DEVELOPMENTS IN THE ANNEALING OF SHEET STEELS

DEVELOPMENTS IN THE ANNEALING OF SHEET STEELS

Proceedings of an international symposium
sponsored by the TMS Ferrous Metallurgy Committee
and held at the 1991 Fall Meeting in Cincinnati, Ohio,
October 22-24

Edited by

R. PRADHAN
Bethlehem Steel Corporation
Bethlehem, Pennsylvania

and

I. GUPTA
Inland Steel Company
East Chicago, Indiana

A Publication of

A Publication of **The Minerals, Metals & Materials Society**
420 Commonwealth Drive
Warrendale, Pennsylvania 15086
(412) 776-9024

The Minerals, Metals & Materials Society is not responsible for statements or opinions and absolved of liability due to misuse of information contained in this publication.

Printed in the United States of America
Library of Congress Catalog Number 92-50174
ISBN Number 0-87339-181-0

Authorization to photocopy items for internal or personal use, or the internal or personal use of specific clients, is granted by The Minerals, Metals & Materials Society for users registered with the Copyright Clearance Center (CCC) Transactional Reporting Service, provided that the base fee of $3.00 per copy is paid directly to Copyright Clearance Center, 27 Congress Street, Salem, Massachusetts 01970. For those organizations that have been granted a photocopy license by Copyright Clearance Center, a separate system of payment has been arranged.

© 1992

PREFACE

Significant process developments continue to occur in both continuous-annealing and batch-annealing technologies, as well as, in the metallurgy of the products processed via these technologies. These developments were the focus of an international symposium, sponsored by the TMS Ferrous Metallurgy Committee, and held during the 1991 Fall Meeting, October 22-24, in Cincinnati, Ohio.

The symposium was dedicated to the many technical contributions of Dr. Hiroshi Takechi (Nippon Steel Corporation) to the field of sheet steels. The willingness of Dr. Takechi to present the keynote address is gratefully acknowledged.

Thanks are due to the following for serving as Session Chairmen:

A.J. Boucek, LTV Steel
S.R. Goodman, US Steel div. of USX
V. Leroy, Centre de Recherches Metallurgiques, Belgium
I. O'Reilly, Dofasco, Canada
S. Satoh, Kawasaki Steel Corporation, Japan
H. Takechi, Nippon Steel Corporation, Japan
A.E. Wilson, Armco

                                        R. Pradhan
                                        Bethlehem Steel Corporation
                                        Bethlehem, Pennsylvania

                                        I. Gupta
                                        Inland Steel Company
                                        East Chicago, Indiana

                                        February 1992

The TMS Ferrous Metallurgy Committee dedicated this Symposium to the recognition of the many invaluable technical contributions of Dr. Hiroshi Takechi to the field of Sheet Steels.

Dr. Hiroshi Takechi

After obtaining his doctorate degree from Kyoto University, Dr. Takechi joined the Yawata Iron & Steel Co. (later, Nippon Steel Corporation) in 1956. At Nippon Steel Corporation, Dr. Takechi has served as General Manager of Kimitsu Works R&D Lab. (1977-81), Yawata Works R&D Lab. (1981-83) and Research Center for Sheet Steels (1983-87). Since 1987, he has been Executive Counselor at Nippon Steel.

TABLE OF CONTENTS

Preface V
Biography, Dr. Hiroshi Takechi VII

I. Keynote Address

Recent Developments in the Metallurgical Technology of Continuous
Annealing for Cold-Rolled and Surface-Coated Sheet Steels 3
    H. Takechi

II. Continuous Annealing: Process Technology

The Equipment and Operations of NKK Fukuyama No. 2
Continuous Galvanizing Line 27
    A. Nakamura, N. Taguchi, H. Yano and K. Takagi

Continuous Annealing Line of Sollac Ste Agathe 43
    P. Louis, C. Deutsch and C. Brugnera

Improvement in First Cooling Technique
(Roll Quench and Water Quench) and the Properties of the Products 53
    M. Okura, H. Makino, Y. Tanaka, J. Iwaya and H. Maeda

Outline of No.3 CAL Water Quench System at NKK Fukuyama Works 65
    N. Matsui, S. Jitsukawa, T. Izushi and M. Yamazaki

New Technology in KM-CAL for Sheet Gage 79
    T. Kaihara, S. Mega, Y. Ida, K. Hirohata,
    T. Nakagawa, K. Kuramoto and T. Fukushima

Reduction Heating Technology of Steel Sheets by Direct Fire 117
    M. Kurihara, A. Honda, S. Uchino and M. Shoji

Development of the Strip Temperature Control Technique
for a Continuous Annealing Line 133
    K. Taya, I. Ueda and M. Honjoh

Tension and Elongation Control for Continuous Annealing Lines 143
    E.A. Cook and R. Mieloo

III. Continuous Annealing: Product Metallurgy
A. Sheet Steels

Production of Hot-Dip Galvanized High Strength Steel Sheets
with Improved Formability 151
    A. Lankila and A. Ranta-Eskola

Effect of Carburizing during Continuous Annealing
on Mechanical Properties of Interstitial-Free Sheet Steels 159
    M. Kitamura, S. Hashimoto, M. Matsumoto and T. Inoue

Carbide Dissolution in Interstitial-Free Steels
during Continuous Annealing 177
    S. Satoh, M. Morita, T. Kato and O. Hashimoto

Recrystallization of Interstitial-Free Steels 189
    D.O. Wilshynsky-Dresler, G. Krauss and D.K. Matlock

Effect of Precipitate Size and Dispersion on Lankford Values
of Ti Stabilized Interstitial-Free Steels 219
    S.V. Subramanian, M. Prikryl, B.D. Gaulin, M. Koch and S. Benincasa

Effects of Manganese Content and Hot-Band Coiling Temperature
on Deep Drawability of Continuous-Annealed Al-killed Sheet Steels 247
    N. Mizui and A. Okamoto

Metallurgical Investigation for Producing Non-Aging
Deep-Drawable Low-Carbon Al-killed Steel Sheets
by Continuous Annealing 261
    K. Ushioda, O. Akisue, K. Koyama and T. Hayashida

Continuous Annealing of ULC-Ti Ferritic Hot-Rolled Strips 287
    P. Messien, J.C. Herman, V. Leroy, Ph. Harlet, F. Beco and L. Renard

Influence of Thermal History Prior to Hot Rolling on Mechanical Properties
of Continuously Annealed High-Strength Sheet Steels
containing Titanium, Manganese and Phosphorus 305
    K.G. Chin, H.J. Kang and S.K. Chang

Cold-Rolled, Intercritically Annealed,
and Isothermally Transformed Sheet Steels 321
    Y. Sakuma, D.K. Matlock and G. Krauss

Annealing Conditions for Hot-Rolled Strip of 13Cr Stainless Steel 339
    K. Miura, S. Satoh, K. Yoshioka and O. Hashimoto

Influence of Steel Type and Surface Condition on Galvanizing Reaction 351
   K. Nishimura, K. Kishida and H. Odashima

Design and Application of a Continuous Annealing Simulator 369
   D.M. Haezebrouck, J.W. Sinclair and D.A. White

III. Continuous Annealing: Product Metallurgy
B. Tin Plate

Metallurgy for Hoogovens' New CA Line for Tin Plate 383
   Th.M. Hoogendoorn and A.J. van den Hoogen

Advanced Manufacturing Process for Tin Mill Black-Plate
with All Temper Designations by Continuous Annealing.............................397
   H. Kuguminato, T. Kato, T. Sekine, A. Tosaka,
   C. Fujinaga, Y. Shimoyama, H. Ohno and R. Asaho

Hardness and Recrystallization Behavior of Continuously Annealed
Soft Temper Blackplates containing Zirconium or Niobium 411
   J.H. Kwak and S.K. Chang

IV. Batch Annealing: Process Technology

Surface Improvements of Steel Strip Annealed
in Ebner Hicon/H_2 ® Batch Furnaces 427
   H. Lochner

Modern Concepts of Process Control and Optimization
in Hydrogen Batch Annealing Shops 443
   M. Bock, K. Nolte and P. Wittler

Development of a Theoretical Annealing Model
for Hydrogen Annealing of Steel Sheets 463
   S. Ramasamy, R.L. Simmons, A.P. DeVito and K.G. Brickner

V. Batch Annealing: Product Metallurgy

Experience with High Convection
Hydrogen Batch Annealing at Dofasco 481
   W.F. Gasse and S.J. Thomas

Gas-Metal Reactions during 100% H_2 Batch-Annealing 493
   J.M. Mataigne, M. Lamberigts and V. Leroy

Selective Oxidation of Cold-Rolled Steel
during Recrystallization Annealing 511
    J.M. Mataigne, M. Lamberigts and V. Leroy

Characteristics of 100% Hydrogen Annealing Furnace 529
    S. Tajima and M. Shirouzu

Subject Index 539

Author Index 543

I. Keynote Address

Recent Developments in the Metallurgical Technology of Continuous Annealing for Cold-rolled and Surface-Coated Sheet Steels

Hiroshi Takechi

Nippon Steel Corporation

1. Introduction

About 20 years have passed since the first continuous annealing and processing line for cold-rolled sheet steels in Japan started operation in 1972. During this time 45 continuous annealing lines have been constructed worldwide, as shown in Table 1, of which 32 are for cold-rolled sheet steels, seven are for tin plate and the remaining six are for both cold-rolled sheet steels and tin plate.

Table 1 Continuous annealing lines in operation or under construction in the world as of Sep. 1991 (published)

Technology (Owner)	Inside technology owner			Outside technology owner			Total
	Line	Starting year	Product	Line	Starting year	Product	
NSC-CAPL (Nippon Steel)	Kimitsu No.1	1972	Sheet	Sumitomo No.1	1981	Sheet	22 lines (7 + 15)
	Yawata No.1	1979	Sheet	SIDMAR	1981	Sheet	
	Nagoya No.1	1982	Sheet	CSN	1986	Sheet	
	Hirohata (FIPL)	1982	Sheet	Nisshin	1986	Sheet	
	Yawata No.2	1982	Tinplate	Sollac/Basse-Indre	1988	Tinplate	
	Kimitsu No.2	1991	Sheet	ISCOR No.1	1988	Sheet	
	Nagoya No.2	(1991)	Tinplate	British Steel	1988	Tinplate + Sheet	
				Baoshan	1989	Sheet	
				ISCOR No.2 (revamp)	1989	Tinplate	
				I/N Tek (FIPL)	1990	Sheet	
				Hoogovens	1991	Tinplate	
				ILVA	1991	Sheet	
				A. H. V.	(1991)	Tinplate	
				ENSIDESA	(1991)	Tinplate + Sheet	
				Sumitomo No.2	(1992)	Sheet	
NKK-CAL (Nippon Kokan)	Fukuyama No.3	1987	Sheet	SSAB	1982	Sheet	12 lines (2 + 10)
	Fukuyama No.2 (revamp)	1991	Tinplate	Novolipetsk	1982	Sheet	
				Kobe Steel	1982	Sheet	
				Inland Steel	1983	Sheet	
				Bethlehem Steel	1983	Sheet	
				Hoesch Stahl	1986	Sheet	
				Sollac/Montataire	1986	Sheet	
				Posco No.3	1987	Sheet	
				China Steel	1988	Sheet	
				LTV	(1991)	Sheet	
KM-CAL (Kawasaki Steel)	Chiba No.2	1980	Tinplate + Sheet	Rasselstein AG	1984	Tinplate + Sheet	10 lines (5 + 5)
	Mizushima No.1	1984	Tinplate + Sheet	Posco No.2	1986	Sheet	
	Chiba No.3	1988	Sheet	Sollac/Florange	1988	Sheet	
	Chiba No.4	1990	Tinplate	Posco No.4	1988	Sheet	
	Mizushima No.2	1991	Sheet	U. P. I.	1989	Tinplate + Sheet	
CRM-HOWAQ (Cockerill)	Jemeppe	1985	Sheet				1 line
Sum							45 lines

For the cooling system after annealing, the HOWAQ system, the water-gas mixture cooling system and the roll cooling system were developed from the initial gas jet system and the water quenching system and are now being actually used. Starting from the gas jet system for slow cooling of about 10°C/sec and the water quenching system for ultrarapid cooling of about 1,000 °C/sec, the water-gas mixture cooling system and the roll cooling system for medium-rate cooling of about 50-400°C/sec are at present coming to be mainly used. In the case of Al-killed steels, this is based on metallurgical reasons with consideration given to the balance between ductility and strain aging property. It is the most important task in continuous annealing technology to precipitate the largest possible amount of solute carbon in the form of iron carbides without deteriorating the ductility of the products during the short-time overaging treatment of Al-killed steels. The results of recent studies on this subject will be reported later in this paper.

As a matter closely related to continuous annealing technology, I must mention the recent remarkable technical development and quantitative expansion of IF steel. Figure 1 shows the increase in annual IF steel production ratio at Nippon Steel. As seen from this figure, IF steel production has recently shown a phenomenal increase, and the absolute quantity has entered the 2 million tons/year/steelmaker age in Japan. The greatest reason for this rapid growth of IF steel is its unparalleled press formability capable of meeting the automotive industry's wide-ranging requirements. The use of IF steel, in which the solute C and N are fixed by Ti and Nb, is essential for manufacturing the deep drawing grade in the hot-dip galvanizing line, which is not provided with an overaging furnace for precipitating solute C and N. Compared with cold rolled steel sheets, galvanized sheets are said to be one or two ranks lower in press formability, and the use of IF steel sheets as matrix steel sheets can cover this inferiority.

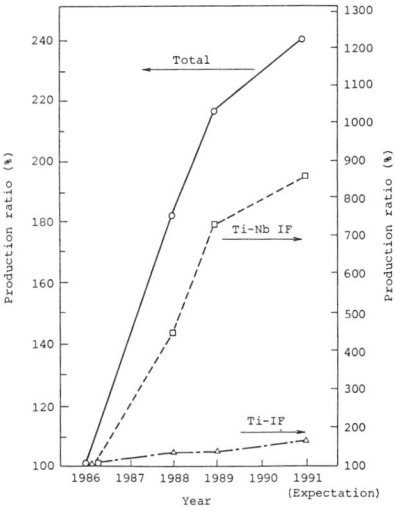

Fig. 1 The periodic change in the production ratio of IF steel in Nippon Steel Corp.

IF steel has characteristics that it is not influenced by the heating rate during annealing and the soaking temperature should be high enough for its higher recrystallization temperature. Due to these characteristics, IF steel is extremely suitable for production in the continuous annealing line and the hot-dip galvanizing line. A combination of IF steel with high-temperature annealing in the continuous annealing line has enabled the steel producers to meet the automotive industry's demand for such new products as super formable steel sheets and bake hardenable steel sheets of deep drawing quality.
With these circumstances as the background, the excellence of IF steel has come to be widely recognized and its use is now rapidly spreading.

The enactment of the CAFE (Corporative Average Fuel Economy) regulation for curbing the greenhouse effect of the earth is being discussed, and it was concluded that the weight reduction of automobiles is essential for decreasing the exhaust CO_2 volume. Therefore, high-strength steel (HSS) came to attract serious attention once again. Continuous annealing technology facilitate the production of various types of HSS such as IF-HSS, BH-HSS, DP-HSS and TRIP-HSS. The ratio of the use of HSS in the white bodies of some Japanese automobiles reaches 40 percent, and in near future, will reach 50%. Although aluminum and plastics are also eagerly studied as lighter materials, they have problems yet to be solved. Therefore, great expectation is still entertained of HSS. A comparison of the characteristics of TRIP-HSS and aluminum sheets will be made later. Continuous annealing technology will be used more actively in the future for the development of new types of HSS and applied also to the development of hot-dip galvanized HSS.

CAPL technology made it possible to produce matrix steel plates for softer tin plate up to temper 1 grade using Al-killed steels. I shall describe later the metallurgical concept of the manufacture of matrix steel plates for softer tin plate using Al-killed steels and the continuous annealing furnace.

Comparative studies of continuous annealing technology and H_2-BAF technology have come to be made worldwide in recent years. These two technologies must be compared from the aspects of the equipment construction cost, operation cost, product characteristics and their uniformities, capability to develop new products to meet new market needs and so forth. In comparing the technologies, the conditions of the comparison must be clarified, and attention must be given to the fact that the criteria for emphasing the comparison items are not always the same among countries or among companies. One thing that all people recognize is the point that compared with the BAF process, the continuous annealing process is more capable of quickly meeting market needs. In other words, the continuous annealing process is a market-driven process. By taking this opportunity to talk about the recent situation of continuous annealing technology, I would like to touch upon the comparison of the two processes later as a reference of my talk.

Regarding the continuous annealing technology I have outlined, I shall now speak about some of the latest topics.

2. New heat cycle for the overaging effect of deep drawing Al-killed steel sheets

After the completion of recrystallization by annealing, the steel sheets are cooled and enter the overaging zone, where the iron carbides are precipitated. This precipitation treatment, which is called overaging treatment, should precipitate the largest possible amount of solute C in the form of iron carbides in several minutes of isothermal holding and grow the precipitated iron carbides as much as possible. The solute C is precipitated to improve the strain aging property of the product, and the iron carbides are grown to secure the ductility of the product. If the steel sheets are rapidly cooled as by water quenching, as shown in Figure 2 (A), after recrystallization annealing and heated again and then isothermally held for 5 minutes in the overaging furnace, many fine iron carbides are found precipitated in each grain, as shown in Photo 1 (A).[1] In this case, as shown in Figure 3, the restoration of the yield point elongation after artificial aging, which is a measure for the strain aging property, is the smallest, while the deterioration of the ductility is the largest. Conversely, if the steel sheets are slowly cooled as by gas cooling, as shown in Figure 2 (C), after overaging treatment, iron carbides are precipitated mostly coarse on each grain boundary and very few iron carbides are precipitated in each

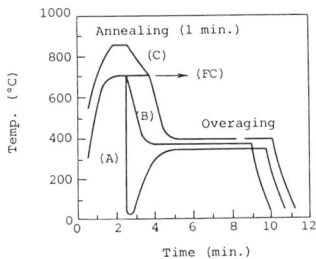

Fig. 2 Heat cycles tested for carbide distributions in overaging

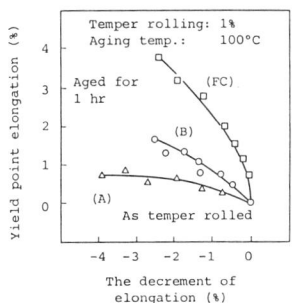

Fig. 3 The change in yield point elongation and total elongation during strain aging due to the change of heat cycles shown in Fig. 2

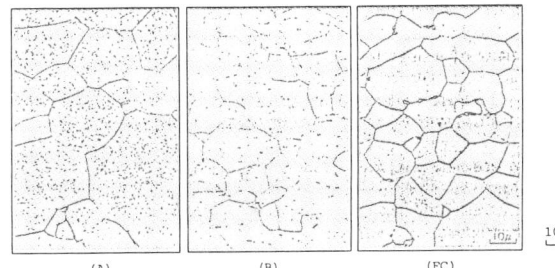

Photo 1 The change in carbide distributions due to the change of heat cycles for overaging shown in Fig. 2

grain. As a result, the restoration of the yield point elongation is larger than in the case of water-quenched steel sheets, while the deterioration of the ductility is the smallest, as shown in Figure 3. If the amount of the supersaturated C is increased by rapid cooling, the driving force for solute C precipitation increases. But since the precipitation is rapid, many fine iron carbides occur in each grain, and the ductility is deteriorated. The reverse is the case with slow cooling. Therefore, the cooling rate at which a compromise is made between strain aging property and ductility was sought, and a cooling rate of about 100°C/sec was selected. To attain this cooling rate, the water-gas mixture cooling system and the roll cooling system are mostly used now.

However, a study was made on a new heat cycle which can further improve the absolute values of strain aging property and ductility while maintaining the compromise between them in a short overaging time of not more than 3 minutes. [2,3] As shown in Figure 4, steel sheets were subjected to recrystallization annealing at 800°C for 60 seconds, were slowly cooled to 675°C and then to various temperatures between 200°C and 350°C at a rate of 100 °C/sec and then were subjected to overaging treatment for not more than 5 minutes at various temperatures between 350°C and 250°C.[2] The size and distribution of iron carbides after overaging treatment changed as shown in Photo 2 according to the change in the degree of supercooling indicated by the temperature difference ΔT between the quench-stopping temperature and the overaging temperature of 350°C. The size and distribution of iron carbides also changed as shown in Photo 3 when the overaging time was changed variously between zero and 5 minutes.

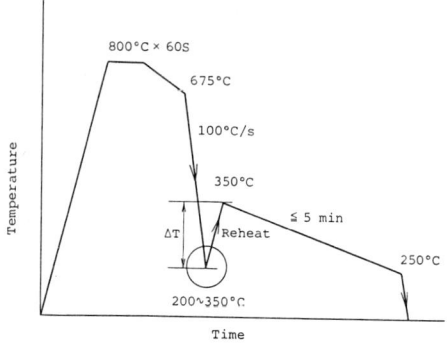

Fig. 4 The new heat cycle for accelerating the overaging of Al-killed, deep drawing sheet steel

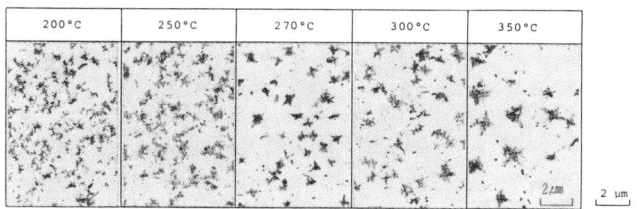

Photo 2 The effect of quench-stopping temperature on the size and the number of iron carbides

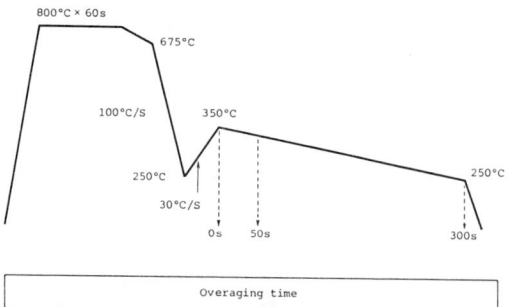

Photo 3 The effect of the overaging time on the size and the number of iron carbides

It was confirmed both empirically and theoretically that the amount of solute C is changed as shown in Figure 5 by changing the quench-stopping temperature and the overaging time. So it was made possible to control the conditions for simultaneously satisfying both strain aging property and ductility by properly selecting the ΔT and the overaging time. In order to prove this fact industrially, we installed an electric induction heating-type reheating device between the primary cooling zone (water-gas mixture) and the overaging furnace in the Hirohata CAPL and tested the above new annealing cycle using this system. As a result, excellent features were achieved as shown in Table 2. Figure 6 shows the makeup of the Kimitsu No. 2 CAPL which is the newest line having the reheat-overaging system above. This line started operation on August 1, 1991.

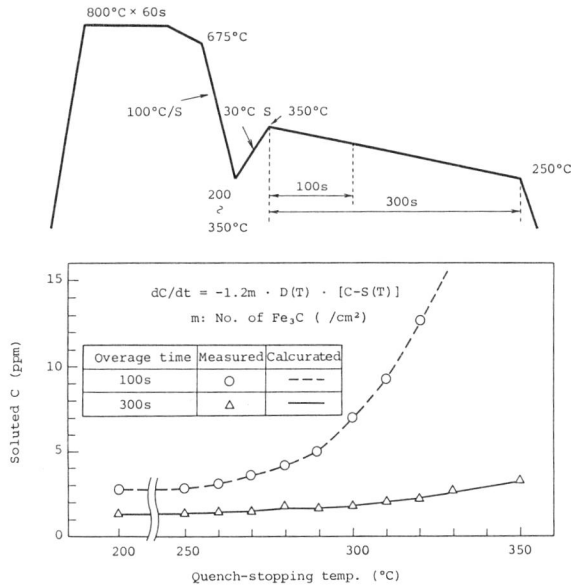

Fig. 5 The effect of quench-stopping temperature on the quantity of soluted carbon during overaging

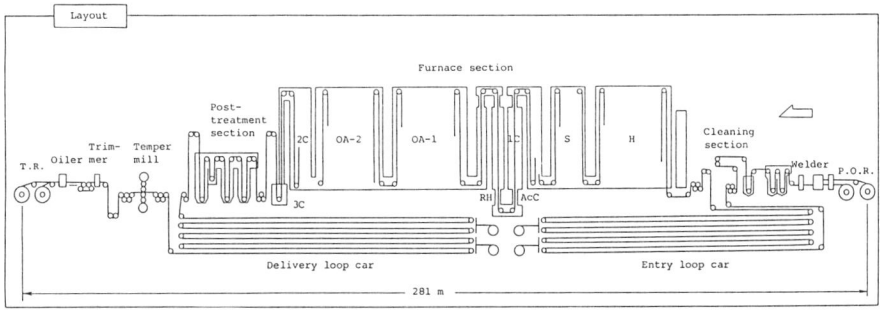

Fig. 6 The layout of Kimitsu No. 2 CAPL which has an induction heater (RH) for new heat cycle after water-gas mixture cooling (AcC)

9

Table 2 Line test results using the new heat cycle for overaging (Hirohata)

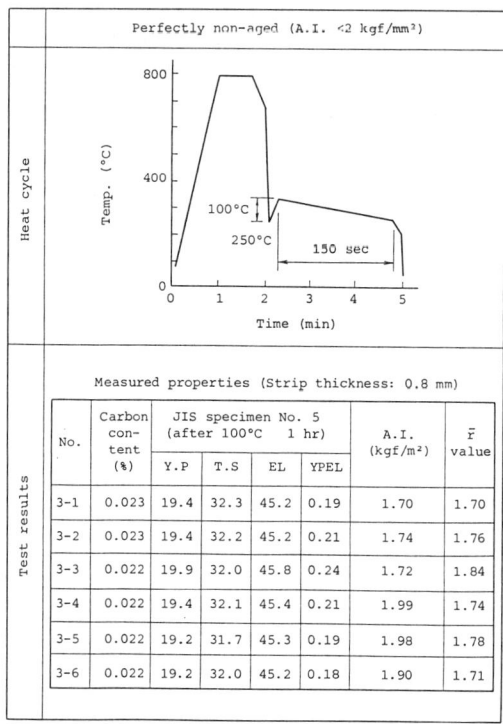

3. Manufacture of super-formable sheet steel using IF steel

Cold-rolled sheet steels were formerly classified by their r̄ value and total elongation (Eℓ) into 4 grades, CQ (commercial quality), DQ (drawing quality), DDQ (deep drawing quality) and EDDQ (extra deep drawing quality) as shown in Figure 7. Recently, however, Super EDDQ grade has come to be in demand, as shown in the figure, due to increasingly severe customer requirements. Surprisingly high formability of over 2.0 in r̄ value and over 50% in Eℓ are required of Super EDDQ. To clear these mechanical properties, IF steel which is coiled at a high temperature of over 700°C at the time of hot rolling should be annealing at a high temperature of over 800°C after cold rolling. For this purpose the continuous annealing process is very advantageous because it is free from the trouble of thermal sticking at the time of annealing. Figure 8 shows the relationship between the annealing temperature and the r̄ value.[4] The importance of the annealing temperature for mechanical properties is quite same in the case of Al-killed steel as shown in Fig. 9.[5] Table 3 shows the manufacturing conditions and mechanical properties of Super EDDQ manufactured in the commercial line by combining IF steel with continuous annealing.

Kimitsu No. 2 CAPL line, for example, is capable for heating cold rolled strips which are 6 feets wide up to 900 °C where the engineering and the operational developments were also achieved.

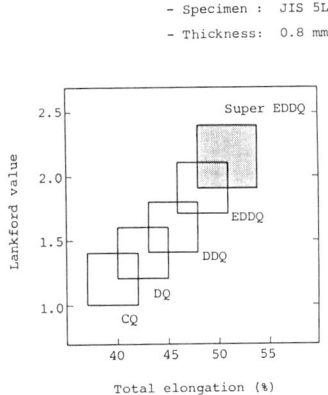

Fig. 7 Super formable steel sheets

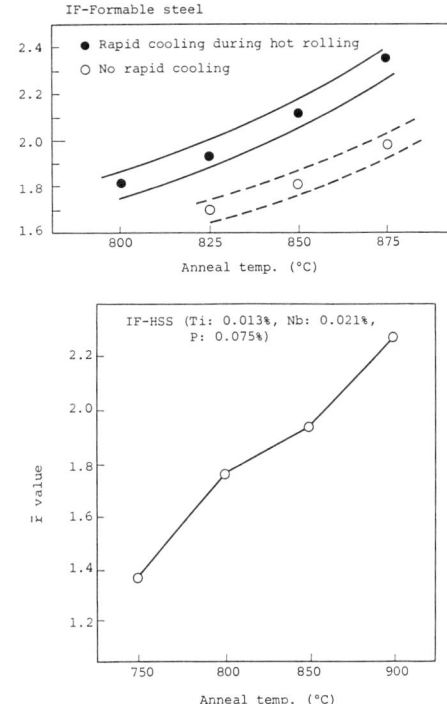

Fig. 8 The effect of annealing temperature on $\bar{r}$ value of IF-Formable steel and IF-HSS

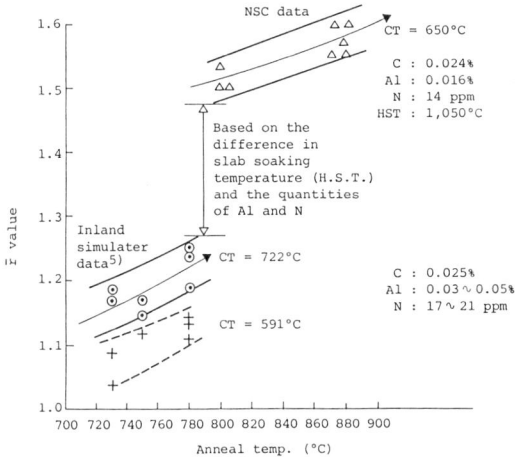

Fig. 9 The relationship between anneal temp. and $\bar{r}$ value of Al-killed steel treated by continuous annealing

Table 3 An example of Super EDDQ produced with IF steel

Chemical composition (%)							Hot roll			CAPL		Mech. properties			
C	Mn	P	S	Ti	B	N	Slab soak (°C)	Coil (°C)	Ann. (°C)	OA (°C)		YS (kgf/mm²)	El (%)	n	$\bar{r}$
0.0014	0.13	0.008	0.003	0.050	0.0003	0.0015	1,150	700	850	350		12.4	52.6	0.27	2.50

Let me now present one recent topic regarding the texture control of IF steel. High-cleanliness steels like IF steel easily coarsen during slab soaking before hot rolling, and consequently the grain size in the hot-rolled strip tends to become large. The fact that the grain size before cold rolling largely influences the texture after cold rolling and annealing which in its turn influences the value was reported by Abe et al.[6] Figure 10 is a {200} pole figure after cold rolling and annealing (700°C - 1 hr) with the grain size before cold rolling changed from 25 to 300 μm. The figure clearly shows that if the initial grain size is large the Goss orientation exists, and that if the initial grain size is small the Goss orientation disappears and the accumulations of the {111} <110> + {111} <211> orientations become strong. This is because of the difference that when fine-grained steel sheet is deformed by cold rolling, constraint due to grain boundaries is strong and multiple slip occurs in many zones, but when coarse-grained steel sheet is deformed by cold rolling, many deformation bands are formed in each grain due to simple slip. In the case of IF steel, also, the grain size in the hot-rolled strip should preferably be as small as possible.

It was reported that if water spray cooling is started soon after strip has passed through the final stand of the hot strip mill, the grain size number of the hot-rolled strip decreases by 2, as shown in Figure 11, with the result that the $\bar{r}$ value of the product is improved by about 0.2.[7]

This may be called a good example of the blossoming of a basic study of the rolling and recrystallization texture into industrial IF steel manufacturing technology.

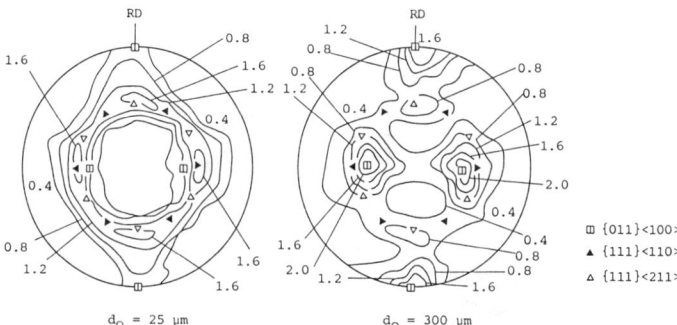

Fig. 10 The effect of the grain size in hot rolled steel sheets (do) on the recrystallization texture obtained after cold rolling and annealing

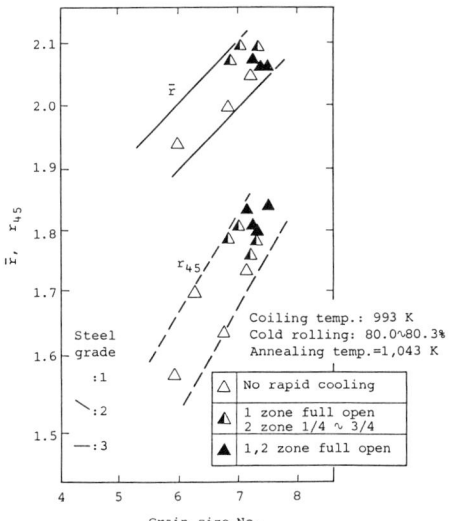

Fig. 11 The effect of grain refining in hot rolled steel sheets by means of the rapid cooling started just after the hot-roll finishing on the r value of products

4. Metallurgical approach to the manufacture of soft grade tin plate by continuous annealing

For the manufacture of matrix steel sheets for softer tin plate of up to temper 1 grade using Al-killed steels, rapid precipitation treatment of solute C by selecting a suitable cooling rate after annealing and by imparting overaging property is necessary as in the case of deep drawing cold-rolled sheet steel. In the case of thin sheets like matrix steel sheets for tin plate, a considerably high cooling rate can be secured by atmospheric gas jetting. Recently it has become possible to narrow the gap between strip and gas nozzles to 50 mm by controlling the flattering of strip by supporting rolls. This system, called the high cooling rate GJC system (H-GJC), made it possible to obtain a cooling rate of 100°C/sec in the whole thickness range of strip for tin plate. Figure 12 shows the relationship between the quantity of solute C after overaging treatment and the overaging time when the H-GJC and the conventional gas jet cooling system (GJC) are used.[8] As seen from this figure, the precipitation of solute C particularly under the short-time overaging condition is conspicuously accelerated by the H-GJC. Such an increase in the cooling rate remarkably improved the fluting property after annealing and 1% skinpass rolling.

What is important for softening the product is to prevent as much as possible the precipitation of AlN during heating in continuous annealing. This is because the fine AlN particles precipitated during heating interfere with the grain growth during soaking, as shown in Figure 13.

There are three possible ways to prevent the precipitation of AlN during heating.
The first is to set the slab soaking temperature at the time of hot rolling as low as possible and prevent the resolution of AlN already precipitated in this stage. Figure 14 shows the relationship between the quantity of N

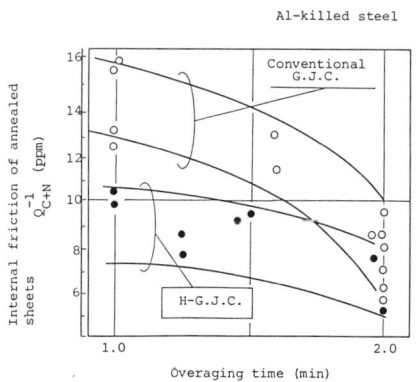

Fig. 12 Change in solute [C] in blackplate between before and after start-up of H-GJC in Yawata No. 2 CAPL

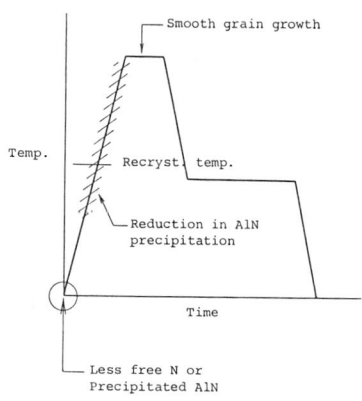

Reduction in AlN precipitation during continuous annealing is necessary for smooth grain growth.

Fig. 13 Schematic diagram of heat cycle for softer tinplate

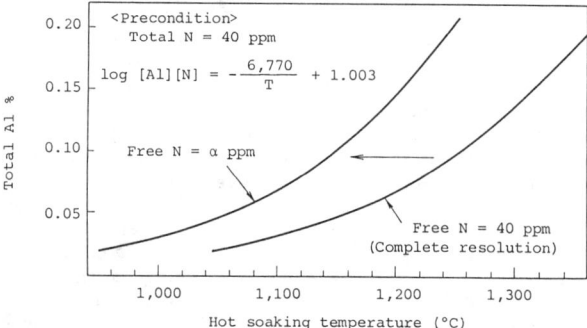

Agglomeration of AlN precipitates prior to annealing (1)

Prevention of AlN dissolution during slab soaking
(Sufficient Al for N content at each slab soaking temp.)

Fig. 14 Relationship between H.S.T. and Al content to keep N in solution at a certain value during soaking

which dissolves during heating according to Leslie's equation and slab soaking temperature. It is said that in order to obtain temper 3 grade, the critical quantity of solute nitrogen should be controlled to below 15 ppm.[9]

The second method is to set the coiling temperature at the time of hot rolling rather high so that even if a considerable quantity of solute N remains at the end of hot rolling, the nitrogen is precipitated as AlN. Figure 15 shows the relationship between coiling temperature in hot rolling and HR30T hardness after cold rolling and annealing.[10] When the slab soaking temperature cannot be set sufficiently low, it is effective in preventing the precipitation of AlN to set the coiling temperature rather high. The quantitative relationship between both temperatures is important.

The third is to reduce the quantity of AlN to be precipitated at the time of annealing by decreasing the quantity of Al and the quantity of N as much as possible in the steelmaking stage. If Al and N to be precipitated do not exist, precipitation hardening naturally does not occur. Even when the second method is adopted, there is no need to set the coiling temperature so high even if the slab soaking temperature is sufficiently high, provided the total quantity of Al and N is small. In this case, the gradient of the curve in Figure 16 becomes gentle, and the dependence of the HR30T hardness of the product steel sheets after annealing upon the coiling temperature at the time of hot rolling decreases.[10]

Combinations of the three methods described above made it possible to manufacture sheet steels of up to temper 1 grade by continuous annealing using Al-killed steels, as shown in Table 4.[11] Of temper 1 grade products, however, temper 1 C grade for which stretcher-strain characteristics must be guaranteed still cannot be manufactured. Although temper 1 C grade can be manufactured without problems if IF steel is used, approval for using Ti and Nb must be obtained for the manufcture.

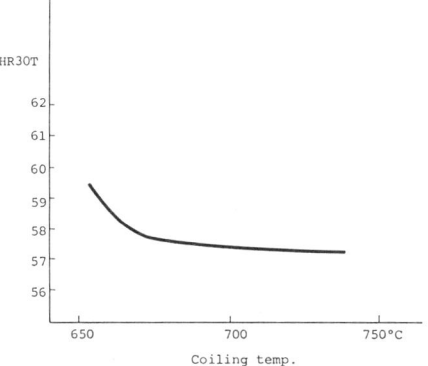

Fig. 15 The effect of coiling temp. on HR30T of continuously annealed steel sheet (Center at middle of a coil)

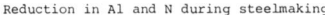

Fig. 16 Relationship between HR30T and coiling temp.

Table 4 Present situation of soft temper grade production from Al-killed steel in Yawata No. 2 C.A.P.L.

<In early stage of Y-2 C.A.P.L. operation>

Temper grade \ Workability rank	A Hardness only	B Resistance to fluting	C Resistance to stretcher strain	D Special application (D&I etc.)
T-1				
T-2				
T-2.5	120S	120S		
T-3	60S	120S		

Introduction of integrated metallurgy + H-GJC ⟹

<In recent stage of Y-2 C.A.P.L. operation with H-GJC>

Temper grade \ Workability rank	A Hardness only	B Resistance to fluting	C Resistance to stretcher strain	D Special application (D&I etc.)
T-1	90S	120S (High ST)		90S
T-2	90S	90S		
T-2.5	90S	90S	120S (High ST)	90S
T-3	60S	60S	120S	

☐ : Low-[C], high purity Al-killed steel
▨ : Normal-[C], Al-killed steel

5. Manufacture of IF-BH and TRIP sheet steels by continuous annealing

The question of how the weight reduction of automobiles can be realized is posed anew because the requirement for the reduction of exhaust CO_2 gas volume which is a large issue in the CAFE regulation becomes urgent and automobiles weight tends to increase for installing many functional apparatuses. Therefore, it is the automotive industry's desire to use high-strength steels (HSS) as much as possible. This problem is closely related to competition between steel and other materials like aluminum and plastics and particularly to the scrap recycling problem which has now become a serious worldwide problem. In Japan, over 10,000 lives are lost every year in traffic accidents. HSS plays an important role in securing drivers' safety at the time of collisions.

Many types of HSS were developed in the past, as shown in Fig. 17. As cold-rolled HSS, rephosphorized steel and BH-rephosphorized steel are mainly used for exposed and unexposed panels and bainite steel and dual-phase steel are used for reinforcement.
The HSSs which are attracting attention at present are IF-HSS with further improved strength and formability for use as exposed and unexposed panels,

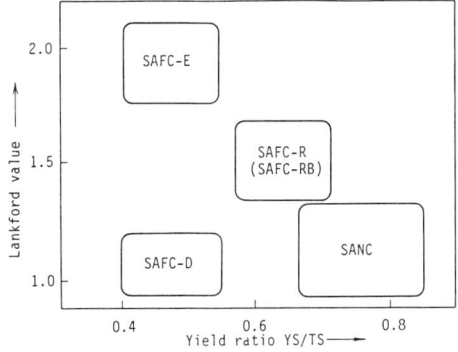

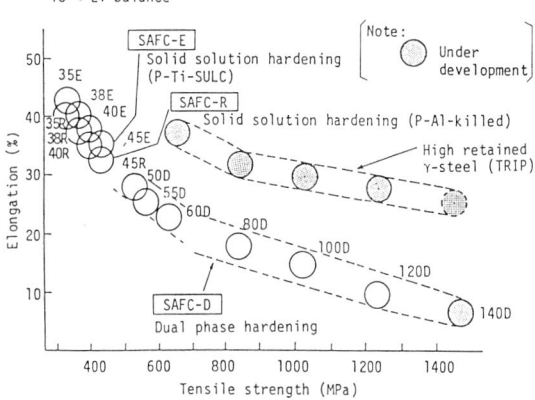

Fig. 17 High strength steel sheets produced in NSC

IF-BH-HSS which is IF-HSS with the BH property and TRIP steel. IF-HSS is IF steel strengthened by adding P and Mn. Let me omit describing the manufacturing process for this steel, as it is much the same as that for IF steel.
IF-BH-HSS has solute C and N left in steel sheets by partly dissolving the carbides of Ti and Nb by selecting the annealing conditions.[12),13)] However, a sufficiently high $\bar{r}$ value can be obtained because the matrix texture is formed before the carbides are dissolved. Table 5 shows an example of the composition of this steel continuous annealed, in which excellent mechanical properties and a BH value of 4-5 kgf/mm^2 are obtained. Figure 18 shows the influence of continuous annealing conditions on bake-hardenability of IF steel.

Among super high strength steels with a tensile strength of over 80 kgf/mm^2, dual-phase steel was formerly the highest in ductility. However, the total elongation of dual-phase steel at a tensile strength of 100 kgf/mm^2 is a maximum of 20% under the conditions that thickness is 0.8 mm and the gauge length of tensile test specimens is 50 mm. Nevertheless, my customers require a total elongation of over 20% for HSS at a tensile strength of 100-140 kgf/mm^2. What can meet this requirement is TRIP steel utilizing transformation-induced plasticity. It is possible to manufacture TRIP steel using two linearly arranged continuous annealing furnaces so as to

Table 5 Bake Hardenability of Sheet Steel Produced by CAPL

Chemical composition (wt. %)								Thickness	Annealing temperature (°C)	Mechanical properties					
C	Si	Mn	P	S	Al	N	Addition			YP (Kgf/mm²)	TS (Kgf/mm²)	EL (%)	r_0	$\bar{r}$	BH (Kgf/mm²)
0.015	0.02	0.10	0.06	0.014	0.031	0.0030	B: 0.0021	0.8	850	20.8	35.7	40.0	1.86	1.72	5.5
0.06	0.05	0.25	0.06	-	-	-		0.8	800	27.0	40.3	40.0	-	1.35	4.5
0.002	0.01	0.2	0.07	0.005	-	-	B: 0.015	0.7	830	20.4	35.7	44.5	-	2.1	4.6
0.0026	0.01	0.15	0.008	-	0.032	0.0024	Nb: 0.008 Ti: 0.007	0.8	780	15	30	49	-	2.1	4.0

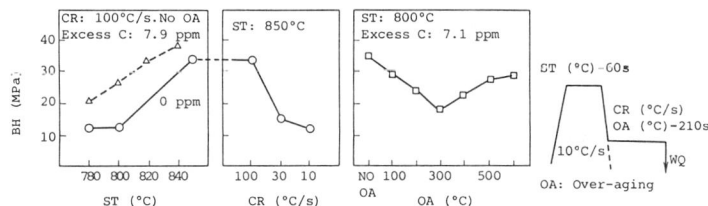

Fig. 18 Effect of annealing temperature (ST), cooling rate (CR) and over-aging (OA) on bake-hardenability

obtain the two stage heat treatment.

The use of expensive alloying elements is not permitted for mass-produced products like automotive body parts. If a steel having the chemical composition shown in Table 6 for instance is treated in the heat cycle shown in Figure 19, material containing about 20% retained austenite in the annealed steel sheets can be obtained depending on the soaking conditions in the bainitic zone.[14] The trip effect of the retained austenite produced in such a large quantity makes it possible to obtain such a remarkable value of total elongation as 30% even at the high tensile strength of 100 kgf/mm². Photo 4 shows the cylindrical cup drawing test results for TRIP steel and dual-phase steel both of which have a tensile strength of 100 kgf/mm². While the dual-phase steel cracked in the initial stage of deformation, the TRIP steel was drawn like mild steel.[15]

Metallurgically, the following phenomena were observed.[16] The austenite produced by annealing the steel sheets in the dual-phase zone where austenite and ferrite coexist is reduced to below 200°-300°C in the Ms point due to the concentration of C and Mn. Carbon is further concentrated in the untransformed area at the time of bainitic transformation, and the Ms point drops to below room temperature. This is related to the formation of bainitic ferrite by retarding the precipitation of iron carbides due to the presence of Si, and the 8-20% retained austenite exists not only between laths but also along ferrite grain boundaries in a globular form. This retained austenite is transformed into martensite during deformation at room temperature, and the transformation-induced plasticity brings the large elongation of 40-30% at the high strength of 60-100 kgf/mm².

When competition between steel and lighter materials is considered, TRIP steel is felt to have a great significance. Table 7 compares the character-

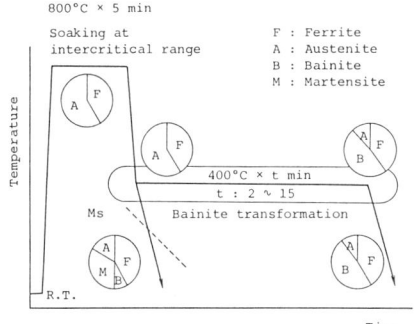

Fig. 19 Manufacturing principle of cold-rolled TRIP steels in the C.A.P.L.

Photo 4 Deep-drawability of the new and conventional steels for the 100 kgf/mm² class

Table 6 The chemical composition of TRIP steels (%) and properties

C	Si	Mn	Sol. Aℓ	T.N.	A_{C1}/A_{C3}	Tensile strength (kgf/mm²)	Eℓ (%)
0.4	1.50	1.20	0.045	0.0050	750/810	130	30
0.3	1.20	1.20	0.045	0.0050	750/830	115	30
0.2	1.20	1.20	0.045	0.0050	745/840	100	30

Table 7 The comparison of mechanical properties between TRIP sheet steel and Aℓ sheet

	Yield strength (kgf/mm²)	Tensile strength (kgf/mm²)	Elongation (%)	n value	r̄ value	Young's modulus (kgf/mm²)
Aℓ (5182-0)	15	30	25	0.3	0.7	7,000
TRIP	80	100	30	0.3	0.9	21,000

istics of 5,000-series aluminum alloy and TRIP steel. Macroscopically, both are nearly the same in press formability, while especially Young's modulus of TRIP steel is much higher than that of aluminum alloy.

Figure 20 shows the relationship between the degree of weight reduction and the material cost for HSS and aluminum sheets which are changed in thickness

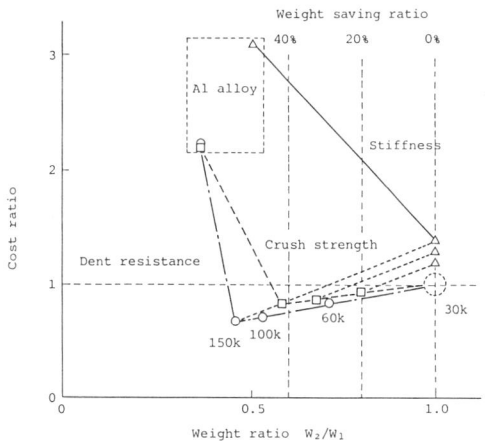

Fig. 20 The relationship between weight ratio and cost ratio of various materials used for weight reduction

to have the same stiffness, dent resistance and crush strength.[17] As you can see from this figure, HSS will have a competitive degree of weight reduction at an incomparably low cost against aluminum, though excluding stiffness, if a highly formable HSS at a tensile strength of 150 kgf/mm^2 is developed. Stiffness can be solved by locally lining HSS with plastics or like material after automotive bodies are assembled. This may be called an example of the great future possibilities of continuous annealing in such a new market area.

6. Conclusion

The use of the continuous annealing process is spreading worldwide as a highly sophisticated sheet steel production system, and its outstanding performance can fully meet the requirements of the Japanese automakers which are most severe in the world. At the same time, continuous annealing technology is now a key technology in the modernization of the integrated process at the steelworks.

In view of the current uncertain and unstable social situation with many environmental and other problems to be solved throughout the world, we consider that it is to be of great significance for the world's steel industry to possess a high-performance and flexible production process capable of meeting any competition with other materials.

(Supplement)

This lecture was delivered mainly on the metallurgical aspect of continuous annealing technology.

For the commercial application of continuous annealing technology as a means of production, however, it was necessary to make remarkable developments from the aspects of equipment technology and operation technology for the continuous annealing line in order to meet metallurgical requirements.

Since continuous annealing technology shows its maximum superiority when controlled chemical composition and high cleanliness of steels are achieved, various technical requirements were put forth for the steelmaking process. Technical requirements were also proposed for the hot rolling process from the aspect of hot rolling temperature and for the cold rolling process from the aspect of cold reduction ratio. These technical requirements for the upstream processes were fulfilled by way of the modernization in each process.

Thus the continuous annealing process fully displayed its effectiveness as the result of the coordinated development of the integrated process from steelmaking to annealing. Therefore, continuous annealing technology may be called the driving force in the modernization of the integrated process at steelworks.

I should have explained the recent development of continuous annealing technology from all these aspects, but time did not permit me to do so during the lecture. I believe it fair to cite now some examples of the progress in non-metallurgical matters other than those few which I touched upon in the lecture.

Allow me to speak about examples at Nippon Steel. In 1990, the continuous casting ratio reached 98.6%, the continuous cold rolling ratio 58%, and the continuous annealing ratio 66%. With these achievements as the background, the product yield from hot-rolled coils reached 97%, as shown in Figure 21.

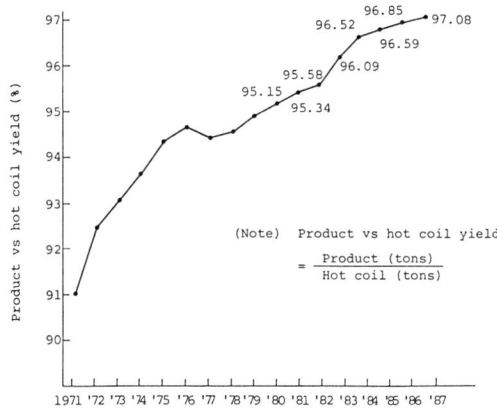

Fig. 21 Progress in product vs hot coil yield in NSC

The production ratio for IF steel is high as I said already. As shown in Table 8, on the other hand, the average working ratio of the continuous annealing line is 98%, and strip breaks inside the furnace, which need a long time to recover, occur only about once a year.

Table 8 Actual data on working ratio in NSC's Nagoya No.1 CAPL

Item Month	Working ratio (%)	Strip breakage (times/month)	
		inside surface	outside furnace
May. '90	98.52	0	1
Jun.	94.55	0	1
Jul.	98.69	0	0
Aug.	94.88	0	0
Sep.	92.39	0	0
Oct.	99.66	0	0
Nov.	99.86	0	0
Dec.	100.00	0	0
Jan. '91	99.84	0	0
Feb.	100.00	0	0
Mar.	100.00	0	0
Apr.	100.00	0	0
May	97.66	1	0
Jun.	98.56	0	0
Jul.	100.00	0	0
Aug.	99.29	0	1

(Note)

Working ratio = $\frac{\text{(Scheduled operating time)} - \text{(Unscheduled stoppage)}}{\text{Sheduled operating time}}$,

where: Scheduled operating time
= (Calender time) - (Scheduled stoppage)*

* incl. Maintenance time

Lastly, let me refer to the comparison between the conventional BA and H_2-BA processes and the continuous annealing (CA) process, the comparison of which has recently been made frequently due to the widespread use of the H_2-BAF.[18),19),20)]

I would like to show one example of operation cost comparison as follows. As Nippon Steel does not employ an H_2-BAF as a means of production, I cannot compare the relative merits of CA and H_2-BA in operation costs based on the company's own data. So I first compared the conventional BA and CA using Nippon Steel's data.[18)] Then, I compared H_2-BA and the conventional BA using the data in reference.[19)] Finally I compared CA and H_2-BA in operation costs, synthesizing the results of both comparisons above.[20)] Figure 22 shows the results of the final comparison. Although CA and H_2-BA both have advantages and disadvantages according to individual items, CA is generally more advantageous in operation cost than H_2-BA at a ratio of 73:91. However, I must add that the results shown in Figure 22 are for your reference only, because the unit prices used in the comparison items differ in different countries and in different companies, as stated in the introduction.

It is impossible to treat all sizes of sheet steels in the CAPL. The CA process and the H_2-BA process each have their own advantages and disadvantages. Therefore, it is also an appropriate idea to use them both in combination. It is preferable in that case to increase the investment efficiency of the CAPL, which requires a large investment though it has a great capability to meet the market demand, by maximizing its production capacity and using the H_2-BAF instead of the conventional BAF to cover the remaining capacity required.

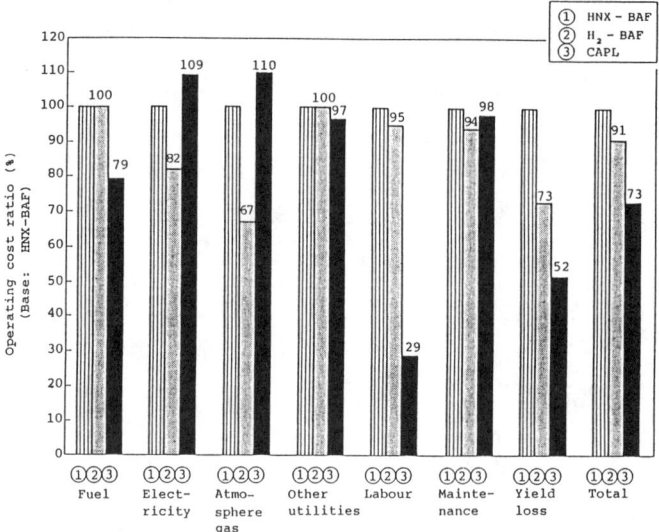

Fig. 22 Comparison of operating cost among HNX-BAF, H_2-BAF and CAPL processes

- partly estimated

< Reference >

1) H. Takechi: No.88 Nishiyama memorial lecture: ISIJ (1983), p.47

2) T. Hayashida & S. Sanagi: CAMP-ISIJ, 1 (1988) p.1721

3) K. Koyama, H. Katoh & M. Naguma: Tetsu-to-Hagané, 72 No.7 1986, p.823

4) N. Matsudzu, K. Koyama, A. Itami, T. Takahashi, H. Ohhashi & M. Shibata: CAMP-ISIJ, 3 (1990) p.1816

5) D. Bhattacharya: Private communication

6) M. Abe, Y. Kokabu, Y. Hayashi & S. Hayami: Trans. JIM, 23 (1982) p.718

7) N. Kino, Y. Matsumura, H. Tsuchiya, Y. Furukawa, H. Akagi & S. Sanagi: CAMP-ISIJ, 3 (1990) p.785

8) K. Maruoka: Private communication

9) ibid.

10) ibid.

11) K. Maruoka, S. Sanagi & T. Kawano: Annual report (1990) R & D Activities at Nippon Steel, p.8

12) M. Yamada, Y. Tokunaga & K. Itoh: Seitsu-Kenkyu No.322 (1986) p.90

13) K. Ohsawa, S. Satoh, T. Katoh, N. Katayama, K. Nishimura & J. Mano: CAMP-ISIJ. 4 (1991) p.1934

14) O. Matsumura, Y. Sakuma & H. Takechi: Trans. ISIJ, 27 (1987) p.570

15) H. Takechi: Proceedings HSLA '90 (1990) ASM. To be published

16) Y. Sakuma, O. Matsumura & H. Takechi: Met. Trans. 22 A (1991), p.489

17) M. Usuda: Private communication

18) H. Takechi, T. Asamura & K. Sakurai: Steel Times Int'l, July (1990)

19) H-T Junius, W. Bleck, W. Müschenborn & C. Strassburger: Stahl u. Eisen, 108 (1988) Nr.20 p.931

20) F. Weber & H-T Junius: Stahl u. Eisen, 111 (1991) Nr.4 p.65

II. Continuous Annealing: Process Technology

The Equipment and Operations of

NKK Fukuyama No.2 Continuous Galvanizing Line

A.NAKAMURA, N.TAGUCHI, H.YANO, K.TAKAGI,

FUKUYAMA WORKS NKK CORPORATION
1 KOKAN-CHO FUKUYAMA-CITY
JAPAN

NKK constructed a brand-new hot dip galvanizing line in the Fukuyama Works in April 1990. This line called No.2 Continuous Galvanizing Line is installed for the purpose of producing exposed panels for automobile. In order to satisfy this demand, many new techniques are introduced into the design of No.2 CGL process.

Major technological points are as follows:

1) A vertical type annealing furnace with a reduction type direct-fired furnace provides good mechnical properties and good surface quality.
2) Dross defects can be improved by double pots.
3) Induction heating equipment of galvannealed furnace is very useful to suppress powdering of the galvannealed products.
4) An Iron-Zinc alloy electroplating equipment is useful to produce couble layer coating products for automotive exposed panels.

Owing to the effective equipment, No.2 CGL has been working efficiently achieving high quality products.

1. Introduction

NKK constructed No.2 Continuous Galvanizing Line (CGL) at Fukuyama Works in April, 1990, in response to the increasing demand for hot dip galvannealed steel sheets for automobiles.

This line was installed particularly for manufacturing outer panels for automobiles. Various new techniques were adopted in the design of No.2 CGL to realize this aim.

The major technological features of this line are an improved annealing furnace with a reduction type direct-fired furnace, double zinc pot, induction heating furnace and iron-zinc alloy electroplating equipment for the production of double layer coated products.

The equipment installed are extremely effective in the realization of good surface quality and good mechanical properties, and are also very useful in suppressing powdering. NKK has been successful in manufacturing top quality galvannealed steel sheets owing to the advanced technology incorporated in the line.

2. Equipment features

The equipment for the new construction of Fukuyama No.2 CGL was planned with the objectives of satisfying the following conditions:

(1) A manufacturing line that can manufacture outer galvannealed steel sheet panels for automobiles, and one that is capable of responding to more sophisticated levels in future.
(2) A new and powerful line that incorporates new technology and new equipment.
(3) A line in which a perfect quality control system has been established. The most important points in terms of technology, for satisfying the fundamental concepts mentioned above, are as follows:
  (a) A vertical annealing furnace, which is an improved annealing reduction type direct-fired furnace, with good mechanical characteristics and surface quality.
  (b) Elimination of dross defect, which is a major defect in surface quality, by installing double pots.
  (c) Improvement in the powdering resistance of galvannealed products (peeling off resistance during press forming) and improvement in surface quality by utilization of an induction type galvannealing furnace.
  (d) Improvement in press processability of outer automobile panels and improvement in surface quality by using iron-zinc alloy plating in the upper galvannealed layer.
  (e) On-line measurement control and off-line check control utilizing several quality control equipment and with a line configuration such that

external inspection of both the top and bottom surface can be simultaneously carried out.

(f) Construction of an extremely efficient and highly functional computer control system by optimum functional allocation to satisfy the rigid quality control needs and to respond to the demands of automation.

3. Equipment outline

3.1 Main specifications

The main specifications of the Fukuyama No.2 CGL are shown in Table 1.

Table 1 Main specifications of No.2 CGL

Items			Specifications
Prodution capacity		t/month	20 000
Strip size	Thickness	mm	0.40～1.60
	Width	mm	610～1 880
Products	Pure Zinc		KZ, PZ
	Iron Zinc		PZA, PZM, PZD
Continuous furnace	Type		Vertical continuous annealing furnace
	Capacity	t/h	62
Galvannealing furnace	Type		Induction heater
	Capacity	kW	3 000
Entry coil	Weight	t	45.0
	OD	mm	2 650
Delivery coil	Weight	t	25.0
	OD	mm	2 250
Line speed	Entry	mpm	160
	Center	mpm	120
	Delivery	mpm	180

3.2 Equipment configuration

The layout of Fukuyama No.2 CGL is shown in Fig.1.

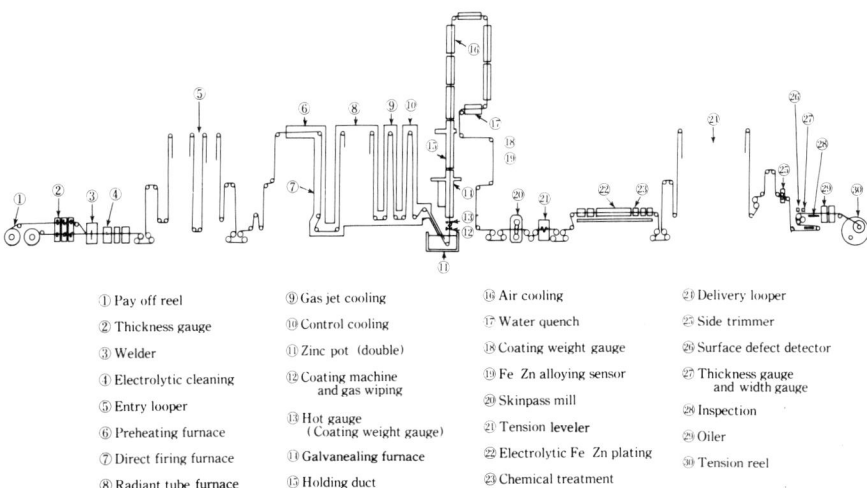

① Pay off reel
② Thickness gauge
③ Welder
④ Electrolytic cleaning
⑤ Entry looper
⑥ Preheating furnace
⑦ Direct firing furnace
⑧ Radiant tube furnace
⑨ Gas jet cooling
⑩ Control cooling
⑪ Zinc pot (double)
⑫ Coating machine and gas wiping
⑬ Hot gauge (Coating weight gauge)
⑭ Galvanealing furnace
⑮ Holding duct
⑯ Air cooling
⑰ Water quench
⑱ Coating weight gauge
⑲ Fe Zn alloying sensor
⑳ Skinpass mill
㉑ Tension leveler
㉒ Electrolytic Fe Zn plating
㉓ Chemical treatment
㉔ Delivery looper
㉕ Side trimmer
㉖ Surface defect detector
㉗ Thickness gauge and width gauge
㉘ Inspection
㉙ Oiler
㉚ Tension reel

Fig.1 Lauout of Fukuyama No.2 CGL

(1) Entry equipment

The entry equipment consists of two pay of reels, top and bottom end processing equipment and one welding machine (direct current type), electrolytic cleaning equipment and looper. Continuous nandling of coils is mostly automatic and the width of the coil and diameter is measured in the automatic charging machine, which is designed to avoid mixing of wrong materials.

(2) Furnace equipment

Furnace equipment consists of the Pre-Heating Furnace (PHF), Direct Fired Furnace (DFF), Radiant Tube Furnace (RTF), Gas Jet Cooling equipment (GJC) and Controlled Cooling (CC) equipment.

(3) Coating equipment

The coating equipment consists of the zinc pot and, gas wiping and coating machine for regulating the galvannealed coating weight, the galvannealing furnace for alloying the surface and cooling equipment.

(4) Central equipment

The central equipment consists of skinpass mill and tension leveler which improves the mechanical qualities, surface properties and shape of steel sheets. Upper layer electroplating equipment improves the appearance of galvannealed material and press processability, and chemical treatment equipment improves chromate treatment.

(5) Exit equipment

The exit equipment consist of looper, trimming equipment, top and bottom side inspection equipment, oiler, tension real equipment (Carousel type), tail end accuracy inspection equipment and binding/weighting equipment.

3.3 Main equipment outline

3.3.1 Annealing Furnace Equipment

(1) Direct heating furnace

A vertical direct fired reduction heating burner was adopted for heating the annealing furnace. This equipment has very good economic advantages because of the improvement in threading, good heat response and compactness in construction. The main product of this line is wide sheets for outer automobile panels. Quality verification tests of the already existing lines were carried out repeatedly as a part of the investigation of equipment specifications in pursuit of the optimum direct heating furnace. Some of the characteristics are given below:

(a) Adoption of NKK's original new direct-fired reduction heating burner (Toroidal burner) and improvement in reduction performance by expansion of reduction zone.

(b) Optimum burner disposition and trimming function enabling uniform heating in the transverse direction of plate, taking into account the stabilized threading of materials with large width.
(c) Adoption of dynamic air ratio control corresponding to the fluctuations in fuel properties.
(d) Installation of optimum after-combustion control system and gas flow control systems for restricting the oxidation on the surface of steel sheets in the furnace.

In addition, NKK technology has also been introduced to the equipment of CGL in many places before and after the direct heating furnace (Preheating furnace, Roll room, etc). Fig.2 shows the construction of the direct heating furnace.

The combustion control zone is divided into four zones, with Selas nozzle mixing burners arranged in zones No.1 and 2 and burners for reduction heating arranged in zones No.3 and 4. The load balance of combustion control zone is designed in the last stage, attaining a large combustion capacity. In zone 3, the nozzle-mix type new direct-fired reduction heating burner developed by NKK has been adopted, and this has resulted in large scale improvement in the reduction performance.

Complete non-oxidized heating is attained up to a steel temperature of 720°C because of this ideal combustion pattern.

(2) New direct-fired reduction heating burner

The schematic sketch of the construction of the nozzle-mix type new direct-fired reduction heating burner is shown in Fig.3. A mixing capacity which is equivalent to the conventional pre-mix type burner is obtained on blowing fuel gas radially from the center part of the burner, and the air for combustion creates a swiveling flow from the tiled wall all around. Owing to this technology, a uniform reduction flame zone is formed after the burner tile exit.

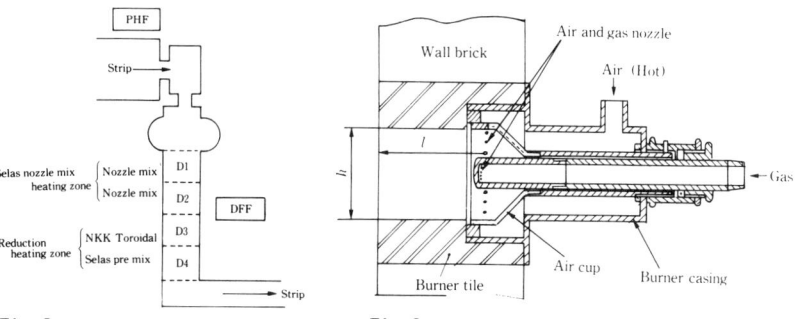

Fig.2　　　　　　　　　　　　Fig.3
Direct-fired furnace arrangement　　Construction of NKK toroidal burner

(3) Atmosphere heat treating furnace

The steel sheets that have been heated nearly up to the annealing target temperature in the direct-fired furnace are uniformly heated and cooled in the atmosphere heat treating furnace in the next stage. Gas Jet Cooling (GJC), which is part of the cooling equipment consisting of double pass, restrains the cooling speed to a low value and can control the air quantity in the transverse direction of the plate, thus preventing deformation of the steel sheet during cooling. The Controlled Cooling (CC) zone adopts air cooled tube system. The GJC cooler is built in the furnace and the internal pressure in the CC air cooled tube is negated, as a countermeasure for preventing the mixing of oxygen in this furnace. It also minimizes any adverse effects should a malfunction occur.

As a countermeasure for preventing minute flaws, the atmosphere heat treating furnace has stainless steel sheets on all the internal walls. This prevents contamination within the furnace due to refractory products.

Furthermore, all the hearth roll surfaces in the furnace have been ceramic coated, and occurrence of dents in steel sheet have been eliminated.

For sheet temperature measurement, radiation thermometers using multi-reflection effect (which is not influenced by emissivity of steel) developed by NKK are widely used. High accuracy in sheet temperature measurement has become possible, using this thermometer. This has also contributed considerably to stabilization of quality.

3.3.2 Galvanizing pot

A ceramic pot is used as the galvanizing pot and there are two inductors for zinc solution and for temperature maintenance, with a 300kW capacity in each pot. There are two pots, and by shifting these pots off-line by air bearing, dross can be eliminated at an early stage. This arrangement also contributes to large scale improvement in appearance.

3.3.3 Gas wiping equipment

One-sided blowing type, large width, acute angled nozzles are adopted with the aim of splash improvement, simplification of devices and for increasing gas wiping ability. Oxygen and nitrogen are interchangeable as gases used for gas wiping. Furthermore, for improving the strip shape, submerged rolls (as shown in Fig.4) are provided with a C warp correcting roll.

3.3.4 Zinc coating control

In CGL, the gas wiping nozzle and coating weight gauge are separated, causing achievement of high accuracy to be difficult throughout the length of the coil by feedback control only. Therefore, NKK has developed and achieved a coating control system with improved setting accuracy by adopting the coating weight control model.

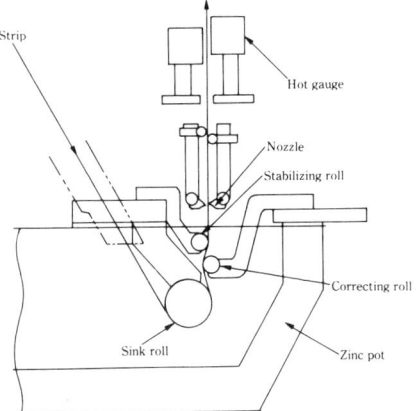

Fig.4 Gas wiping and coating machine of No.2 CGL

The features of this system are listed below.
(1) The setting calculations are carried out by the control model shown below.

$Ps = a \times Cw^b \times Dc \times Vd \times \exp(-eV)$

where,

Ps : Pressure setting value
Cw : Target coating weight
a-e : Model constants
D : Nozzle spacing
V : Line speed

(2) The coating weight condition changes according to factors which have not been taken into account (such as the type of steel), but by changing the model constants because of these factors, value setting errors are minimized and high accuracy control is realized.

(3) To facilitate tuning of model constants, a mini computer specially meant for coating weight control, capable of collecting, analyzing and editing a large volume of data, has been installed.

3.3.5 Galvannealing furnace

The galvannealing furnace makes use of an induction heating system (rated output 3000kW), the first in Japan, with the aim of improving the powdering resistance and surface appearance quality. The advantages of this furnace, based on the principles of induction heating system, are:

(1) Uniform heating in the transverse direction is possible because it is unaffected by changes in emissivity of strip surface or changes in strip shapes.

(2) The response to changes in size and coating quantity is extremely fast.

(3) Repeatability is possible because there is no thermal inertia.

Good results are being obtained on account of these advantages. An edge heater is also installed as a measure to prevent alloying of the strip edge. The soaking furnace for alloying adopts the electrical heater heating system and retains it for a sufficient holding time.

Furthermore, the air cooling system is adopted for cooling after the soaking zone.

3.3.6 Alloy coating analyzer

An alloy coating analyzer has been installed for on-line measuring of the alloying rate (iron content) in the coating. This device makes use of a monochromatic excitation X-ray fluorescence method with two optical systems. This has been independently researched and developed by NKK. As shown in Fig.5, the coating weight and the level of alloying can be determined by the X-ray values measured by the low and high angle optical systems. Fig.6 shows the results of the measurement accuracy test check. It is evident from this figure that the target value of below 0.5 Fe percent has been attained.

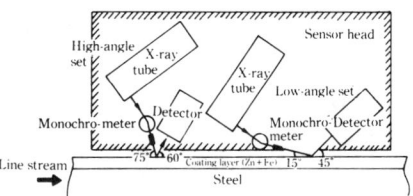

Fig.5 Configuration of sensor head

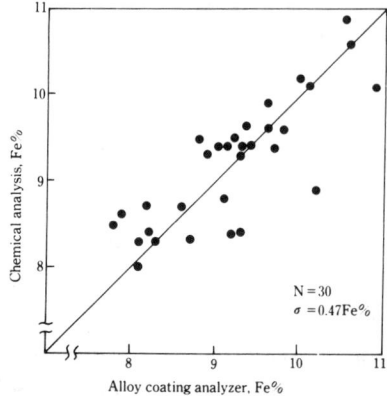

Fig.6 Accuracy of alloy coating analyzer

Now, this device is becoming indispensable (as a quality control instrument) for the stable operation of alloying galvanized steel sheets and the improvement of quality control. Fig.7 shows the system configuration of coating weight control system and alloying control systems.

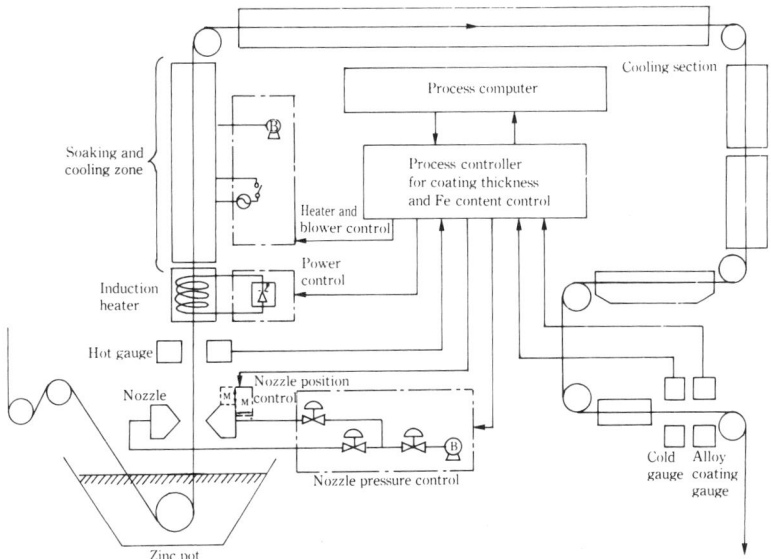

Fig.7 System configuration of coating weight and iron content control

3.3.7 Skinpass mill

(1) Skinpass mill

The skinpass mill has been installed for improving the mechanical properties and surface appearance of the product. The main specifications of Skinpass Mill are as follows:

Type   : 4Hi hydraulic push up control
Rolling load   : 7840kN maximum
Drive system   : Twin back-up roll independent drive system
Shape control   : Upper back-up roll equipped with TP roll
                  Increase and decrease work roll bender

When the welding point is passing through, two modes can be selected, the open mode which opens the mill roll and soft rolling mode which opens the mill gap value and reduces the push-up power. These are automatically corresponded by tracking the welding point.

For shape control, in addition to the increasse-decrease bender roll, a TProll (a system in which a piston with taper is provided in the outer tube sleeve, the piston is moved in the longitudinal direction of the roll body and the BUR crown is varied) is used in the upper BUR, so that all the materials can be rolled with one type of work roll crown.

3.3.8　Upper layer electroplating equipment

　　The upper layer electroplating equipment aims at press processability and improvement of surface appearance quality. It is used for electroplating iron and zinc and has 3 plating tanks with two rectifiers, each with a capacity of 25000A per tank.

3.3.9　Quality assurance equipment

　　Perfection of quality assurance system is very important for responding to the needs of consumers. This equipment makes use of high quality control systems including strip temperature control and coating weight/alloying control systems for obtaining stabilized quality. For quality assurance of appearance, various types of quality assurance devices, have been installed. In addition, tail-end accuracy inspection equipment is also provided to detect minute flaws for perfecting the quality assurance controlsystem. The quality assurance devices installed for this equipment are shown in Table 2.

Table 2　Facilities for quality assurance

Facilities	Type	Usage
Thickness gauge (Entry side)	γ-ray	Cutting the off gauge portion
Hot gauge	X-ray	Adjustment of the coating weight
Coating weight gauge	X-ray	Judgement
		Control of the coating weight and logging it's data
Alloy coating analyzer	X-ray	Control of the iron content in the coating layer and logging it's data
Thickness gauge (Delivery side)	γ-ray	Judgement
		Feed forward and logging the off gauge data
Width gauge	Image sensor	Judgement
		Feed forward and logging the off gauge data
Surface defect detector	Flying spot	Judgement
		Feed forward and logging the top and bottom surface defect data

　　Deviation from the tolerances of coating weight, strip thickness and strip width are verified with the order reports and recorded verification reports, along with the positions in the longitudinal direction. The deviations from the target values are also similarly recorded using the alloying gauge. The type of surface defect, degree of flaw and its longitudinal and transverse position are also recorded using the surface defect detector, and the flaw is automatically ranked. This serves as a back-up for the inspector during evaluation.

　　Moreover, during the pay out of product coil from the line, in addition to the coil sizes and weight, the elongation rate, inner diameter, coil winding direction, etc., are also automatically verified with the orders, therefore avoiding operation mistakes.

3.3.10 Computer control system

The configuration of computer control system used in the No.2 CGL is shown in Fig.8.

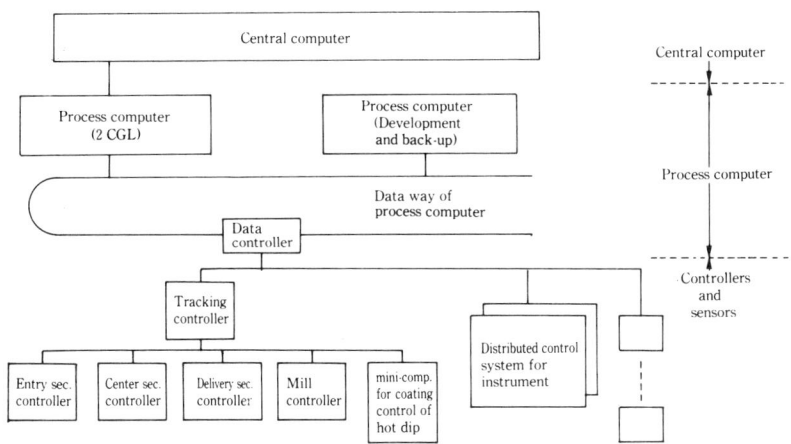

Fig.8 Configuration of computer control system

As is evident from this figure, this system consists of three groups of computers, namely the central computere for production process control, the process computer (which handles overall process control and management), and the control system group that executes various kinds of measurement control, handling and operation of lines. The classified functions of the computer control system are shown in Table 3.

Table 3 Classified functions of computer control system

Level	Component	Main function	Process control function
1	Central computer	· Production planning and scheduling · Material flow control · Result data management	· Master table management
2	Process computer	· Coil tracking · Preset and data gathering · Yard management	· Line speed management · Strip temperature control · Coating control of electrolytic galvanizing
3	Process controllers and measuring devices	· Line drive · Sequence control · Measurement and instrument	· Coating control of hot dip · Mill and leveller control

As mentioned before, the No.2 CGL is a large scale complex process line, incorporated with several processes and an annealing furnace. By means of system configuration and optimum distribution of functions mentioned above, all the equipment and data in the No.2 CGL, starting from complex line controls to quality assurance controls and total yard controls, are effectively managed.

4. Operating status

4.1 Results of production

The No.2 CGL, which is a process line integrating the hot dip galvanizing equipment and electroplating equipment, has been successfully operating since the commencement of operations on April 27, 1990, with the monthly production after seven months reaching 20000 tons, and has continued to operate at a high production level.

Furthermore, after optimization of coating weight control and alloying conditions, optimization of upper layer electroplating operation conditions, the operation conditions were established in the early stage, with the result that the manufacture of double layer Iron-Zinc alloy galvanized steel sheets "PZM", (the main product of this equipment), has also progressed smoothly. Since September 1990, the production per month has reached 4000 tons, demonstrating that the functions have been fully and effectively utilized.

4.2 Quality characteristics

4.2.1 Mechanical Properties

This is a line focusing mainly on the manufacture of outer panels for automobiles, but to respond to the users diversified needs in terms of material quality, the line has equipment which enables the manufacture of products with various grades, from the ultra deep drawing grade to the high ten grade. The annealing furnace has a longer path for the heating zone and heating is uniform; it can also maintain the optimum sheet temperature at 870°C. This sheet temperature is controlled with good accuracy by the process computer. The maximum rolling load of the skinpass mill is 7840kN and elongation is controlled appropriately to suit the material grade. The typical characteristic values of mechanical properties presently being marketed are shown in Table 4.

Table 4 Examples of mechanical properties produces on No.2 CGL

Grade		Mechanical properties				r	2%BH* N/mm^2	Kind of steel
		YP N/mm^2	TS N/mm^2	El %	n			
Commercial quality		255	355	38	0.21	—	—	Low carbon steel
Drawing quality		215	325	44	0.22	—	—	〃
Deep drawing quality		165	305	46	0.23	1.5	—	Ultra-low carbon steel
Ultra-deep drawing quality		155	295	48	0.24	1.7	—	〃
Hi-ten	35^{kilo} BH	235	350	40	0.20	1.5	35	Ultra-low carbon P-added steel
	40^{kilo} class	315	410	36	0.20	—	—	Low carbon steel
	45^{kilo} class	345	450	34	0.20	—	—	Medium carbon steel

*: BH property means raised quantity of yielding point after 170°C × 20min. treatment after giving 2% pre-strain.

4.2.2 Uniformity of coating

The most fundamental and important point of a galvannealed product is the uniformity of zinc coating weight in the longitudinal and transverse direction. This is because the uniformity has a large influence on the corrosion resistance, weldability and powdering resistance.

The conditions for uniformity of coating are as follows:
(1) Optimization of slit gap profile in the gas wiping nozzle transverse direction.
(2) Flattening of steel strip shape at the nozzle position and vibration prevention.
(3) Accurate measurement of distance between nozzle and steel strip and repeatability.

To resolve these problems the following investigations were made. An investigations of dynamic pressure distribution and coating weight distribution with actual nozzles; perfection of stability conditions for steel strip shape by appropriate combination of submerged rolls disposition, roll diameter and roll pressing quantity; and improvement in repeatability of front/rear nozzle pitch setting values. Presently, these conditions and coating weight gauge have been linked and an almost satisfactory coating distribution has been obtained. Furthermore, adjustment of the hot gauge, installed directly above the gas wiping nozzle, is also being carried out and further improvements in accuracy are anticipated.

4.2.3 Powdering resistance

One of the most important advantages of the galvannealed steel strip is the powdering resistance. Particularly in the case of pressing such steel sheets as the outer panels of automobiles, if powdering should occur, it leads to such defects as press dents and inability of continuous pressing. The most important methods for improving the powdering resistance are as follows:
(1) Uniformity of coating weight in the transverse and longitudinal directions.
(2) Uniformity of iron concentration in the transverse and longitudinal directions.
(3) Control of alloying temperature and heat pattern after the galvannealing furnace.

Item(1) has already been discussed before but for items(2) and(3), the conclusions arrived at in laboratory tests were verified in the actual lines. The results showed that the induction heating system alloying furnace demonstrated the anticipated performance in terms of high response speed, uniform heating, and ease in changing the heat cycles compared to the conventional gas furnace. Moreover, it also demonstrated the feasibility of manufacturing products with good powdering resistance. Furthermore, even for strips with varying steel compositions, because of the ease in presetting and control, the variation in iron concentration can be restricted to a minimum. In addition, iron concentration is always checked by means of the on-line alloying gauge and the concentration is being controlled so that it falls within the optimum range.

The topics for the future are the establishment of automatic control between this alloying gauge and the loading power of galvannealing furnace and steel sheet temperatures.

The typical coating weight distribution and iron content distribution in the coating layer is shown in Fig.9. Furthermore, Fig.10 shows the relation between coating weight and iron content in the coating layer, and the powdering resistance.

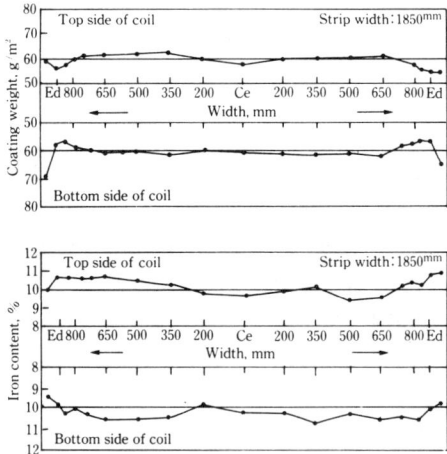

Fig.9 Coating Weight and Iron Content Profile

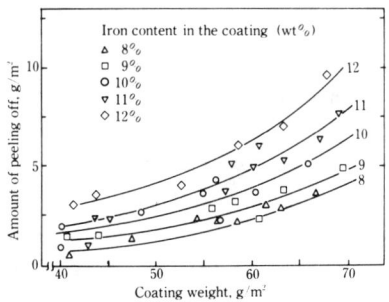

Fig.10 Effects of coating weight on the amount of peeling off by drawbead test

5. Conclusion

The Fukuyama No.2 CGL was newly installed at the Fukuyama Works in April, 1990. This CGL was installed to maintain high productivity and mainly for automobiles, particularly for manufacturing high quality products to suit outer panel applications, and is a new and powerful line equipped with the latest high technology equipment. Since the commencement of production and marketing, work has stabilized in a short period of time and operations are proceeding smoothly amidst favorable environment of large scale demands for surface treated steel strips, mainly for automobile applications.

In order to fully utilize the merits of this line and aim to satisfy the high level needs of consumers, NKK shall improve existing equipment and aim for improvement in operating technology, concentrating also on the rationalization aspects, such as cost reduction.

CONTINUOUS ANNEALING LINE

OF SOLLAC STE AGATHE

P. LOUIS, C. DEUTSCH, C. BRUGNERA

SOLLAC - 57191 FLORANGE - FRANCE

ABSTRACT

The continuous annealing line of SOLLAC Ste Agathe has been built in 1987/1988 for both tin plate and sheet production. Since february 1989 this line produces only sheets but some tests are being performed to produce tin plate again.

In a first part of this report, the technology used to produce coïls with an excellent uniformity is presented, especially the cleaning section with a system of quick changing (SOLLAC Patent) which allows the maintenance while the line is running.

In a second part of this report, the properties of various commercial qualities are presented and the results of industrial production from the bake-hardening DQ quality which is now developed on this line are particulary discussed.Otherwise, for this last quality, the influence of process conditions (baking time, baking temperature and prestain after temper-rolling) on the bake-hardening values has been also studied.

I - INTRODUCTION

The continuous annealing line from Ste Agathe (SOLLAC - FLORANGE) has started in June 1988. This plant has been designed to produce some steels for packaging (tin plate, tin free steel) as well as cold rolled sheets. Since February 1989, this line produces sheet only. Some tests are being performed to produce tin plate again.

Its main characteristics are :

- thin sheet :   thickness   : from 0.35 mm to 1.4 mm
                 width   : from 700 mm to 1700 mm
                 furnace speed : 230 m/mn max.

- tin plate :   thickness   : from 0.17 to 0.5 mm
                 width   : from 700 to 1300 mm
                 furnace speed : 600 m/mn max.

- total length of the line : 335 meters

- rated production : 103 tons/hour, max

- annual rated production :   450 000 tons of sheet

          or   300 000 tons of sheet and
              100 000 tons of tin plate

Eighteen months after the beginning the rated capacity has been reached (figure 1 : evolution of the productivity since the beginning).

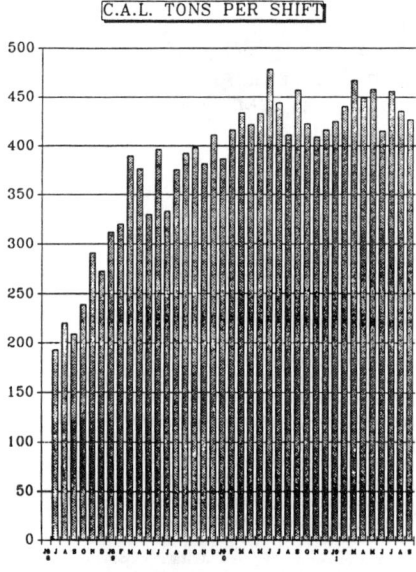

Figure 1 : EVOLUTION OF PRODUCTIVITY

II - TECHNOLOGICAL CHOICES

The study of the dual purpose continuous annealing line has been lead in the constant care of building an equipment allowing to produce coils of a perfect quality from the first to the last meter. Each technological choice has been made to reach this purpose.

Today, the rate of sheet for car outer panel is 20 to 25 % of the total production. Among this production, the downgrade rate due to the continuous annealing line is below 1 %.

The technological choices that guarantee this quality are :

1 - Entry section

A horizontal cleaning section designed for sheet degreasing at a maximal speed of 800 m/mn.

A brusher with three brushes per face coupled to a demineralized water rinsing over three pairs of drying rolls secures the sheet cleanliness. A system of quick changing (SOLLAC Patent) allows the replacement of the brushes or the drying rolls while the line is running. The advantages of such a system are :

- it is not necessary to wait for the next maintenance period to ensure the replacement,

- the cleanliness of the strip is maintained to its better level always using the brushes to the maximum.

2 - Furnace section

The ceramic coating of the furnace rolls helps to get rid of any risks of dents or pick up. On the other hand this coating limits the abrasion of the rolls roughness and so it contributes to the good tracking of the strip in the furnace.

Using the stretching rolls before the top rolls of the heating and soaking sections, delays the appearance of heat buckles that allows the production of particularly fragile sizes.

The rapid cooling is performed by a cooled HN gas blowing at high speed (KAWASAKI patent). This technology allows to keep a good flatness at the exit of this section and also a rapid cooling process from 650°C to 400°C with cooling rates reaching 80°C/sec. Using guiding rolls in this section avoids all risks of scratches that are inappropriate to the quality of the product.

Respecting the annealing cycle is secured by the use of seventeen pyrometers installed all along the furnace. The temperature set points of the heating and soaking zones are defined according to a mathematical model. The use of this model allows to manage the temperature transitions better and so it limits the strip lengths affected by these phenomena.

3 - Exit section

The skin-pass equipment (quarto), with a flatness regulation driven by a planometer-roll ASEA, secures a perfect flatness without any operator.

The use of CVC (continuous variable crown) cylinders (SMS patent) for the temper rolling of electrical sheet (elongation from 2 to 7 %) allows to control the strip flatness with a single type of cylinder crown.

A mathematical model of pressure gives the predetermined temper rolling parameters for the next coil. The application of these predetermined parameters is made automatically while the weld runs through the skin-pass. This model allows :

- to avoid a line strop in the skin-pass,
- to limit the transitional periods,
- it doesn't need the attendance of an operator during the change of product.

The control of the strip aspect is performed by an employee of the Quality Service standing at an inspection post especially designed for this purpose. A detection system of defects with a laser beam (SICK system) increases the control efficiency of the surface aspect.

III - METALLURGY

1 - Qualities produced on the continuous annealing

Steel metallurgy with the continuous annealing depends on the chemical analysis and also on the various processing conditions all along the manufacturing process.

For each stage of production, the important parameters are :

a) Steel making plant :   the metallurgy will be according
to the quality aimed

- either aluminium-killed steels
- or interstial-free steels
- or niobium low alloyed steels
- or rephosphorized steels.

b) Hot rolling : the aluminium killed metallurgy requires a hot coiling (more than 700°C). To limit the effects of the cooling on the coil extremities, it is imperious to make a selective coiling according to the following graph.

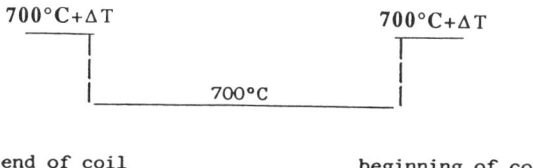

end of coil   beginning of coil

c) Continuous annealing : the standard annealing cycle, made on Ste Agathe Line, is represented on the graph below. It is composed of six phases : heating - soaking - slow cooling - rapid cooling - overaging - final cooling.

The main parameters that define our annealing cycles are :

- the soaking temperature,
- the overaging time that defines the maximum line speed,
- for certain qualities the rapid cooling rate.

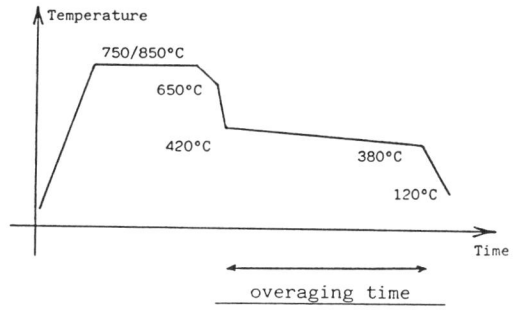

Standard cycle on the Ste Agathe CAL

The characteristics obtained for the different qualities industrially produced are given in the Table 1.

QUALITY	YIELD STRENGTH (MPa)	TENSILE STRENGTH (MPa)	ELONGATION (%)	r	n
CQ	220	325	39	2.0	0.185
DQ	205	315	40	2.0	0.195
DDQ (AlK)	200	315	41	2.2	0.215
DDQ (IFS)	170	315	42	2.3	0.235
EDDQ	160	315	42	2.5	0.235
Rephosphorized steel	285	380	34	1.9	0.180
Steel for enamelling	225	330	39	1.9	0.195
HSLA (Soldur 340)	370	435	28	1.1	0.155
HSLA (Soldur 380)	410	475	25	1.0	0.140

Table1 : MECHANICAL PROPERTIES FOR THE MAIN QUALITIES

The mix product is divided as follows :

Quality		
	CQ	20 %
	DQ	25 %
	DDQ (Alk)	5 %
DDQ and	EDDQ (IFS)	25 %
	HSLA	5 %
Steels for enamelling		5 %
Rephosphorized steels		5 %
Electrical steels		10 %

2 - Metallurgical developments

The metallurgical developments mainly deal with :

- Interstitial free steels with high tensile strength.
The developed qualities are the following.

IFHR 350 : Rm ≥ 350 MPa

IFHR 380 : Rm ≥ 380 MPa

- Steels with a warranted bake-hardening level obtained with different means

- aluminium-killed steels,
- rephosphorized steels,
- non stoichiometric niobium-bearing steels ($\frac{Nb}{c} < 1$)

The bake-hardening DQ quality can be achieved with aluminium-killed or non stoichiometric niobium-bearing steels. Next, only the results on bake-hardening DQ quality obtained with aluminium-killed steels will be discussed.

2a - Results of industrial production

Development of bake-hardening qualities was limited to batch annealed steels and the bake-hardening values were rather weak.

With its two continuous annealing lines dedicated to sheet products, SOLLAC took on the development of these bake-hardening qualities with this new process.

In order to obtain a bake-hardening DQ quality with low yield strength and high Lankford value through Alk metallurgy, two initial conditions are required :
- high coiling temperature (700/750°C),
- high annealing temperature (800/850°C).

When all the nitrogen has precipitated with aluminium, the solute carbon content, which yields bake-hardening, is controlled by the overaging conditions after soaking.

The industrial production of bake-hardening DQ quality shows :

- an excellent homogeneity of the results for the yield strength (variation from 200 to 230 MPa with a mean value of 216 MPa) (fig. 2 histogram of yield strength),

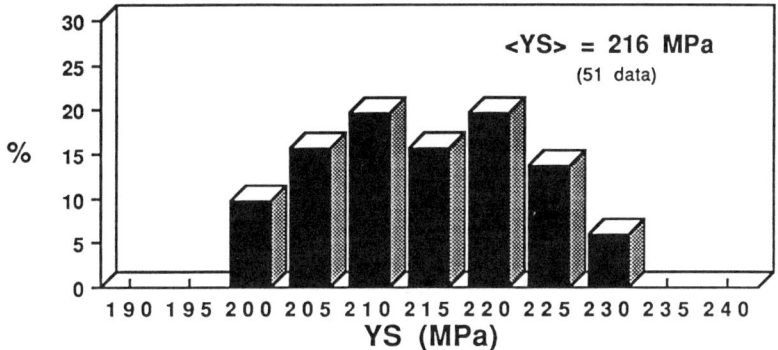

Figure2: BAKE-HARDENING DQ QUALITY: RESULTS ON YIELD STRENGTH

- a bake-hardening mean value of 41 MPa with only 2 % of the measurements less than 30 MPa (fig. 3 bake-hardening histogram),

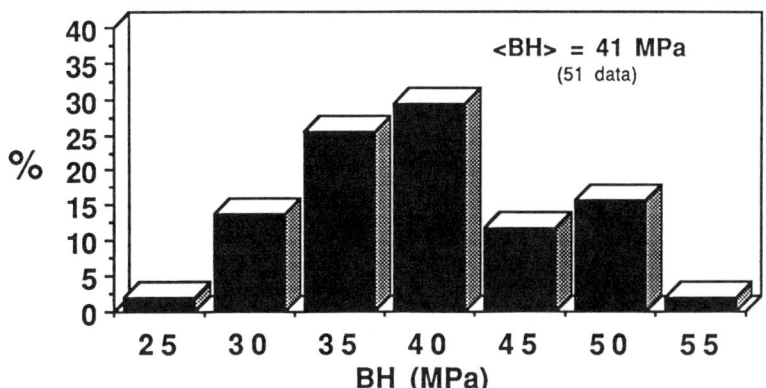

Figure3: BAKE-HARDENING DQ QUALITY : RESULTS ON BH VALUES.

- the sum work-hardening plus bake-hardening is centered on 72 MPa and 90 % of the results are superior to 65 MPa (fig. 4 histogram of the sum work-hardening and bake-hardening).

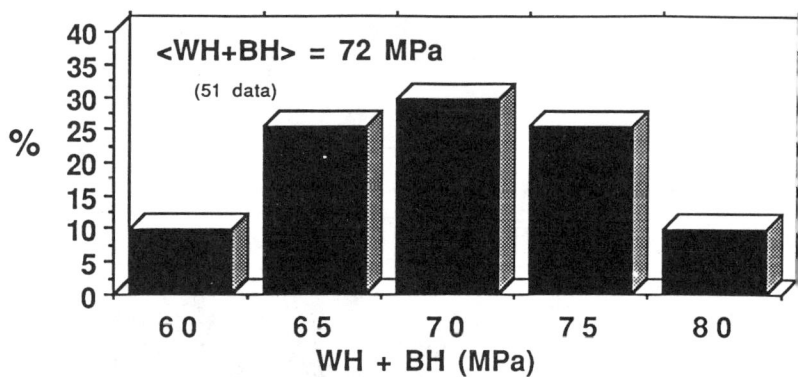

Figure4: BAKE-HARDENING DQ QUALITY : RESULTS ON "WH+BH" VALUES.

2b - Influence of process conditions on bake-hardening

2b.1 - Influence of temperature and baking time on bake hardening

A laboratory study has been done on bake-hardening DQ quality. The metal has been annealed and temper rolled under the same conditions as those used on the CAL Line. Four temperatures (120, 140, 160 and 170°C) and two baking times (10 and 20 minutes) have been tested.

The results (fig. 5 bake-hardening according to the baking temperature and time) indicate that the bake-hardening values are steady in the range of temperature and baking time studied.

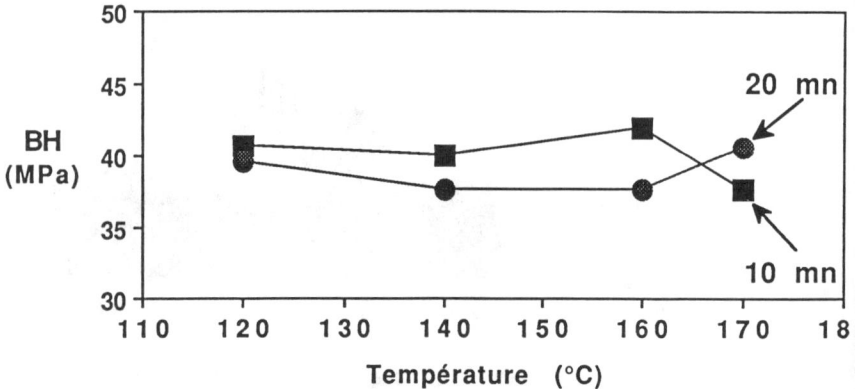

Figure5: INFLUENCE OF TIME AND TEMPERATURE OF BAKING ON THE BAKE-HARDENING

2b.2 - Influence of prestrain

Some laboratory tests have proved that :

- bake-hardening values slightly decrease as the prestrain increase. Yet, this loss is largely compensated by the work-hardening increase. Overall, as the prestrain is increased (fig. 6 work-hardening and bake-hardening according to the predeformation level),so is the bake-hardening.

- bake-hardening values after 0,4 - 0,5 % prestrain (uniaxial tensile test) are typically at the same level that those obtained for more important prestrains.

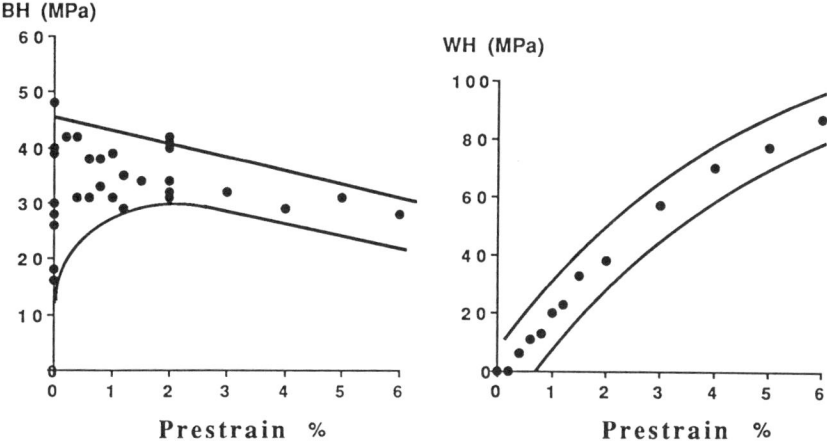

Figure 6 : BAKE-HARDENING DQ QUALITY:INFLUENCE OF PRESTRAIN

2b.3 - Influence of time between drawing and paint baking

One point that is particularly sensitive for car builders, is the influence of time between the process of drawing and cataphoresis on the bake-hardening.

The influence of aging before the paint baking has been studied as follow :

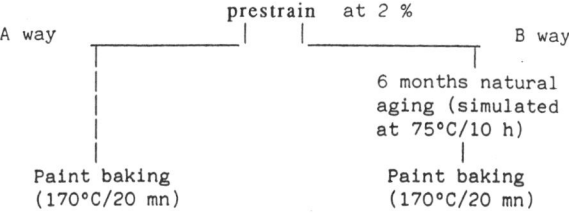

The results observed indicate :

A way	: bake-hardening level	: 48/50 MPa
B way	: natural aging level	: 30/35 MPa
	bake-hardening level	: 15/20 MPa
	total hardening after baking	: 45/50 MPa

2c - Conclusion

The bake-hardening steels used by car builders for outer panels allow :

a) for a constant thickness, an increase of the dent resistance,

b) for an equivalent dent resistance, a decrease of thickness,

c) for constant thickness and dent resistance a better drawability.

SOLLAC development on the bake-hardening DQ quality aims to achieve these purposes.
The present results show a good making of this quality and an interesting behaviour of the bake-hardening phenomenon in the operating conditions.

IV - CONCLUSION

The continuous annealing line of SOLLAC Ste Agathe has been built in 1987/1988 for both tin plate and sheet production. Since february 1989 this line produces only sheets but some tests are being performed to produce tin plate again.

The technology used allows the production of coïls with an excellent uniformity.The cleaning section with a system of quick changing (SOLLAC Patent) which allows the maintenance while the line is running increases the productivity.

Various commercial qualities are yet producted with this continuous annealing line and bake-hardenable steels have been developped.For the bake-hardening DQ quality, industrial results show that the bake-hardening values are superior to 30 MPa for 98% of the production. Otherwise, for this last quality we have studied the influence of process conditions on the bake-hardening values and we have show that baking time between 10 and 20 minutes and baking temperature between 110 and 170 °C don't have any influence on the bake-hardening values. Other results show that the bake-hardening values reached after 0,4/0,5 % of prestrain are similar to those reached after 2 % prestrain and that time between drawing and painting does'nt have any significant influence on the total hardening.

Improvement in First Cooling Techinique (Roll Quench and Water Quench) and the Properties of the Products

Mineki Okura, Hidetada Makino, Yoshiki Tanaka, Jiro Iwaya, Hiroyuki Maeda

Kobe Steels Ltd
1, Kanazawacho, Kakogawa, Hyogo, 675-01, JAPAN

Abstract

A continuous annealing line with two types of quenching methods, Roll Quenching and Water Quenching, for steel sheets started up in 1982, at Kakogawa Works of Kobe Steel Ltd. Cold rolled mild sheets are produced by Roll Quenching system. Various types of high strength steel sheets are also available from the Water Quenching system. This paper discusses the Roll Quenching characteristics that are mainly affected by the quenching roll surface conditions, and the Water Quenching technology, such as modifications of equipments and improvements of operational conditions for producing high strength cold rolled dual phase steel sheets.

Introduction

A continuous annealing line (CAL) with two types of quenching methods, Roll Quenching and Water Quenching, for steel sheets started up in 1982 at kakogawa works of Kobe Steel Ltd. In this CAL, mild steel sheets are produced by the roll quenching system, and high strength steel sheets are primarily produced by the water quenching system.

This paper discusses the Roll Quenching technology supporting the homogeneous quenching conditions, and the Water Quenching technology for producing high strength cold rolled dual phase steel sheets.

Layout and Specification of the Continuous Annealing Line (CAL)

Figure 1 shows the layout of the CAL and Table 1 lists the main specifications. The roll quenching system is suitable for producing cold rolled mild steel sheets with a cooling rate of 100~400°C/second and no reheating energy. The water quenching system is characterized by a very quick quenching rate of 1000~2000°C/second; this produces high strength dual phase steel sheets with excellent drawablity without expensive alloying elements. Presently we have the chance of using water quenching cycle once a month. It requires about four hours of down-time to change the roll quenching cycle to the water quenching cycle and vice versa.

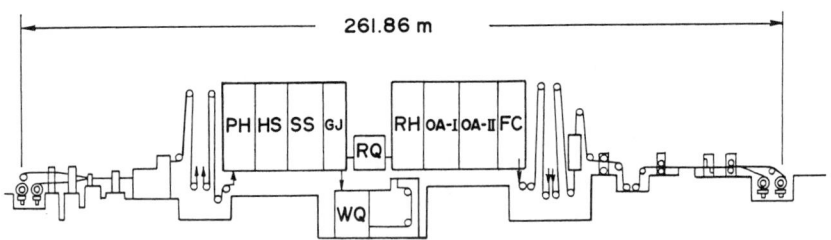

Figure 1. Layout of CAL

Table 1. Specification of CAL

Capacity	t/month	45000
Productivity	t/h	max. 108
Coil Thickness	mm	0.40~2.30
Coil Width	mm	600~1620
Coil Weight	t	max. 50
Line Speed	m/min	Entry, Delivery Section 340 Center Section 250

The roll quenching system can control the cooling rate, by changing the contact length from 0m to 5.4m by moving the four quenching rolls of 1400mm in diameter. Strip tension is controlled by bridle roll units at the entry and the delivery sides of the roll quenching system. The quenching temperature of the water quenching system is measured to control mechanical propreties at the entry of the quenching section. After quenching, steel sheets shall be pickled by hydrochloric acid to remove the oxidized surface and then tempered in the re-heating section. Tempering temperature is controlled with the pyrometer on the delivery side. Figure 2 shows the layout of the first cooling section.

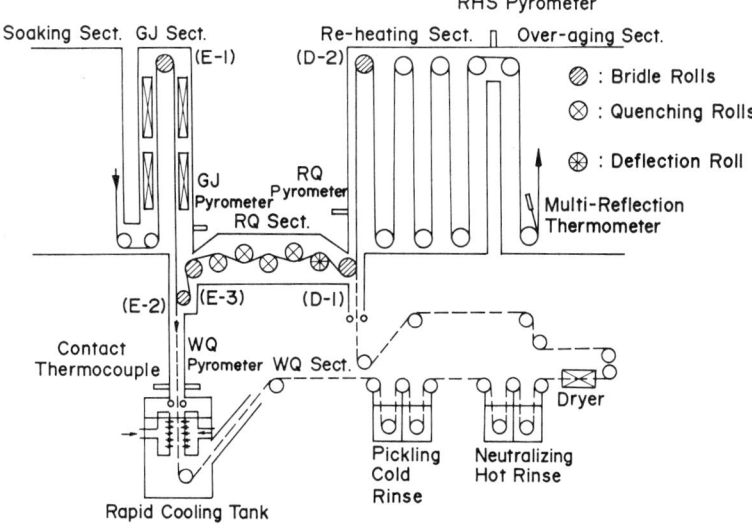

Figure 2. Layout of first cooling section

Figure 3 shows the typical heat patterns of the first cooling systems, roll quenching and water quenching.

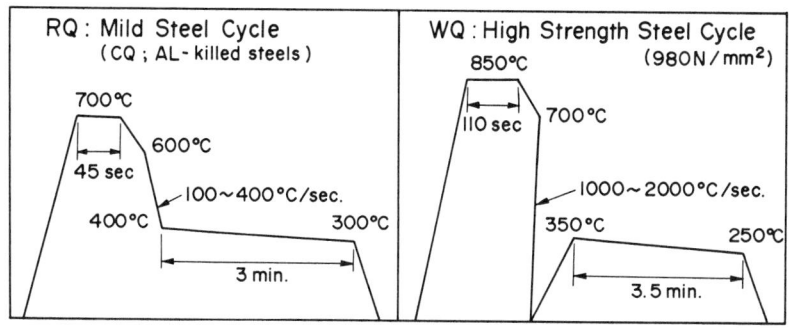

Figure 3. Typical heat patterns

Homogeneous quenching technology for a roll quenching system

It is important to control appropriate contact pressure and heat transfer coefficients for roll quenching by homogenizing the contact heat transfer between quenching roll and strip. Improper condtions for contact pressure and heat transfer coefficient bring unbalanced ,and uneven widthwise cooling, which have a bad effect on material properties and strip flatness. Cooling buckles occasionally occur in the case of unbalanced cooling. It is necessary for preventing unbalanced cooling to control strip tension and quenching roll surface condition.

Strip tension

In the roll quenching section, the strip doesn't have the prefect flatness and the homogeneous widthwise temperature distribution. Consequently strip tension has an unbalanced widthwise distribution. In case of a great unbalanced distribution, the contact condition between strip and quenching rolls becomes worse and results in bad cooling condition. Appropriate strip tension is required to prevent bad contact condition. Strip tension is controlled properly by the bridle rolls at the entry and the delivery sides of roll quenching system.

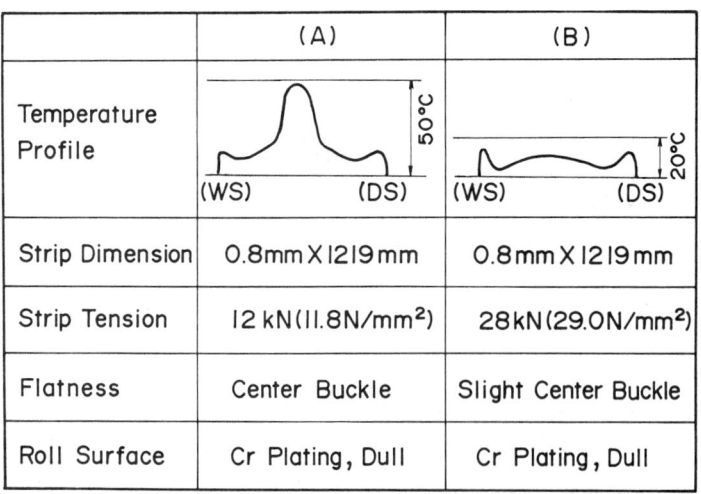

Figure 4. Effect of strip tension on temperature profile in width direction

Figure 4 shows the temperature profile at the delivery side of roll quenching system using chromium plating roll. Figure 4(A) shows the uncontacted condition for the center of the strip due to the shortage of strip tension and results in improper heat transfer. Figure 4(B) shows how the increase of tension contributes to the flatness of the temperature profile.

Condition of quenching roll surface

Local abrasion of chromium plating roll with dull surface ; Strip profile is largely improved with the increase of strip tension. But then the strip temperature of the center and both edges, after roll quenching, are higher than other areas. This phenomenon is due to the increase in the contact area between quenching roll and strip, because of local abrasion of peak at surface profile in chromium-plated dull surface roll, and the consequest increase of heat transfer. Figure 5 shows the abrasion rate of the chromium quenching roll surface. Kobe Steel's strip width distribution treated in the CAL results in a slight slip on the strip contact at about 400mm from roll center. And then it is necessary for the roll surface material to be high wear resistant and undiscerning of the variation of heat transfer coefficient with the abrasion.

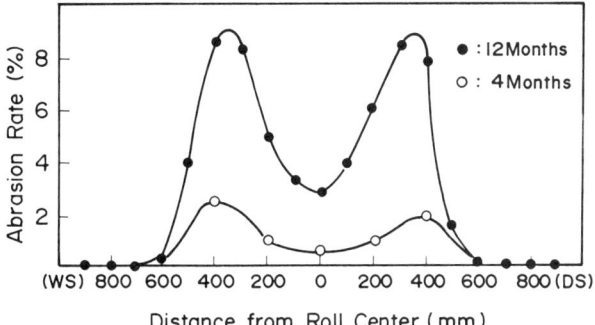

Figure 5. Abrasion rate profile of RQ roll surface

Roll surface material ; Figure 6 shows various characters of heat transfer with the surface materials applied for quenching roll.

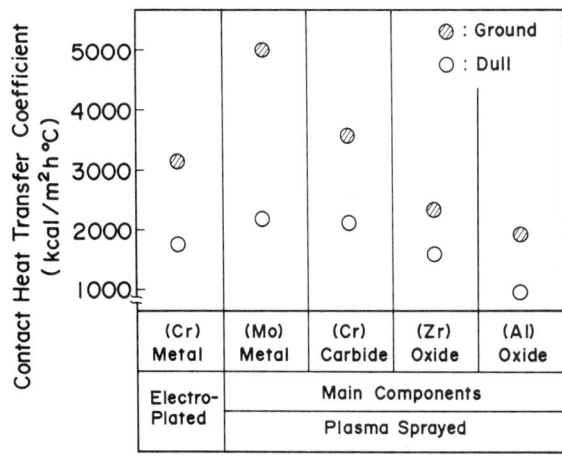

Figure 6. Contact heat transfer coefficient of various surface coating (at 29.0N/cm^2)

The variation of the heat transfer coefficient with surface conditioning is smallest in metallic oxide coating surface, so this coating is suitable for quenching roll. At present Kobe Steel applies the aluminum oxide coating surface for the quenching roll, resulting in good wear resistance. Table 2 lists chemical composition and figure 7 shows applicable condition.

Table 2. Chemical composition of RQ roll surface

Material	Al₂O₃	ZrO₂
Ratio	75 %	25 %

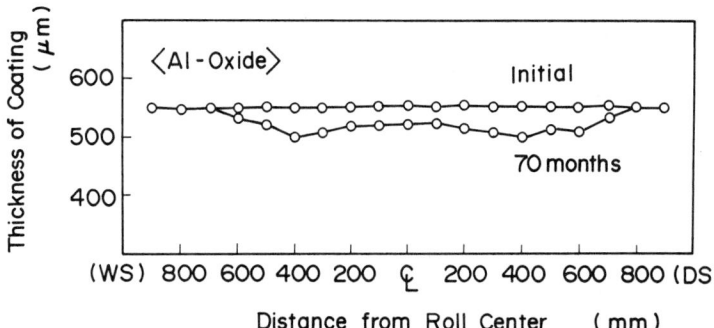

Figure 7. Wear of metal oxide plasma sprayed RQ roll surface

Aluminum oxide coating surface can be used successfully for 70 months. This is a remarkable extension of life, much longer than chromium plating surface (whose life is 2~3 months).

Temperature Profile	≈ flat profile with small peaks at edges (20°C scale)
Strip Dimension	1.0 mm × 1070 mm

Figure 8. Improvement of temperature profile in width direction

And temperature profile in width direction is improved by using aluminum oxide coating rolls. Figure 8 shows improved temperature profile measured at the delivery side of roll quenching system.
Consequently these improvements for roll quenching technology bring the quite stable operation of the CAL without cooling buckles.

Table 3 lists mechanical properties of cold rolled mild steel sheets produced with roll quenching system.

Table 3. Mechanical properties of mild steel sheets

Material			Mechanical properties				
Grade	Thickness mm	YS N/mm²	TS N/mm²	El %	r	n	
Mild steel CQ	0.8	225	335	44	1.3	0.23	Al-killed steels
Mild steel DQ	0.8	150	305	48	1.8	0.26	I.F. steels
Mild steel DDQ	0.8	145	305	49	1.9	0.26	I.F. steels

Homogeneous quenching technology for a water quenching system

In a water quenching system, it is very important to control the quenching temperature for the transformation volume of martensite under a regulating cooling rate over 1000°C/s. The faltness of the strips is generally required for commercial use. However these are normally produced under specialized quenching conditions, because strip flatness is largely changed by heat contraction with quick quenching rate and transformation.

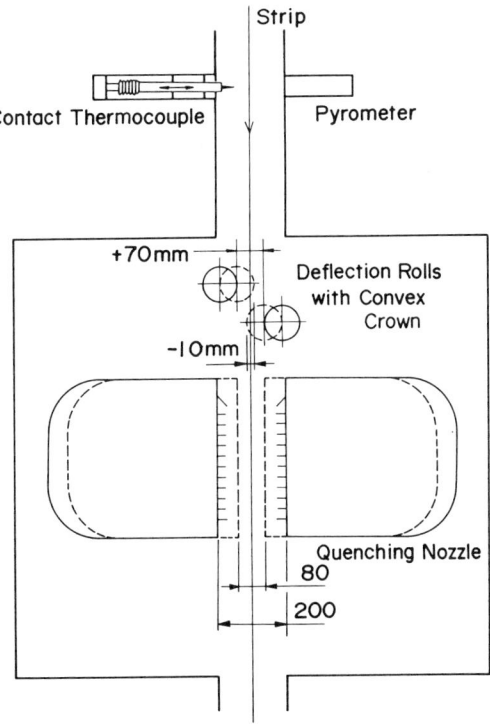

Figure 9. Layout of water quenching section

Temperature of quenching and tempering

High strength steels are susceptible to oxidation on the strip surface during the addition of a great amount of alloying elements. This oxidized surface makes it is impossible to measure the true temperature of the strip by general pyrometer, because of the variation of emissivity from the surface and slight variation of temperature causes change to material properties with high heat susceptibility.

Quenching temperature : Material properties are unstable because of the uncertain cooling quantity of strip temperature from the pyrometer to the quenching point, when the pyrometer for controlling quenching temperature is installed far from quenching point. Consequently it is desirable that thermometer for controlling the quenching temperature is installed in front of the quenching point. A contact-type thermocouple is set up at same point as the pyrometer, and is occasionally used to calibrate strip surface emissivity by contacting the measure temperature in the remote control system. The guide roll of a deflection type is set up near the contact-type thermocouple, cannot fail in measuring the temperature of a fluttering strip. Figure 9 shows the layout of a water quenching system. Table 4 lists the strip emissivity of typical materials, and shows large differences of emissivity among steel grades.

Table 4. Strip emissivity of typical high strength steels

Grade (N/mm^2)	WQ Pyrometer	WQ Contact Thermometer	Strip Emissivity
440	627 °C	546 °C	0.57
980	628 °C	527 °C	0.50
1470	593 °C	462 °C	0.37

Tempering temperature : When mild steels are produced, the reheating section is used as an over-aging zone. But when high strength steels are produced, it is used as a tempering zone operating with the heating system. The tempering thermometer has the same problem in that it can't measure accurately the temperature because of the emissivity variation. To improve this measurement, we installed a multi-reflection-type thermometer, but excluded the effect of strip emissivity at the delivery side of re-heating section, and succeeded in controlling the tempering temperature. Figure 10 shows the results measured by conventioal a pyrometer and the multi-reflection-type thermometer. Multi-reflection type thermometer is more stable toward the variation of emissivity than the conventional pyrometer.

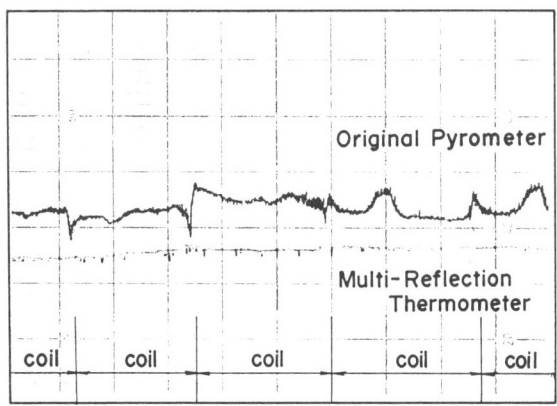

Figure 10. Effect of tempering temperature with multi-reflection thermometer

Strip flatness

Strips are deformed largely during the complicated stress balance of quick cooling and transformation which occurs in the quench from about 700°C to the normal water quenching temperature. Figure 11 shows the schematic condition of strip flatness under the effect of calculated heat contraction in water quenching. Deformation of strips in the water quenching section result in nonconforming flatness for commercial use, and the scratch mark and strip walking in the reheating section. It is is required for perfect flat strips production to control strip tension, water spout and flatness before quenching.

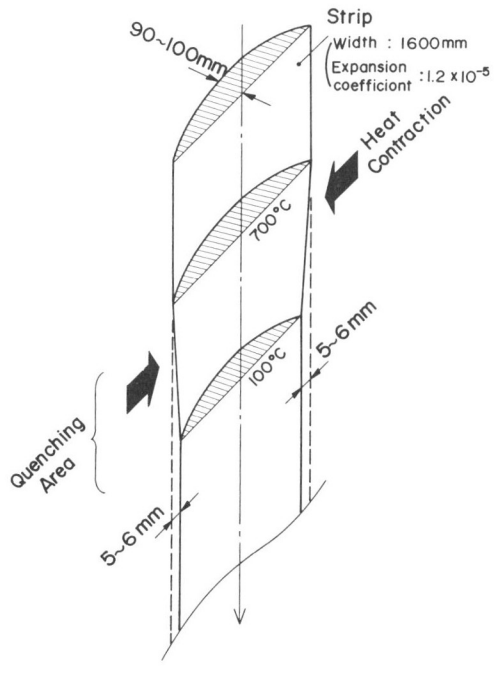

Figure 11. Crossbow with heat contraction in water quenching section

61

Equipment and operation technology ; When the strip quenches into water at high strip temperature, unbalanced quenching occurs due to film boiling condition and this results in bad flatness and nonuniform material properties. Film boiling is prevented by spouting water near the strip and increasing water quantity in water quenching. Consequently, it is necessary for the preventing of strip fluttering to detain the strip with deflection rolls. However, when the strip is detained by flat profile deflection rolls, edge wave occurs. So for a good profile of the strip edge, it is required to select the appropriate crown and to off set the deflection-roll center to fit the quenching condition. These improvements can result in stable cooling condition due to the closeness of the quenching nozzle to the strip. In addition it is required basically to adjust the strip tension and quantity of water the spout for ensuring the effect of the deflection rolls. Kobe Steel produces high strength steels under the typical condition of ~ $22N/mm^2 \sim 25N/mm^2$ ~ for strip tension , ~ $1500m^3/hour \sim 1600m^3/hour$ ~ for quantity of water spout, and ~ $0mm \sim 10mm$ ~ for the off set of deflection-roll center. Figure 12 shows height of the crossbow of high-strength steels after these improvements for the $1470N/mm^2$ steel grade.

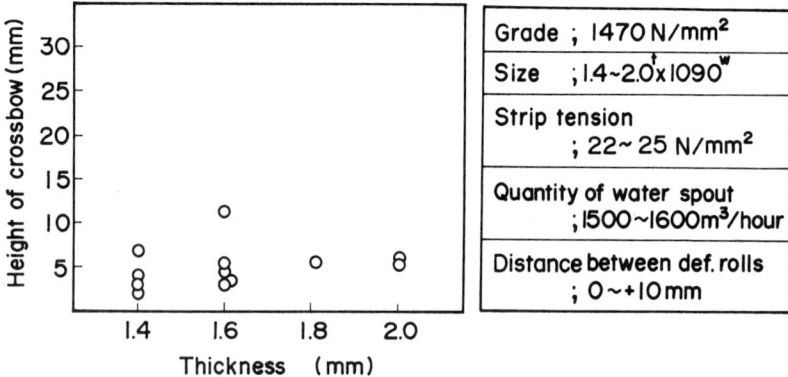

Figure 12. Height of crossbow after improvement

A water quenching system is suitable to produce high strength steels with excellent drawablity. However it is indispensable to improve equipment and operation technology along with metallurgical development for establishing the technology of new products. Kobe Steel has already completed water quenching system menus for producing high strength steels from $440N/mm^2$ to $1470N/mm^2$. And table 5 lists the typical mechanical properties of high strength steels.

Conclusion

Kobe Steel produces mild steels with ultra drawablity (Interstital Free steel etc) to ultra high strength steels by appling the excellent functions of quenching system, roll quenching and water quenching. Especially high strength steels are required to minimize the weight of automobiles which results in good fuel economy, and a safe structure. Kobe Steel positively attempts to develop high strength steelswith excellent drawablity in correspoudence to the needs of the customer. We will continue to develop new products, steadily and perfectly by appling the quenching technology of Kobe Steel Ltd.

Table 5. Mechanical properties of continuous-annealed products

Material			Mechanical properties			
Grade	Type	Thickness, mm	YS, N/mm^2	TS, N/mm^2	El, %	λ, %
KBCF 440D	Draw	1.2	300	475	34	-
KBCF 490D	Draw	1.2	315	515	32	-
KBCF 590D	Draw	1.2	380	630	27	-
KBCF 780D	Draw	1.4	515	805	19	-
KBCF 980D	Draw	1.4	645	1020	18	30
	Bend	1.4	780	1000	14	70
KBCF1180D	Draw	1.4	800	1195	13	27
	Bend	1.4	1010	1215	11	68
KBCF1370D	Bend	1.4	1275	1400	9	40
KBCF1470D	Bend	1.4	1320	1510	7	30

Outline of No.3 CAL Water Quench System at NKK Fukuyama Works.

N.Matsui* , S.Jitsukawa **, T.Izushi*** , M.Yamazaki****

Fukuyama Works, NKK Corp, Kokan-cho 1 Fukuyama 721 JAPAN

NKK installed the Water Quench (WQ) system on September 1990 in addition to the existing Roll Quench (RQ) system at No.3 Continuous Annealing Line (hereinafter referred to as CAL) in Fukuyama Works. No.3 CAL is the nucleus facilities of the new cold rolling mill started in 1987. The WQ system is installed for the purpose of increasing the demand of heavier or wider gauge of high tensile strength steel sheets such as for CAFE, etc.

No.3 CAL is combination line for the WQ process and RQ process by installation of the WQ system. Thus, it makes possible to produce the wide range of cold-rolled steel from a soft cold-rolled steel to the high tensile strength.

* Superintendent, Cold Strip Mill, Hot and Cold Strip Dept., Fukuyama Works.

** Chief manager, Operation Technology Sec., Hot and Cold Strip Dep., Fukuyama Works.

*** Manager, Operation Technology Sec., Hot and Cold Strip Dep., Fukuyama Works.

****Manager, Plant Construction and Engineering Dept.

1. Introduction

NKK Fukuyama Works installed a new cold strip mill plant which consists of a combined pickling - cold rolling mill[1] NK-PPCM(NKK Pickling and Profile Control Mill), a continuous annealing line, No.3 CAL[2], a recoil line and a coil packing line in 1987. The plant has improved yield and remarkably reduced manufacturing time by supplying hot rolled coils from HDR[3] (Hot Direct Rolling) process at the adjacent No.5 CCM(Continuous Casting Mill) - No.2 HSM(Hot Strip Mill).

No.3 CAL is the nucleus facility at the new cold strip mill plant which has adopted the direct fired method for a heating equipment, the Roll Quench system(RQ) for fast cooling equipment. It is a large-scale CAL capable of handling approximately 80,000 tons per month. This system has played a large role in improvement of yield and productivity and reduction of manufacturing time for the cold-rolled sheet since operation began in August, 1987.

NKK Fukuyama works first built No.2 CAL[4] as the genuine NKK-CAL process in 1976. No.2 CAL was equipped with WQ system for fast cooling and had been producing high tensile strength steel sheets from 390 MPa grade to 1180 MPa grade. However, it was hard to meet the demands of thicker and the wider gauge of high tensile strength steel sheets required by CAFE in recent years because of restrictions of No.2 CAL specifications. In view of these demand trends, No.3 CAL was planned at the new cold strip mill plant so as to provide equipment which is capable of producing the high tensile strength steel sheets.

No.3 CAL features the change of process on the line using both WQ process and RQ process in short time without cutting the strip. Additionally, various new mechanisms are introduced in No.3 CAL. The following deals with No.3 CAL WQ equipment, operating conditions and the quality of steel sheets.

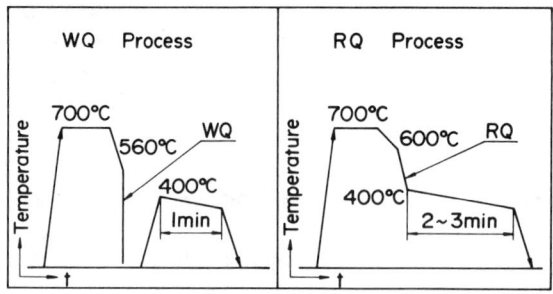

Figure 1 Heat cycle of NKK - CAL (CQ grade)

2. No.3 CAL equipment

No.3 CAL was started up only with RQ cooling process.

2.1 Features of WQ process and RQ process

The typical heat cycles of WQ and RQ processes in NKK-CAL process are shown in Figure 1.

After heating the cold rolled steel sheet above the temperature of recrystallization, the strip is rapidly cooled. Thus, an excellent cold rolled steel sheet with anti-aging properties can be produced by precipitating super saturated solute [C] with an overaging treatment in a short time.

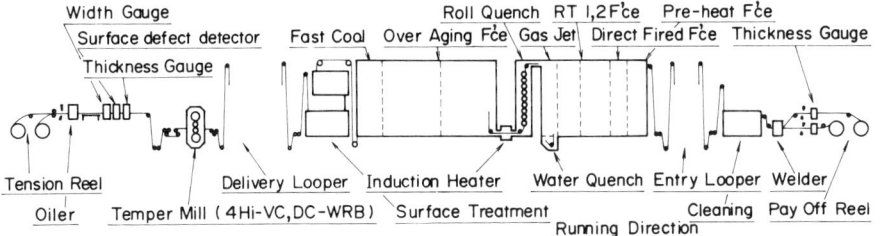

Figure 2   Layout of the Fukuyama No.3CAL

Table I.- Main specifications of No.3CAL

Items			Specifications
Production capacity		t/month	80000
Strip size	Thickness	mm	0.35～2.30
	Width	mm	600～1650
Coil weight	Entry	t	45.0
	Delivery	t	34.0
Line speed	Entry	mpm	480
	Furnace	mpm	370
	Delivery	mpm	515
Furnace capacity	RQ	t/h	210 at CQ
	WQ	t/h	60 at 590MPa

WQ process differs from RQ process on the cooling speed of the fast cooling section. The former is 2000 ℃/sec and the latter 200 ℃/sec. Since the super saturation of cooled solute [C] varies by different cooling speeds, the overaging time of 1 minute is enough on WQ process when the soft cold rolling steel is produced. Two or three minutes are required on the RQ process.

In WQ process, with the strip cooled by water, a oxidizing film is generated on the steel sheet surface. Pickling equipment to eliminate this oxidizing film and re-heating equipment to raise the strip to overaging temperature are required. As a result, the equipment becomes a complicated structure. However, WQ system features the production of the high tensile strength steel sheet from the economical steel which is a low alloy element because WQ process allows an austenite to transform into a martensite by cooling the steel from a two-phase region of the austenite and a ferrite, utilizing the high cooling speed to produce the high tensile strength steel sheet of the dual-phase structure in which the martensite and ferrite coexist.

In the RQ process, the strip is not oxidized while cooling and the strip temperature at the outlet of RQ can be controlled, thus eliminating the pickling and re-heating equipment. This is an energy saving type and manufacturing costs are reasonable. Consequently, RQ system is adopted for producing the soft cold rolled steel.

2.2 No.3 CAL equipment

The layout of No.3 CAL is shown in Figure 2, and main specification is shown in Table I.

The equipment on the entry side consists of two coil charging equipment and rewinding reel, crop cut shear, welding machine, surface cleaning section and looper. The furnace also consists of PHF(Pre Heat Furnace), DFF(Direct Fired Furnace), RTF(Radiant Tube Furnace), GJ(Gas Jet cooling Furnace), OA(Over Aging Furnace) and FC(Fast Cool). in the furnace, the first adopted two-path direct fired furnace in NKK-CAL process makes it possible to pass the strip through a recrystalization temperature region which is apt to getting worse strip shape with 1 path, thus resulting in stable high speed traveling in the furnace. In RQ equipment, nine cooling rolls are vertically arranged to save space and enable the roll to be changed in a short time without cutting the strip.

The equipment on the delivery section consists of the looper, temper rolling mill, surface defect detector, thickness gauge, width gauge, inspection table, electro-static oiler, shear, two coil winding reels and coil discharging device. The strip shape is assured with 4Hi-1 stand temper mill equipped with VC back-up rolls and double-chock work roll bender. The surface defect detector performs high accurate automatic inspection combined with the dry temper rolled strip surface.

3. Arrangement and construction of WQ system

   3.1 Arrangement and specifications of new equipment

   For the newly-installed WQ equipment on No.3 CAL, the following two points were considered.
   (1) Process change between WQ process and RQ process can be easily performed in a short time.
   (2) Both the existing and the new equipment offer the soft cold rolled sheet by the RQ process and the high tensile strength steel by the WQ process.

   To realize the above production, the WQ equipment are continuously arranged at the front RQ equipment and the pickling equipment ST(Surface Treatment section) are arranged at the outlet of furnace section.

   In conventional NKK-CAL, process change between the RQ process and WQ process is performed by cutting the strip with the RQ and WQ parallel arrangement systems. Because the strip cutting and connection works are complicated, air is easily entered into the furnace when the strip in the furnace is moved. Consequently, gas in the furnace must be replaced for a long period of time. As a result, much time is required for process change.

   In the No.3 CAL, the strip cutting and connection work are eliminated. The clamp devices between furnaces installed at the front and rear WQ equipment protect the furnace from air entering, thus reducing the change time by 30% in comparison with the conventional layout.

   In the WQ process, induction heating equipment has been installed at the OA furnace inlet to prevent the strip walking on reheating furnace when a part of OA furnace is used as reheating. Table II shows the specifications of the newly installed equipment.

   3.2 High tensile strength steel producing equipment

   3.2.1 WQ section arrangement

   Figure 3 shows the arrangement of WQ section.

   Spray nozzles to make a water quench, wringer roll unit for squeezing after quenching and dryer are installed in WQ section. Two sets of rolls and

Table II.- New installed equipment

section	equipement	
WQ section	WQ nozzle	x 1set
		Cooling capacity 60t/h
		flow rate MAX 1500t/h
	Dryer	x 1
		Combuster 35×10⁴Kcal/h
	Wringer roll unit	x 3sets
	WQ seal roll	x 2sets
OA section	Induction heater	x 1set
		Capacity MAX 1000kw
ST section	Pickling tank	x 2
		Rectifier MAX 8000A
	Water spray tank	x 1
	Wetting duct	x 1
	Alkaline tank	x 1
	Scruber tank	x 1
		Brushing roll x 2sets
		Wringer roll x 2sets
	Rinse tank	x 1

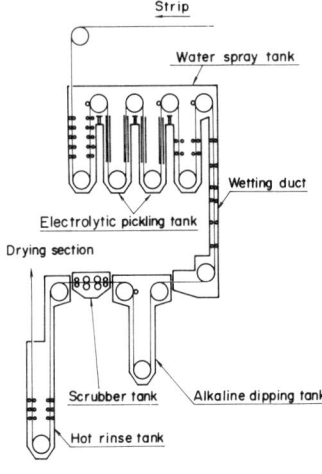

Figure 4 ST section arrangement

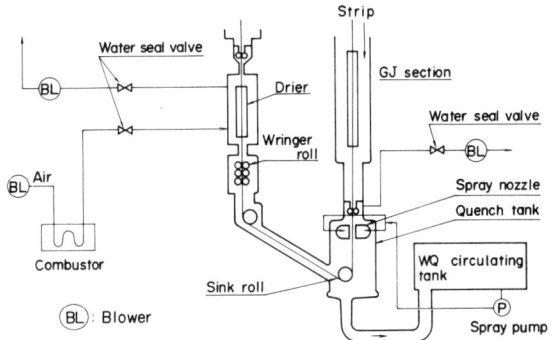

Figure 3 WQ section arrangement

exhaust equipment are provided on the upper quench tank and dryer to prevent the raising of due point in the furnace because of steam generation in a quench tank or high temperature dryer air entering in the furnace. Sink roll provided in the quench tank and the quench tank are existed in the high temperature inert gas on RQ process operation and underwater of ordinary temperature on WQ process operation. The heat-protecting plate is provided on the wall in the tank and an gas seal method is adopted for the shaft of sink roll to prevent heat deformation of the quench tank and defective seal of the sink roll shaft.

3.2.2 Surface treatment (ST) section arrangement

Figure 4 shows the ST section arrangement.
The ST section consists of such tanks as pickling, water cleaning, neutralization, scrubber and rinse.

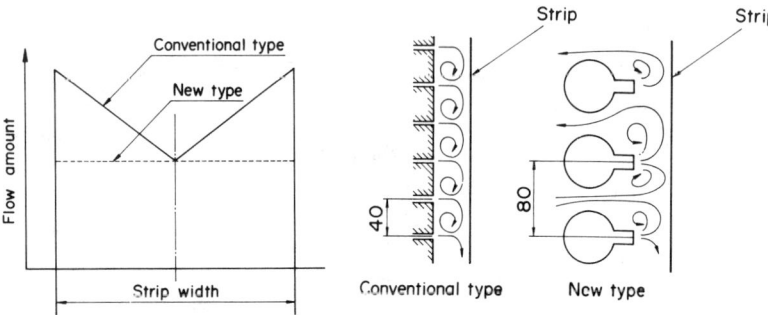

Figure 5 Flow amount distribution Figure 6 Flow pattern of each nozzle

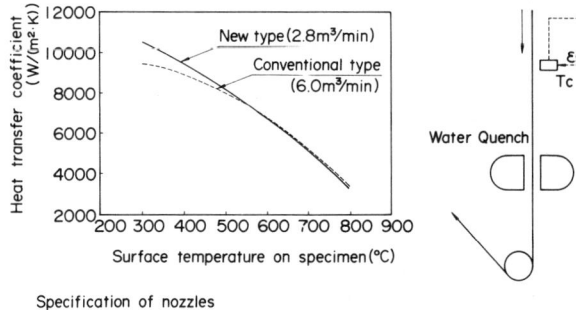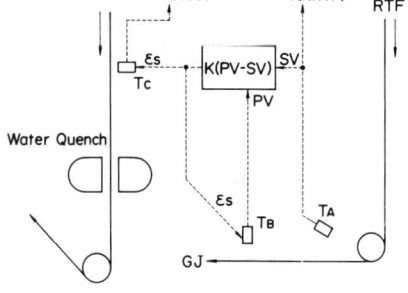

Specification of nozzles

Nozzle width ; 600 mm
pitch ; New type ; 80 mm
Conventional type ; 40 mm
number ; New type ; 7
Conventional type ; 13

T_A ; Thermometer (Multi reflection type)
T_B, T_C ; Thermometer (Conventional type)
ε_s ; Strip emissivity
T_s ; Strip temperature (Section)

Figure 7 Heat transfer coefficient   Figure 8 Strip temperature measurement

The electrolytic pickling method by sulfuric acid is applied in the pickling section so as to descale stably in wide line speed variation and to obtain good surface quality. After pickling the strip surface is washed through high pressured water (60 MPa) to remove the acid, then dipped through Alkaline tank for neutralizaiton, thus the good quality of strip surface is obtained. Water sprays between tanks on ST zone are provided to prevent the strip surface from defective stains by drying the strip at an intermediate path.

3.2.3 WQ spray nozzle

New spray nozzles have been installed on WQ process for the purpose of uniformly cooling across the strip width and improving the cooling efficiency. In the conventional nozzles [5], slit nozzles are arranged on one face in a box. The water collided with the strip flows out through the space between the spray nozzles and the strip. As a result, cooling water flow rate increases from the strip center to the strip edge, causing an over cooled strip edge as shown in Figure 5. New nozzle, header type, is independently and multistagedly arranged for the purpose of uniforming cooling water distribution width so that water colliding with the strip will flow to rear spray nozzles through space between nozzles.

Figure 6 shows each flow nozzle pattern confirmed with an air bubble tracer by experiment. With the conventional nozzles, a stable vortexes are formed across the strip width and water flows to the strip edge from the strip center. With new type nozzles, it is observed that the stable vortexes are not formed on the strip surface and water colliding with the strip flows smoothly to the rear nozzles.

Figure 7 shows the comparison of surface heat transfer coefficient for each nozzle found with the experiments. It is found that the cooling efficiency is improved since new nozzles can obtain the equivalent heat transfer coefficient with the flow rate of 1/2 of the conventional nozzles for reducing water flow pressure loss.

WQ spray nozzles on No. 3 CAL have been designed on the basis of these experimental results. As comparison with the conventional nozzles, cooling water is reduced by about 40%.

3.2.4 Strip temperature measurement

In the high tensile strength steel sheet manufactureing, it is very important to control a quenching temperature since the tensile strength of sheet depends upon this temperature.

The strip temperature measurement system of the quenching temperature is shown in Figure 8.

The ordinary radiant thermometer is arranged on the top quench tank inlet to precisely measure the strip temperature.

The detecting energy G of the ordinary radiant temperature is indicated with the following formula.

$$G = \varepsilon E(Ts) + a(1-\varepsilon) E(Tw) \tag{1}$$

Where

$E(T)$: Energy emitted from black body of temperature T
Ts : Strip temperature
Tw : Furnace wall temperature
ε : Emissivity of strip
a : Influence coefficient of background

As known from the formula (1), there are the following problems when measuring the strip temperature by the ordinary radiant thermometer.
(1) It is hard to confirm the strip emissivity.
(2) The influence of background cannot be disregarded.

The following two countermeasures are executed to solve these problems.

The ordinary radiant thermometer is installed in addition to the multi-reflection type thermometer[6] for the strip temperature control already installed on RT zone outlet. As shown in Fig.8, the strip emissivity which is obtained from these two thermometers is preset to the thermometer on WQ inlet in this system.

To eliminate the influence of background noise, the area around the optical axis of the thermometer is enclosed with a water-cooled cylinder so that radiant energy from the wall furnace does not reach the thermometer.

Under these countermeasures, the strip temperature has been precisely measured without varying strip emissivity by difference of steel grade or surface quality.

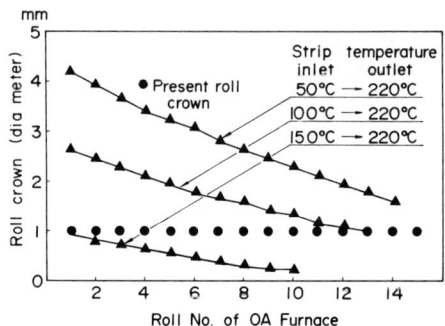

Figure 9 Required initial roll crown in OA Furnace

3.3 Countermeasure of strip walking in reheating section

By installing WQ equipment, No. 3 CAL system is operated using both as WQ process and RQ process. Consequently, two kinds of heating cycles for WQ and RQ processes must be used.

The heat cycle of WQ process for the soft cold rolled steel sheet is described in paragraph 2.1. In the heat cycle for high tensile strength steel, the reheating treatment in lower temperature is also required for purpose of improving the elongation by precipitating the solute [C] in the ferrite or to control bake-hardnability by tempering. Accordingly, in No. 3 CAL, it is necessary to use the half of OA furnace as the reheating section for tempering on the WQ process and as the slow cooling zone for overaging treatment on the RQ process.

If the strip is reheated by only a radiant tube during WQ process operation, a minus heat crown forms because the roll temperature of the strip is higher than that of on the strip by the radiant heat from the furnace. For this reason, it is presumed that the strip walks in the furnace. For this countermeasure, providing a large initial crown roll is considered. However, when the thinner and wider material of the soft cold-rolled sheet passes in RQ process, there is a possibility that heat buckling occur because of large crown roll.

In the front OA furnace, an induction heating equipment is provided at the OA furnace inlet, to prevent the strip walking in the WQ operation and the heat buckling in the RQ operation by the current small roll crown.

Figure 9 shows the relationship between OA furnace inlet strip temperature and proper roll crown obtained by FEM analysis in basis of operation data on No. 3 CAL.

During WQ operation, the strip cooled down to a normal temperature after water quenching is raised to 150 °C by the induction heating equipment and it is also raised to target strip temperature with the radiant tube. Thus, the variation of roll crown is prevented so the strip passes stably.

3.4 Process change (WQ/RQ) mechanism

Process change between RQ and WQ processes at the WQ section is shown in Figure 10.

When changing the operation mode from the WQ process to RQ process, the spray nozzles and wringer rolls are extracted to the outer furnace and the piping for

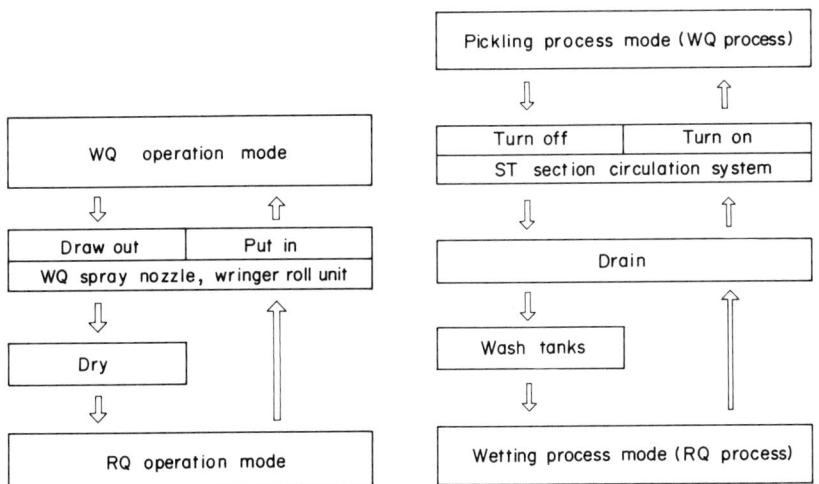

Figure 10 WQ section process change   Figure 11   ST section process change

cooling water circutation is separated from the furnace to dry the inside furnace.

After drying, the inside of the furnace is replaced with nitrogen. Simultaneously, each piping connected to the furnace is sealed with a water seal valve to finish the change.

Reversely, to change from the RQ process to WQ process, spray nozzles and wringer roll are provided to the inside of the furnace to connect the cooling water circular piping.

In the ST section, the operating process is changed without cutting the strip also. For this reason, the system is operated with the pickling mode on the WQ process, while operated with a wetting mode with all tanks filled with water on the RQ process. Figure 11 shows how to change at ST section.

To change from the WQ process to RQ process, the solution in the tank is drained and the tank is cleaned after each circulating system is stopped before supplying water to move to the wetting mode.

Reversely, to change from the RQ process to WQ process, each circulating system is operated after draining water in the tank to move to the pickling mode.

The all existing control instruments are fully used to perform these process change work for labor saving in a short time.

In the process change at the WQ section, working procedures are indicated on the CRT in the delivery pulpit as operators' guidance. Simultaneously, valve operations required in the drying of quenching tank, purging of nitrogen in the furnace, etc., are fully automated. Also, in ST section, valves and pumps are automatically operated and the process change is finished in about one hour.

4. Operation and quality

4.1 Production

This equipment was installed, stopping the line for 8 days in January, 1989,

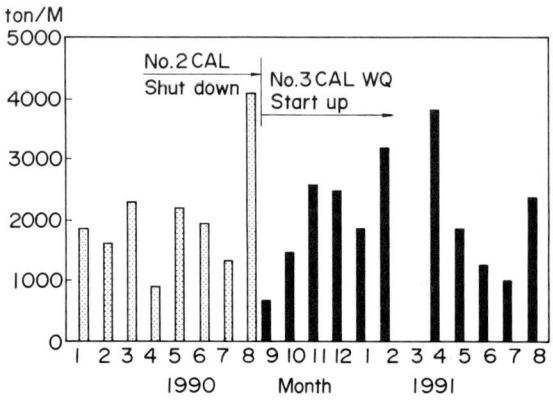

Figure 12 Production volume of high tensile strength steel

10 days in June and 6 days in January, 1990. Hot run performed every month to confirm the quality of all high tensile strength steel sheets from 390 MPa grade to 1470 MPa grade. The equipment started commercial production on September, 1990. Figure 12 shows the productional transition of the high tensile strength steel sheet.

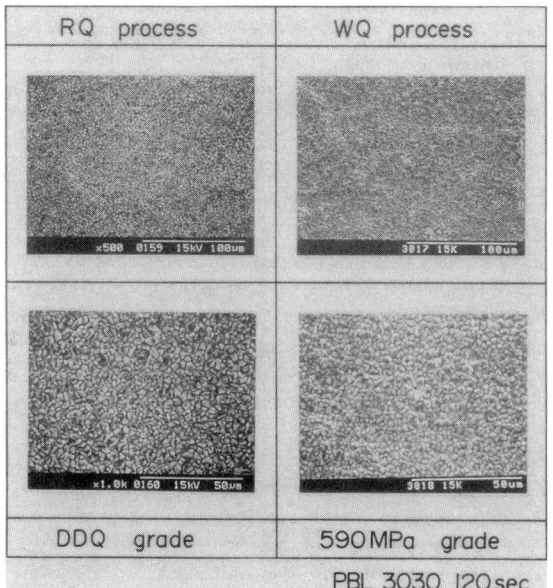

Figure 13 SEM image of the phosphate film

4.2 Quality

Figure 13 shows the SEM image of phosphate film of EDDQ by RQ process and 590 MPa grade high tensile strength steel sheet by WQ process.

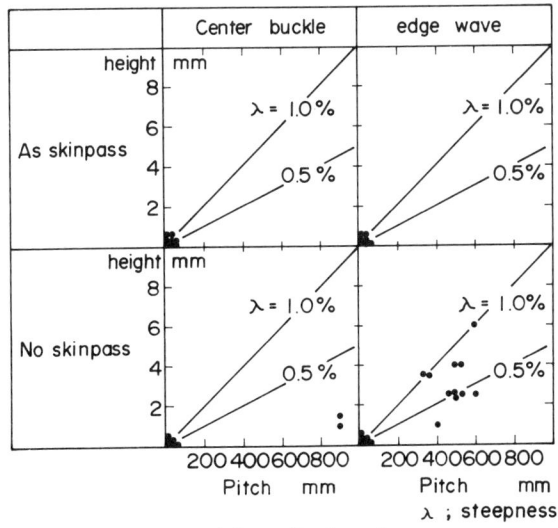

Figure 14 Strip shape

λ ; steepness

Table III.- Example of manufacturing condition

Category	Grade (MPa)	Chemica composition (%)					HRM C.T.(°C)	Annealing temp.(°C)	
		C	Si	Mn	P	S		Heating	Re-heating
High-bakehardnability type	390	0.05	0.02	0.23	0.012	0.010	560	810	220
	440	0.06	0.02	0.36	0.013	0.011	560	810	210
	490	0.05	0.16	0.38	0.058	0.010	560	810	210
	540	0.07	0.20	0.52	0.048	0.009	560	800	200
	590	0.10	0.20	0.75	0.068	0.005	560	810	200
Ultra high-strength type	780	0.11	0.42	1.25	0.016	0.004	560	830	180
	980	0.15	0.35	1.50	0.015	0.006	560	830	210
	1180	0.15	0.35	1.50	0.015	0.006	560	860	230
	1470	0.20	0.40	1.76	0.013	0.002	560	850	150
Low yield-ratio type	490L	0.05	0.02	1.40	0.013	0.007	560	800	150
	590L	0.08	0.51	1.40	0.012	0.005	560	800	150

Table IV.- Example of mechanical properties

Process	Category	Grade (MPa)	YP (MPa)	TS (MPa)	El. (%)	BH (MPa)	Thickness (mm)
RQ	Soft steel	CQ	209	324	46	-	0.8
		DQ	178	319	47	-	0.8
		DDQ	162	307	48	-	0.8
		EDDQ	142	305	49	-	0.8
WQ	High tensile strengh steel	390	265	403	37	88	1.2
		440	316	451	34	88	1.0
		490	362	510	32	88	1.2
		540	392	568	30	88	1.2
		590	406	603	27	78	1.2
		780	487	811	21	58	1.2
		980	631	1021	14	-	1.2
		1180	864	1229	9	-	1.6
		1470	1166	1546	8	-	2.3
		490L	304	548	34	29	1.0
		590L	313	598	34	29	1.2

The uniform phosphate film is formed with both materials, showing adequate surface quality.

The strip shape of the high tensile strength steel sheet produced by WQ process is shown in Figure 14. The good strip shape is obtained from materials of 590 MPa or less with skinpass, and also it is obtained the steepness of 1% or less from no-skinpass materials of 780 MPa to 980 MPa grade.

Table III shows the manufacturing conditions of the high tensile strength steel sheet. The high tensile strength steel sheet which is mainly strengthened by dual-phase with water quench is produced. Steel sheets from 390 MPa to 1470 MPa grade are produced by controlling a maltensite volume fraction with heating and quenching temperatures applied to the steel chemical composition.

Table IV shows mechanical properties of each grade.

The WQ equipment makes it possible to produce a wide range of cold-rolled steel sheets from the soft steel sheets to the high tensile strength steel sheets.

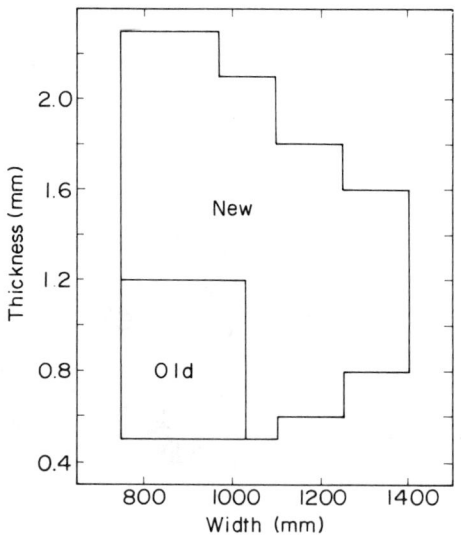

Figure 15 Production capability
(980 MPa grade)

4.3 Manufacturing strip size of high tensile strength steel

No. 3 CAL is capable of producing high tensile strength steel sheets in wider size range compared with the size range of No. 2 CAL.

The enlarged high tensile steel sheets of 980 MPa grade are shown in Figure 15.

5. Conclusion

NKK newly installed the water quench system (WQ) in   No. 3 CAL started in 1987 at Fukuyama Works and the line was available for both RQ process and WQ process. This system enabled to product the wider strip dimension and grade of high tensile strength steel sheets which are expected increasing demands.

We established the wide range of the cold-rolled steel sheet manufacturing system from the soft steel sheets by the RQ process to the high tensile strength steel sheets by the WQ process in No.3 CAL.

In this reconstruction, WQ and RQ equipment are arranged in a series and the pickling equipment is also arranged at the furnace outlet. As a result, NKK-CAL features at first the change of the RQ process and WQ process without cutting the strip. Furthermore, new various mechanisms such as common use of RQ and WQ processs, short time change between processes have been introduced.

This system started the commercial production of the high tensile strength steel sheets in September, 1990. Furthermore, we will improve production capability such as the strip dimension and the steel grade.

References

1) Iwadoh,S et al, NKK Technical Review No.54, P69-78(1988)
2) Kanetoh,S et al, NKK Technical Review No.56, P49-57(1989)
3) Taniguchi,I et al, Nippon Kokan Technical Report Overseas No.46, P128-136(1986)
4) Naemura,H et al, Nippon Kokan Technical Report, No.73(1977), P47-55
5) Nakaoka,K et al, Tetsu-to-Hagane ,62(1976)6,P634
6) Yamada,T et al, Nippon Kokan Technical Report Overseas No.41, P126-134(1984)

NEW TECHNOLOGY IN KM-CAL FOR SHEET GAGE

T.Kaihara, S.Mega* ,Y.Ida *, K.Hirohata, T.Nakagawa, K.Kuramoto and T.Fukushima**

Kawasaki Steel Corporation
Mizushima,Kurashiki 712, Japan
* Chiba 260, Japan

**Mitsubishi Heavy Industry
Hiroshima 733,Japan

Abstract

Two new type KM-CALs (Kawasaki Steel Multipurpose Continuous Annealing Line) for sheet gage started their operation. One is Chiba NO.3CAL(started in February 1988) and the other is Mizushima NO.2CAL (started in May 1991). These lines mainly produce extra low carbon steel sheets for automotive use.

For mass-producing annealing line for high temperature operation, several technical innovations are achieved : (1) radiant tubes and burners for high temperature operation, (2) high speed and stable processing at high temperature operation (preventing "buckles" and "mistracking"), (3)quick cooling technology, (4)high accuracy in strip temperature control.

1. Technical problems of continuous annealing process in the manufacture of cold-rolled extra low carbon steel sheet

Cold-rolled steel sheet that was required to exhibit good press formability and high cosmetic beauty in the manufacture of automotive parts, used to be made by means of the steps to be listed in the following:

(a) Pickling for removal of surface oxide layer peculiar to hot-rolled steel sheet
(b) Cold-rolling
(c) Degreasing of the surfaces of steel sheet by means of electrolytic cleaning
(d) Improvement of workability of steel sheet by means of batch annealing
(e) Rust-preventive cooling down to room temperature
(f) Conditioning of mechanical characteristics, improvement of form, and adjustment of surface grain through tempering rolling
(g) Finishing of product by conducting inspection and finishing steps

Starting in the 1970s, a continuous annealing processing system for cold-rolled steel sheet combining the steps (c) through (g) above into one single process was developed, thereby markedly improving the production efficiency (labor productivity, yield, etc.) [1],[2].
At Kawasaki Steel, too, a multi-purpose continuous annealing furnace for processing different products such as sheet for making tinplate, electrical sheet and common cold-rolled steel sheet was worked out, and operation using the new system was started (Chiba No. 2 CAL) at the cold-rolling mill, the Chiba Works, in July, 1980.[3]
In the face of market demand for products offering better workability and strength, there were limitations to the conventional cold-rolled steel sheet made by the continuous annealing process on the basis of the existing low carbon-aluminum killed steel. It was for this reason that the above-mentioned new product was developed and introduced on the basis of extra low carbon steel.[4]
It was essential that problems such as those given below had to be solved first in order to generate the above-mentioned new product in large quantities and in a stable manner:

(1) Durability of burners and radiant tubes to be used for high-temperature annealing (850 to 900° C).
(2) Processing across the furnace of the soft type extra low carbon steel while avoiding formation of "buckles" (heat buckle and cooling buckle).
(3) Quenching technique for imparting mechanical characteristics to the product.
(4) Control of the annealing temperature of the product.

For the purpose of solving those problems, the continuous annealing process for the extra low carbon steel was worked out by placing major developmental emphasis on the following points:

① Mass-producing annealing furnace for high temperature operation

In this furnace, strip is indirectly heated by radiant tubes in a non-

oxidizing atmosphere (1 - 3%H_2 - balance N_2) so as not to oxidize the surface of the strip.
For achieving the high temperature annealing (850 to 900° C), the furnace temperature is raised up to 950 to 1,000° C, and the conbustion load required by each radiant tube is as large as 150,000 to 160,000 kcal/hr (as against the conventional level at 100,000 to 130,000 kcal/hour).
The following problems were sustained as a result:

(1) Need for stable combustibility under an extensive fluctuation in combustion load.
(2) Need for suppressing the generation of NOx (to meet the standards for environmental restrictions).
(3) Durability of the radiant tubes under high temperature.

A solution to those problems has been achieved through development of burners and radiant tubes for use under high temperature.

② Establishment of the technology for high speed and stable processing of sheet at high temperatures

On the continuous annealing system, by means of friction and tension provided by the turning rolls, the steel strip is processed at line speed that runs at 100 to 420 m/min. For the purpose of causing the strip to run straight without wavering on the line, roll crown and tension are provided, but unless they are given at right levels, the strip will sustain buckles commonly known as heat buckles. Particularly, in processing soft type extra low carbon steel at high temperatures, the following measures were taken:

(1) To decide on the conditions required for designing the equipment, as well as, for running the operation, after quantifying the limits up to which no heat buckles are generated.

(2) To improve the accuracy in tension control, so as to enable operations possible within the range of tension where no strip wavering or heat buckles will be caused.

(3) For attaining stable processing of the strip, to control thermal crown (crown caused to develop as a result of heat distribution taking place within the rolls).

③ Development of quick cooling technology

A rate of cooling ranging from 30 to 100° C per second is required for the purpose of manufacturing bake-hardening type highly workable steel sheet. In view of such cooling rate, the gas jet cooling that has been employed extensively is somewhat deficient in terms of capacity (about 50 ° C per second in handling steel strip with a thickness of 0.7mm). While the cooling rate was adequate in roll cooling (a system whereby steel strip is wound on a hollow roll through which cooling medium is circulated), the roll cooling of the conventional type was inadequate for cooling extra low carbon steel because of the development of buckles (cooling buckles) as a result of uneven heat distribution across the soft extra low carbon steel.

Aiming at the settlement of the above-mentioned problem, a new RGCC system (roll and gas jet combined cooling system) was developed by combining the roll cooling and the gas jet cooling, thereby making it possible to be successfully applied to the manufacture of extra low carbon steel by putting an end to the problem of uneven heat distribution during the cooling process.

④ High accuracy in strip temperature control

In operating a continuous annealing furnace, the time constant of the process becomes large as a result of thermal inertia of the heating furnace. In the face of diversification of types of products, on the other hand, the frequency of changes in scheduling for heating load (strip thickness, strip width and target strip temperature) has been on the increase. Use of a static model for the furnace has been inadequate to accommodate such high frequency in scheduling changes. Toward a solution to such problems, a system for high accuracy control of the strip temperature was developed by introducing a method of strip temperature control on the basis of a dynamic model for expressing heat conductance behavior of the process, combined with development of a "chance free section" which is capable of promptly responding to changes in the heating load.

2 Development of mass-production annealing furnace for high temperaure operations

2-1 Technology required for high temperature annealing

For the prevention of surface oxidation, thin steel sheet is required to be heated in a non-oxidizing atmosphere (such as one consisting of 3% hydrogen and 97% nitrogen). For a continuous annealing furnace calling for such an atmosphere control, a heating system often used is composed of radiant tubes. This system, as illustrated in Figure 2-1, is an indirect type radiant heating system where the fuel is burned inside the radiant tubes and the heat thus generated is radiated from the surface of radiant tubes for heating the strip.

The radiant tubes employed for the continuous annealing furnace are available in two types, U type and W type as shown in Figure 2-2. One of the two is selected depending on the maximum width of the work as well as on the combustion load of the burner. In processing the work in a large volume, in most cases the W type is used because the thermal efficiency would be poor if the U type is employed when the burner in use is of a large combustion load. In the following, therefore, what will be explained will be the W type radiant tube. The mean temperature of the radiant tubes needs to be brought to 950 to 1,000 ° C in order to quickly heat the steel sheet up to 900° C through radiant tube heating.

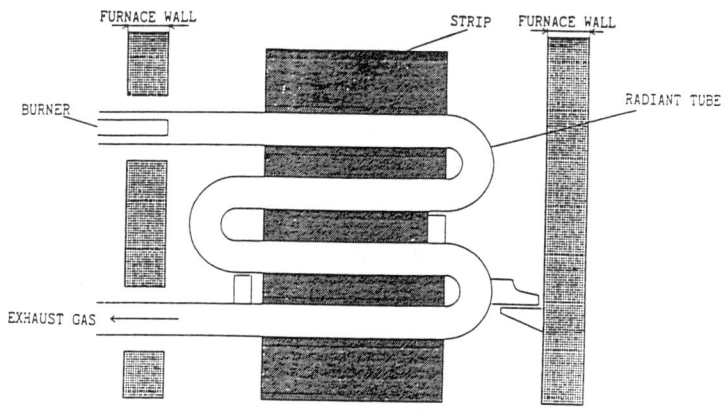

Fig.2-1 Radiant tube heating

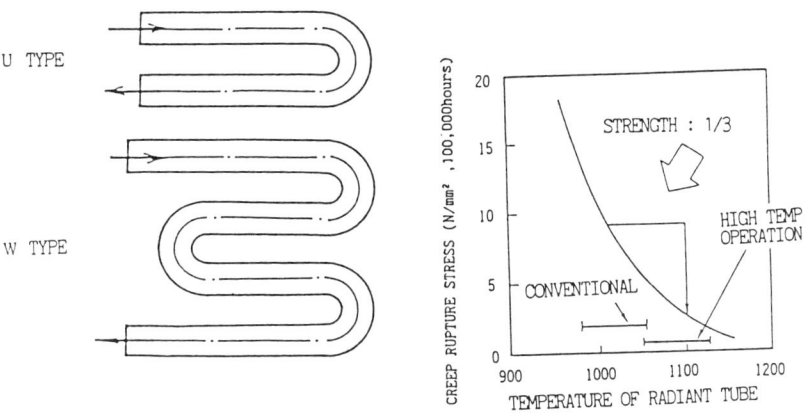

Fig.2-2 Types of radiant tubes   Fig.2-3 Heat resistance of radiant tube

In such a case, use of the conventional radiant tubes and burner will pose a problem related to the strength of the radiant tubes as shown in Figure 2-3 because the temperature of the radiant tubes will reach

1,100 °C at peak. Moreover, a furnace with a length being larger than that of the conventional furnace will be required for annealing in large quantities at high temperature because of the limited combustion capacity of the burner. It was therefore necessary to come up with a burner having a capacity being larger than that of the conventional burners (120,000 kcal/hr) so as to obtain a compact furnace and to attain reduced equipment cost.

Since it was clear as was so far described to attain the target annealing temperature and processing capacity by use of the conventional radiant tubes and burner, Kawasaki has developed the "high temperature radiant tubes" and the "burner for high combustion load" as a new technical solution enabling a high temperature annealing for a large volume production process.

2-2 Development of radiant tubes for high temperature use

(1) Radiant tubes

The life of the radiant tubes is dependent on the deformation of the straight pipe section and on the cracking developing near the burner combustion tube rather than on the life peculiar to the material resulting from deterioration of the material under high temperature oxidation. In such a case, the limits of heat resistance of the radiant tubes can be evaluated on the basis of the high temperature strength characteristics of the radiant tube material, the difference between peak temperature and average temperature of the radiant tubes (hereinafter called " ΔT" as shown in Figure 2-4)) as well as of the level of stress to which the radiant tubes are exposed.

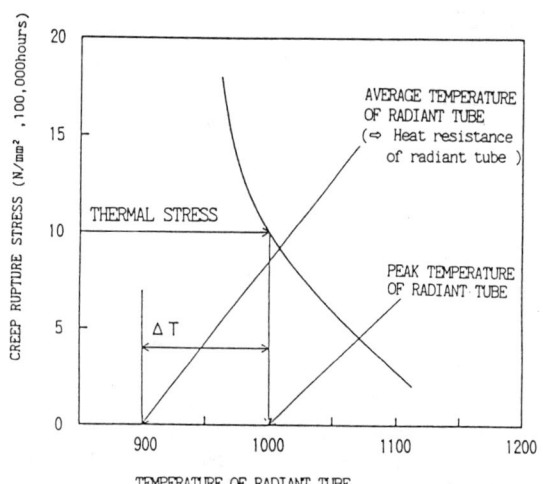

Fig.2-4 Heat resistance of radiant tube

In improving the heat resistance of the radiant tubes, in other words, there are three possible means, namely, the method of improving the high temperature strength of the radiant tube material, the method of reducing the $\triangle T$, and the method of reducing the stress to which the radiant tube is exposed.

As a material making up the radiant tubes, many a time an alloy steel mainly consisting of chromium and nickel is used, and use of some other material in search of improvement in the high temperature strength of the radiant tubes can result in a great increase in cost for limited improvement in the strength. An attempt of improving the limits of heat resistance of the radiant tubes could not be considered advantageous for the continuous annealing process that calls for a large number of radiant tubes.

The $\triangle T$ can be reduced by the two methods;
1) by lowering the surface thermal load by enlarging the inside diameter of the radiant tubes of the burner, and 2) by lowering the local heating through optimization of the burner.

The stress applied to the radiant tube can also be reduced through optimization of the method of upholding the radiant tube and of the structure of the radiant tube itself.

(2) Basic principle

In the case of the conventional radiant tube, all the straight pipes are normally made equal in inside diameter and the furnace body is upheld at one point when the burner is installed on the straight pipe, whereas in the case of the radiant tube developed this time, it features, as shown in Figure 2-5, in that the inside diameter of those pipes on which the burners are installed is made larger and in that the furnace body is upheld at three points.

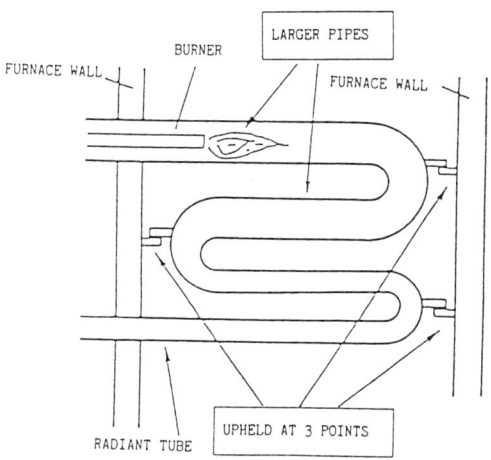

Fig.2-5 Radiant tube for high temperature

The basic purpose of this radiant tube is to improve its heat resistance by means of reducing the heat stress in the vicinities of the burner by bringing down the peak temperature of the radiant tube as well as by preventing deformation resulting from high temperature creep through reduction of the stress applied to the straight pipe.

(3) Thermal stress occurring in the radiant tube
With the purpose of clarifying the cause of stress developing in the radiant tube and the behavior of radiant tube deformation, the elastic plasticity creep was calculated and the hot deformation was tested, and they were followed by verification of those characteristics.

As shown in Figure 2-6, the stress occurring in the radiant tube as the temperature of the furnace is raised from room temperature to 900 °C by use of a conventional burner, and it is concentrated near the combustion tube of the burner, in each bend, and in the end of the pipe, and what is predominant is the stress in the cross-sectional direction.

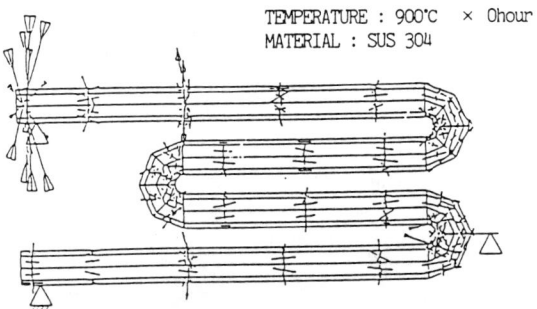

Fig.2-6 Distribution of stress in the radiant tube calculated by finite element method

The above phenomenon can be explained as follows:
The end of the pipe is greater than the burner combustion tube in terms of temperature gradient at 0.5° C/mm, whereas the temperature gradient between the burner combustion tube and No. 1 bend is small at 0.1° C/mm, and therefore a stress resulting from the difference in expansion in the cross-sectional direction is present near the burner combustion tube where the temperature gradient varies.

By way of the means for the prevention of cracking and deformation near the burner combustion tube, what is important is to reduce the $\triangle T$ and to lower the temperature gradient in the radiant tube in the longitudinal direction.

It is conceivable that the expansion of each bend is small and its stress level is high because the movement of each bend caused by its expansion is restricted, and therefore the strength of this portion needs to be improved by using bends of greater thickness.

(4) Behavior of creep deformation

As illustrated in Figure 2-7, the stress occurring in the radiant tube in 100 hours after it has been raised to the temperature of 900 °C is seen to increase two- to three-fold in the No. 2 bend and No. 3 bend. This is because the No. 1 straight pipe is made to descend due to the creep deformation, thereby applying a load on the No. 2 bend and the No. 3 bend.

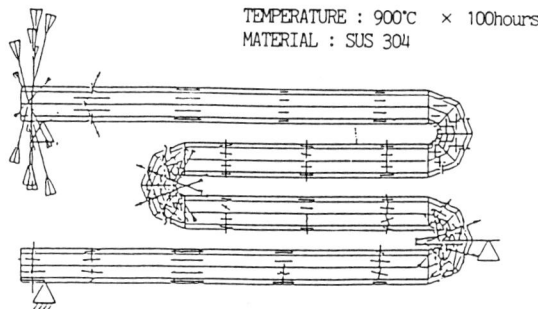

Fig.2-7 Distribution of stress in the radiant tube calculated by finite element method

It is understandable that the radiant tube will undergo a large deformation when the No. 1 straight pipe descends due to its creep deformation, thereby inducing deformation of the other straight pipes.

The extent of deformation owing to the above-mentioned mechanism is seen to coincide with the result of the deformation tests on stainless steel SUS 304, as shown in Figure 2-8. It has been known therefore that the most effective method for the prevention of the radiant tube from deforming is to support all the bends from the furnace body.

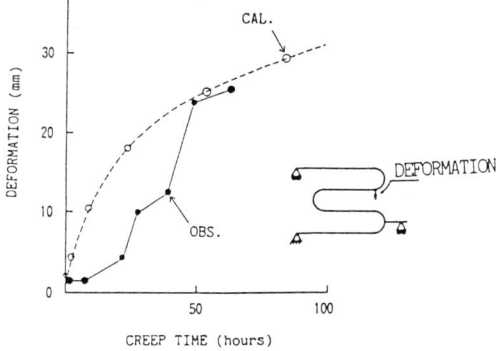

Fig.2-8 Creep deformation of radiant tube

2-3 New type radiant tube burner

① Characteristics required of a radiant tube burner
In addition to the characteristics required of a combustion equipment, the radiant tube burner must have the characteristics that will not pose adverse effect on the life of the radiant tubes, and those requirements may be summarized as follows:

(a) To cope with large fluctuations on the load on the furnace, it is necessary that the flame remain stable over a wide range of combustion load and the vibrating combustion is free from combustion noise.
(b) To better meet the environmental requirements, NOx generation needs to be as low as possible.
(c) To better protect the radiant tubes, it is necessary that no hot spots are caused to the radiant tubes and that the uniformity in heat temperature distribution is high in the radiant tubes in the longitudinal and circumferential directions.
(d) To better protect the structure of the burner, no hot spots must be caused to the combustion tube, etc.

In the case of the conventional radiant tube burner, however, due to its structural restrictions, it was difficult to accomplish high load combustion with good temperature distribution in the longitudinal and circumferential directions of the radiant tubes while suppressing the generation of NOx, and the combustion capacity of the burner in the case of the W type radiant tubes with an inside diameter of 7 inches, was 120,000 kcal/hr at the maximum.

It is well known that most of the NOx generated from the combustion in the radiant tube burner derives from a high temperature chemical reaction between nitrogen and oxygen in the combustion air which is commonly known as the Zeldovich's mechanism[5]. Because of the mechanism of this chemical reaction, the generation rate of NOx which is affected by the temperature, oxygen concentration as well as the retention time under high temperature, will markedly rise under high temperature especially in excess of 1,300° C. For suppressing the generation of NOx, it is important that the flame temperature is reduced and further it is effective if a two-stage combustion and recycle of exhaust gas are employed.

The above-mentioned two-stage combustion is also effective in equalizing the temperature distribution in the longitudinal and circumferential directions across the radiant tubes.

② Basic principle
The conventional burners are either of one-stage combustion or of two-stage slow combustion, whereas our burner of the recent development as

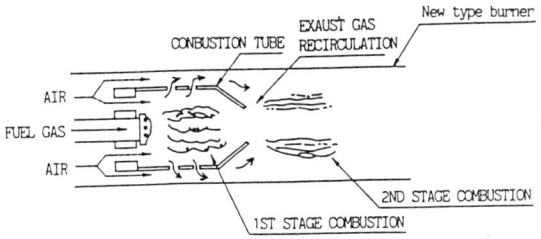

Fig.2-9 New type burner

shown in Figure 2-9 features a two-stage combustion enabling recycle of part of the exhaust gas from quickly reducing combustion gas within the limited space inside the radiant tubes.The purpose of this burner are reducing NOx generation under high load combustion, stabilizing combustion over a wide range, and equalizing the temperature distribution across the radiant tubes by means of the following basic principle:To eject from a multi-holed tube part of the combustion gas deriving from the quickly reducing primary combustion occurring inside the combustion tube, to put the above-mentioned part of the combustion gas to a complete combustion using the secondary combustion air supplied from outside the combustion tube, and to conduct a secondary combustion by mixing the combustion exhaust gas with the remaining reducing primary combustion gas.

③ Performance of the developed burner
The burner that Kawasaki has developed was tested for checking and verifying its basic performance. The burner was found to have the performance to be detailed below at a combustion capacity of 160,000 kcal/hr.

(a) Stability of the combustion
As shown in Figure 2-10, the stability of the combustion can be seen to be high over a wide range of conbustion load.

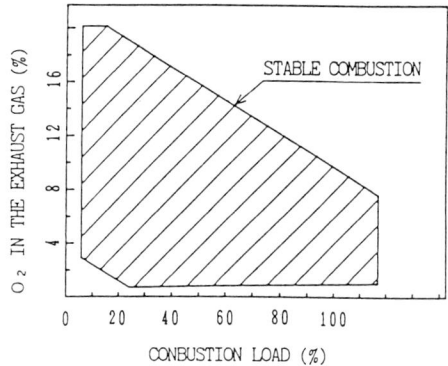

Fig.2-10 Stability of the conbustion of the new type burner

(b) Low NOx generation
It was verified that as compared with the conventional burners, the NOx level in the exhaust gas of this burner was found largely lower as is clear from Figure 2-11.

(c) Temperature distribution across the radiant tubes
The temperature distribution in the radiant tubes in the longitudinal direction is as shown in Figure 2-12. It is understandable from the figure that the $\triangle T$ of the burner introduced here is 40 to 50° C and its temperature gradient in the longitudinal direction of the radiant tubes is 0.2° C/mm. These numbers are only half of those of the conventional burners, thus indicating that its temperature distribution in the radiant tubes in the longitudinal direction has been made even.

(d) Temperature of the burner combustion tube
No hot spots occurring in the burner combustion tube have been found, and the maximum temperature of the combustion tube was 1,005° C, at the average radiant tube temperature of 1,000° C, thus meaning that there is a decline of 100° C or more in comparison with that of the conventional burners.

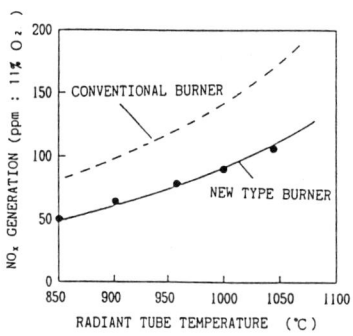

Fig.2-11 The generation of NOx

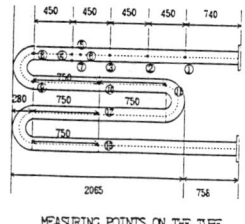

MEASURING POINTS ON THE TUBE

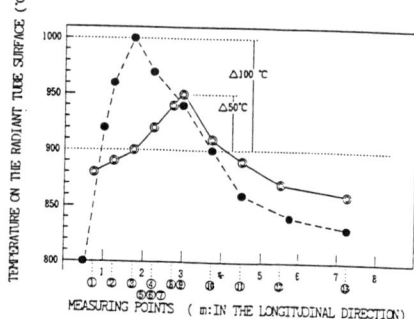

Fig.2-12 The temperature distribution in the tubes in the longitudinal (combustion load = 70% , R.T.temperature = 900°C (AVE.))

3 Establishment of the technology of high-temperature and high-speed processing of steel strip

3-1 Stable steel strip processing technology by use of a continuous annealing furnace

A continuous annealing system is made up of a heating zone for raising the temperature of the strip up to an annealing temperature level and a cooling zone for bringing it down to room temperature. For stable processing of the strip through continuous annealing, what is most critical is to work out a stable processing technology particularly in the heating zone. In general, for the purpose of causing the strip to run straight in the heating zone, while the strip is still retained cold, the strip is prevented from mistracking by imparting tension to the strip by providing a hearth roll with a crown (hereinafter called the initial crown) as illustrated in Figure 3-1 (a). When the steel strip is run hot, however, the central portion of the hearth roll (part A) is cooled by the strip as seen in Figure 3-1 (b) becoming lower in temperature than the edge of the hearth roll (part B), thereby causing a reverse crown (hereinafter called the thermal crown). When the width of the steel strip is narrow and the form of the strip is not good enough in the state shown in Figure 3-1 (b), the strip will begin to mistracking as there will be an imbalance in the mistracking correcting force as shown in Figure 3-2.

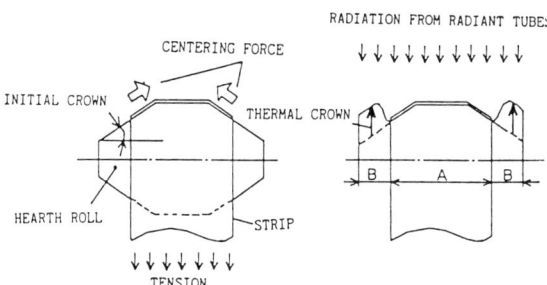

Fig.3-1 Hearth roll profile in the heating section

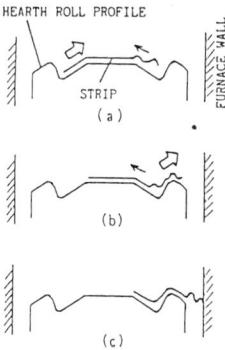
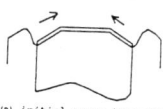
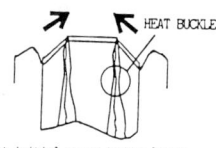

Fig.3-2 Mistracking of the strip in the heating section

Fig.3-3 Heat buckle of the strip in the furnace

As soon as the mistracking strip gets on the thermal crown, the strip will be made to waver toward the wall of the furnace owing to the mistracking correcting force posed by the thermal crown, thus resulting in a problem of the strip contacting the wall of the furnace or of breakage of the strip itself.

The method that has been practiced as a solution to the above problem is to enlarge the initial crown of the hearth roll. However, if and when the steel strip is small and large in width, as the initial crown of the hearth roll is made too large, the mistracking correcting force provided will be excessively large as seen in Figure 3-3; there would be a problem with the steel strip buckling on the shoulder of the hearth roll where most stress is concentrated (hereinafter called the heat buckle). The initial crown of the hearth roll employed in operating the conventional continuous annealing furnace has been so selected that the problems of mistracking and heat buckles may be minimized.

It is known further that the possibility of causing heat buckles is greatest when the tension applied during processing of the steel strip is the highest.

Reducing the tension, on the other hand, the steel strip will begin to waver in the widthwise direction, the operations have been made setting the tension within an appropriate range depending on the dimensions and the type of the strip as well as on the annealing temperature, so that mistracking of the strip and heat buckles may be prevented.

3-2 Problems associated with the high-temperature and high-speed stable processing of steel strip

Comparing with a case where steel strip is processed at a low speed by the conventional technology, in processing the steel strip at a high speed by operating a large capacity continuous annealing system, the heat input required per unit time in the heating zone needs to be increased, thereby enlarging the thermal crown as seen in Figure 3-4 and greatly increasing the danger of mistracking of the steel strip. In high temperature annealing (900 °C), on the other hand, heat buckles will become easier to happen because the yield stress of the strip will decline to approximately one third of the conventional process as shown in Figure 3-5. In the case of the existing technology, to achieve high-temperature and high-speed processing of steel strip, there was the necessity to work out the following technologies:

> I   Development of heat buckle preventing technology
> II Development of thermal crown controlling technology

RADIATION FROM RADIANT TUBES

ENLARGING THE THERMAL CROWN

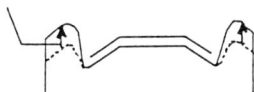

Fig.3-4 Hearth roll profile at high speed operation

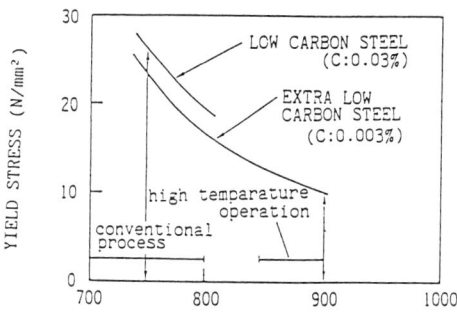

Fig.3-5 Yield stress in the high temparature

3-3 Development of heat buckle preventative technology

(1) Development of heat buckle prediction formula
In an attempt to prevent the occurrence of heat buckles during high-temperature annealing of extra low carbon steel, it is essential that the mechanism of their occurrence is firstly found so as to clarify the factors with the possibility of causing heat buckles. For the purpose of such clarification, the heat buckle prediction formula to be described below was worked out by implementing heat buckle simulation test using aluminum foil at the pilot plant, calculation of stress across the steel strip by means of a finite element method, as well as, by means of calculations of the buckling stress.

1) The symbols used in the prediction formula are as follows as shown in Figure 3-6.

$$T_{cr} = K \cdot \frac{E \cdot t^2}{(B-H)^3 \cdot \theta} \cdot \frac{dT}{d\sigma} \quad (2-1)$$

T_{cr} : Buckling critical tension (N/mm²)
θ : Taper gradient (rad)
T : Tension of steel strip (N/mm²)
E : Young's modulus of steel strip (N/mm²)
t : Plate thickness (mm)
B : Plate width (mm)
H : Flat section length (central portion of roll) (mm)
σ : Compressive stress across steel strip (N/mm²)
K : Material constant (function of yield stress etc.)

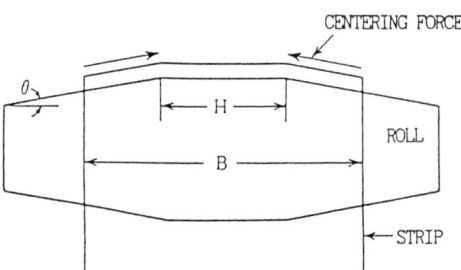

Fig.3-6 The symbols used in the heat buckle prediction formura

The results of verification conducted on the actual production line for the purpose of confirming the accuracy of the heat buckle prediction formula are as indicated in Figure 3-7. The effective means

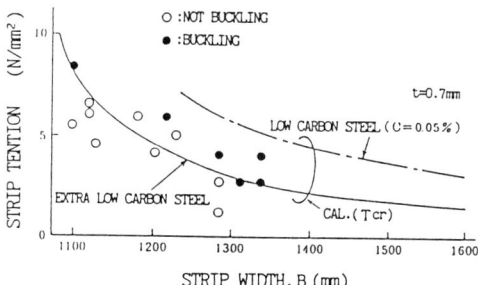

FIG.3-7 Comparison between actual buckling critical tension and calculated curve by the heat buckle prediction formula

worked out for the prevention of heat buckles were the following as understandable from the heat buckle prediction formula (2-1) and Figure 3-7:
(i) Optimization of the hearth roll initial crown (taper θ).
(ii) High-accuracy tension control (for processing of the extra low tension trip).

(2) Optimization of the hearth roll initial crown
It is necessary that the hearth roll crown be optimized so that the occurrence of heat buckles may be prevented. The hearth roll crown was designed based on the heat buckle prediction formula (2-1) and by the flow illustrated in Figure 3-8.

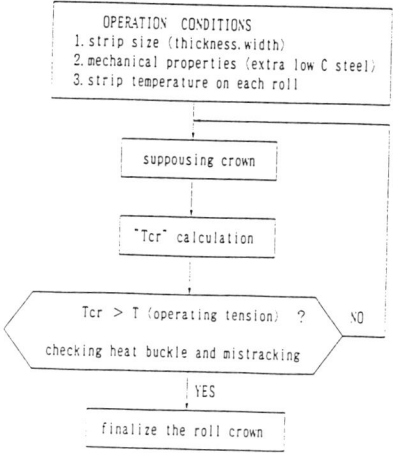

Fig. 3-8 The hearth roll crown design flow

(3) High accuracy tension control

(a) Develolpment of bridle roll control in the furnace

As shown in Figure 3-7, it was understood that reducing the steel strip tension can greatly contribute to reducing preventing the occurrence of heat buckles. The furnace of the continuous annealing system is mainly composed of a heating zone and a cooling zone. In the cooling zone, if the cooling is run by use of gas jet, the steel strip is required to have a high tension or the vibration of the strip can not be reduced. In the heating zone, however, the steel strip must be under low tension for the prevention of heat buckles. To satisfy those two different requirements, as illustrated in Figure 3-9, a new tension control system for inside the furnace was developed where a bridle roll in the furnace was installed in the inlet portion of the cooling zone, and the bridle roll was made the master roll as the reference for controlling the speed in the furnace. Use of this new system has enabled to properly and accurately control the tension of the steel strip in the furnace because the tension control for the section where the strip expands in being heated (the heating zone) and for the section where the strip is shrunk in being cooled (the cooling zone), can be separated over the master roll as the border.

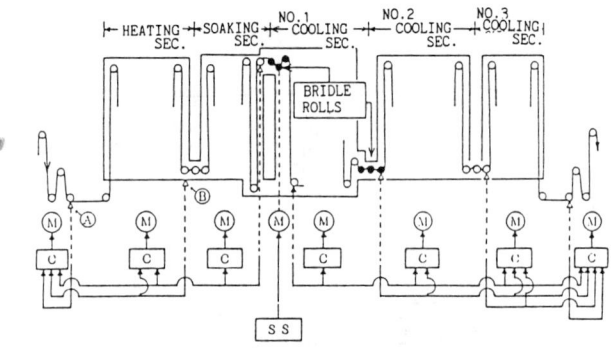

△ : TENTION METER—
Ⓜ :HEARTH ROLL DRIVING MOTOR
C : SPEED CONTROLLER—
(FOR EACH HELPER ROLL)
S S : SPEED REFERENCE FOR THE MASTER ROLL

FIG.3-9 The schematic diagram of the strip tention control with the furnace bridle rolls

(b) Development of high-accuracy tension control technology

In addition to the above-mentioned control of bridle roll inside the furnace, a technology enabling high-accuracy control of the tension has been developed, and its details are as follows:

The fluctuation in the tension of the steel strip in the annealing furnace occurs at the following two times:
i) Changes in the tension of the steel strip in the looper are transmitted into the furanace, and those changes are seen to occur in raising and lowering the looper as a result of changes in the speed for coil changing in the inlet or outlet section.
ii) Changes in the tension are seen to occur because the speeds of the respective rolls are not equal at the time of changing the speed of the furnace section. All those phenomena are caused by uneven speeds of the respective rolls which are to transfer the strip, and the specific causes may be explained as follows:

① When the speed control is of PI control, the drooping function is provided, so that the motor can be prevented from being over-loaded as a result of a roll diameter that may be selected, and the speed command is made to undergo changes in increasing or reducing of the speed due to the drooping function.

② In raising or lowering the speed, the drive roll is exposed to the inertia of the strip, non-driving roll, etc., thereby lowering the response of the motor. (The inertia of the non-driving roll is large particularly in the looper.)

With regard to ① and ② above, because of the differentials of the motors in speed because of their respective mechanical conditions, changes in the tension are seen to develop.
For the purpose of quantitatively grasping the factors leading to changes in the tension inside the annealing furnace, analytical models were structured and a simulation was run. And based on the results of the simulation, the high accuracy tension control system was developed as shown in Figure 3-10.
One feature of the new tension control system was that the following functions were added to the motor control so as to equalize the speeds of the motors under different conditions. (See Figure 3-10.)

CONVENTIONAL TENSION CONTROL SYSTEM

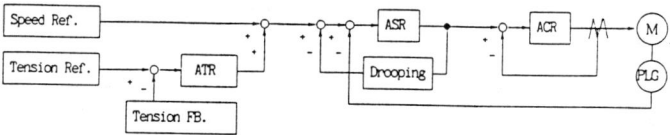

NEWLY DEVELOPED SYSTEM

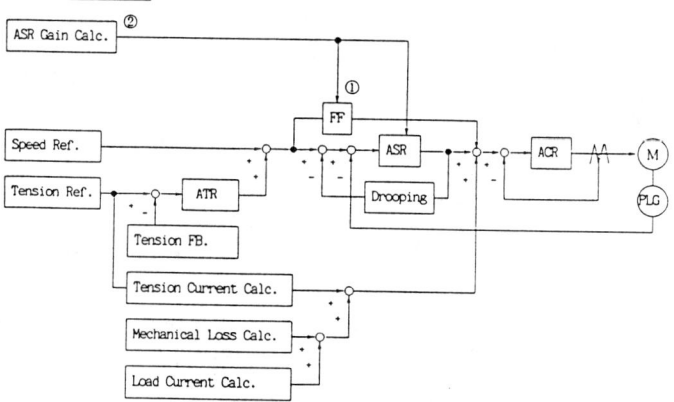

Fig 3-10 Newly developed tension control system

① FF compensation (compensation for feed-forward speed changing current)

Current command is given by calculating the current necessary for speed changes on the basis of the speed command and the inertia to be burdened by the motor. The ASR is for burdening what corresponds to the deviation of the speed resulting from deviation of this compensation current, etc. Therefore, speed changes owing to the drooping at the time of speed change are nil.

② Compensation for ASR gain

Automatic adjustment of the gain for the purpose of rendering the ASR response to be constant at all the time by taking into account not only the load directly connected to the motor, but also the GD^2 of the load (of the strip, non-driving roll, etc.) that the motor is required to burden at the time of speed change, etc.

$$K^* = K_s \cdot \frac{J_M + J_L}{J_M}$$

K* : Corrected value from the host computer
Ks : Adjusted value (gain) when the motor is under no load (connected directly to the roll).
J_M : The sum of the inertia moment of the motor and that of the directly connected roll.
J_L : The sum of the inertia moment of the load to be burdened by the motor at the time of speed change.

(c) Effect of the high-accuracy tension control system

Shown in Figure 3-11 is the effect of the new tension control based on the control of bridle roll in the furnace. The control accuracy attained by this control is ±300N.

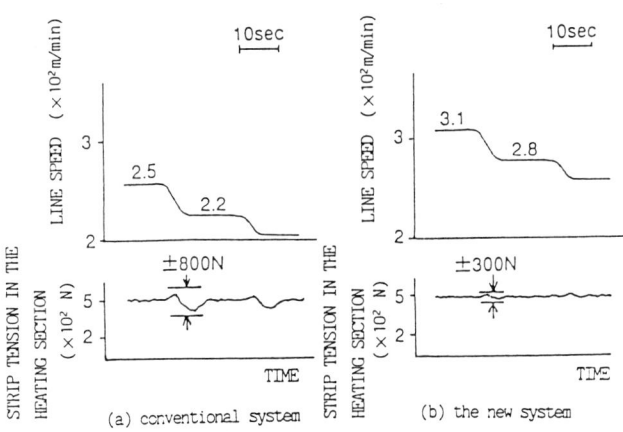

Fig.3-11 Accuracy of the new tention control system

(4) Results of the heat buckle prevention technology

Thanks to the development of the heat buckle prevention technology and of the high-accuracy tension control technology, high-temperature annealing of thin wide sheet made of extra low carbon steel has been made possible, even though such work was difficult, by means of the conventional technology. The results of the new technology are shown in Figure 3-12.

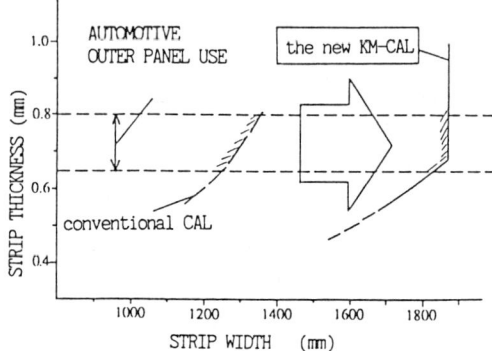

Fig.3-12 Product size of extra low carbon steel sheet

3-4 Development of thermal crown control technology

(1) Development of thermal crown prediction formula

As described in 2-3, as the technology for the prevention of heat buckles, what were effected and sought were low-tension operation through high-accuracy tension control and optimization of hearth roll crown (reducing of crown). However, each of those measures taken was found to go counter to the purpose of preventing the mistracking of the strip. In order to stop mistracking of the strip in processing it at high speed, therefore, it is necessary to reduce the thermal crown itself, that is the main cause of the strip mistracking. For an accurate determination of the behavior of the thermal crown, the temperature distribution in the roll in the widthwise direction of the steel strip was measured by use of the temperature measuring roll shown in Figure 3-13. Shown in Figure 3-14 is an example of the measuremnts.

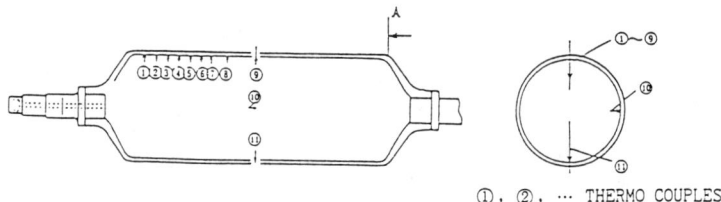

①, ②, ⋯ THERMO COUPLES

Fig.3-13 Temperature measuring roll

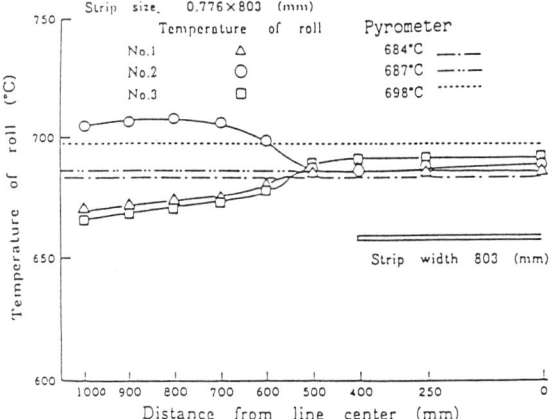

Fig.3-14 An example of the measurements

Furthermore, on the basis of the results of measurement made using the tempearature measuring roll, the heat balance across the circumference of the hearth roll was analyzed. Its analytical model is as illustrated in Figure 3-15. Through comparisons of the thermal crowns determined this time and in the past in this way, it has been made possible to make an accurate prediction of the thermal crowns.

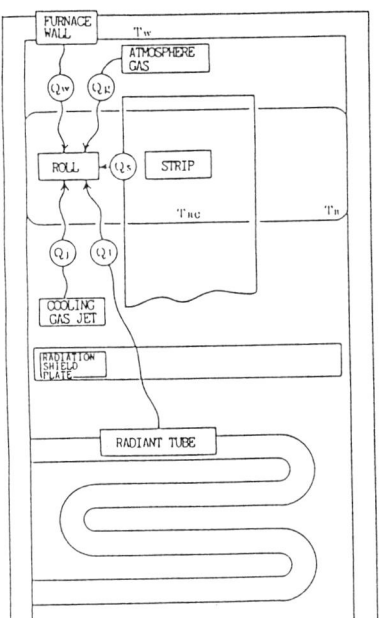

Fig.3-15 The analytical model of the heat balance around the hearth roll

(2) Development of thermal crown control technology

There are two methods for the control of thermal crowns, one being heating of the central portion of the roll that has been cooled by the steel strip, and the other being cooling the edge portion of the roll that has been heated by the radiant tube, etc. The first method has little possibility of being put to use because of necessity for a large heating capacity necessary to bring the temperature of the steel strip up to the temperature of the roll edges, since the roll is cooled down by the strip.
In the case of the method of cooling the roll edges, on the other hand, what is required to cool is no more than the heat input that has been fed to the roll edges, and that cooling quantity has been found by the above-mentioned analytical model to be of a relatively low capacity. On the basis of this finding, a new system of controlling the thermal crown by cooling the roll edges has been worked out.
The system may be outlined as illustrated in Figure 3-16.

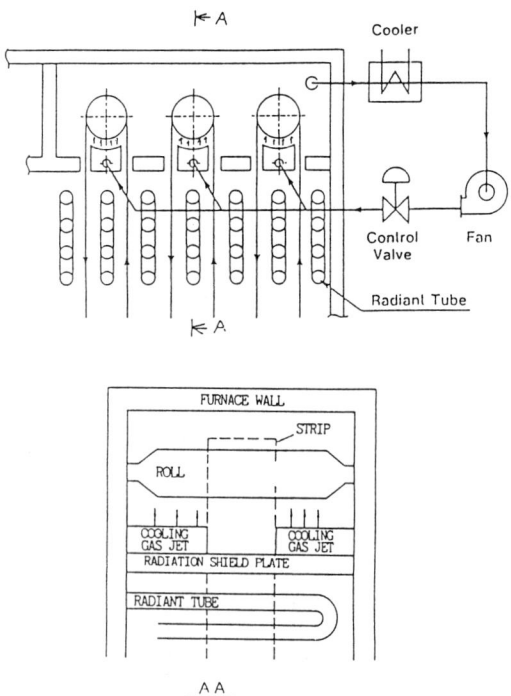

Fig.3-16 The thermal crown control unit (HEATING SECTION)

The makeup of this system consists of a roll chamber for reducing heat input into the hearth roll edges from the outside, a plenum chamber for cooling the roll edges and its circulation system. The thermal crown is controllable by changing the flow rate of blow gas.

(3) Effect of the thermal crown control unit

The effect of the thermal crown control unit is as given in Figure 3-17. The possibility of mistracking of the steel strip is risked when the production capacity P is 100 or higher thereby making it difficult to cause the steel strip to run at high speed, and when, on the other hand, roll edge forced cooling by means of the thermal crown control unit is in use, the possibility of weavering can be avoided even in a high speed range at the production capacity P of 300 (that corresponds to 80,000 metric tons per month). Use of this thermal crown control unit has enabled high-speed, stable processing of the steel strip.

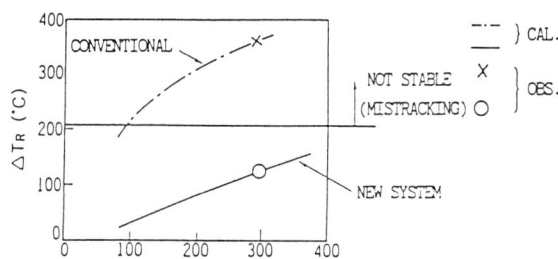

Fig.3-17 The effect of the thermal crown control unit

4 Development of quenching technology

4-1 Quenching in the continuous annealing process

In the case of continuous annealing process of general cooling type steel sheet, the mechanical properties of the product are affected by the primary cooling conducted after heating and soaking. The cooling rate based on the matallurgical requirements in the case of general cold rolled steel sheet for fabrication use and of bake hardening type steel sheet, is in the range from 30° C to 100° C per second.

The strip that has undergone annealing is required to be flat in form, and it is also important that it remain good in its surface quality so as to receive chemical processing after the annealing step.
For this primary cooling, one of the following four methods is used according to the characteristics of the line (capacity and type of product): gas jet cooling, roll cooling, mist cooling and wter cooling,

and Figure 4-1 compares those methods of strip cooling.

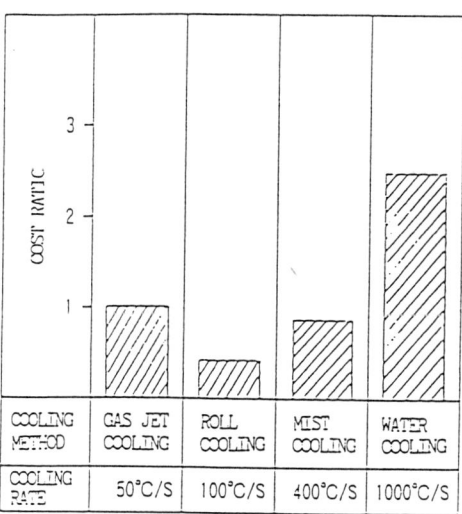

Fig. 4-1 Comparison of consumption of the cooling methods

The roll cooling is a method whereby cooling medium is caused to flow inside the roll and the strip is cooled by causing it run while contacting the surface of the coo roll, and this method offers the following advantages:
① Low running cost (Figure 4-1)
 The cost of power required for running blowers is high in the case of the gas jet cooling, whereas by the water cooling and the mist cooling, oxide film is formed on the surface of the steel sheet during the cooling process, calling for pickling treatment as an additional post processing. Applying the water cooling, on the other hand, reheating is required immediately after the cooling step if the product in the making calls for over-aging treatment.

In the case of the roll cooling, the energy cost is extremely low because the heat exchange between the cooling medium and the strip is carried out by way of the roll shell.

② Excellent surface quality
 In the roll cooling, since cooling is conducted in the ambient gas (nitrogen and hydrogen), the surface conditions of the strip are no different from those obtainable by the gas cooling, and thus posing no problems for the chemical treatment. Additional processing of pickling becomes necessary in the case of the water cooling and the mist cooling where formation of oxide film on the surface of the strip is inevitable.

Furthermore, compared with the strip that has undergone treatment in a non-oxidizing atmosphere (without pickling), the one that has been subjected to the pickling process is slightly inferior in terms of chemical processability. (See Photos 4-1 and 4-2.)

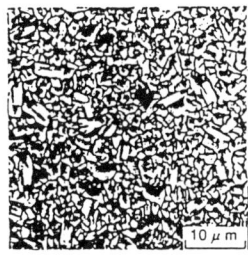

Photo.4-1 Strip surface with chemical treatment
(CAL in a non-oxidizing atmosphere ⇨ chemical treatment)

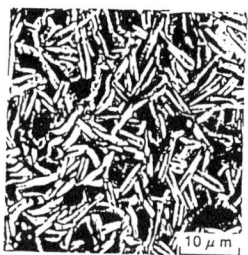

Photo.4-2 Strip surface with chemical treatment
(CAL ⇨ pickling ⇨ chemical treatment)

③ Appropriate rate of cooling
In the manufacture of the coating bake and cure type steel sheet, as already referred to in 4-3 above, the cooling rate required for the CAL for cold-rolled steel sheet is 30 to 100° C per second, whereas the maximum cooling rate in the gas jet cooling is about 50° C per second if the sheet thickness if 0.7mm.
In consideration of the above conditions, Kawasaki has developed an RGCC (roll and gas jet combined cooling) system, putting the roll cooling and the gas jet cooling combined.

4-2 RGCC system

The outline of the RGCC system is shown in Figure 4-2.
In this system, as seen in the figure, gas is blown onto the back surface of the strip which is in contact with the cooling roll. Because the opening between the strip (which is fixed to the roll) and the gas nozzle can be made small, the power required for the gas blower can be small enough. By the conventional roll cooling system, as a result of improper temperature distribution in the strip in the widthwise direction, steel sheet with non-satisfactory form is produced.
By means of this RGCC system, use of the gas jet placed opposite the roll

has made it possible to make even the contact between the roll and the strip thereby preventing occurrence of improper heat distribution in the strip. (See Figure 4-3.) In this way, the roll cooling is now possible to be used also for extra low carbon steel and thin steel sheet (thickness: 0.4 mm), even though this roll cooling system has been found difficult to apply to the type of use mentioned above.

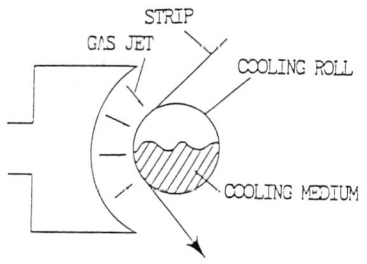

Fig. 4-2 Outline of RGCC system

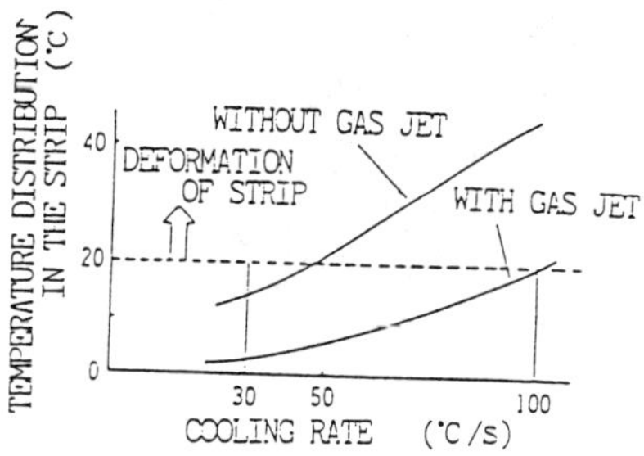

Fig. 4-3 The effect of RGCC system

5 Establishment of high-accuracy strip temperature control technology

5-1 Makeup and features of the strip temperature control technology

The purposes of the strip temperature control in the continuous annealing system lie in the following:

(1) Achievement of the product with stable quality by controlling the temperature of steel strip at the outlet of the continuous annealing system within the target range strip temperatures.
(2) Automation of furnace operations
(3) Improvement in the productivity and in the specific consumption of fuel

With the tendency intensifying in the recent yearstoward the production of small volumes of product in many different types and grades, the frequency of schedule changes (in strip thickness, width and target strip temperature) has been on the incraease. On the other hand, as a result of use of a large capacity heating furnace, the time constant of the process is long at 10 to 20 minutes due to the thermal inertia, and by simply changing the operation volumes in steps, high-accuracy control of strip is impossible because of need for long time for stabilizing the strip temperature.

The method conventionally used for the control of the temperature of steel strip has been based on the static characteristic model of the furnace, and therefore the consideration extended has been inadequte for the non-routine operations such as increasing or decreasing the speed of the production line and schedule changes.

To solve those problems, the high-accuracy steel strip temperature control has been developed and applied on the basis of a dynamic characteristic model of the steel strip temperature.

The makeup of the steel strip tempearature control system is shown in Figure 5-1.

CONVENTION MODEL

HIGH-ACCURACY STRIP TEMPERATURE PREDICTION MODEL

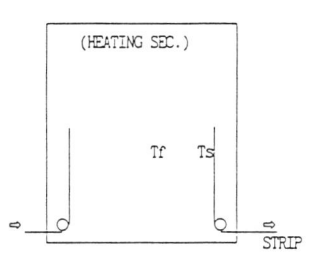

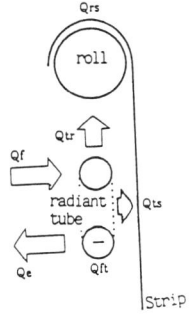

$\dot{Q} = \phi_{cc} \sigma (T_f^4 - T_s^4)$

Q : strip heat absorption
ϕ_{cc} : Overall coefficient of heat transfer
σ : Stefan-Boltzmann constant
Ts : Strip temperature
Tf : Furnace temperature

Q=f(Qf, Qft,Qts,Qtr,Qrs,Qe)

Q   : strip heat absorption
Qf : input heat of fuel gas
Qft : heat flux (fuel gas ⇒ radiant tube)
Qts : heat flux (radiant tube ⇒ strip)
Qtr : heat flux (radiant tube ⇒ roll)
Qrs : heat flux (roll ⇒ strip)
Qe : heat of exhaust gas

Fig. 5-1 Highly accurate strip temperature control system

5-2 High-accuracy strip temperature prediction model

Mathematical models based on the theory of heat conductance are used as the steel strip temperature models on which the strip temperature control of the heating zone and the soaking zone is based, and (1) Radiation heating formula and (2) Heat balance formula are made the basic models.

Radiation heating formula:

$$\frac{dT_s}{dt} = \frac{d\sigma}{C \cdot \rho \cdot D} \cdot \phi_{CG} (T_f^4 - T_s^4) \qquad (1)$$

Ts : Strip temperature (K)
Tf : Furnace temperature (K)
σ : Stefan-Boltzmann constant 4.88 x 10-8 (kcal/h·m²·k⁴)

C   : Specific heat (kcal/kg·K)
ρ : Density (t/m³)
ϕ_{CG} : Overall coefficient of heat transfer
D   : Strip thickness (m)
t   : Time (h)

Heat balance formula:

$$Ls \cdot D = \frac{[Qc - (1-\mu)Qg]F - QL}{\rho \cdot Qs \cdot W} \quad (2)$$

Ls : Line speed (m/h)
D : Strip thickness (m)
Qc : Fuel calorific value (kcal/Nm³)
μ : Recuperator efficiency
Qg : Sensible heat of exhaust gas (kcal/Nm³)
F : Flow rate of fuel (Nm³/h)
QL : Fixed heat loss (kcal/h)
ρ : Density (t/m³)
Qs : Strip heat absorption (kcal/t)
W : Strip width (m)

For the verification and accuracy improvement of the strip temperature model, extensively utilized are the strip temperature gauges installed at a number of principal positions and a temperature measuring roll which is a helper roll in the furnace in which thermocouples are embedded.

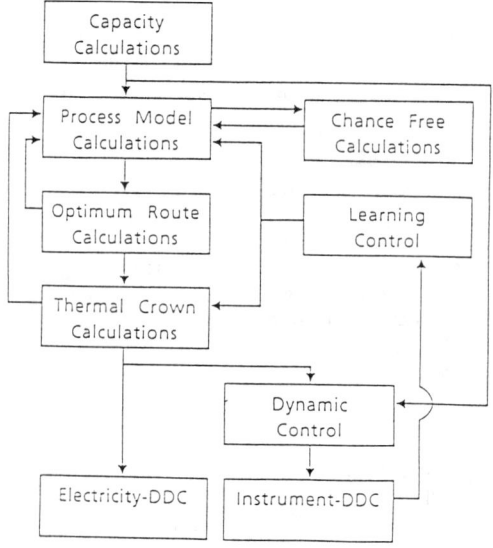

Fig. 5-2 High-accuracy strip temperature prediction model

5-3 Optimal route control

① Summary

By use of a dynamic model formula representing the relationship between the strip temperature, strip thickness and the line speed obtained by formulas (1) and (2), and the flow rate of fuel, calculations are made to obtain the path of strip temperature following up the target strip temperature of each coil with the minimum of deviation as well as the time series of the fuel flow rate, and this time series value of the fuel flow rate is set at as the fuel flow rate for feed forward (FF). What is more, high accuracy control of the strip temperature can be realized by means of feed back (FB) control using the path of strip temperature as the strip temperature setting signal.
Further descriptions will be given in this section regarding optimal route control for the FF control. As for the FB control, it will be explained in the following section as dynamic strip temperature control.

② Optimal calculations

By solving the optimization problem for minimizing the function of evaluation of formula (A) under the restricting conditions of formula (B), the value of time series of optimal fuel gas flow, the value of time series of line speed and the path of strip temperature can be calculated.

$$cTs_{HS}(k) \geq oTs_{HS}(k)$$

$$MIN \leq V_{MT}(k) \leq MAX \quad (A)$$

$$J = \sum_{K=1}^{N} (cTs_{HS}(k) - oTs_{HS}(k))^2 \rightarrow MIN \quad (B)$$

where,
$cTs_{HS}(k)$: Calculated value of strip temperature at the heat outlet
$oTs_{HS}(k)$: Target strip temperature
$V_{MT}(k)$: Fuel flow rate
N : Number of data

③ Prediction of strip temperature

The calculated value of strip temperature used in the function of evaluation can be obtained on the basis of the strip temperature prediction model shown in Figure 5-2. In other words, using the strip temeprature prediction model, dynamic changes in the strip temperature in the furnace can be predicted and calculated on basis of change in the strip thickness, and the line speed and the flow rate of fuel gas that change with time.

④ Results of use of the optimal route control
The results of use of this control are as shown in Figure 5-3. Compared with the conventional system, the extent of change in strip temperature at the timing of schedule change has been reduced to approximately 1/3, and the effect of this system is high in terms of improvement in the control accuracy of strip temperature.

Using this system, it is also possible to calculate the optimum pattern of the line speed, but the line speed has its restrictions in connection with troubles such as of the operation conditions on the inlet and outlet sides, mistracking of the strip in the furnace, and buckling of the strip. Applications of this system will be tried after having attained improvement in the level after having taken those factors into account.

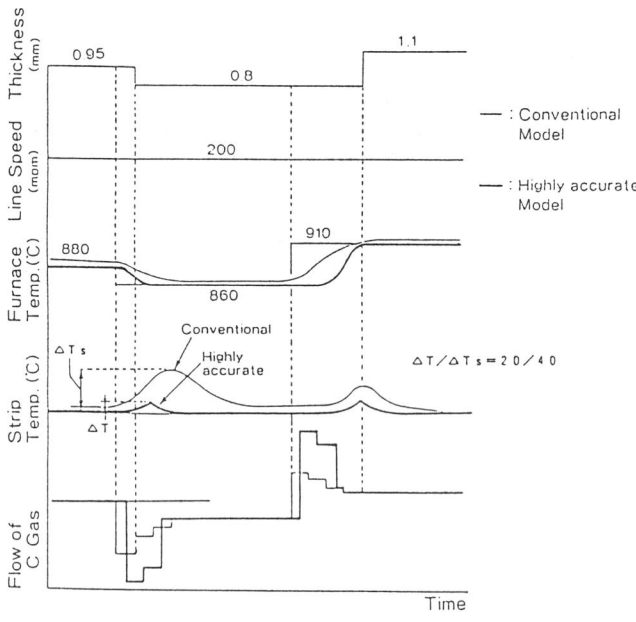

Fig. 5-3 The result of the optimal route control

5-4 Dynamic strip temperature control

① Makeup of the function

The purpose of the function is to control the actual strip temperature in accordance with the target strip temperature, and a model expressed in a status equation is used for considering the dynamic characteristics of the plant, and identifying structure for real time modification is provided. The function, furthermore, comprises a status prediction unit for predicting the strip temperature on the basis of the results of the identification, as well as, a control unit for deciding on an extent of changing the fuel flow rate on the basis of the quantity of status.

② Control model formulas

$$\Delta X_K = P \cdot \Delta X_{K-1} + Q \cdot \Delta U_{K-1}$$

$$\Delta Y_K = C \cdot \Delta X_K$$

ΔX_K : Extent of change in the status at the point of time k
ΔU_K : Extent of change in the input at the point of time k
ΔY_K : Extent of change in the output at the point of time k

③ Results of use of the on-line control

Shown in Figure 5-4 are the results of use of the on-line control. Even under load fluctuation up to about 30%, a good accuracy in strip temperature control has been attained at 20° C. Shown in Figure 5-5 are the results of control that have been achieved by the conventional system whereby the furnace temperature is set in advance, and since the strip temperature is changing up to the level of the target plus about 40° C, the effect of improvemnt in the strip tempeature control by this system is very high.

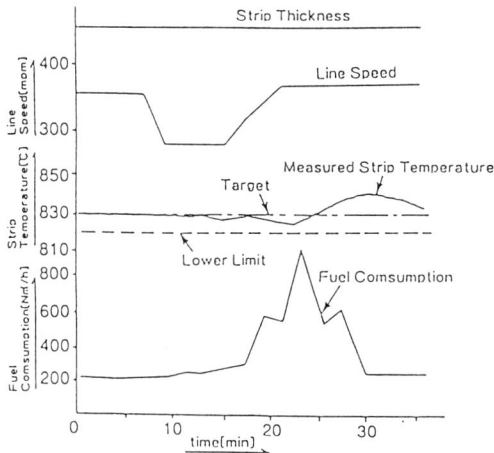

Fig. 5-4 The result of the dynamic control

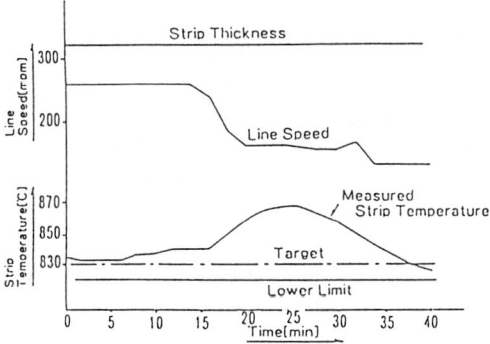

Fig. 5-5 The result of conventional the control system

5-5 Chance-free control

① Function of the chance-free control

The purpose of the function is to improve the controllability of strip temperature at the point of change in heating load such as changes in the target strip temperature, and changes in the strip thickness. At the point of joint -- representing the concept of chance-free control as shown in Figure 5-6 -- between the preceding steel strip (strip thickness: 0.8 mm and target temperature: 830° C) and the following steel strip (strip thickness: 0.48mm and target temperature: 830 ° C), in process - ing the preceding steel strip (strip thickness: 0.8mm) aiming at the target temperature, the following steel strip, immediately after the point of joint, will unavoidably go up excessively in its strip temperature.

The chance-free section, which is positioned half way in the heating section, is for minimizing possible overshoot of the strip temperature on the outlet side of the heating section by blowing cooling gas immediately at after having passed the point of joint.

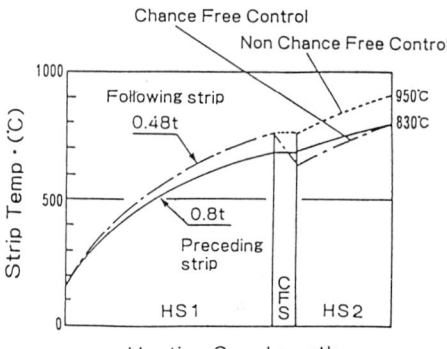

Fig. 5-6 The concept of chance-free control

② Outline of the equipment

An outline of the equipment of the chance-free section is as shown in Figure 5-7. The main cooling fan is run at all times, and the gas is circulated by-passing the furnace body while the chance-free section is not in use.

While the chance-free section is in use, the damper of the by-pass system is closed and the damper leading to the gas chamber opens so as to cool the the strip. In this case, the set value of pressure calculated on the basis of the strip temperature control model is set on the pressure control system (indicated to be PIC in the figure) from the host computer (indicated to be P/C in the figure), and depending on the set value, the revolutions of the fan are changed for adjusting the extent of cooling. In a high temperature atmosphere, as the cooling fan is run, a temporary decline in the pressure in the furnace is caused resulting from the shrinkage due to drop in the temperature of the atmosphere, and the steps to be given in the following are taken for the prevention of the above-mentioned consequences.

(1) By running the main cooling fan at all times (by-passed when not in use), decline in the furnace pressure is prevented at the time of starting the fan.

(2) For the prevention of sudden changes in the gas temperature across inside the chance-free section, the temperature of the atmosphere is retained low at all times by use of an auxiliary cooling fan.

(3) Back-up atmosphere gas is injected in case of drop in the furnace pressure.

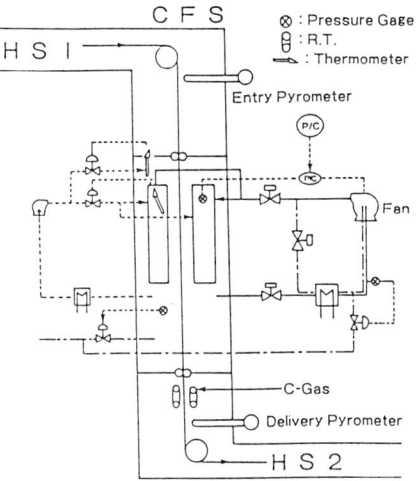

Fig. 5-7   The outline of the equipment of the chance-free section

③ Results of application of the chance-free control

An example of use of the chance-free control is illustrated in Figure 5-8. At the joint point where strip thickness is changed between the preceding strip (strip thcikness: 0.96 mm and target temperature: 780° C) and the following strip (strip thickness: 0.8 mm and target temperature: 780 ° C), by providing gas blast immediately after the point of thickness change, over-shoot of temperature could be limited to 5° C. (The overshoot is 40° C when no control is applied is a calculated value).Use of this chance-free control has made it possible to bring the accuracy of strip temperature control to ±10° C when the change in the load is ±20% (such as in changing the strip thickness from 1.0 mm to 1.2 mm).

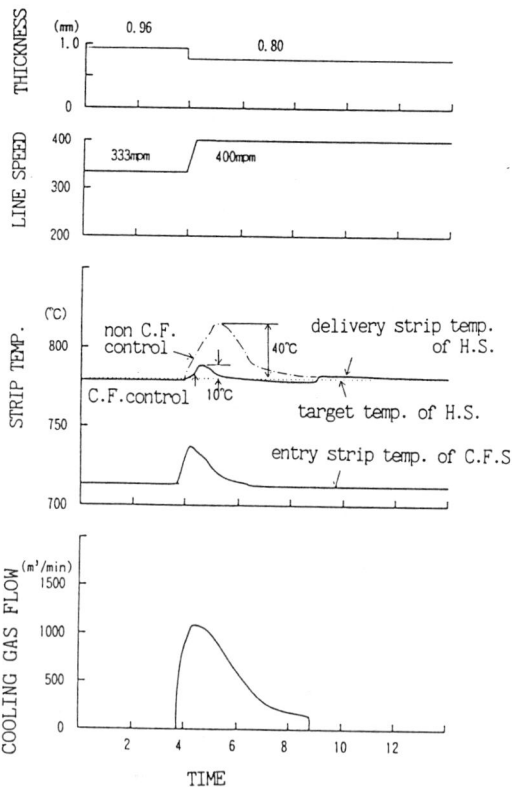

Fig. 5-8 An example of use of the chance-free control

6 Conclusions

In this paper, new technologies in KM-CAL for sheet gage are described. Using these technologies, a great deal of extra low carbon steel sheets are produced in Chiba NO.3CAL and Mizushima NO.2 CAL at Kawasaki Steel. These lines achieved high productivity and very stable operation at very high temperature operation.

References

[1] K.Toda et al:Production of drawing quality steel sheet by continuous annealing and processing ,Steel metal industries,51(1974)9, 586-603.
[2] M.Matsumoto : NO.2 continuous annealing line for drawing quality cold strip at NIPPON KOKAN,FUKUYAMA WORKS,Steel times,205(1977) 3,149-163.
[3] F.Yanagishima et al.:Characteristics and operation of multipurpose continuous annealing line at Chiba works,Kawasaki steel technical report,2(1981)3,1-13.
[4] K.Tsunoyama et al.:Development of Extra-Deep-Drawing cold rolled steel sheets for integrated parts,Kawasaki steel technical report,24(1991)3,84-90.
[5] J.Zeldovich:Acta Physicochimica U.R.S.S.,21(1946),577.

REDUCTION HEATING TECHNOLOGY OF STEEL SHEETS BY DIRECT FIRE;

Masanori Kurihara, Akiyoshi Honda

Shuzo Uchino and Masahiro Shoji

NKK Corporation
FUKUYAMA Works
1 Kokan-cho, Fukuyama-city, 721, JAPAN

Abstract

NKK has adopted reduction type direct-fired furnace for continuous annealing process of steel sheets including CGL (Continuous Galvanizing Line). A direct-fired furnace method has many advantages such as better tracking of steel sheets, minimizing-heat inertia of the furnace and compact equipment. NKK has further advanced improvement and development concerning this direct fired heating technology, thus establishing direct fired heating equipment which has high level reduction heating performance. Major technological characteristics are as follows.

(1) Nozzle-mixing type direct reduction heating burner provided with powerful reduction performance and energy-saving system.
(2) Two-pass direct-fired furnace and its optimum combustion control technology corresponding to large type equipment and high speed.
(3) Automatic control system of high accuracy steel sheets temperature.

NKK has applied the direct-fired method to No. 3 CGL in KEIHIN works, No. 3 CAL (Continuous Annealing Line) and No. 2 CGL in FUKUYAMA works, thereby resulting in an effective manufacturing of high quality steel sheets, stable and high speed travel operation, energy saving, etc. NKK has also scheduled application to 3 lines being newly provided.

1. Introduction

Increasing demand for high quality cold-rolled steel sheets has induced Continuous Annealing Process to be enlarged in size, more efficient and to ensure high-grade products.

To cite an example, CAL (Continuous Annealing Line) of increased annual production up to one million tons, and CGL (Continuous Galvanizing Line) of half a million capacity, including high grade products for exposed car bodies, have been installed. Recently, to respond to these tendencies, the steel industry has been placing emphasis on new investment and on research and development in the area of CALs, CGLs and related technologies.

In a coventional CAL, indirect heating by radiant tube was used mainly to prevent surface oxidation.[1] The indirect heating system, however, has inherent problems to accompany the increase in size, namely restricted heating capacity, slow thermal response, strip snaking and so on.

For CGL, the NOF (non-oxidizing furnace) system of mostly horizontal type had been established, in which employed direct firing burners have no reduction effect on the strip surface to permit inevitable oxidation by combusted gas.

These conventional thechnologies hardly overcome some substantial operational problems at the large-sized CAL and high-grade CGL.

As a solution of the problems, NKK took notice of the SELAS type DFF (Direct-Fired Furnace) and introduced the technology from Selas Corporation of America to construct the KEIHIN No.3 CGL in 1983. Because the furnace had redution heating burners [2] and was widely applied for CGLs in production of construction materials, and the superior heating characteristics of direct firing it was considered to be compatible with the required product quality.

After constructing the KEIHIN No.3 CGL in cooperation with SELAS, NKK has been continuing to improve and develop this DFF technology in order to meet new production levels and higher product qualty. Especially, the reduction heating mechanism has been clarified and the most stable and optimum conditions have been determined through research. A result of this research was the development of an unique reduction heating burner giving higher performance than the conventional burner. Over the years, much information has been gathered from basic experiments, operations, various tests and our own heat transfer analysis model of DFF. And this know-how has been applied on a continual basis.

NKK has already adopted this DFF system to our 3 lines, KEIHIN No.3 CGL, FUKUYAMA No.3 CAL and No.2 CGL with positive results. Furthermore, three more lines, namely KEIHIN No.4 CGL, FUKUYAMA No.3 CGL and No.4 CAL, are under design and construction on the basis of this DFF technology.

2. Features of direct-fired reduction heating technology

2.1 Effect of reduction heating DFF

When installed in the heating zone of a CAL or a CGL, instead of the conventional indirect heating furnace or the NOF system, the reduction DFF brings about the following advantages:

(1) Compact line arrangement

The DFF can realize rapid heating up to the target strip temperature, as shown in

Fig.1. It appears that heating capacity of DFF is 10 times greater than that of a radiant-tube furnace and total required path number for heating up can be drastically reduced. This leads to cost reduction, 20% or more for heating furnace only, and good strip tracking with no trouble of buckling and snaking caused by roll contact, The latter can be regarded as a superiority of vital importance for this DFF.

(2) Quick thermal response

The compact furnace can lower thermal inertia,. Therefore control ability of strip temperature is remakably improved because of quick resonse of furnace temperature according combustion rate change.

(3) Optimum heating corresponding to strip width

Optimum heating is possible corresponding to various strip widths and results in energy saving. With multiple burner arrangement in the direction of strip width, several burners can be trimmed to achieve uniform heating and to avoid overload especially while narrow-width strip is processed.

(4) Strip surface quality

Strip surface is so bright at the exit of the DFF that there is no need to reduce the surface by higher hydrogen atmosphere in the following furnace. It also means that roll pick-up, sintered iron oxide accumulation on the roll which causes dents, can be avoided. [3]

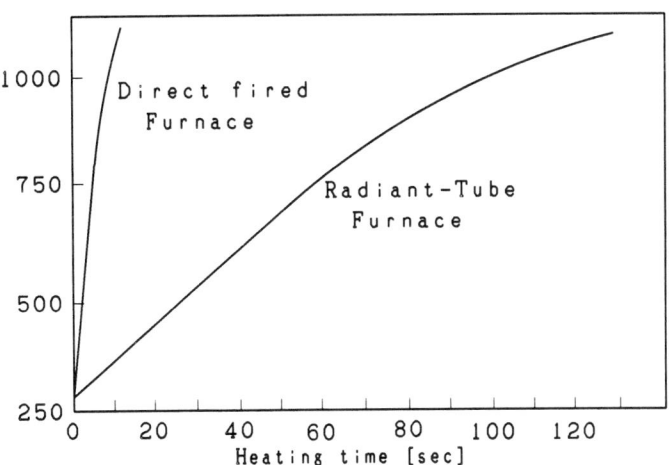

Fig.1 Comparison between direct fired furnace and radiant tube furnace

2.2 Basic characteristics of direct-fired reduction heating mechanism

2.2.1 Non-equilibrium and stationary experiment   Reduction heating of the DFF is in principle obtained by maintaining the strip surface in the region where the combustion flame possesses a reduction force. The reduction force is considered to be due to the presence of intermediate products such as ions and radicals, which occured in the hot gas during immediate combustion process. The strip surface oxidation/reduction mechanism, however, is not verified clearly and optimum stable conditions for sufficient reduction are difficult to find out spedifically.

Therefore, the examination was conducted forcusing on the chemical reaction in heating by the flame impinging on the sample steel sheet, which was maintained at constant temperature (a constant heating condition). Fig.2 shows an outline of the experimental apparatus. In the experiment, a partition was provided between the burner and the sample sheet. N_2 gas was blown against the surface of the specimen and heated to prescribed temperature, taking preventive measures against oxidation. Then, the partition plate was lifted so that the flame would impinge on the surface of specimen. Simultaneously, the surface on the opposite of the burner side was cooled with N_2 gas to stabilize the temperature of the specimen at a certain level. After the flame was applied to the specimen for a specific period of time, the partition was again lowered between the specimen and the burner, and the specimen quenched by N_2 gas blowing. Radiant cup burner using pre-mixed COG(coke oven gas), by-product in complex steel works, was employed in this experiment.

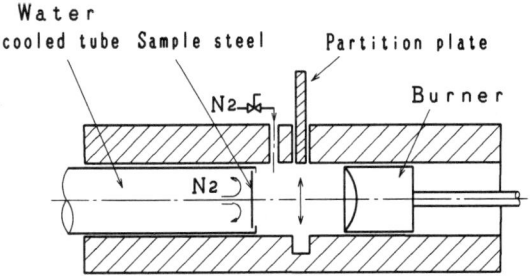

Fig. 2 Experimental apparatus

2.2.2 Experimental result Fig.3 shows the measurement of reduction competion time for the sample oxidized beforehand in atmosphere. The reduction speed is faster at

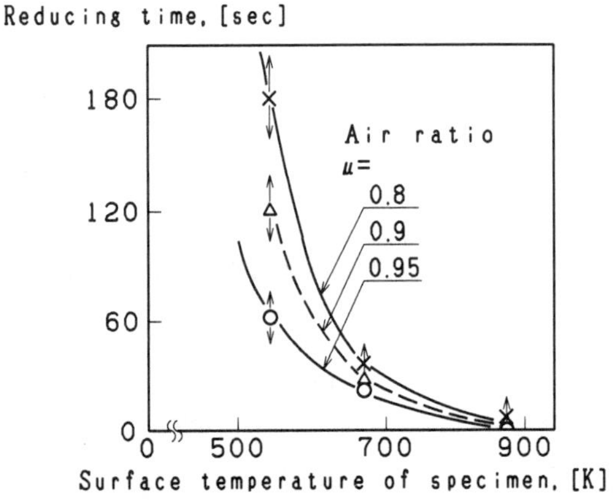

Fig.3 Relation betweeen steel temperature and reducing time

an air ratio of 1.0 or less, but closer to 1.0. Based on these data, it seems reasonable to assume that the activity of the reducible intermediate products increases because of higher flame temperature. As shown in Fig.4, reduction speed increases according to higher combustion rate, as well as higher flame temperature. These data means that total amount of the reducible intermedate products, impinging on sample surface, has also effect on reduction force.

In this experiment, however, reoxidation occurred at high sample temperature of above 1023 K. It is considered that the activity of intermediate products depends on temperature and that oxidizing intermediate products surpass redusing products in activity in the temperature range of higher than 1023 K.

The temperature that starts reoxidization is hereinafter defined as the maximum temperature for reduction heating.[3]

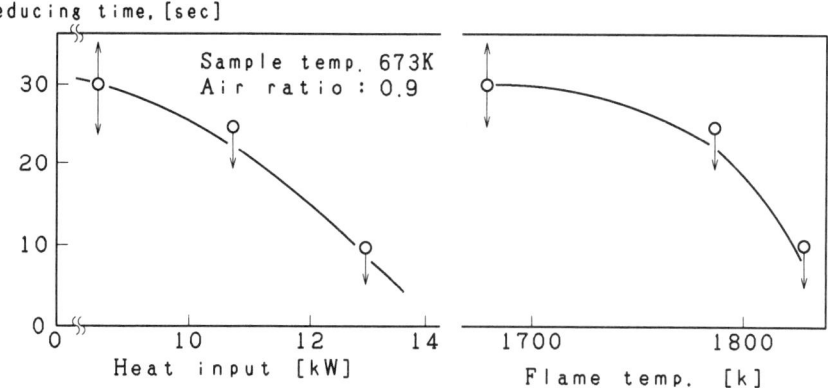

Fig. 4 Influence of heat input and flame temperature on reducing time

3. NKK new technology and its development

3.1 Development of new type burner

The profile of characteristics of reduction heating by pre-mix type burner has virtually been clarified through a series of basic experiments. These experimental data appear to indicate that it is essential to raise the flame temperature for drastic improvement of reduction force.

To raise the flame temperature, the most effective means are to preheat the combustion air. SELAS reduction heating burner of pre-mix type,[4] however, can not accept the preheated combustion air inherently from the point of safety. Therefore, NKK has been developing of a nozzle-mix type burner which has equal or more intense reduction potential than that of the SELAS pre-mix burner. This effort results in so-called 'NKK Toroidal Burner'.

3.1.1 Basic burner construction   In order to form the stable reducing flame, new type nozzle-mix burner was designed, as illustrated in Fig.5. The fuel gas jets radially from the center nozzle and the preheated combustion air jets at a certain angle through many holes arranged inside the burner cup. This strong swirling flow can accelerate the mixing of fuel and air to achieve rapid combustion in the burner tile. Furthermore, the flow forms a strong recirculating eddy (toroidal eddy) in the center of the burner and form a uniform reducing flame downstream from the exit of the tile. A series of tests was conducted on the basis of a small burner of 12kW/Burner to explore the optimum conditions for the relative position of air and gas nozzles, jet hole bore, jet speed etc,. As a result, a lalrge-sized burner of 87 kW/Burner, which Fig.6 shows, was realized for practical use.

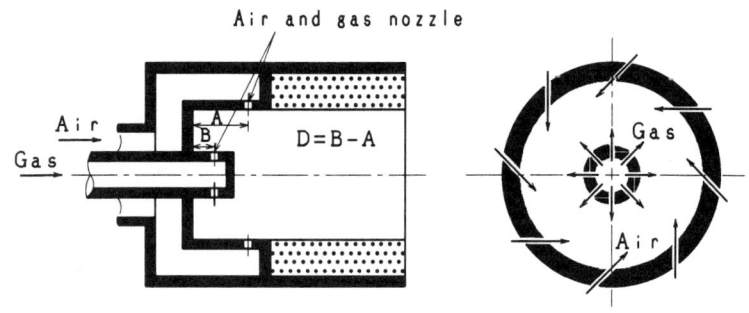

Fig. 5 Concept of NKK new nozzle mix burner

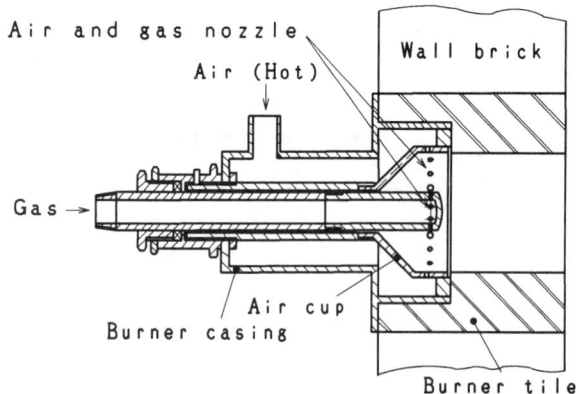

Fig.6 Construction of NKK toroidal burner

3.1.2 Demonstration experiment and results An outline of the experimental furnace with toroidal burners is shown in Fig.7. The following sequence of tests were conducted. After heating the furnace to the prescribed temperature, the specimen was charged into the furnace from the top. The two sides of the specimen were heated by burner flame, then dropped in the lower N_2 cooling chamber and then removed. The specimen was as cold rolled low carbon steel sheet.

Fig.8 and Fig.9 show the experimental results. Fig.8 shows the effect of the preheated air on the maximum temperature for reduction heating, which temperature is raised by using the preheated air. By 623-673K preheated air, the maximum temperature for reduction heating of 1173K is reached within a wide range of air ratio, 0.8-0.95. It is due to a sufficient swirling force by high speed jet flow and high flame temperature by the preheated air, even if air ratio is low.

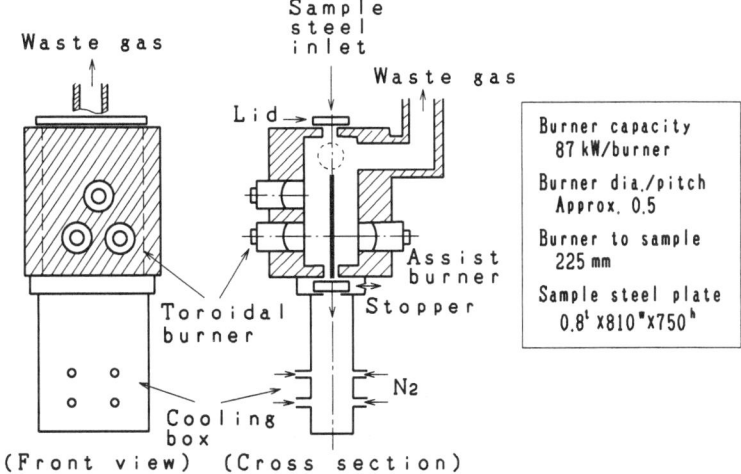

Fig.7 Outline of combustion test furnace

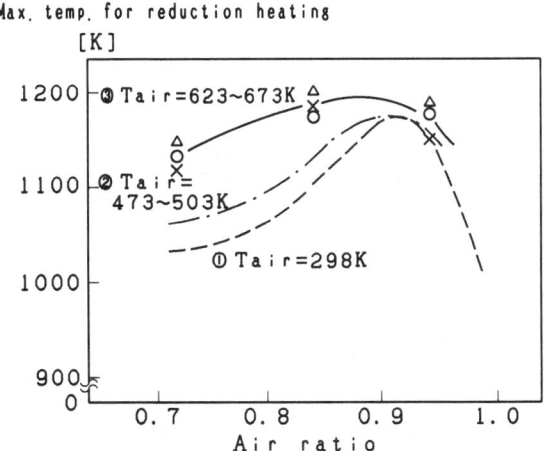

Fig.8 Effects of preheated air on reducing ability

Fig.9 shows the reducing abilities for various combustion conditions of actual operation. The toroidal burner appears stable reducibilty over a wide range because reducing force is strong in comparison with the conventional pre-mix burner.

For this reason, it is clear that this burner allows flexibility of the DFF operation to increase, and higher quality products to be heated. [5]

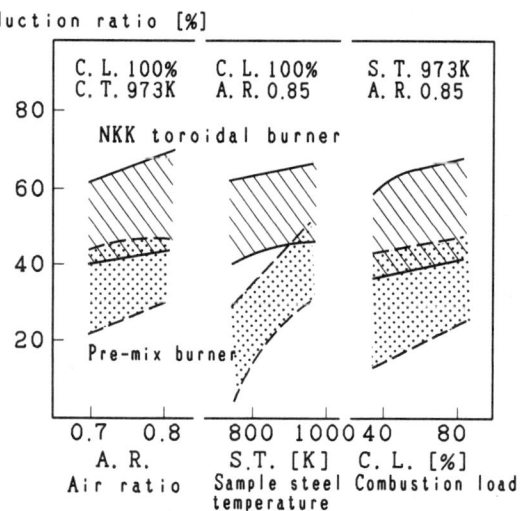

Fig.9 Comparison of pre-mix burner and NKK toroidal burner (Reduction ratio)

3.1.3 <u>Practical application</u> The developed NKK toroidal burners were applied to the DFF No.3 zone of the FUKUYAMA No.2 CGL [6], in which 48 burners were installed with an opposite staggered arrangement as shown in Photo 1. On the other hand, No.1 and 2 zones employ SELAS nozzle-mix burners and No.4 uses SELAS pre-mix burners like other preceding SELAS type DFF. The reason for applying the new burners to No.3 zone only

Photo.1-(a)   NKK toroidal burner arrangement in FUKUYAMA No.2CGL

Photo.1-(b) NKK toroidal burner flames

was that this burner show good reduction performance in a range of moderate strip temperature, 773-823K where the pre-mix burner is weak in reduction.

Combustion air preheated to 723K is supplied to each toroidal burner, through a common header and each flow equalizing valve.

Long hours of operation at the commercial line has served to accumulate know-how and experience of the durability, air ratio unbalance among the burners and its adjusting procedures etc,. Because of successful performance exemplified on product quality and energy saving, it has been decided to apply these burners to the FUKUYAMA No.3 CGL and No.4 CAL, under designing. Especially, No.4 CAL will be a large 2-pass DFF comprising 8 heating zones, of which the last 2 reduction zones will be equipped with these toroidal burners.

3.2 Development of large 2-pass DFF

3.2.1 Outline of facilities The world's first 2-pass DFF was installed in FUKUYAMA No.3 CAL (operating since Aug.1987). Its operation is continuing successfully ever since commercial production was started. In designing the large-scale furnace capable of handling approximately 80,000 tons of steel sheets per month, focus was placed on how to make the coils pass stably through the furnace. In the case of a conventional indirect fired furnace, which is inevitably very long, steel sheets are conveyed on a large number of rolls while they are in the stress relief and recrystalization temperature range which adversely affects shape retention of the sheets, and therefore, their travel becomes extremely unstable.

For this high production rate, NKK designed a 2-pass DFF the height of which is the same as the following furnace section and has developed the optimum equipment and control systems of it. Fig.10 illustrates the arrangement of the DFF on No.3 CAL. Each pass of furnace is divided into four zones, and Zones No.4 and No.8 at the end of each pass are composed of pre-mix type radiant cup burners (for reduction heating). The other zones are composed of energy saving nozzle mixing type radiant cup burners (for non-oxidation heating), in which combustion air can be preheated.

The burners in each zone burn with a higher gas content than the previous one with decreasing air ratio of 1.05 to 0.85. Thus the furnace can effect reduction heating with excellent thermal efficiency.

Exhaust gas from the DFF is utilized to preheats steel sheets in the PHF (Pre heating furnace). It is also used to preheat combustion air for the DFF as well as to preheat air in the dryer attached to the cleaning section on the entry side.

3.2.2 Furnace pressure control method   In realizing the 2-pass DFF, the method of removing exhaust gas is of overriding importance. Since exhaust gas contains much moisture, it is essential to prevent the exhaust gas from flowing into the succeeding atmosphere heat treating furnace. Therefore, the pressure control system was adopted so that each pass can independently draw the exhaust gas in the reverse direction to which the steel sheet is traveling. See Fig.10. DFF-2 furnace pressure is controlled by damper-A to keep the bottom pressure at 10mm H2O. DFF-1 pressure is controlled by damper-B to keep its top pressure the same as that of DFF-2, which varies with the burning rate of DFF-2. When damper-B control is difficult because of a low DFF 1 burning rate, some of the exhaust gas is recycled upstream of damper-B to ensure its suitable opening. Header pressure is controlled by damper-C as well as exhaust fan speed to keep the openings of damper-A and B within the controllable range.

This control method prevents not only the oxidization of steel sheets but also the occurrence of pick up by rolls in the roll chamber of the following furnaces.

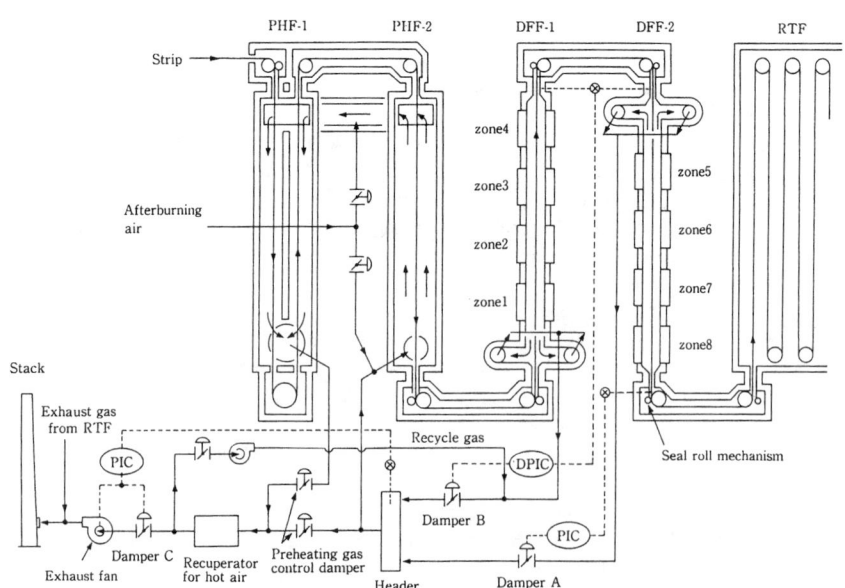

Fig.10 2-pass DFF of FUKUYAMA No.3 CAL

3.2.3 Operation The compact facilities by introducing the DFF have effect on the hearth roll crown in the succeeding furnace which can be minimized and heat buckle can also be eliminated.

Fig.11 shows productivity and production amount, and Fig.12, the fuel consumption. Each control system has been finely adjusted after operation began in Aug., 1987, and various improvements made and optimum operating conditions maintained, thereby resulting in nominal capacity value of 80,000 tons of steel sheets per month after one year. Since then, steel sheets are produced in large quantities and the fuel consumption is securely reduced at approximately 850 MJ/T level.[7]

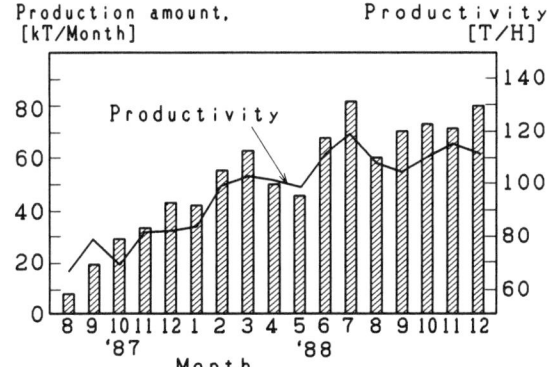

Fig.11 Transition of production amount and productivity

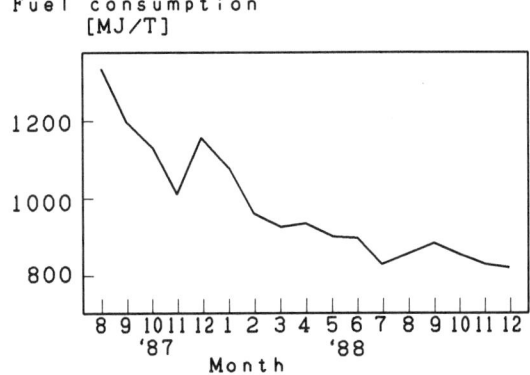

Fig.12 Transition of fuel consumption

3.3 High precision strip temperature control

3.3.1 Multi-reflection thermometer   NKK has developed the thermometer utilized with multi-reflection in which emissivity of steel sheets can be set to 1.0 and it has been adopted at plants since FUKUYAMA No.3 CAL. This method is to set the emissivity of appearance to 1.0 of black body condition by seeing the wedge-shaped

part formed with the hearth roll and the strip. The following effects are obtained by the multi-reflection thermometer.
(1) A standard radiant thermometer can be used.
(2) Overall accuracy level is remarkably improved since the thermometer is not influenced by emissivity surging or back ground light of the strip.
(3) It is unnecessary to set the emissivity value corresponding to steel sheets differing in composition or surface conditions,

3.3.2 Strip temperature control model The dynamic strip temperature control model based on the distributed digital instrumentation system (Fig.14) have been adopted for the No.3 CAL.

A heat transfer model for the DFF is inevitably complex because of the various heat transfer factors present in the furnace, such as radiation from combustion gas, forced convective heat transfer, and convective heat transfer by exhaust gas, as well as radiation from the furnace wall.

In view of this estimation, a simple heat transfer model was developed for on-line control to express heat transfer based on radiation from the furnace wall and combustion gas as shown in equation. However, each radiation heat transfer coefficient is modified, considering the convective heat transfer of combustion gas.

$$ph \cdot \frac{dQs}{dt} = Ks \cdot [U_{Gs}(u \cdot T_G) \cdot \{(T_G + 273)^4 - (T_S + 273)^4\} + U_{ws}(u \cdot T_w) \cdot \{(T_w + 273)^4 - (T_S + 273)^4\}] \quad \cdots\cdots (*)$$

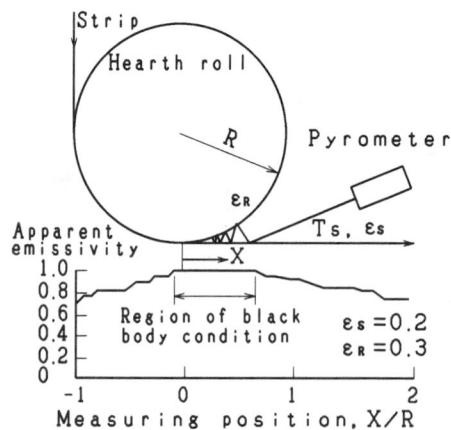

Fig. 13 Multi reflection strip temperature measurement method

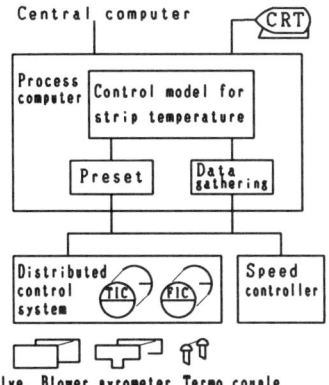

Fig.14 Schematic diagram of automatic temperature control system

Where,
- ρ : Strip density (Kg/m³)
- Qs : Strip heat content (Kcal/Kg)
- u : Combustion gas flow rate (Nm³/h)
- T_w : Furnace wall temperature (℃)
- U_{cs} : Coefficient of heat transfer between gas and steel sheet(Kcal/m²h℃)
- U_{ws} : Coefficient of heat transfer between furnace was and steel sheets (Kcal/m²h℃)
- h : Strip thickness(m)
- K_s : Model parameter
- T_G : Combustion gas temperature (℃)
- T_S : Strip temperature (℃)

This heat transfer model has made it possible to estimate strip temperature in DFF, thereby setting DFF's target strip temperature and determining the rate of line speed change[7].

3.4 Other new technologies

Further improvement, for the preheating furnace, roll chamber, after-burning chamber and other related equipment, are necessary to support the DFF technology obtaining high-grade products. The following new technologies, therefore, have been introduced to the DFF to cope with several precise items:

(1) Dynamic air ratio control

In complex steel works, COG is usually used as the fuel gas for the heating furnaces. For the DFF, NKK has also been using COG. The composition of COG fluctuates depending on operation conditions of the coke oven such as coking coal blending ratio etc,.

The air ratio changes easily by $\pm 5\%$ or over corresponding to the fluctuation of COG composition.

The precise control of air ratio, however, is imperative especially for the pre-mix type burner to be applied in the reducing zone.

The air ratio should be within a range of 0.85-0.90. For this reason, a new control system for the air ratio has been established, in which on-line measurement data of the COG composition is fed back to the air/fuel gas flow rate control.

(2) Prevention of strip oxidation at emergency line stop

The DFF is originally provided with a N_2 gas quenching system. This system blows a large amount of N_2 gas into the furnace to purge out the combusted gas and to cool down the furnace wall to prevent strip breakage and oxidation at an emergency line stop.

This system, however, is not predictable to obtain completely clean strip surface. There has been some oxidation troubles at line stop. The causes were the exhaust gas flow into the following roll chamber and soaking furnace, air invasion at the entrance of the preheating furnace due to temporary negative pressure, and other factors. Thus the following positive countermeasures were executed to that the combusted gas could be immediately exhausted through a regular flue duct;

(a) Additional installation of seal rolls in the roll chambers to avoid combusted gas invasion into the atmosphere section.
(b) N_2 gas supply to the roll chamber when pressure here drops rapidly.
(c) Improved sealing of the preheating furnace inlet and reliable shut-off of the combustion air/after-burning air piping.
(d) N_2 gas supplied to the extinguished burner nozzles and other unused nozzles to avoid combusted gas inflow and water condensation in the pipes.

These measures have the effect of improving stable gas flow and furnace pressure control not only at line stop but during normal operation.

(3) Optimum design based on numerical analysis

In order to investigate an optimum burner arrangement and burner group control corresponding to a wide range of strip width, a heat-transfer model, based on radiative heat transfers among the gas/wall/burner-cup and strip, was made by dividing the furnace wall and the strip into hundreds of fine elements. Each temperature such as gas, furnace wall, burner cup, etc. of this model is calculated based upon a use of each formula of combustion and heat transfer. The various coefficients of the heat transfer in this model were verified by actual operation data and heat balance. This numerical analysis becomes a very useful method to design optimum furnace length/width, and burner arrangement, etc. under various conditions of strip sizes and speeds.[8]

Furthermore, the oxide film thickness for each strip temperature can be estimated with a certain equation, of which coefficients of oxidizing and reducing speeds were derived from basic experiments.

This estimation method is also available for optimum design of the DFF system, in conjunction with the above-mentioned model.

4. NKK-CAL/CGL equipped with direct-fired furnace

Table.1 shows the details of NKK-CAL and CGL equipped with reduction heating type DFFs at the KEIHIN and FUKUYAMA works, including practical applications of the new technologies.

Table. 1 NKK-CAL/CGL with reducible-type DFF

Line name	Production start	Monthly production	Main product use	Max. strip width	Toroidal burner	2pass DFF	Multi-reflection thermometer	Dynamic Air ratio control	Others
KEIHIN No.3CGL	1983.4	30 000t	Structural	1 300mm	-	-	-	-	
FUKUYAMA No.3CAL	1987.8	80 000t	General	1 650mm	-	O	O	O	
FUKUYAMA No.2CGL	1990.5	20 000t	Exposed car body	1 880mm	O D3 zone	-	O	O	New gas seal equip. (PHF, Roof chamber) New After Burning
KEIHIN No.4CGL	1992.5 (planned)	30 000t	Structural	1 300mm	-	-	O	O	New gas seal equip. (PHF)
FUKUYAMA No.3CGL	1992.10 (planned)	40 000t	Exposed car body	1 650mm	O D3 zone	-	O	O	New gas seal equip. (PHF, Roof chamber) New After Burning
FUKUYAMA No.4CAL	1993.4 (planned)	80 000t	General	1 880mm	O D7, D8 zone	O	O	O	New gas seal equip. (PHF)

5. Conclusion

Eight years have passed since NKK introduced the reduction heating DFF technology from SELAS Corporation of America and installed the first furnace with the DFF in Japan, the KEIHIN No.3 CGL. During these years, NKK has already applied the DFF to the six lines (three under construction included), and has achieved smooth startup of operation and the customer's approval for product quality due to close co-operations among the related section. Conforming to the needs of a large-scaled CAL and a high-grade CGL with continued improvements and technological breakthroughs, the DFF is now recognized as the main constituent of NKK-CAL/CGL.

The excellent DFF technology, originated by SELAS, has been further evolved by NKK to be the well-established annealing furnace of steel sheets. We are hoping that in the future the SELAS-NKK type DFF will be applied to other continuous annealing processes.

(Reference)

1) M. Imose "Heating and cooling technologies of the Continuous annealing process," ISIJ : No.88, 89 Nishiyama Memorial Seminar, (1983) 84-89
2) SELAS Corp. of America, "Continuous steel strip heat processing equipment" (Technical report)
3) S. Fukuda et al., "Non-oxidation heating of steel product by direct fire" Industrial Heating, 23(4) (1986) 25-32
4) M. Takusagawa, "Application of the SELAS combustion control system" Industrial Heating, 24(2) (1987) 58-64
5) H. Yoshida et al., "Reduction heating technology by Toroidal burner" NKK Technical Report 1989, no.127; 120-126
6) K. Takaki et al.,"The Equipment and Operation of Fukuyama No.2 CGL" NKK Technical Report 1991, no.135; 34-42
7) S. Kaneto et al., "The Equipment and Operation of NKK Fukuyama No.3 CAL" NKK Technical Report 1989, no.126; 16-23
8) M. Nakayama et al., "Application of Direct Fired Furnace to NKK-CAL/CGL" NKK Technical Report 1991, no.136; 16-23

Development of the strip temperature control technique for
a continuous annealing line

Kohichi taya,Ichiro Ueda,Motoi Honjoh

Sumitomo Metal Industries,Ltd
3 Hikari Kashima-cho Kashima-gun Ibaraki pref. 314
JAPAN

Abstract

A new strip temperature control system for a heating furnace in the continuous annealing line has been developed and tested on No.1 continuous annealing line in the cold strip mill of Sumitomo Metals Kashima Steel Works with satisfactory control performance. The system is characterized by the transition control to optimize the adjustment timing of the furnace temperature and the line speed based on the prediction of the strip temperature transition. The prediction model fully includes the dynamics of the strip temperature transition due to the furnace temperature response and the heat transfer between the strip and hearth rolls.

1. Introduction

Due to recent increases in customers demand for mere deversified and higher quality steel products, high-strength steel and ultra-low-carbon steel products have increased in production. These products are peculiar for their narrow range of temperature tolerance at the exit of the the heating furnace. Given this situation, the operation load for temperature control has increased. Flexibility in dealing with frequent changes in the strip tenperature cycle has been required. In order to solve these problems, a new strip temperature control for the heating furnace has been developed.

2. Description of continuous annealing line

Fig.1 shows the layout of No.1 continuos annealing line at Kashima Steel Workes. Monthly production capacity is 40,000 tons. The maximum center speed is 250 mpm. Thickness·range is from 0.4 to 1.6 mm and width range is 600 to 1425 mm. This furnace consists of a heating furnace, a soaking furnace, a 1st cooling furnace, an over-aging furnace and a final cooling furnace. The heating process in the heating furnace is indirect heating using gas fired radiant tubes. According to strip grade, strip reference temperature at the exit of the heating furnace ranges from 700 to 850 C. Some high-strength steel grades have a reference temperature higher than 800 C as shown in Fig.2.

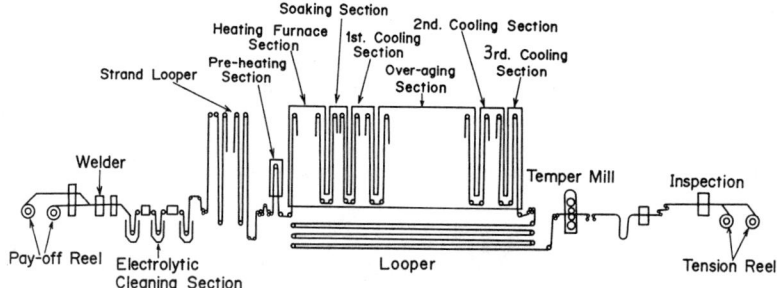

Fig. 1 Layout of No 1 CAL at Kashima Steel Works

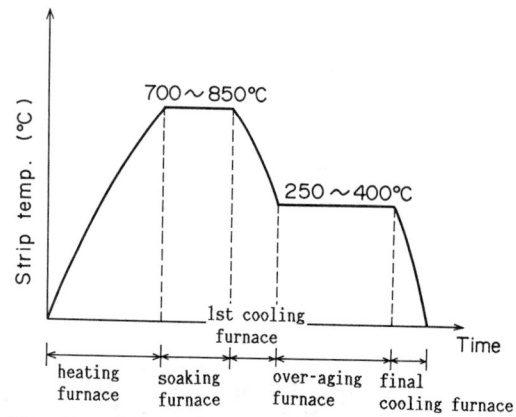

Fig. 2 Strip Temperature Cycle

3. Description of control system

3.1 Strip temperature calculation model

It is necessary to control furnace temperature by predicting strip temperature changes before weld point reaches the heating furnace, because the response time of furnace temperature change is so long. Thus a basic strip temperature calculation model has been developed. Characteristics of that model are as follows;

1. Deviding the strip length into nodes which are 20 m long, the model calculates the strip temperature of each node in real time.
2. The calculation model takes the heat transfer between strip and hearth roll into account, adding it to the heat transfer between strip and radiant tubes, and the heat transfer between strip and atmospheric gas.

The condition of that model is as follows;

1. Strip temperature distribution is uniform toward both thickness and width direction.
2. No heat transfer occurs toward longitudinal ditection.

Above condition is shown in Fig.3.

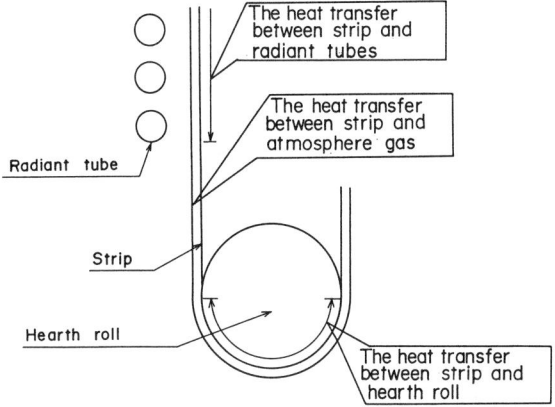

Fig. 3 Calculation model

3.2 Control system overview

Fig.4 shows the control system. Adaptive heat trasfer model is correcting the strip emissivity in real time. Set up model determines the line speed and the furnace temperature in steady-state. The above functions are based on strip temperature calculation models. Transition control determines the optimal adjustment timing of the furnace temperature. This area is based on the prediction model of the strip temperature transition. Dynamic speed control and steady-state control are added. In order to improve strip temperature measurement more precisely, a multireflective thermometer was adopted.

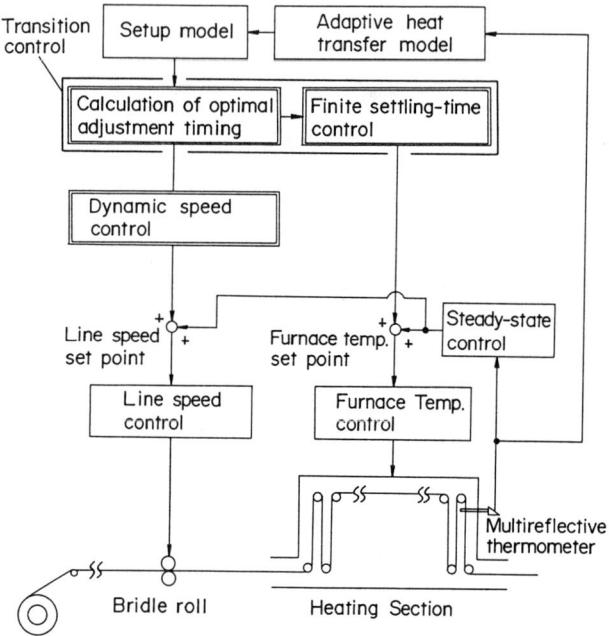

Fig.4   Control System Overview

3.3 Transition control

Strictly speaking, the transition control needs to be calculated over and over using the strip temperature calculation model. But it is not possible to do so, because of the length of time needed for calculating. Therefore instead of strip temperature calculation model, the prediction model approximating the strip temperature transition was needed. In order to establish the prediction model of the strip temperature transition, dynamics of the strip temperature at the exit of heating furnace were examined by the strip temperature calculation model which are shown in Fig.5.

(a) shows the calculated strip temperature when the furnace temperature set point was changed. The response of strip temperature is very slow and is similer to the response of the actual furnace temperature.

(b) shows the calculated strip temperature when the line speed was changed. The response of the strip temperature is a combination of the ramp response caused by the heating time change and the first-order delay response caused by heat transfer between the strip and hearth roll.

(c) shows the calculated strip temperature when the strip thickness was changed. The response of the strip temperature is stepwise at the weld point, then it is similer to the first-order delay response shown in (b).

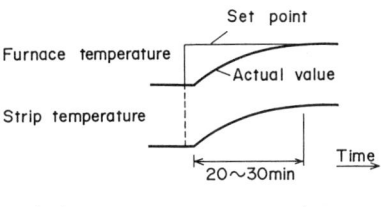

(a) Furnace temperature change

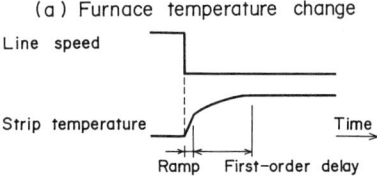

(b) Line speed change

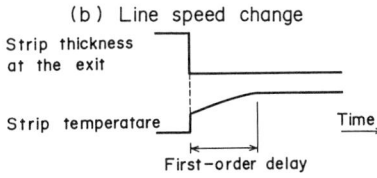

(c) Strip thickness change

Fig. 5 Dynamic response of strip temperture

$$\Delta TS(t;\tau_v,\tau_F) = \alpha(t)\frac{\partial TS}{\partial h}\Delta h + \beta(t-\tau_v)\frac{\partial TS}{\partial V}\Delta V + \gamma(t-\tau_F)\frac{\partial TS}{\partial TF}\Delta TF$$

$$\alpha(t) = \begin{cases} 0 & (t<0) \\ 1-(1-\eta)\exp(-t/T_1) & (t \geq 0) \end{cases}$$

$$\beta(t) = \begin{cases} 0 & (t<0) \\ \eta t/\tau_{VD} & (0 \leq t \leq \tau_{VD}) \\ \eta+(1-\eta)[1-\exp\{-(t-\tau_{VD})/T_1\}] & (t > \tau_{VD}) \end{cases}$$

$$\gamma(t) = \begin{cases} 0 & (t < \tau_{FD}) \\ \dfrac{[1-\exp\{-(t-\tau_{FD})/T_2\}]}{1-\exp(-\tau_P/T_2)} & (\tau_{FD} \leq t \leq \tau_{FD}+\tau_P) \\ 1 & (t > \tau_{FD}+\tau_P) \end{cases}$$

t : time (t=0 : when the weld point is at the exit)
τ_v : line speed adjustment timing
τ_F : furnace temperature adjustment timing
ΔTS : strip temperature transition
Δh : strip thickness change
ΔV : line speed change
ΔTF : furnace temperature change
∂TS/∂h : partial derivative of the strip temperature with respect to the strip thickness
∂TS/∂V : partial derivative of the strip temperature with respect to the line speed

∂TS/∂TF : partial derivative of the strip temperature with respect to the furnace temperature
T_1 : time constant due to the heat transfer between the strip and hearth roll
T_2 : time constant of the furnace temperature response
τ_{VD} : heating time at the furnace
τ_{FD} : dead time of the furnace temperature response
τ_P : target response time of the furnace temperature
η : response parameter due to the heat transfer coefficient between the strip and hearth roll

Fig. 6 Strip temperature transition

137

In accordance with the results, the equation of strip temperature transition can be modeled which is shown in Fig.6. Strip temperature transition is expressed as a linear approximate equation which takes strip thickness, line speed and furnace into account.

3.4 Calculation of the optimal adjustment timing

In order to evaluate strip temperature deviations from temperature tolerances, the cost function was determined as shown in Fig.7. The first and the second terms on the right side of this equation represent the strip temperature deviations. The factor δ is determined based on the relative grade priorty. The factor δ is 1 when the present strip grade is higher than next strip, and 0 is in the opposite case. In case of the same temperature tolerance, a higher reference temperature gets priority. The optimal adjustment timings Tv, Tf can be obtained based on the Flether-Powell method which makes J minimum.

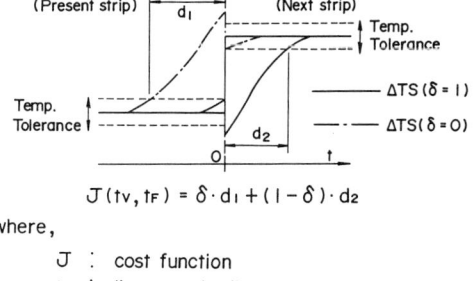

$$J(t_V, t_F) = \delta \cdot d_1 + (1-\delta) \cdot d_2$$

where,

- J : cost function
- t_V : line speed adjustment timing
- t_F : furnace temperature adjustment timing
- δ : relative grade factor ($\delta = 1$ or 0)
- d_1 : temperature deviations of present strip
- d_2 : temperature deviations of next strip
- ΔTS : strip temperature transition

Fig.7 Calculation of the optimal adjustment timing

3.5 Finite settling-time control

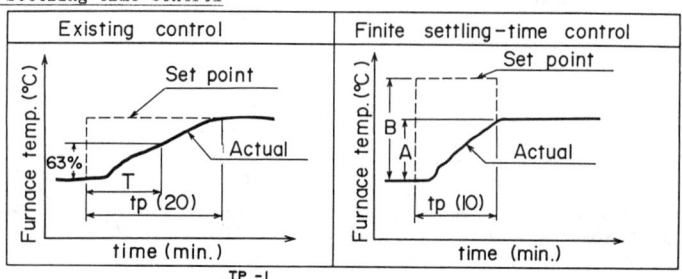

$B/A = K$, $K = (1 - e^{-\frac{T_P}{T}})^{-1}$ (>1)

T = Time constant of the furnace temperature response at stepwise

T_P = Target response time of furnace temperature

Fig.8 Comparison of furnace temperature control between the existing control and finite settling-time control

Finite settling-time control is applied to improve the response of furnace temperature. The furnace set point is biased as K in direction of exceeding that temperature and kept, during the target response time as Tp and then is returned to the steady-state level. The response time is reduced to about ten minutes, which is half of existing control one shown in Fig.8.

3.6 Speed dynamic control

Speed dynamic control is also applied to improve the response time of the strip temperature which is shown in Fig.9. At the same time the line speed is adjusted, the reference speed is biased as Δv and then it is reduced, according to the strip temperature transition based on prediction model. Δv is speed compensation value.

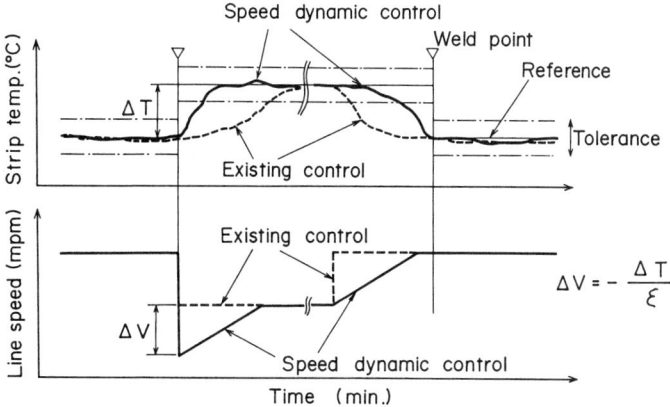

Fig.9 Speed dynamic control

Fig.10 shows the effect of speed dynamic control. When reference temperature changes 40 C in both upward and downward direcrion, speed dynamic control shortens the response time in each occasion.

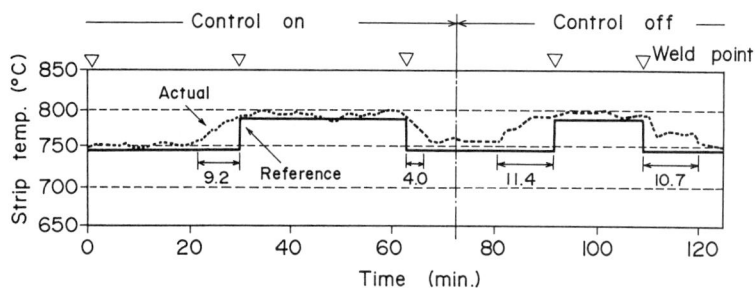

Fig.10 Effect of speed dynamic control

3.7 Multireflective thermometer

In order to make strip temperature measurement more precise, multireflective thermometer has been adopted. The principle of measurement is that the radiation energy reflects many times between the strips and is measured by radiation pyrometer, in this way the strip temperature is accurately measured. It is set between the strip just under the top hearth roll, there two part of the strip are opposite to each other as shown in Fig.11.

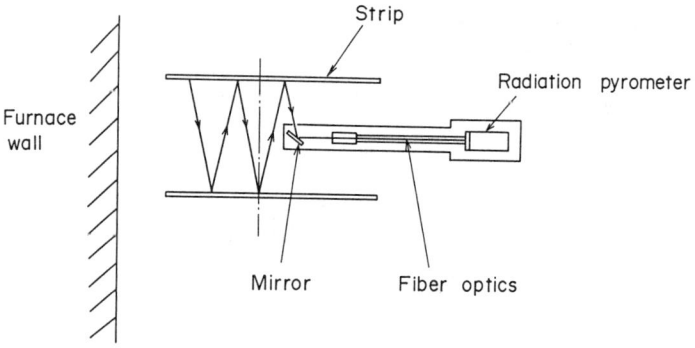

Fig.11 Multireflective thermometer

Fig.12 shows the measurement results of the multireflective thermometer with the comventional radiation pyrometer. Using the radiation pyrometer, temperature shows higher value with high-strength steel, because of the higher emissivity it has. On the other hand, the multireflective thermometer shows stable measurement.

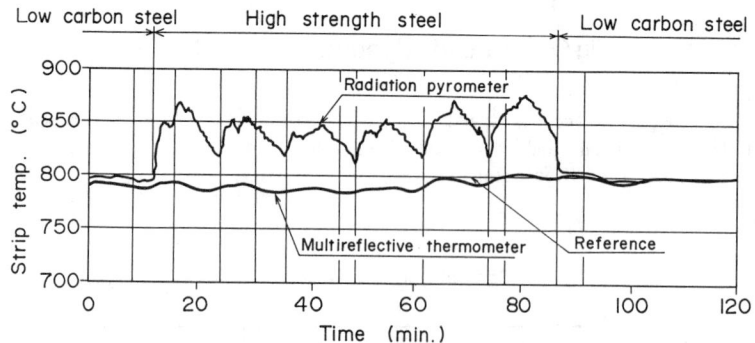

Fig.12 Effect of multireflective thermometer

4. Testing results

Fig.13 shows test results from the No.1 continuous annealing line in Kashima Steel Works. The control system performance shows that changing points were well controled. With the change in strip thickness, the strip reference temperature mirrored the actual temperature and kept the temperature well within the narrow range of temperature range required for higher quality steel strips. Tests were made over a long term. It is shown in Fig.14. In these longer rerm tests the control system performed stably, and the accuracy ratio in the inner strip proved better than that with manual control. The mean valu of the accuracy ratio in the inner strip was 88.69%.

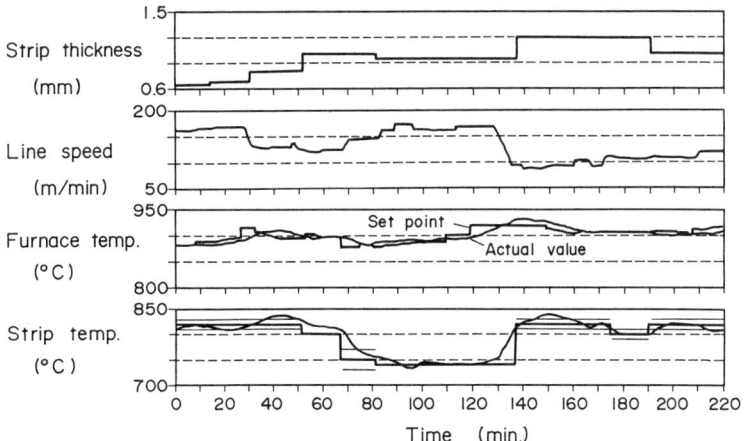

Fig. 13 Actual Control

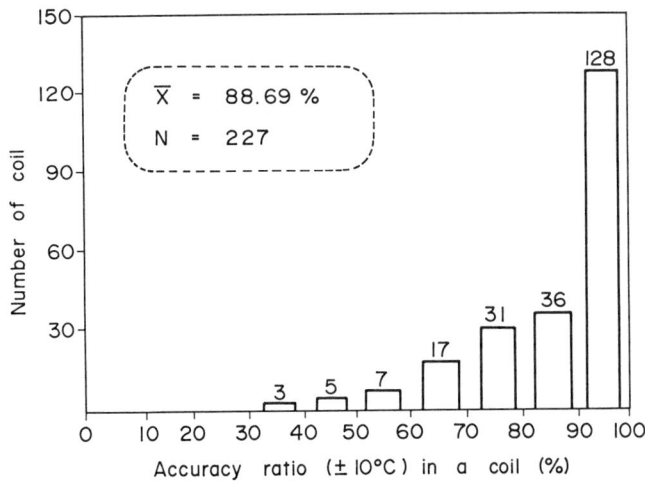

Fig.14 Effect of the temperture control

5.Conclusion

Our latest development in temperature control techniques for the continuous annealing line in the cold strip mill of Sumitomo Metals Kashima Steel Works is discused. Important features of this temperature control system were stable performance over the long term and a grater accuracy ratio than with manual control. As this system is presently only in use for furnace using the radiant tube heating system, in future we will need to develop similer systems for temperature control for other heating processes, such as ,direct fired heating furnaces.

REFERENCES

(1) J.A.Kilpack, E.J.Seeman:
"Computer control of a continuous annealing line", Preprint of IIS 27th Mechanical Working and Steel Processing Conference, Cleveland, October 1985

(2) N.Yoshitani:
"Optical and Aduptive Control of Strip Temperature for a Heating Furnace in C.A.P.L", Preprint of the IFAC Symposium, Tokyo, August 1986

(3) C.D.Kelly, D.Watanapongse, K.M.Gaskey:
"Application of Modern control Theory to a Continuous Annealing Line"

(4) I.Ueda:
"Strip Temperature Control for a Heating Section in CAL", Preprint of the IECON '91 Kobe October 1991

TENSION and ELONGATION CONTROL for

CONTINUOUS ANNEALING LINES

E. A. Cook - Selas Corporation of America
Dresher, PA 19025

Robert Mieloo - Stelco Steel
Hamilton, Ontario, Canada

This paper presents a new concept of strip tension profile control and elongation control for the annealing furnace of a modern C.A.L., including galvanizers. Patents are pending for these concepts, now entering production at the Steel Company of Canada, and in design phase for a plant now under construction in the USA.

Modern CAL's require heat cycles from 550°C for pre-annealed to 900°C for EDDQ and HSLA products. The metallurgy contains both low carbon and ultra-low carbon compositions whereby yield and tensile strengths vary widely. These combinations result in large required tension ranges particularly when the substrate is in the plastic state.

A typical CAL passline is shown in Figure 1. Initially the substrate enters the line from the cold mill where it is reduced up to 85% with very large induced stresses which are not uniform, resulting in irregular flatness across the strip width, and with various frequency of such defect lengthwise of the strip. Flatness may vary from 100 to 400 I-units. The resulting irregular contact with the furnace rolls, coupled with a very high yield point, requires a high tension to avoid slippage and sideways mistracking. The required tension corresponds to stresses of up to 1.4 Kg/mm^2.

As the strip travels in the heating section of the furnace, its temperature increases and some flattening or removal of stresses occur as its yield point lowers due to temperature. When the strip temperature reaches a point where plastic or permanent extension begins to occur, the strain rate (a function of stress, temperature, and time) must be significantly decreased to avoid over-extension which would occur at the strain rates at or near those required at furnace entry. The required tension at this stage corresponds to stresses of 0.15 to 0.5 Kg/mm^2.

To accomplish this large tension change, shown in Figure 2, each of the heating furnace rolls are fitted with individual regenerative static power units designed in a series of multi-roll bridles. The end points

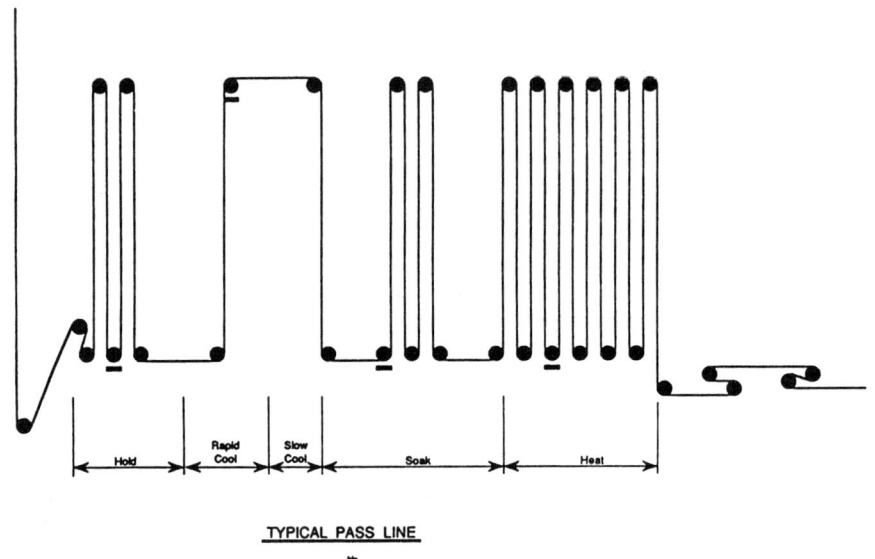

TYPICAL PASS LINE

Fig. 1

STRIP TENSION PROFILE THROUGH THE FURNACE

Fig. 2

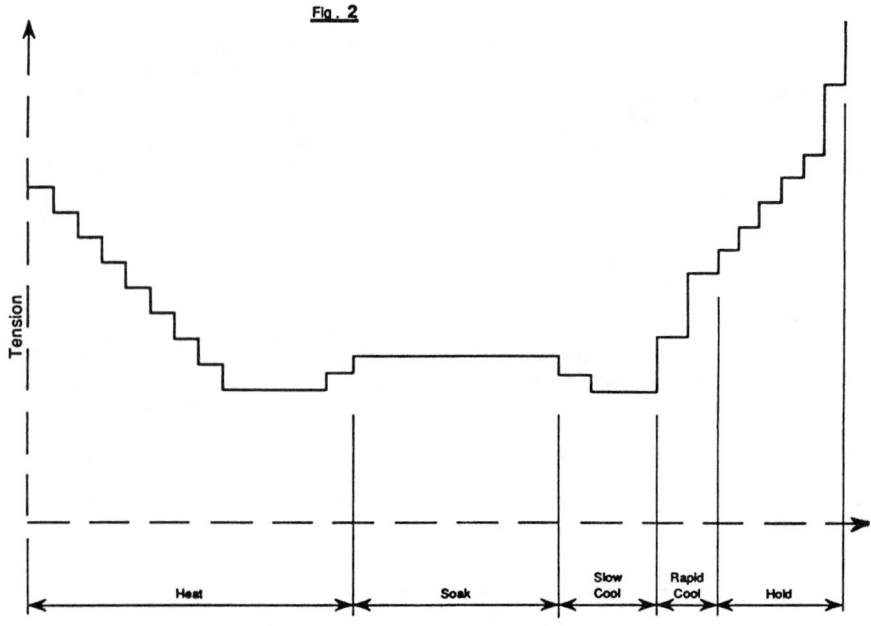

of the desired tension change are fitted with tension controllers using load cells. Deviation from setpoint is fed to speed auctioneering blocks serving each roll. Each roll is powered such that it's contribution in the multi-roll bridle can be changed by the auctioneering block, thereby adjusting of the tension profile as required by operations.

While the strip is in the temperature range where plasticity occurs, final flattening of the strip is obtained by plastic extension of the strip. This extension, however, should be carefully controlled as stresses which are too high can cause heat buckles or strip breaks. Further, over extension of the sheet causes more narrowing than necessary to flatten. Excessive narrowing or necking requires more width at the pickle line and is more difficult to keep in commercial tolerance without later side trimming.

For example, for 200 I-units out of flatness, the minimum permanent elongation must be 0.002 meter/meter or 0.2% extension. The corresponding width reduction,

$$Y = \mu\Sigma \times W \quad (1)$$

is then equal to

$$y = 0.3 \times .002 \times W = 0.0006 W \quad (2)$$

where W is the strip width and μ is Poisson's ratio.

Thus for a 60" wide strip the minimum narrowing would be 0.036 inches to flatten the strip. Further extension or narrowing is unnecessary and tends to promote heat buckles and strip breakage.

These initial parameters can be used to find the minimum desirable stress or tension. Figure 3 shows a qualitative relationship between extension and stress for various strain rates at 850°C. The solid line strain rates are for a typical 0.05 carbon steel and the dotted line rates are for a typical 0.003 C steel.

For the example above with a desired extension of 0.2%, the strain rate can be determined by the time the strip resides in the soak zone.

$$\text{Strain Rate} = \frac{\text{Extension}}{\text{Time}}$$

In our example, if the time is 25 seconds then the strain rate is

$$\text{Strain Rate} = \frac{.002}{25} = 8 \times 10^{-5} \text{ per second}$$

From Figure 3, using extension of 0.2% and strain rate of 8×10^{-5}/second, the stress may be determined. If this stress is below the minimum tracking stress of approximately 250 psi, the extension must be increased until tracking stress is satisfied. In this manner, tension tables for any given product mix can be determined.

It may be seen if a speed change occurs due to strip thickness change or by the operator, the strain rate changes due to time. If the tension or stress is not changed extension may well occur beyond desirable limits. Elongation control senses this deviation with automatic correction.

While the strip is in the furnace soaking section, at constant temperature only plastic change is encountered and this change can be relatively easily measured. On either side of the soaking section, however, the strip is at a temperature where both plastic and thermal change are taking place, and extension measurement is much more difficult without accurate temperature measurement at the beginning and end of these shoulders, plus software to calculate the thermal change and subtract it from the total measured change.

To more easily and accurately control extension and narrowing then, the controlled extension is accomplished mainly in the soak section. The stress in this section is set at a level to assure a high percentage, approximately 75%, of the total extension occurs in this section. The stress in the heating and cooling shoulders is reduced to minimize the plastic extension at a precalculated value.

The rolls of the heating shoulder, the soak section and the cooling section are grouped as individual multi-roll bridles. As in the heating section, each roll is fitted with an individual regenerative static power unit. All rolls are fitted with precision resolvers for speed and consequent elongation measurement.

In the soak section, the primary mode of operation is elongation with the process variable elongation being measured between the first and last roll of the section. Any correction generated by a difference between this process variable and a downloaded setpoint changes the reference to the speed intermediate loop. This reference is then distributed to all rolls of the section through auctioneering blocks thereby adjusting the individual contribution of each roll to the given torque demand change.

The rolls of the shoulder zones are tension controlled in the same manner as for the heating section. However, each zone is further equipped with a single auctioneering block to give a vernier adjustment to the tension in the same proportion as the elongation change in the soak section.

The exit end of the furnace may contain a high speed cooling unit where higher tension are required for a stable non-fluttering passline. In the case of a galvanizing line, the furnace exit tension must correspond to stresses of 1.8 to 2.7 Kg/mm^2 to assure uniform coating regardless of coating weights and speed.

To accomplish this large rise in tension from that of the soak section, the rolls in the exit end are also designed as a series of adjustable multi-roll bridles. Again, each roll is fitted with an individual regenerative static power unit each adjustable from an auctioneering block within the drive group.

Since the critical, low, tension in the soak section is sensitive to transient tension changes usually emanating from line speed changes, the exit end drive group is made the master speed section of the line. Whenever a change of substrate cross section occurs necessitating a line speed change, the danger of over-extension, heat buckles or strip break increases greatly. For example, some overheating of the thinner strip usually occurs. If the tension required for the thicker strip is left applied, the stress and the strain rate will combine to cause a higher extension than allowed by heat buckle mechanism.

To avoid this probability and since such changes can be anticipated, a mathematical model is highly desirable, particularly if combined with a furnace strip temperature control model. Such a model works with weld tracking for timing and consists of a list of events initiating speed changes, with tension and elongation setpoint downloaded progressively as the weld travels through the furnace. The timing functions are, of course, dependent on the change; that is, with strip change going thick to thin or vice versa.

The model is also used to control the tensions and protect the strip during line stops, starts and slow downs.

In summary, the major features of this system are:

1. Extension control of the strip rather than tension control in the critical plastic stages. This concept provides the most advantageous heat buckle index, minimizing heat buckles and strip breaks as well as excessive necking.

2. Use of all furnace rolls as bridle combinations with adjustable contribution. This arrangement provides all the possibilities to control the strip elongation over the full range of heat cycles and strip shapes entering the furnace.

3. Use of the rolls in the furnace exit section as the master speed control bridle of the line. This device minimizes line transients into the critical soak section, while providing for high exit tension at constant speed for coating in a galvanizing line whatever the tension level in the soaking section.

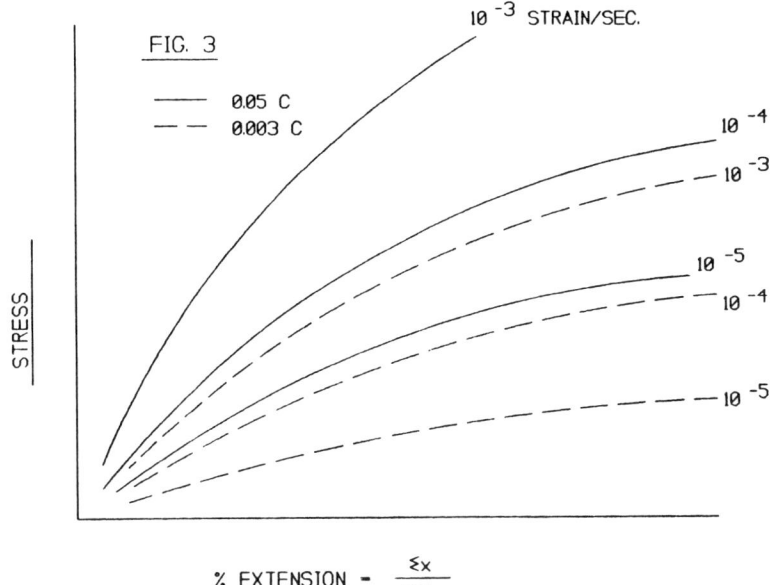

FIG. 3

—— 0.05 C
— — 0.003 C

10^{-3} STRAIN/SEC.

10^{-4}
10^{-3}
10^{-5}
10^{-4}
10^{-5}

STRESS

% EXTENSION - $\frac{\varepsilon x}{100}$

III. Continuous Annealing: Product Metallurgy

A. Sheet Steels

PRODUCTION OF HOT-DIP GALVANIZED HIGH STRENGTH

STEEL SHEETS WITH IMPROVED FORMABILITY

A.Lankila and A.Ranta-Eskola

Rautaruukki Oy, Hämeenlinna Works
13300 Hämeenlinna, Finland

Abstract

Three types of hot-dip galvanized steels with yield strengths exceeding 300 MPa are compared. The steels are conventional C-Mn steels, dual phase steels and rephosphorized overaged steels. The effect of chemical compositions, annealing practices and microstructures on the strength and formability is studied. Although the total elongation of structural C-Mn steel is relatively good, this steel type suffers from quench aging and discontinuous yielding. Dual phase steels are characterized by their low yield strength in comparison to tensile strength, high work hardening rate in the early stages of plastic deformation, and good ductility during forming. Rephosphorized steels overaged in a batch-annealing furnace exhibit significant improvement in formability compared to conventional hot-dip galvanized structural steels.

Introduction

The demand of formable high strength hot-dip galvanized steel sheet for structural applications is increasing. The principal user of formable steel sheets is the automotive industry, but good forming properties are needed also in certain construction applications. The formability of hot-dip galvanized steels is normally inferior compared to cold rolled steels because of differences in annealing cycle. The annealing time in hot-dip galvanizing line is short (1 to 2 minutes) and the temperature is high (700 to 850 °C). Quench ageing due to fast cooling after annealing restricts formability.

There are a number of strengthening methods available in hot-dip galvanizing. Yield strength levels between 300 and 400 MPa are achieved by solid-solution hardening, precipitation and grain refinement. Conventional structural qualities are carbon-manganese steels having a maximum of 0.2 %C and 1,5 %Mn. High pearlite content of these steels impairs formability. Bendability can be improved by reducing carbon and manganese contents. Niobium or phosphorus, for example, are useful as strengthening agents in this case.

Dual phase steels offer several advantages over C-Mn steels. Their high strain propagation ability gives good shape fixability for shallow parts and good stretch formability. Spring back of these steels is small for a large bending radius because of low yield strength, but is fairly large for a small bending radius because of high work hardening rate (1). Also dent resistance is good because of high work- and bake-hardenability (1,2).

Hot-dip galvanized dual phase steel sheets can be produced if the hardenability of the steel is sufficient for the available coolig rate. Hardenabity is normally increased by alloying manganese, chromium and molybdenum. Alternatively, the Zinquench-process can be used to increase cooling rate (3).

Soluble carbon in hot-dip galvanized steels leads to ageing and bake-hardening in coil coating lines. The reason for the rather high soluble carbon content in hot-dip galvanized steels is rapid cooling after annealing. Carbon is precipitated during cooling at grain boundaries, inclusions (such as MnS) and precipitates. Because of smaller grain size of high strength steels compared to drawing qualities, soluble carbon content after cooling is smaller (4).

Ageing resistant and formable steel can be produced by overageing coils in batch-annealing furnaces (5). Interstitial-free high-strength steels have also been developed (6).

In this work several hot-dip galvanized steels having yield strengths exceeding 300 MPa are compared. The steel types are conventional structural steels, dual phase steels and rephosphorized overaged steels. The effect of chemical composition, annealing practice and microstructure on the strength and formability is studied.

Experimental Procedure

The experimental steels were produced in LD-KG converter and continuously cast into slabs. These were hot rolled using a reheating temperature of 1250°C and a finishing temperature of 850 - 900°C. Hot mill coiling temperature was 640 - 680°C. Tandem reductions of 60 - 80 % were used to produce cold band gauges from 0.50 to 0.75 mm. The chemical compositions of the steels are shown in Table I.

Table I. Chemical compositions (wt %)

Steel	C	Mn	Si	P	S	Al	Cr
C-Mn	0.12	0.80	0.02	0.01	0.015	0.035	-
Mn-P	0.06	0.70	0.02	0.08	0.009	0.037	-
Mn-Cr	0.09	1.34	0.02	0.01	0.014	0.043	0.47

The effects of annealing temperature and zinc quenching temperature were studied in a sendzimir type galvanizing line. Rephosphorized steels were overaged in a batch annealing furnace at 280°C. All the results are from production trials.

The effect of tempering was studied using laboratory salt bath furnaces. The samples from production trials were annealed in a laboratory furnace at temperatures 160, 230 and 300 °C for 20, 60 and 180 seconds.

Tensile tests were carried out using transverse specimens, 20 mm in width and of 80 mm gauge length.

Results

The effect of annealing temperature on the mechanical properties of steels C-Mn and Mn-P is quite small (Fig. 1 a and b). C-Mn steels are tested after hot dip galvanizing and tension leveling. The tensile strength increases slightly and yield strength decreases with increasing annealing temperature. Mn-P steels are tested after overageing in batch-annealing furnaces and skin passing. The tensile and yield strengths decrease when the annealing temperature increases.

The effect of zinc quenching temperature on the mechanical properties of steels C-Mn and Mn-Cr is presented in Figures 2 a and b. When the quenching temperature of C-Mn steel reaches 700°C, the tensile strength increases and yield strength decreases. The strength of Mn-Cr steel increases with increasing quenching temperature. The total elongation of both steels decreases with increasing quenching temperature.

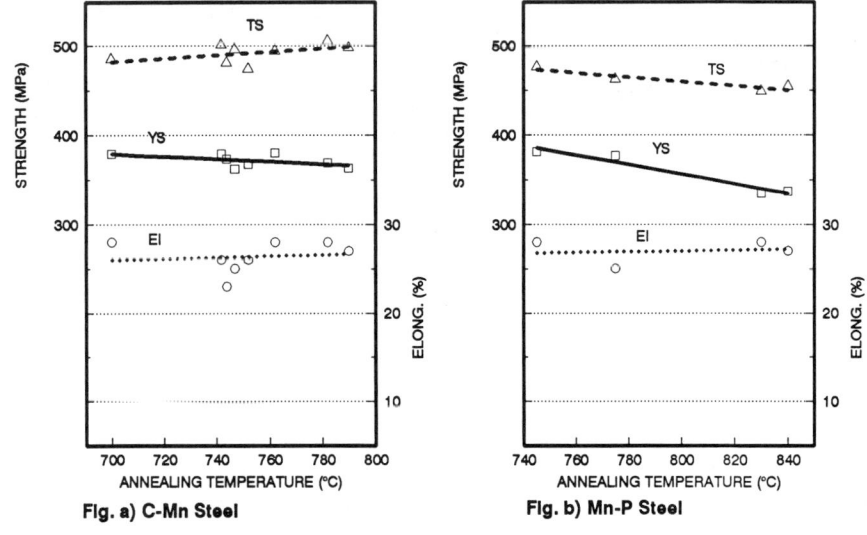

Figure 1 - The effect of annealing temperature on mechanical properties.

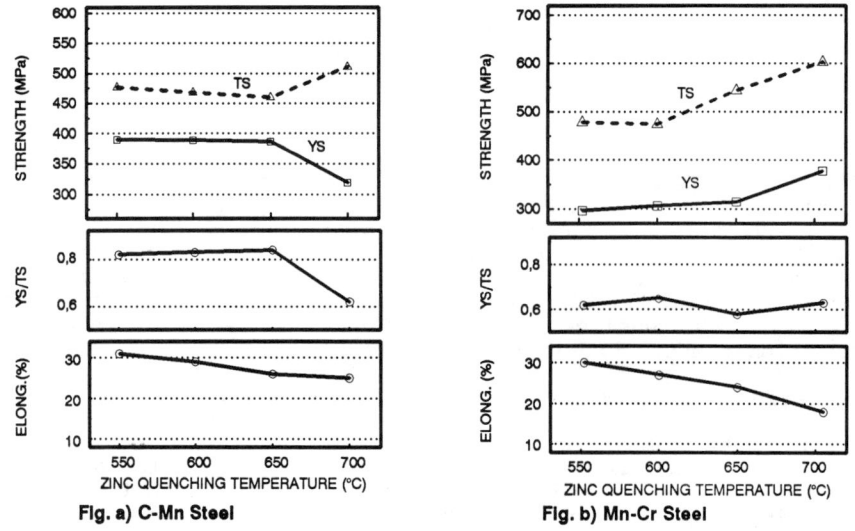

Figure 2 - The effect of zinc quenching temperature on mechanical properties.

Typical mechanical properties of the steels are presented in Table II. The strengths of C-Mn and Mn-P are approximately equal, the yield strength (YS) of Mn-Cr steel is lower and the tensile strength (TS) higher compared to other steels. Heat treatment at 170°C for 20 minutes after 2 % deformation increases the yield strength of C-Mn and Mn-Cr steels by more than 100 MPa (WH + BH). Overaged Mn-P steel does not show any bake-hardenig.

Table II. Typical mechanical properties

Steel	YS (MPa)	TS (MPa)	El %	WH + BH (MPa)
C-Mn	365	475	26	110
Mn-P	381	475	28	0
Mn-Cr	300	520	28	120

Differential n-values are seen in Figure 3. After a prestrain of about 0.03, the differential n-value level is around 0.15 for C-Mn steel and around 0.22 for Mn-P steel. The n-value of Mn-Cr steel is more than 0.3 at low strains, but it is impaired as the strain increases.

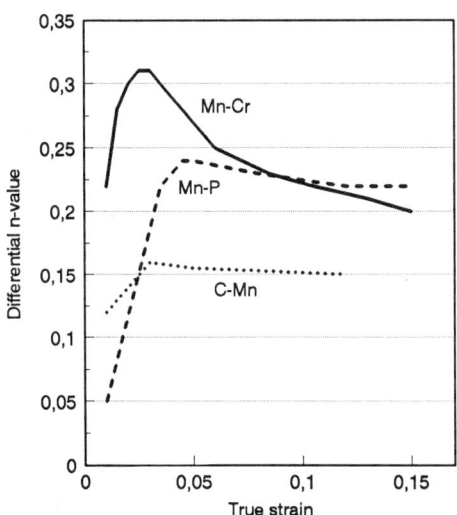

Figure 3 - Differential n-values of the steels.

The effect of tempering was studied using laboratory salt bath furnaces. The mechanical properties of steels C-Mn and Mn-Cr are presented in figure 4. Tempering at 160°C does not change the yield strength of Mn-Cr steel, but the strength of C-Mn steel increases as the tempering time increases. At the same time the total elongation of both steels is impaired.

At 230°C the yield strengths and yield point elongations of the steels increase as the tempering time increases. At 300 °C the maximum ageing is achieved in 20 second and the properties stays at constant level while the tempering time increases.

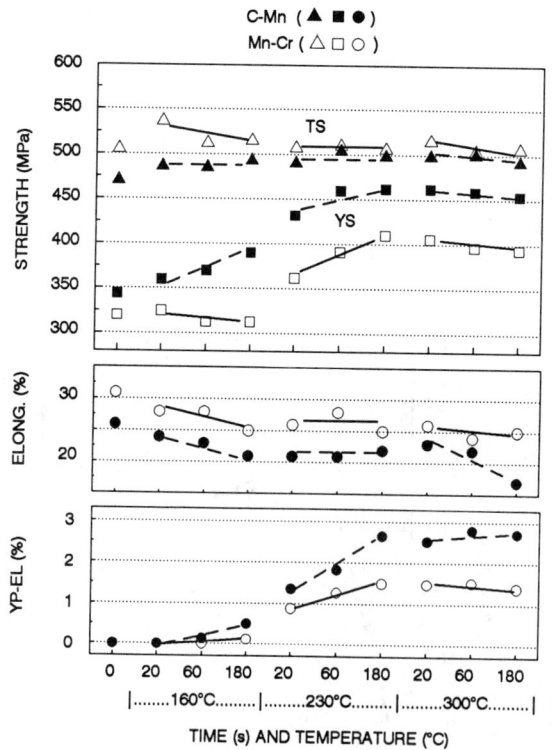

Figure 4 - The effect of tempering temperature and time on mechanical properties.

Discussion

Yield strengths exceeding 300 MPa are economically achieved by using C-Mn steels. Conventional hot-dip galvanizing process gives a good combination of strength and total elongation to these steel grades. The effect of annealing temperature on mechanical properties is quite small. The properties are dependent on ferrite grain size, pearlite content and the amount of soluble carbon. The growth of grain size is restricted by austenite, which is easily formed because of high carbon content.

The production of rephosphorized overaged steels is complicated compared to C-Mn steels. Their advantages are better formability and good ageing resistance. The strength of these steels decreases slightly with increasing annealing temperature due to growth of the grain size.

Dual phase steels have higher alloying and are more expensive than C-Mn steels. High work hardenig rate and other special properties are useful in certain applications. The Zinquench-process makes it possible to adjust cooling rate in hot-dip galvanizing, which improves flexibility in production. The hardenability of Mn-Cr steel is high and quenching temperatures above 650°C lead to impaired formability.

Conclusions

The strength of structural C-Mn hot-dip galvanized steel is high compared to a batch annealed steel having the same composition. Although the total elongation is relatively good, this steel type suffers from quench ageing and discontinuous yielding.

Dual phase steels are characterized by their low yield strength in comparison to tensile strength, high work hardening rate in the early stages of plastic deformation, and good ductility during forming. Bake hardening combined with work hardening (2 % prestrain) is very strong.

Rephosphorized steels overaged in a batch-annealing furnace exhibit significant improvement in formability, and especially ageing resistance compared to conventional hot-dip galvanized structural steels.

REFERENCES

1. K.Nakagawa and H.Abe, "Press formability of high strength cold rolled steel sheets," Memoires Scientifiques Revue Metallurgie, 4 (1980), 475-484.

2. E.-J.Drewes und B.Engl, "Contigeglühte höherfeste Stahlsorten für Feinbleche," Stahl und Eisen, 108 (21) (1988), 47-54.

3. A.Lankila, "Production experience and properties of hot-dip galvanized dual-phase steels," in Proc. of the 16th Biennal IDDRG Congress, Borlänge, 1990, 355-362.

4. H.Katoh, K.Koyama and K.Kawasaki, "The Kinetics of Precipitation of Cementites During Over-Aging in Continuous Annealing," in Technology of Continuously Annealed Cold-Rolled Sheet Steel, ed. R.Pradhan, (Warrendale, PA: The Metallurgical Society, 1984), 79-94.

5. W.Bleck, D.Giesel, M.Menne and W.Müschenborn, "Development of Hot-Dip Galvanized Higher-Strength Steels with Good Cold Forming Properties," Thyssen Technische Berichte, 2 (1989), 197-211.

6. H.Takechi, "Developments in High-Strength Hot- and Cold-Rolled Steels for Automotive Applications," in Hot- and Cold-Rolled Sheet Steels, ed. R.Pradhan and G.Ludkovsky, (Warrendale, PA: The Metallurgical Society, 1988), 117-138.

Effect of Carburizing during Continuous Annealing on Mechanical

Properties of Interstitial-Free Sheet Steels

M.Kitamura, S.Hashimoto, M.Matsumoto and T.Inoue

Iron & Steel Research Laboratories, Kobe Steel Ltd.,
Chuo-ku, Kobe, 651 Japan

Abstract

Interstitial-free (IF) sheet steels are currently used for automotive panels because of their excellent formability and shape-fixability. However, due to reduced amount of carbon in solid solution, insufficient bake hardenability (BH) and low resistance to cold-work embrittlement put limitations on the extensive use of IF steels. In this paper, it is shown that carburizing during continuous annealing can improve BH and the resistance to cold-work embrittlement without deteriorating deep drawability. The carbon atoms in the carburized layer exist in the form of both solute atoms and carbides such as TiC and/or NbC. The carbon content in solid solution of carburized steels is controlled by the composition of carburizing atmosphere, the initial Ti or/and Nb content of the steels and the heat cycle for annealing such as cooling rate.

Introduction

IF sheet steels, which are ultra low carbon steels with carbonitride forming elements such as Ti and Nb, are extensively produced for the demand of deep drawability in automotive panel sheets because of their high Lankford-value (r-value) and non-aging property.

It is known that intergranular fracture occurs during cold-working after deep drawing in decarburized cold rolled sheet steels(1). Furthermore, in IF steels, the intergranular fracture tends to occur as well as decarburized steels because the grain boundaries are weak due to extremely reduced carbon content and the addition of carbonitride forming elements such as Ti and Nb.

On the other hand, the demand for high strength sheet steels has become greater in order to improve the fuel efficiency by reducing automotive body weight. IF steels with high phosphorus content are produced for this demand. However, phosphorus encourages the cold-work embrittlement to occur due to intergranular fracture(2). Boron is effective in inhibiting the cold-work embrittlement(3,4), however, the addition of boron causes deterioration in r-value(4,5,6) and rise of recrystallization temperature(4,6).

In addition, the demand for bake hardenable sheet steels has increased in recent years. Bake hardenability (BH) is a property that increases yield strength due to strain aging mainly by the interaction between dislocation and carbon in solid solution during paint baking after press forming.

Many studies were performed to produce bake hardenable sheet steels in the past. The bake hardenable Al-killed steels were produced under the adjustment of chemical composition of the steel and annealing condition in box-annealing process(7). In the case of Ti- or/and Nb-added ultra low carbon steels, some processes, such as the decomposition of TiC or NbC by high temperature annealing(8), the retardation of the formation of TiC by lowering sulfur content(9), or the suppression of precipitation of TiC by the control of chemical composition(10), were proposed. However, Al-killed steels has a low r-value compared with IF steels. Even in Ti- or Nb-added ultra low carbon steels, the presence of carbon in solid solution before cold rolling brings about the inhibition of the favorable texture development for deep drawability(11).

In this study, the effect of carburizing during continuous annealing on mechanical properties such as the resistance to cold-work embrittlement and bake hardenability was investigated. Besides, the state of entered carbon by carburizing and the carbon concentration profile after carburizing during continuous annealing were investigated.

Experimental procedure

The materials used in this experiment were vacuum-melted ingots of 90kg. Chemical compositions of steels are shown in Table I. The excess Ti, Ti^*, was calculated by equation (1), since most of sulfides, nitrides, carbides and oxides in hot bands are thought to be stabilized as TiS, TiN, TiC and Al_2O_3.

$$Ti^*(mass\%) = Ti(mass\%) - 1.5S(mass\%) - 3.43N(mass\%) - 4C(mass\%) \qquad (1)$$

On the other hand, the excess Nb was calculated by equation (2), since there

Table I Chemical compositions of steels used. (mass%)

Steel	C	P	S	Ti	Nb	N	Ti*	Nb*
T 1	0.0030	0.010	0.0056	0.033	—	0.0033	0.0013	—
T 2	0.0025	0.010	0.0064	0.050	—	0.0030	0.0201	—
T 3	0.0025	0.010	0.0058	0.066	—	0.0035	0.0353	—
T 4	0.0022	0.010	0.0051	0.086	—	0.0034	0.0579	—
T 5	0.0041	0.085	0.0046	0.054	—	0.0031	0.0201	—
N 1	0.0021	0.010	0.0053	—	0.029	0.0031	—	0.0127

Si=0.01%, Mn=0.20%, Al=0.025%
$Ti^{*}(\%) = Ti(\%) - 1.5S(\%) - 3.43N(\%) - 4C(\%)$
$Nb^{*}(\%) = Nb(\%) - 7.75C(\%)$

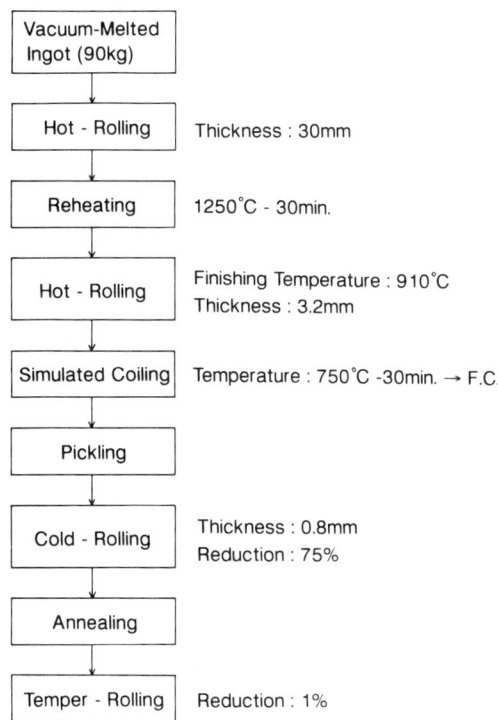

Figure 1 - Schematic illustration of experimental procedure.

was the little formation of NbS, NbN and NbO.

$$Nb^{*}(mass\%) = Nb(mass\%) - 7.75C(mass\%) \qquad (2)$$

The experimental procedure is shown in Figure 1. These steels were hot-rolled to 30mm thickness after soaking at 1150℃ for 1h. The specimens,

120mm wide and 160mm long, were soaked at 1250℃ for 30min and hot-rolled to 3.2mm thickness in 3 passes under a finish temperature of 910℃ by the laboratory mill. After hot rolling, the hot bands were held at 750℃ for 30min and cooled to room temperature in the furnace for a simulation of hot strip coiling.
After pickling, they were cold-rolled to 0.8mm thickness with a reduction of 75%. The cold-rolled steels were then cut into test pieces in 70mm wide and 200mm long. The test pieces were annealed using continuous annealing line (CAL) simulator, infrared-ray heating furnace capable of controlling the composition of atmosphere such as dew point (D.P.).

Figure 2 shows the heat cycles and atmospheres for annealing. In the heat cycle (a), the test pieces were heated to 800 or 850℃ at a rate of 10 or 30℃/s and held for intended times (30-90s) in the reduced atmosphere, ($5\%H_2-N_2$), or in the carburizing atmosphere, ($0.5\%CO-5\%H_2-N_2$). (H_2-N_2) atmosphere is conventional for the continuous annealing process.
In the heat cycle (b), the test pieces were heated to 850℃ in ($5\%H_2-N_2$) atmosphere at a rate of 10℃/s, and were immediately cooled to room temperature at a rate of 80℃/s, and then, those were reheated to 850℃ at a rate of 50℃/s and held for 40s in ($0.5\%CO-5\%H_2-N_2$) atmosphere and cooled to room temperature at a rate of 80℃/s. The heat cycle (b) simulates the process that carburizing is limited in the soaking section during continuous annealing process.
After annealing, they were temper-rolled with a reduction of 1.0%.

Microstructural observation of annealed steels was performed in nital etching by optical microscope.
Figure 3 shows the procedure of cold-work embrittlement test. The specimens were stamped out to 68mm diameter and drawn at a deep-drawing ratio of 2.7. After that, the cups were trimmed to the height of 35mm (drawing ratio=2.5) and pushed them over the conical punch having an apex of 34 degrees after holding at temperatures ranging from -70℃ to -140℃ for 10min. The resistance to cold-work embrittlement was evaluated by the temperature that frequency to exhibit cracking is equal to 50%. That temperature was defined as transition temperature of cold-work embrittlement.
The fatigue test was performed under completely reversed plane bending. The fracture surface was observed by Scanning Electron Microscope (SEM).
Bake hardenability was measured by the difference between the flow stress at 2% elongation and the lower yield stress after aging at 170℃ for 20min.
The test piece size for measurement of r-value was 12.5mm wide and 100mm long for longitudinal and diagonal, 70mm long for transverse directions to rolling direction. The r-value was measured after straining of 15% for 3 directions. The average r-value ($\bar{r}$) was calculated by equation (3).

$$\bar{r}=(r_L+2r_D+r_T)/4 \tag{3}$$

The pole densities at 50μm depth from surface and the half plane in the direction of thickness were measured by X-ray diffraction.
The quantitative analysis of precipitates was performed by electrolytic extraction. And carbon concentration profile was measured by chemical analysis and Secondary Ion Mass Spectroscopy (SIMS) analysis.

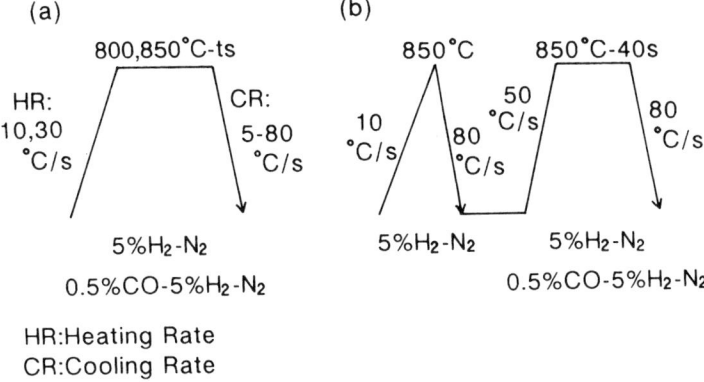

Figure 2 - Heat cycles and atmospheres for annealing.

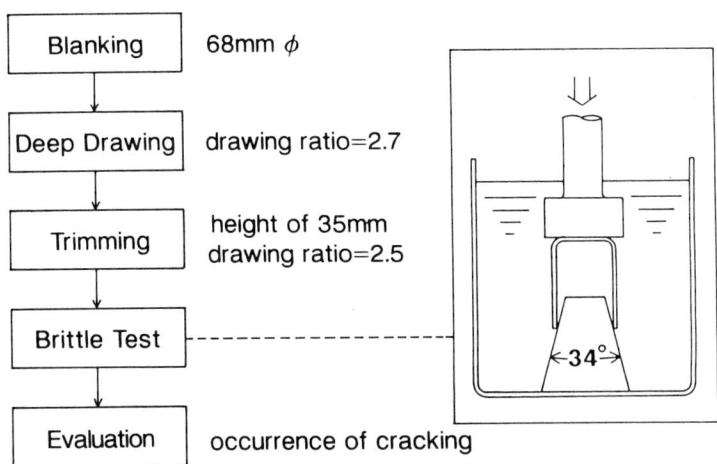

Figure 3 - Procedure of the cold-work embrittlement test.

Results and Discussion

Microstructure of annealed steels

Figure 4 shows the effects of carburizing and holding time at 850℃ on microstructures of steel T2. The grain size of steels annealed in carburizing atmosphere was slightly finer than that of steels annealed in conventional atmosphere. It is considered that the grain refinement was brought about due to the pinning effect of increased precipitates by carburizing. However, there was not a marked change in microstructure by carburizing. Furthermore, the grain size of the steels annealed in the both atmospheres slightly became larger with increase in holding time.

Resistance to cold-work embrittlement

Figure 5 shows the effect of carburizing on the transition temperature of cold-work embrittlement. The transition temperature rose with increase in

Heat cycle (a) : HR=30℃/s, CR=20℃/s

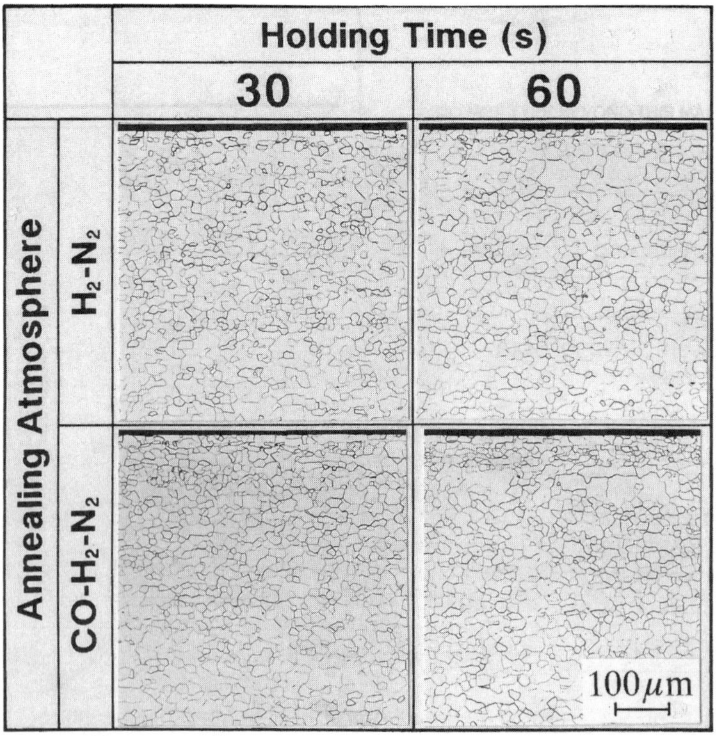

Figure 4 - Optical micrographs of steel T2 annealed at 850℃ for 30s and 60s in conventional and carburizing atmospheres.

tensile strength. The transition temperature of carburized steels was about 50 ℃ lower than that of steels annealed in conventional atmosphere. So, it is clear that carburizing improves the resistance to cold-work embrittlement. However, from SEM observation, the morphology change of brittle fracture surface was not clearly observed in this experiment.

Figure 6 shows the SEM micrographs of fatigue fracture surface. In both steels annealed in conventional and carburizing atmospheres, intergranular fracture was observed. It was recognized that the area of intergranular fracture of carburized steels decreased compared with the steels annealed in conventional atmosphere. From these results shown in Figures 5 and 6, it may be concluded that entered carbon by carburizing segregated to grain boundaries raised the intergranular strength by itself(12) or prevents the segregation of phosphorus at grain boundaries due to the site competition between carbon and phosphorus(13,14).

Bake hardenability

Figure 7 shows the effects of carburizing and cooling rate on bake hardenability. Bake hardenability increased due to entered carbon by carburizing. And it increased with increase in cooling rate for Ti-added and Nb-added steels. Bake hardenability of steels annealed in conventional atmosphere was obtained mainly due to carbon in solid solution resulting from dissolution of carbide (TiC), but the amount of carbon in solid solution was decreased by reprecipitation in cooling process. In addition, for the carburized steels, precipitation of carbides by combining entered carbon with excess Ti, Nb is prevented with increase in cooling rate. Irie

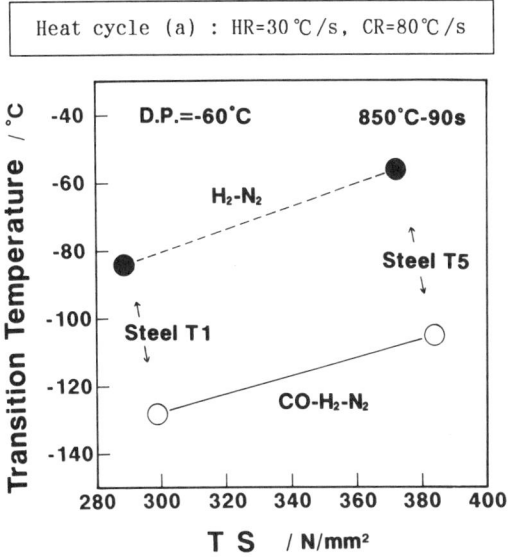

Figure 5 - Effect of carburizing on the transition temperatute of cold-work embrittlement.

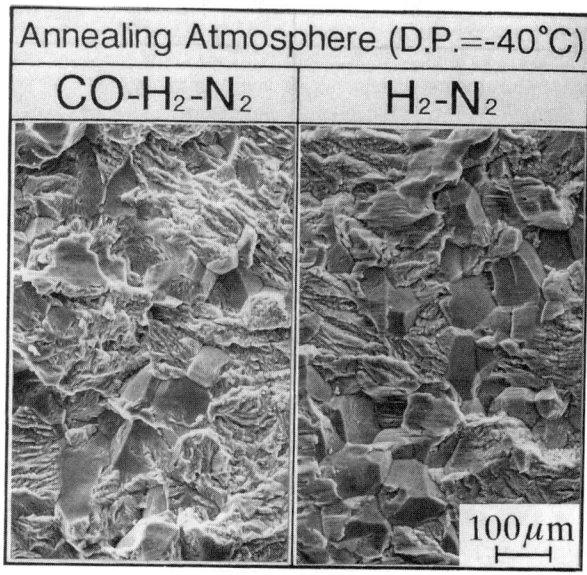

Figure 6 - SEM micrographs of fatigue fracture surface of steel T1 annealed at 800 °C for 60s.

et al.(8) reported that the amount of carbon in solid solution decreased with decrease in cooling rate for Ti-added and Nb-added steels, and both steels have the same cooling rate dependence of the amount of carbon in solid solution. In this experiment, for the carburized steels, there was a difference in cooling rate dependence of bake hardenability between Ti-added steel and Nb-added steel. The cooling rate dependence of bake hardenability in Ti-added steel was slightly larger than in Nb-added steel. TiC tends to precipitate compared with NbC in cooling process because TiC is more stable than NbC in ferrite region from the viewpoint of equilibrium solubility product. And Nb-added steel had a higher bake hardenability compared with Ti-added steel. Compared with Ti-added steel, the entered carbon in Nb-added steel tends to remain in solid solution for the above reason and the grain size of Nb-added steel is finer. Bake hardenability increases with decrease in grain size(15).

Figure 8 shows the effect of dew point in carburizing atmosphere on bake hardenability. Bake hardenability decreased with increase in dew point because of the decrease in the carbon activity through the following reaction (4).

$$CO + H_2 \rightleftarrows C + H_2O \qquad (4)$$

Dew point had a large effect on bake hardenability. So it is important to control the dew point in the furnace to obtain a constant bake hardenability.

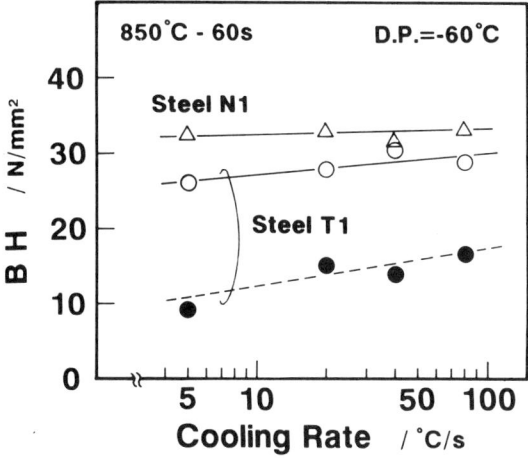

Figure 7 - Effects of carburizing and cooling rate on bake hardenability.

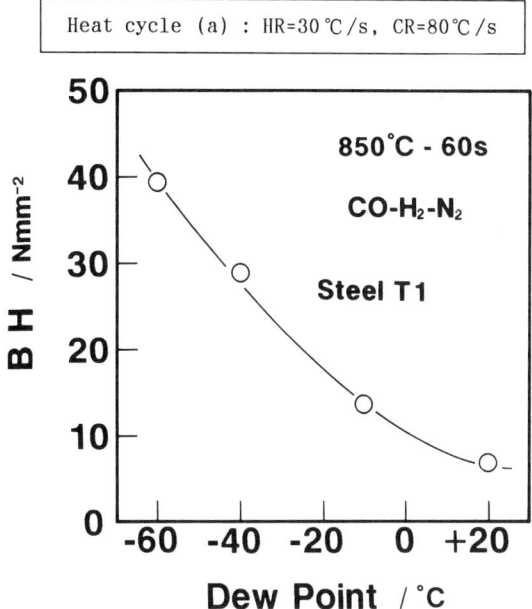

Figure 8 - Effect of dew point in carburizing atmosphere on bake hardenability.

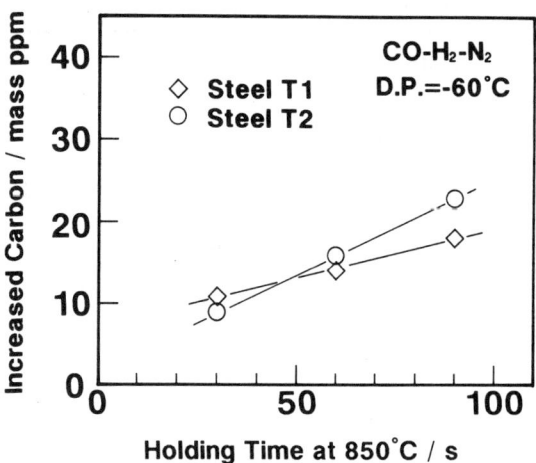

Figure 9 - Change in the amount of increased carbon by carburizing with holding time at 850 °C.

Figure 9 shows the change in the amount of increased carbon by carburizing with holding time at 850 °C. The amount of increased carbon increased with holding time. The increased carbon content of steel T1 was as almost same as that of steel T2 for the same holding time. Figure 10 shows the change in bake hardenability of carburized steels with holding time. Bake hardenability increased with holding time for steels T1 and T2. However, bake hardenability of steel T1 was higher than that of steel T2. There was a difference in bake hardenability between steel T1 and steel T2 for the same amount of increased carbon. Figure 11 shows the relationship between bake hardenability and excess Ti. Bake hardenability decreased with increase in excess Ti. For the carburized steels, the grain size slightly became finer or almost unchanged with increase in excess Ti. Because bake hardenability increases with decrease in grain size (15), it is considered that the amount of carbon remained in solid solution decreased by combining entered carbon with excess Ti.

Figure 12 shows the change in average amount of Ti as compound between surface and 40 μm depth with holding time. The amount of Ti as compound in steel T2 increased with holding time, but that in steel T1 almost remained unchanged. It is considered that the amount of compound increases due to increase in TiC by combining excess Ti with entered carbon by carburizing. So it resulted in Figure 12 because the excess Ti in steel T2 was about 0.02%, but that in steel T1 was almost equal to 0%. Furthermore, the variation in the amount of Ti as compound in the direction of thickness is shown in Figure 13. The amount of compound in the steel annealed in conventional atmosphere was uniform in the direction of thickness. However,

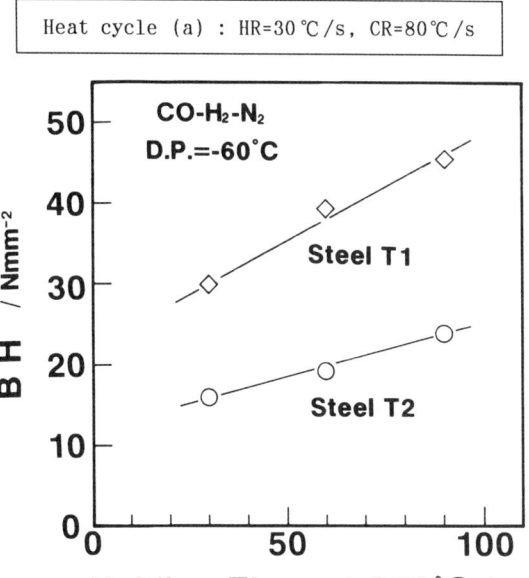

Figure 10 - Change in bake hardenability of carburized steels with holding time at 850℃.

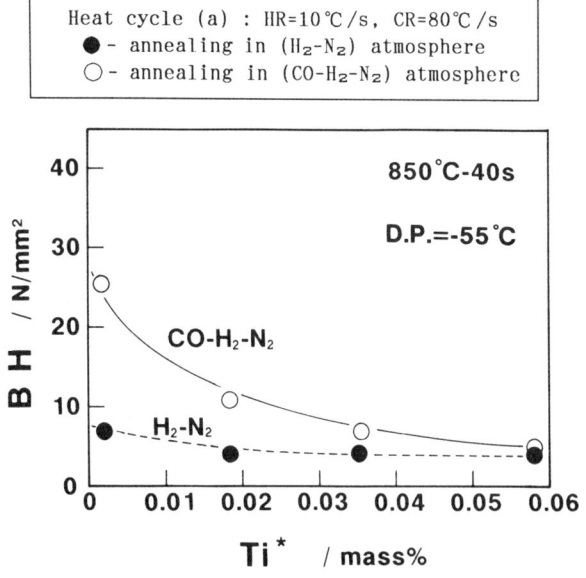

Figure 11 - Relationship between bake hardenability and excess Ti.

that of carburized was nonuniform. The amount of Ti as compound near the surface was equal to initial Ti content for carburized steel, and it decreased from the surface to center. It is speculated that this distribution was due to the carbon concentration in the direction of thickness by carburizing.

Figure 14 shows the carbon concentration profile after carburizing. Both results by chemical and SIMS analyses were approximately on the same line. The carbon concentration decreased from the surface to center through diffusion of entered carbon by carburizing.

Figure 15 shows the distribution of carbon in solid solution of carburized steel T2. The amount of carbon remained in solid solution was estimated by taking the carbon as TiC from the carbon concentration. It took maximum value at a certain depth from the surface. However, the distribution curve varies with Ti content of the steel, the composition of carburizing atmosphere and heat cycle.

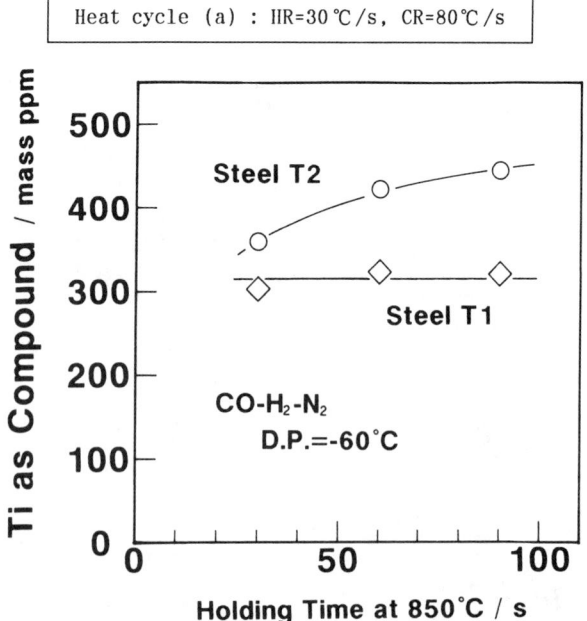

Figure 12 - Change in average amount of Ti as compound between surface and 40 μm depth with holding time at 850°C.

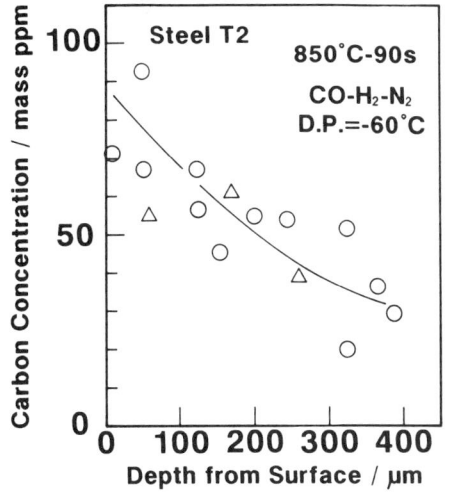

Figure 13 - Variation in the amount of Ti as compound of annealed steel T2 as a function of depth from the surface.

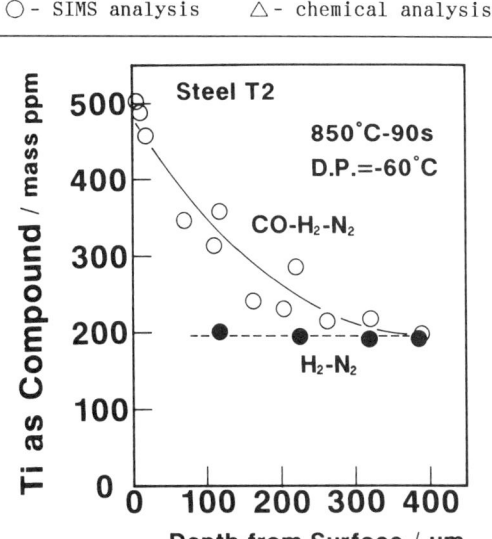

Figure 14 - Carbon concentration of carburized steel T2 versus depth from the surface.

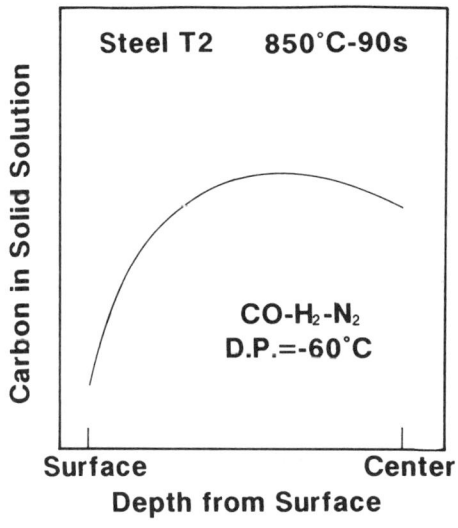

Figure 15 - Schematic distribution of carbon in solid solution of carburized steel T2.

r-value

Figure 16 shows the effects of carburizing and excess Ti on r-value. The r-value increased with excess Ti and leveled off above a certain Ti content as reported in the past(5). The Ti content must exceed the amount required to stabilize carbon, sulfur and nitrogen to obtain sufficiently high r-value. In this study, the r-value leveled off above 0.03% of excess Ti. The amount of excess Ti that r-value levels off varies with carbon, sulfur and nitrogen content of the steel. In addition, the r-value was deteriorated by carburizing in heat cycle (a). It is caused mainly by the entered carbon before recrystallization by carburizing. It is known that carbon in solid solution during annealing suppresses the development of {111} texture(16). Figure 17 shows the effect of carburizing on pole densities at surface and center of thickness for the steels with various Ti content. At center, both steels annealed in conventional and carburizing atmospheres had the almost same recrystallization texture. At surface, for carburized steels, {222} intensity was lower and {110} intensity was higher compared with the steels annealed in conventional atmosphere. This change in texture is thought to be mainly due to the presence of carbon in solid solution during recovery and recrystallization. In short, it is considered that the carburizing during heating retarded the recovery and affected the nucleation of recrystallization. So the annealing was performed in heat cycle (b) to carburize after recrystallization. Figure 18 shows the r-value of carburized steels in heat cycle (b) with that of steels annealed in conventional atmosphere in heat cycle (a). The effect of carburizing

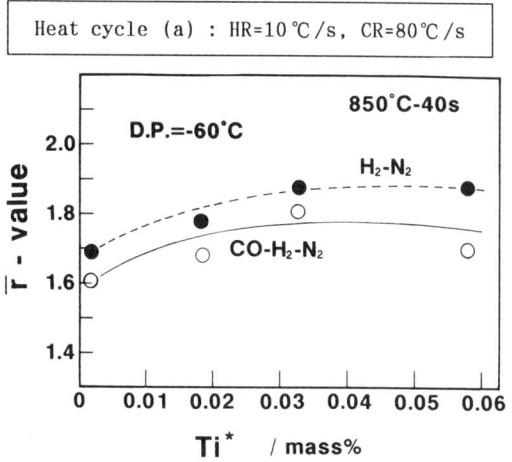

Figure 16 - Effects of carburizing and excess Ti on r-value.

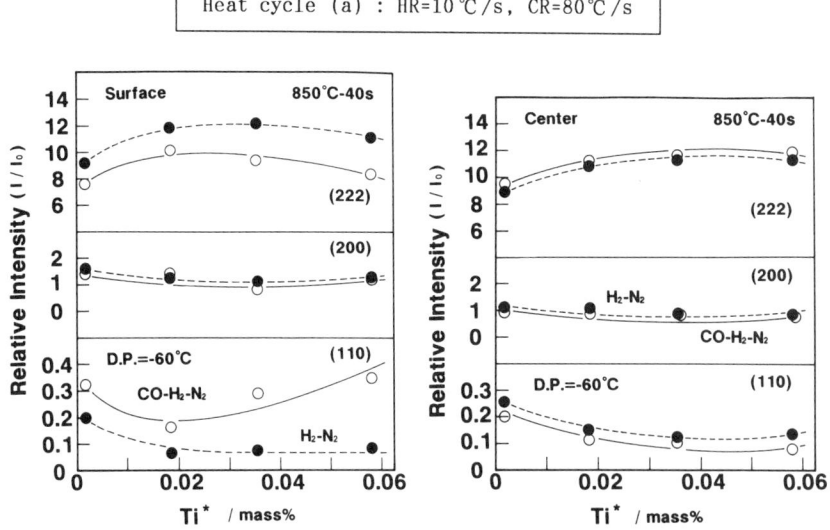

Figure 17 - Effect of carburizing on pole densities at surface and center of thickness for the steels with various Ti content.

after recrystallization on r-value was small. This suggests that the effect of carbon in solid solution after recrystallization on texture is small.

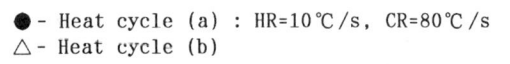
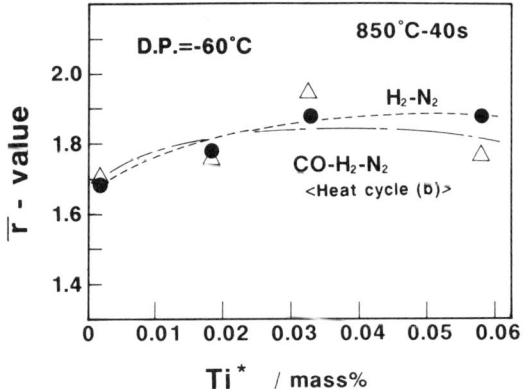

Figure 18 - Effect of carburizing after recrystallization on r-value.

However, it is supposed that the increased carbides such as TiC and NbC by carburizing slightly retards the favorable texture development for r-value due to inhibition of grain growth.

Conclusions

The effect of carburizing during continuous annealing on the resistance to cold-work embrittlement, bake hardenability and r-value of Ti- or Nb-added IF sheet steels were studied in laboratory. The following results were obtained:
1. The resistance to cold-work embrittlement was improved by the segregation of entered carbon by carburizing to grain boundaries.
2. Bake hardenability higher than $30N/mm^2$ was obtained due to entered carbon remained in solid solution without combining with excess Ti, Nb.
3. The carbon in solid solution had a nonuniform distribution in the direction of thickness due to the diffusion of carbon and precipitation of carbides by combining the entered carbon with excess Ti, Nb. The distribution varies with Ti or/and Nb content, the composition of carburizing atmosphere and heat cycle.
4. The development of {111} texture at surface was suppressed by the presence of carbon in solid solution before recrystallization and an increase in precipitates such as TiC and NbC by carburizing, and the inhibition of {111} texture development brought about the deterioration in r-value. However, if the carburizing is performed after recrystallization, that is, limited in soaking section in actual continuous annealing process, we can minimize the deterioration in r-value to nearly 0 compared with the steels annealed in conventional atmosphere.

References

(1)M.Konishi, N.Ohashi and H.Yoshida, "Study on the Brittle Fracture of Decarburized Steel Sheet after Press-forming," Kawasaki Steel Technical Report, 6(1974), 305-323.

(2)M.Konishi, T.Obara and T.Tanaka, "Fracture at Grain Boundary on the Denitrided and Decarburized Sheet Steels and Relation Phosphorus and Carbon Content on the Fracture," Tetsu-to-Hagane, 63(1977), S874.

(3)N.Takahashi, M.Shibata and Y.Furuno, "Production of Super Deep Drawing Quality High Strength Cold Rolled Steel Sheets," Tetsu-to-Hagane, 66(1980), S1127.

(4)K.Tayama et al., "Improvement of Resistance to Brittle Fracture in Extra-Low-Carbon Niobium Stabilized Steel after Press-forming. (Development of Ultra-Deep Drawing Cold Rolled Steel with Less Tendency for Brittleness-I)," Tetsu-to-Hagane, 69(1983), S1365.

(5)N.Fukuda and M.Shimizu, "Effect of Titanium Addition on the Improvement of r Value in Cold Rolled Sheet Steel," Journal of the JSTP, 13(1972), 841-850.

(6)I.Tsukatani, "Effects of Titanium, Manganese and Boron Contents on the Mechanical Properties in Extra-low Carbon Sheet Steels by Room-temperature Coiling Process," Tetsu-to-Hagane, 75(1989), 774-781.

(7)A.Okamoto et al., "Bake-hardenable Al-killed Steel Sheet by Box-annealing Process," Tetsu-to-Hagane, 68(1982), 1369-1377.

(8)T.Irie et al., "Development of Deep Drawable and Bake hardenable High Strength Steel Sheet by Continuous Annealing of Extra Low-Carbon Steels with Nb or Ti," Metallurgy of Continuous-Annealed Sheet Steel, TMS-AIME, Dallas, Feb. 1982, 155-171.

(9)K.Tsunoyama et al., "Effect of Lowering Sulfur Content in Ti-added, Deep Drawable Hot- and Cold-Rolled Sheet Steels," Hot and Cold Rolled Sheet Steels, TMS-AIME, Cincinnati, 1987, 155-164.

(10)N.Mizui and A.Okamoto, "Control of Bake-hardenability in Ultra-low Carbon Ti-added Cold Rolled Sheet Steel. (Development of Bake-hardenable Galvannealed Sheet Steel -1)," CAMP-ISIJ, 3(1990), 1814.

(11)N.Ohashi et al., "Effects of Phosphorus and Solute Carbon on (111) Texture Formation in Extra-Low Carbon, Titanium- or Niobium-Added Cold Rolled Steel Sheets," Textures of Materials, Proc. ICOTOM 6, Tokyo, 1(1981), 195-208.

(12)S.Suzuki et al., "Prevention of the Intergranular Fracture in Iron-Phosphorus Alloys by Carbon," Tetsu-to-Hagane, 70(1984), 2262-2268.

(13)H.Erhart and H.J.Grabke, "Equilibrium Segregation of Phosphorus at Grain Boundaries of Fe-P, Fe-C-P, Fe-Cr-P, and Fe-Cr-C-P Alloys," Metal Science, 15(1981), 401-408.

(14)S.Suzuki et al., "Effect of Carbon on the Grain Boundary Segregation of Phosphorus in α-Iron," Scripta Metallurgica, 7(1983), 1325-1328.

(15)M.Kinoshita and A.Nishimoto, "Effect of carbon content and grain size on bake hardenability or aging property of extra-low carbon cold-rolled steel sheet," CAMP-ISIJ, 3(1990), 1780-1783.

(16)H.Kubotera et al., "Effect of the High-Temperature-Coiling on the Properties of Continuously Annealed Product," Tetsu-to-Hagane, 62(1976), 846-855.

CARBIDE DISSOLUTION IN INTERSTITIAL-FREE STEELS DURING CONTINUOUS ANNEALING

S. Satoh, M. Morita, T. Kato* and O. Hashimoto*

Technical Research Division
Kawasaki Steel Corporation
Mizushima, Kurashiki 712, Japan
*Chiba 260, Japan

Abstract

To clarify the reason for the drawability of interstitial-free steels being influenced by the carbide size prior to cold-rolling, the stability of the carbides at the beginning of recrystallization was investigated. The specimens aged for a short period contained fine and coherent carbides, while coarse and incoherent carbides were observed in the specimens aged for a long period. The amount of solute C distinctly increased in the specimens containing fine carbides at the beginning of recrystallization, although the specimens contained little C in solution before cold-rolling. The calculation of the change in free energy based on the micromechanics has supported the carbide dissolution due to the elastic interaction between coherent carbides and dislocations induced by cold-rolling.

Introduction

Interstitial-free (IF) steels are extensively used as the material for press-forming sheet steels manufactured by the continuous annealing and hot-dip galvanizing processes. Practical IF steels contain strong carbide-forming elements such as Ti and Nb, the C content being less than 0.005mass% (1-7). IF sheet steels produced by cold-rolling and annealing exhibit excellent mechanical properties, especially extra-deep drawability.

It is well known that C in solution retards the development of the { 111 } recrystallization texture favorable for deep drawability (8, 9). To stabilize C in the form of stable carbides in IF steels, a sufficient amount of Ti and/or Nb for the C content is added. It is important to produce the carbides in hot bands prior to recrystallization annealing. Therefore, the drawing property of IF sheet steels is affected by hot-rolling conditions such as the slab reheating temperature (SRT) and coiling temperature (CT) (2-4, 10).

Under normal processing conditions, coarsening of Ti- or Nb-carbides in hot bands is effective for improving the deep drawability of IF sheet steels. This metallurgical principle has been applied to develop extra-deep drawing quality (EDDQ) steels with Lankford values higher than 2.0 (11, 12). Furthermore, bake-hardenable EDDQ steels have been produced by dissolving NbC during high-temperature continuous annealing, thereby coarsening the NbC in the hot band by high CT (13-15). Not enough research has been done on the reason why coarse carbides in hot bands are advantageous for improving the deep drawability of IF sheet steels.

Some previous study clarified that the dispersion of carbides in IF hot bands influenced the recrystallization texture (16-18), most C being stabilized as carbides. Satoh et al. (18) extensively changed the size distribution of carbonitrides in Nb-IF hot bands by controlling the hot-rolling conditions. The hot band containing fine and dense precipitates showed a lower intensity of { 111 } texture after cold-rolling and annealing, compared with the hot band containing coarse precipitates. Both hot bands contained little solute C. No distinct differences in the ferrite grain size and the texture in either hot bands were recognized. They emphasized the contribution of carbides to the growth of { 111 } recrystallized grains. They, however, suggested that fine carbides in the hot band became unstable after cold-rolling and subsequent heating to the stage of recovery or the beginning of recrystallization.

In this study, we investigated the stability of carbides in IF steels during annealing after cold-rolling, thereby changing the size of the carbides existing before cold-rolling. The stability of the carbides is discussed in relation to the elastic interaction between coherent carbides and dislocations induced by cold-rolling.

Experimental Procedure

The chemical composition of a vacuum-melted steel used is shown in Table I . The ingot was homogenized at 1523K for 3.6ks and hot-rolled to a 30mm thick sheet bar. This sheet bar was reheated to 1523K and then hot-rolled to a 3mm thick hot band, the finishing temperature being 1173K. The hot band was cut into small

Table I Chemical composition of steel used (mass%)

C	Si	Mn	P	S	N	O	Al	Ti	Ti* (at%)/ C (at%)
0.0021	0.01	0.01	0.001	0.001	0.0018	0.0020	0.018	0.053	5.4

Ti*(mass%)=Ti (mass%)-(48 / 32) S (mass%)-(48 / 14) N (mass%)

specimens, which were machined to 2mm thick sheets, solution annealed at 1573K for 7.2ks under an Ar gas atmosphere, and quenched in water. The subsequent treatment is illustrated in Fig. 1. The quenched specimens were aged at 873K to 1073K for 120s to 3.6ks. The aged specimens were cold-rolled to 0.5mm thick sheets (reduction : 75%). The aged and cold-rolled specimens were then soaked at 673K to 1073K for 180s, and again quenched in water.

To evaluate the stability of TiC, the aging index (AI) was measured by a tensile test. In this report, AI is defined as the stress difference between the flow stress at a 2% tensile strain and the lower yield stress after aging at 373K for 1.8ks. AI is also sensitive to solute N. However, we confirmed by chemical analysis that almost all the N had been stabilized as TiN at the stage of solution annealing and subsequent quenching.

An AI of 40MPa corresponds to approximately 5 mass ppm C in solution inside the grains when the density of dislocations is low (19). We determined AI of specimens heated to the recovery stage, containing a high density of dislocations, after cold-rolling. Figure 2 demonstrates the change of AI before cold-rolling, and after cold-rolling and soaking at 773K for 180s. The solute C content of the hot bands before cold-rolling was measured by the internal friction method. The steels used were plain steels with no strong carbide-forming elements. Specimens cold-rolled and heated to 773K showed lower AI values than the hot bands, in which the solute C content might not have changed between the two stages. We concluded that a 10MPa AI value at the stage of recovery or at the beginning of recrystallization corresponds to approximately 3 mass ppm of solute C.

The TiC precipitates were observed with a transmission electron microscope (TEM), operating at an accelerating potential of 200keV.

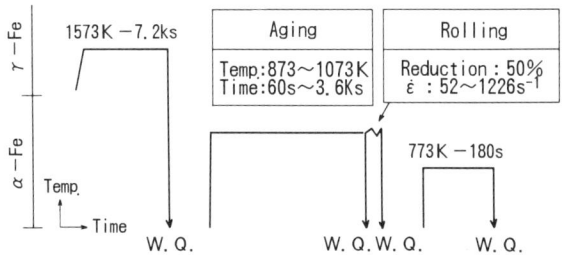

Fig. 1 Schematic illustration of heat treatment process.

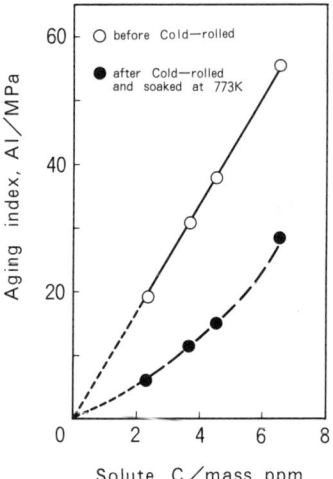

Fig. 2 Change in aging index before cold-rolling, and after cold-rolling and soaking at 773K for 180s in plain steels with no strong carbide-forming elements.

Results

Aging indices (AI) of the solution-treated and aged specimens are shown in Fig. 3. The solution-treated specimen contained approximately 5 mass ppm of solute C. Aging at 973K and 1073K produced very low AI values independent of aging time, while a small change in AI was seen in the specimens aged at 873K. The distinct decrease in the AI of the specimens aged at 973K and 1073K corresponds to the precipitation of TiC.

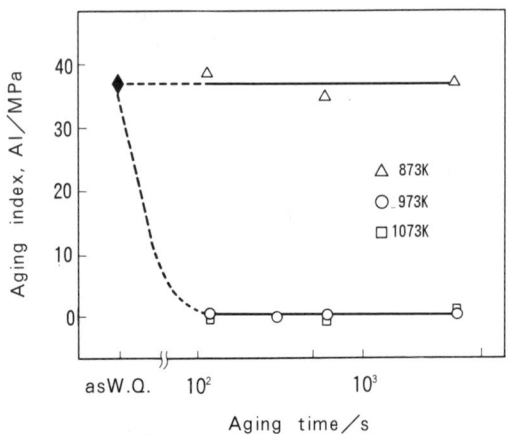

Fig. 3 Aging indices of solution-treated and aged specimens.

Figure 4 shows a transmission electron micrograph of the specimen aged at 973K for 120s. Very fine precipitates were recognized, their being spherical or ellipsoidal. Micrographs taken from specimens with other crystal orientations showed the same shape. The image of the fine precipitates exhibited the strain contrast typically associated with a coherent spherical particle with a dilational misfit strain (20). The average radius of precipitates was estimated to be approximately 2nm, although measuring the size accurately was difficult due to the strong strain ontrast.

Figure 5 shows a transmission electron micrograph of the specimen aged at 973K for 3.6ks. Many more precipitates had coarsened than in the short-term aged specimens (see Fig. 4). The strain contrast around these precipitates was distinctly

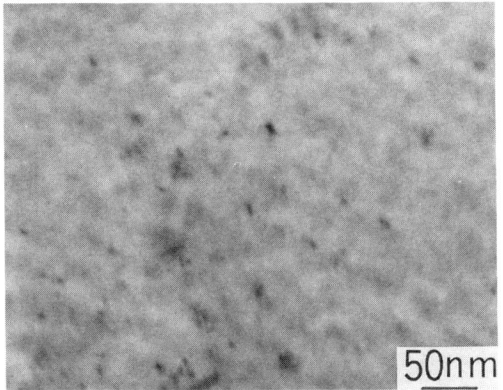

Fig. 4   Transmission electron micrograph of the specimen aged at 973K for 120s.

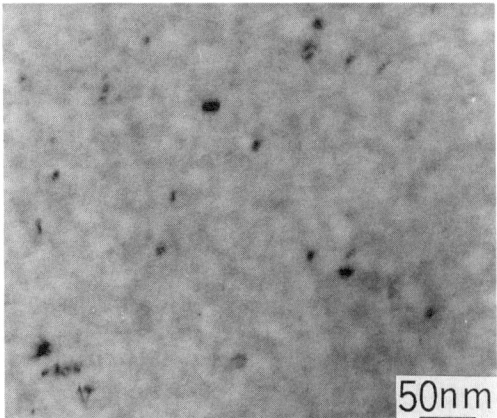

Fig. 5   Transmission electron micrograph of the specimen aged at 973K for 3.6ks.

weaker than that around the fine precipitates shown in Fig. 4. The precipitates in the specimens aged for 120s and 3.6ks were electrolytically extracted and then analyzed by the X-ray diffraction technique. Extracted precipitates were determined to be cubic TiC.

The specimens aged at 973K were cold-rolled by 75% reduction, then soaked at 673K to 1073K for 180s, and finally quenched in water. Figure 6 shows the hardness of the specimens as a function of soaking temperature. The change in hardness shows that recrystallization started at approximately 800K, independent of the aging time prior to cold-rolling. Figure 7 represents the effect of cold-rolling

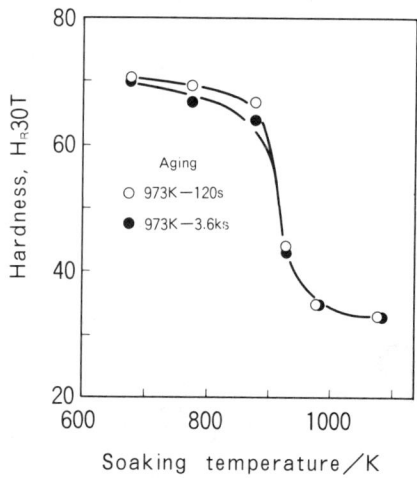

Fig. 6   Relationship between soaking temperature and hardness of specimens aged at 973K for 120s and 3.6ks.

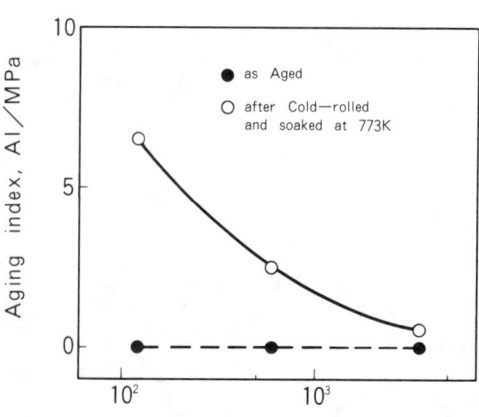

Fig. 7   Effect of cold-rolling and soaking on aging indices of specimens aged at 973K.

and soaking on the aging indices (AI) of the specimens aged at 973K. The soaking temperature was chosen as the beginning of recrystallization as shown in Fig. 6. AI of the as-aged specimens shows that they had a very low solute C content prior to cold-rolling. After cold-rolling and soaking at 773K, the specimens aged for a short time showed higher AI than the specimen aged for longer. This indicates a distinct increase in solute C in the specimen containing fine TiC after cold-rolling and heating at the beginning of recrystallization.

Discussion

The reasons for the increase in solute C content after cold-rolling and heating at the beginning of recrystallization, especially in the specimens aged for a short time, include :

(i) A temperature increase due to deformation. The temperature of the deformed sample would have increased during cold-rolling, and the temperature reached might have resulted in the dissolution of TiC. However, it was confirmed that the temperature increment was less than 20K under our experimental conditions.

(ii) Binding between dislocations and C at interfaces such as grain boundaries. A high density of dislocations close to grain boundaries might have resulted in a high incidence of binding between such dislocations and C at the grain boundaries, producing high AI. However, the binding energy between dislocations and C is 30 to 50kJmol^{-1} (21), while the segregation energy of C at a grain boundary is estimated at 80kJmol^{-1} (22). Thus, the possibility of binding between dislocations and C at grain boundaries is very low. Furthermore, it is difficult to explain the difference in solute C content in the specimens containing fine and coarse TiC by this mechanism.

(iii) Dissolution of TiC due to elastic interaction between dislocations and fine TiC. Fine TiC precipitates can be expected to be coherent with respect to the α-Fe matrix, as shown in Fig. 4. The presence of coherent particles generates an elastic stress field as well as dislocations. Dissolution of TiC might have occurred when the interaction between the stress fields in a region with a high density of dislocations and coherent TiC particles exceeded a critical level. In the following discussion, this interaction energy is estimated and compared with the system free energy.

First, we will consider the change in free energy (ΔG) between the initial state, consisting of a coherent TiC particle and a group of dislocations, and the state in which TiC has dissolved, as shown in Fig. 8. ΔG is given by

$$\Delta G = G^D - (G^P + G^D + G^{P-D})$$

$$= \Delta G^P + \Delta G^{P-D} \qquad (1)$$

where ΔG^P and ΔG^{P-D} are defined by

$$\Delta G^P = -(\Delta G^{chem} V_p + \gamma A_p + E^P V_p) \qquad (2)$$

$$\Delta G^{P-D} = -\lambda E^{P-D} \qquad (3)$$

ΔG^{chem} and γ denote the chemical free energy of a Ti-C system in α-Fe, and the interfacial energy, respectively. The volume and the interfacial area of a particle are represented by V_p and A_p. The self elastic strain energy per unit volume of a coherent precipitate is denoted as E^P. The free energy change of elastic interaction between a coherent particle and dislocations is expressed by Eq. (3), where λ and E^{P-D} denote a constant and the elastic interaction energy between a TiC precipitate and one dislocation, respectively. A TiC particle dissolves when Eq. (1) becomes a negative value.

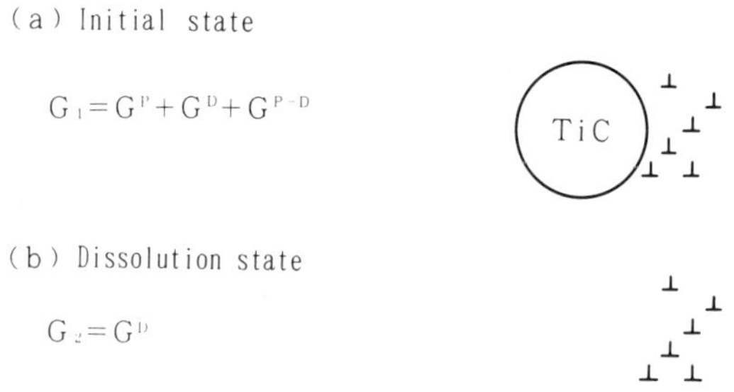

(a) Initial state

$G_1 = G^P + G^D + G^{P-D}$

(b) Dissolution state

$G_2 = G^D$

Fig. 8   Schematic illustration showing (a) initial state and (b) TiC-dissolution state for calculating free energy change.

We will next calculate the elastic interaction energy (E^{P-D}) based on micromechanics (23, 24). Table II presents the elastic constants for α-Fe and TiC (25) at room temperature. The table contains the Zener anisotropy ratio (ω) in terms of the elastic constants for both. The elastic constants for α-Fe and TiC are very different. The anisotropy of α-Fe is great. Therefore, the subsequent calculation for the elastic interaction energy was made for an inhomogeneous and anisotropic system.

We will assume a system comprised of an isolated coherent TiC particle and an edge dislocation surrounded by stress-free external boundaries. The particle possesses a uniform stress-free transformation strain, ε^T, and elastic constants, C^P, different from those of the matrix, C^M. The coordinate system for the calculations based on an α-Fe matrix is shown in Fig. 9. The coordinate axes of x_1, x_2, and x_3 correspond to crystalline orientations [100], [010] and [001] in α-Fe , respectively.

Table II   Elastic constants used in calculations ($10^4 \mathrm{MNm}^{-2}$)

	C_{11}	C_{12}	C_{44}	ω
α-Fe	22.6	14.0	11.6	2.7
TiC	50.0	11.3	17.5	0.9

$\omega = 2C_{44} / (C_{11} - C_{12})$

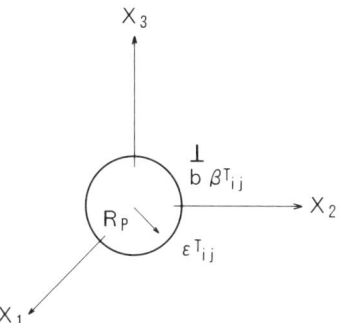

Fig. 9 Coordinate system based on α-Fe matrix for calculating elastic interaction energy.

The TiC particle is assumed to be spherical with the radius, R_p, as suggested by the TEM observation. The straight-edge dislocation parallel to the x_1 axis is located outside the particle. Its stress-free transformation strain is expressed by β^T.

The elastic interaction energy is given by

$$E^{P-D} = -\int_V \sigma_{ij}^P(r)\beta_{ji}^T(r)dr \qquad (4)$$

where σ^P denotes the stress induced by the coherent TiC precipitate. The integration is performed over the system volume, V. We assume the following relative orientation of the TiC particle and the α-Fe matrix (26) :

(001)TiC // (001)α-Fe
[100]TiC // [110]α-Fe
[010]TiC // [110]α-Fe

Using this orientation relationship, the elastic moduli of TiC, C^P, were coordinate-transformed on the α-Fe coordinate system. The stress-free transformation strain, ε^T, can be expressed on the α-Fe coordinate system by using the lattice parameters of α-Fe (0.2866nm) and TiC (0.4380nm). Detailed calculation procedures for the elastic interaction energy, Eq. (4), will be presented in our paper (27).

We will next determine the change in free energy, ΔG, described by Eq. (1) to clarify the possibility of TiC dissolution caused by elastic interaction. In the calculation, we utilized the following chemical free energy of the Ti-C system in α-Fe (28) :

$$\Delta G^{chem} = -188754 + 22.504T \qquad (5)$$

where T denotes the temperature (K). We used 1 Jm^{-2} as the interfacial energy (γ) for calculation purposes (29). Figure 10 shows ΔG (R_p : 2nm, T : 673K to 873K) as a function of λ in Eq. (3). We used the maximum interaction energy of E^{P-D} in Eq. (3), with the edge dislocation located in the region of $x_3<0$ close to TiC (27). ΔG was normalized by the chemical free energy, and exhibits a negative value when λ exceeds approximately 8. This means that a coherent TiC might dissolve in the case of elastically interacting with a high density of dislocations.

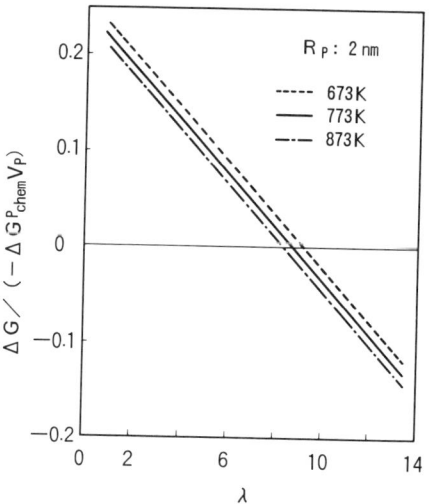

Fig. 10 Free energy change as a function of λ in Eq. (3) (R_p : 2nm, T : 673K to 873K).

Conclusions

To clarify the reason for the drawability of IF steels being affected by the size of carbides prior to cold-rolling, the stability of the carbides at the beginning of recrystallization was investigated.

(1) The amount of solute C at the beginning of recrystallization was estimated by the aging index, utilizing strain-age hardening in a tensile test.

(2) With short-time aging after solution treatment, very fine carbides might have had coherency with respect to the α-Fe matrix, while coarse and incoherent carbides were observed in the specimens aged for a long time.

(3) The amount of solute C distinctly increased in the specimens containing fine carbides at the beginning of recrystallization, although the specimens contained little C in solution before cold-rolling.

(4) This phenomenon can be explained by the carbide dissolution due to elastic interaction between the coherent carbides and dislocations induced by cold-rolling. This mechanism was supported by a calculation of the change in free energy based on micromechanics.

References

1) N. Fukuda and M. Shimizu, "Effect of Titanium Addition on the Improvement of r-value in Cold-rolled Sheet Steel", Journal of Japan Society for Technology of Plasticity, 13 (142) (1972) 841-850.

2) J. A. Elias and R. E. Hook, "Interstitial-free Steels", 348-368 in Proceedings of the 13th Mechanical Working and Steel Processing Conference, ISS-AIME, 1971.

3) O. Hashimoto, S. Satoh and T. Tanaka, "Development of Low Carbon-Niobium Steel with Ultra-Deep Drawability", Testu-to-Hagané, 67 (11) (1981) 1962-1971.

4) S. Satoh, T. Irie and O. Hashimoto, "Effect of Hot-Rolling on Mechanical Properties of Extra-Low Carbon Cold Rolled Steel Sheet Containing Niobium", Tetsu-to-Hagané, 69 (2) (1983) 283-290.

5) M. Yamada and Y. Tokunaga, Tetsu-to-Hagané, 71 (1985) S640.

6) K. Sakata, K. Hashiguchi, O. Hashimoto and S. Okano, Tetsu-to-Hagané, 71 (1985) S1363.

7) K. Tsunoyama, T. Obara, S. Satoh, H. Abe, O. Shibasaki and N. Uesugi, "Development of Extra-Deep-Drawing Cold-Rolled Sheet Steels for Integrated Parts", Kawasaki Steel Technical Report, No. 24 (1991) 84-90.

8) A. Okamoto and M. Takahashi, "Effect of a Small Amount of Carbon on the Recrystallization Texture Formation in Steel", The 6th International Conference on Textures of Materials, Tokyo, Sep. 1981.

9) N. Takahashi, M. Abe, O. Akisue and H. Kato, "Metallurgical Basis to Design a Continuous-Annealing Process for Producing Deep Drawing Steel Sheet", Metallurgy of Continuous-Annealed Sheet Steel, ed. B. L. Branfitt and P. L. Mangonon, Jr. (TMS-AIME, Warrendale PA, 1982) 51-81.

10) S. Satoh, T. Obara, M. Nishida and T. Irie, "Effects of Alloying Elements and Hot-Rolling Conditions on the Mechanical Properties of Continuous-Annealed, Extra-Low-Carbon Steel Sheet", Technology of Continuously Annealed Cold-Rolled Sheet Steel, ed. R. Pradhan (TMS of AIME, Warrendale PA, 1984) 151-166.

11) K. Tsunoyama, S. Satoh, Y. Yamazaki and H. Abe, "Recent Advance in Interstitical-Free Steels for Formable Cold-Rolled Sheet Applications", Metallurgy of Vacuum-Degassed Steel Products, ed. R. Pradhan (TMS of AIME, Warrendale PA, 1990) 127-142.

12) N. Kino, M. Yamada, Y. Tokunaga and H. Tsuchiya, "Production of Nb-Ti-Added Ultra-Low-Carbon Steel for Galvannealed Application", ibid., 197-213.

13) T. Irie, S. Satoh, A. Yasuda and O. Hashimoto, "Development of Deep Drawable and Bake Hardenable High Strength Steel Sheet by Continuous Annealing of Extra Low-Carbon Steels with Nb or Ti, and P", Metallurgy of Continuous-Annealed Sheet Steel, ed. B. L. Bramfitt and P. L. Mangonon, Jr. (TMS of AIME, Warrendale PA, 1982) 155-171.

14) M. Kurosawa, S. Satoh, T. Obara and K. Tsunoyama, "Age-Hardening Behavior and Dent Resistance of Bake-Hardenable and Extra Deep-Drawable High Strength Steel", Kawasaki Steel Technical Report, No. 18 (1988) 61-65.

15) S. Satoh, S. Okada, T. Kato, O. Hashimoto, T. Hanazawa and H. Tsunekawa, "Development of Bake-Hardening High Strength Cold-Rolled Sheet Steels for Automobile Exposed Panels" to be published in Kawasaki Steel Technical Report.

16) R. E. Mould and J. M. Gray, "Plastic Anisotropy of Low-Carbon, Low-Manganese Steels Containing Niobium", Met. Trans., 3 (1972) 3121-3131.

17) O. Akisue and K. Takahashi, "Effects of Nb Addition on Textures in Low Carbon Steel Sheets", Journal of Japan Institute of Metals, 36 (11) (1972) 1124-1130.

18) S. Satoh, T. Obara and K. Tsunoyama, "Effect of Precipitate Dispersion on Recrystallization Texture of Niobium-added Extra-Low Carbon Cold-rolled Steel Sheet", Trans. ISIJ, 26 (1986) 737-744.

19) Y. Yamazaki, S. Matsuoka, S. Satoh and T. Kato, "Relation between Solute Carbon and Aging Index in Extra-Low Carbon Sheet Steels", Current Advances in Materials and Process, (ISIJ, Tokyo, 1991) 827.

20) M. F. Ashby and L. M. Brown, "Diffraction Contrast from Spherically Symmetrical Coherency Strains", Phil. Mag., 8 (1963) 1083-1092.

21) K. Furusawa and K. Tanaka, "Amplitude Dependent Internal Friction of Iron Containing a Small Amount of Carbon and Nitrogen", Journal of Japan Institute of Metals, 33 (1969) 985-991.

22) H. J. Grabke, W. Panlitchke, G. Tauber and H. Viefhans, "Equilibrium Surface Segregation of Dissolved Nonmetal Atoms on Iron (100) Faces", Surf. Sci., 63 (1977) 377-389.

23) J. D. Eshelby, "The Determination of the Elastic Field of an Ellipsoidal Inclusion, and Related Problems", Proc. Roy. Soc. (A), 241 (1957) 376-396.

24) A. G. Khachaturyan, Theory of Structural Transformations in Solids, Wiley, Berlin (1983).

25) J. J. Gilman and B. W. Roberts, "Elastic Constants of TiC and TiB_2", J. Appl. Phys., 32 (1961) 1405.

26) R. Uemori, M. Saga and H. Morikawa, "Characterization of Ultra-fine Region in Steels", Bulletin of the Japan Institute of Metals, 30 (1991) 498-505.

27) S. Satoh and T. Kato, submitted to Trans. ISIJ.

28) M. Hasebe, private communication with authors.

29) W. Kesternich, "Dislocation-Controlled Precipitation of TiC Particles and Their Resistance to Coarsening", Phil. Mag., 52 (1985) 533-548.

RECRYSTALLIZATION OF INTERSTITIAL-FREE STEELS

D.O. Wilshynsky-Dresler, G. Krauss, D.K. Matlock

Advanced Steel Processing and Products Research Center
Department of Metallurgical Engineering
Colorado School of Mines
Golden, Colorado 80401

Abstract

The recrystallization kinetics of interstitial free steels stabilized with Ti, Ti+Nb, or Nb, were evaluated and compared to an unstabilized IF steel with the same base composition. Recrystallized volume fractions were determined as a function of annealing time and temperature. The rates of recrystallization were severely retarded by the stabilizing element additions. Fine precipitates were identified by scanning transmission electron microscopy. The activation energies for recrystallization in the stabilized IF steels were greater than the value for the self-diffusion of iron, and the differences in activation energies with respect to pure iron were interpreted based on analysis of interface restraint by precipitates and the effect of solute atoms on the self-diffusivity of iron during recrystallization.

Introduction

Interstitial-Free (IF) are a class of steels with carbon levels equal to or less than 0.008 wt pct (1-4). Following cold-rolling and recrystallization annealing, IF steels develop high normal anisotropy, r_m; often with values greater than 2.0 (1-4). Many factors which produce high r_m values and good formability of IF steels have been identified (5). However, there are still questions regarding the role alloying elements such as niobium and titanium play in optimizing recrystallization behaviour and texture formation. Previous work by the authors (6), on a set of IF steels with and without stabilizing additions, indicated that with a cold reduction of 90%, and at an isothermal annealing temperature of 815°C, all steels recrystallize very rapidly, within 15 seconds. Even at this very short time for recrystallization there are differences in recrystallization behaviour. However, the very brief times makes it difficult to more closely examine differences and similarities in recrystallization behaviour. This study examines the recrystallization of an experimental series of IF steels at lower annealing temperatures in order to identify differences in kinetics and mechanisms of recrystallization as a function of stabilizing additions.

Experimental Approach

Material Selection

Steels with four different chemistries were examined in order to compare the effects of Nb and Ti on recrystallization behaviour of IF steel. The chemical analyses are listed in Table 1. All four steels have the same nominal composition of 0.004 wt pct carbon, 0.2 wt pct manganese, and between 0.022 and 0.047 wt. pct. aluminum. The major difference in composition is the type and amount of carbon stabilizing element, Ti and/or Nb. As listed in Table I, the atomic ratio of $(Ti^* + Nb)/C$ ranges from 0 to 3.77; Ti^*, the effective Ti content, is defined in Table I. This ratio specifies the degree of stabilization of interstitials. For example, a ratio of 1 indicates that the exact stoichiometric amount of stabilizing element is present to combine with carbon. Therefore, the ferrite matrix should be free of carbon and {111}<110> texture formation should be enhanced (2); however, analysis of precipitate composition, density and size is necessary to rigorously construct this ratio.

Table I. Chemical Analysis of IF Steels (in wt pct)
Analyses provided by the Bethlehem Steel Company

Steel	C	Mn	P	S	Al	Ti	Nb	N	$(Ti^*+Nb)/C$
AK	0.004	0.26	0.005	0.006	0.047	---	---	0.0060	0
+Ti	0.004	0.19	0.005	0.003	0.022	0.087	---	0.0065	3.77
Ti+Nb	0.004	0.19	0.005	0.003	0.039	0.035	0.049	0.0065	1.99
+Nb	0.004	0.20	0.004	0.003	0.036	0.003	0.045	0.0060	1.45

$Ti^* = Ti - (48/32)\%S - (48/14)\%N$

The alloys were melted in a laboratory-scale vacuum-induction furnace to produce 230 mm by 230 mm, 227 kg ingots. These ingots were slabbed to a thickness of 19 mm at a slab reheating temperature of 1260°C, and hot rolled to 3 mm (0.120"). The finishing temperature was 900°C, in the austenite phase. The coiling temperature was set at a "low" value of 565°C, or a "high" value of 730°C for each steel by quenching the steels in a bath of varying concentrations of a polymer quench; once at the coiling temperature, the steels were slow cooled in a furnace at a rate of approximately 30°C/hour to simulate coil cooling. With this series of steels, the effect of the type and amount of stabilizing addition on recrystallization kinetics can be assessed. Also, the effect of coiling temperature on recrystallization behaviour can be clarified for IF steels.

Experimental Program

In order to compare the effects of the stabilizing additions in the steels on recrystallization kinetics, recrystallization curves were generated for the series of IF steels at temperatures between 550°C and 750°C. Partially recrystallized specimens were examined by light microscopy and transmission electron microscopy (TEM). Large precipitates, greater than 0.1 μm, were examined with a JEOL JXA-840 scanning energy dispersive (EDX) microanalyser. Fine precipitates (50 to 1000 Å) were identified with a VG 501HB dedicated Scanning Transmission Electron Microscopy (STEM) operated at 100 kV. Carbon extraction replicas were used for chemical microanalysis of fine precipitates.

Recrystallization Evaluation

Recrystallization kinetics were determined by a series of isothermal heat treatments of cold-rolled material. The as-received hot-band was pickled in a 50 vol pct solution of hydrochloric acid in water at approximately 70°C until the scale was dissolved. The 3 mm thick pickled and oiled hot band was cold-rolled 75% on a laboratory scale 2-Hi Fenn mill with liberal application of lubricant. The cold-rolled material was sheared into coupons, 19 mm by 9.5 mm, which were immersed in a 50% barium chloride salt bath at temperatures between 650°C and 750°C for the stabilized steels, and at temperatures between 550°C and 700°C for the unstabilized IF steel, for various times and then water quenched. All times were measured from the first immersion in the salt baths and include the time to reach temperature. After these heat treatments, longitudinal sections of the coupons were mounted, polished, and sequentially etched in 4 pct picral, Marshall's reagent, and 1:1 4 pct picral-4 pct nital combination etch. The progression of recrystallization was determined by point counting in accordance with ASTM E562-83; the magnifications used varied between 500x and 1000x, and the point counting grids were spaced either 12.7 mm or 19 mm depending on the mean feature size.

Thin foils for transmission electron microscopy were prepared from the center-line thickness of the cold-rolled and partially recrystallized specimens, parallel to the rolling plane. The specimens were first mechanically thinned to approximately 0.1 to 0.13 mm by wet grinding on 320, 400, and 600 grit SiC paper. Discs with 3 mm diameter were punched out, and wet ground on 600 grit SiC to approximately 0.075 - 0.123 mm. The discs were then polished at room temperature to perforation in a Fischione twinjet electro-polisher with a electrolyte of 5 pct perchloric and 95 pct acetic acid (by volume) at a polishing current of approximately 80 ma at room temperature. The thin foils were examined in a Phillips EM 400 TEM operated at 120 Kv.

Results and Discussion

Recrystallization Behaviour

Results in the form of fraction recrystallized versus annealing time for the isothermal heat-treatments of the 75% cold-rolled stabilized IF steel specimens at 650°C , 700°C, and 750 are shown in Figures 1,2, and 3 for both low and high coiling temperatures. Figure 4 shows the recrystallization curves for the unstabilized IF steel isothermally annealed at 500, 550, and 620°C. The data in Figures 1 to 3 indicate that the recrystallization kinetics of stabilized IF steels are independent of hot mill coiling temperature. At each temperature, the unstabilized IF steel recrystallizes the fastest, followed by the Ti-stabilized steel, and finally by the Nb-stabilized and the Ti+Nb-stabilized IF steels. As the isothermal annealing temperature decreases from 750°C to 650°C, the incubation time for recrystallization for the Ti+Nb-stabilized IF steel increases relative to that for the Nb-stabilized IF steel; at 750°C, the Ti+Nb-stabilized IF steel and the Nb-stabilized IF steel take about the same time to recrystallize, at 650°C the Ti+Nb-stabilized steel recrystallizes the most sluggishly.

The recrystallization kinetics for the series of IF steels can be partly ascribed to their precipitate distributions. Figure 5, shows typical precipitate distributions in partially recrystallized unstabilized steel coiled at 730°C; the density of precipitates is low. A typical recrystallized grain growing away from its center of curvature into the unrecrystallized matrix is shown in Figure 5a. A higher magnification, Figure 5b, shows the paucity of fine precipitates in the unstabilized IF steel. is seen. Figures 5c and 5d, other typical fields, illustrate the low fine-precipitate density in the unstabilized IF steel. The precipitates that are present in the unstabilized hot band coiled at 730°C were identified as AlN and MnS as shown in the annular dark field (ADF) images and accompanying x-ray spectra shown in Figures 6 and 7, respectively. The MnS precipitates are arranged on what may have been prior γ grain boundaries, whereas the AlN precipitates appear to have precipitated within the α grains.

In the Ti-stabilized steel, there are two distinct precipitate types: large cubic precipitates, greater than 1000 Å, and finer precipitates that average around 200-300 Å in diameter. The Ti-stabilized IF steel has the largest hot band grain size (6); on this basis, one would expect that the Ti-stabilized IF steel should manifest the most sluggish recrystallization behaviour (7). However, the large cubic precipitates act as preferential nucleation sites for recrystallization, as shown in Figure 8, and accelerate recrystallization. This observation is consistent with previous studies which have shown that large particles (diameter greater than 1 μm) act as nucleation sites for recrystallization (8). The large precipitates are identified as TiN. In some of the large precipitates, Mn substitutes for Ti, as shown in the ADF image, Figure 9a, and the accompanying x-ray spectrum, Figure 9b.

The fine precipitates in the Ti-stabilized hot band coiled at 730°C are all Ti-rich: TiS, TiC, TiN, and possibly Ti(C,N) and $Ti_4C_2S_2$. The use of carbon replicas made it impossible to accurately define the stoichiometry of the precipitates. However, qualitatively, the distribution of C,N, and S does not seem to be a function of size of the fine precipitates. Figure 10 shows a typical ADF image and spectra for the fine precipitates. The distribution of the fine precipitates is nonuniform, as shown in the bright field images of partially recrystallized Ti-

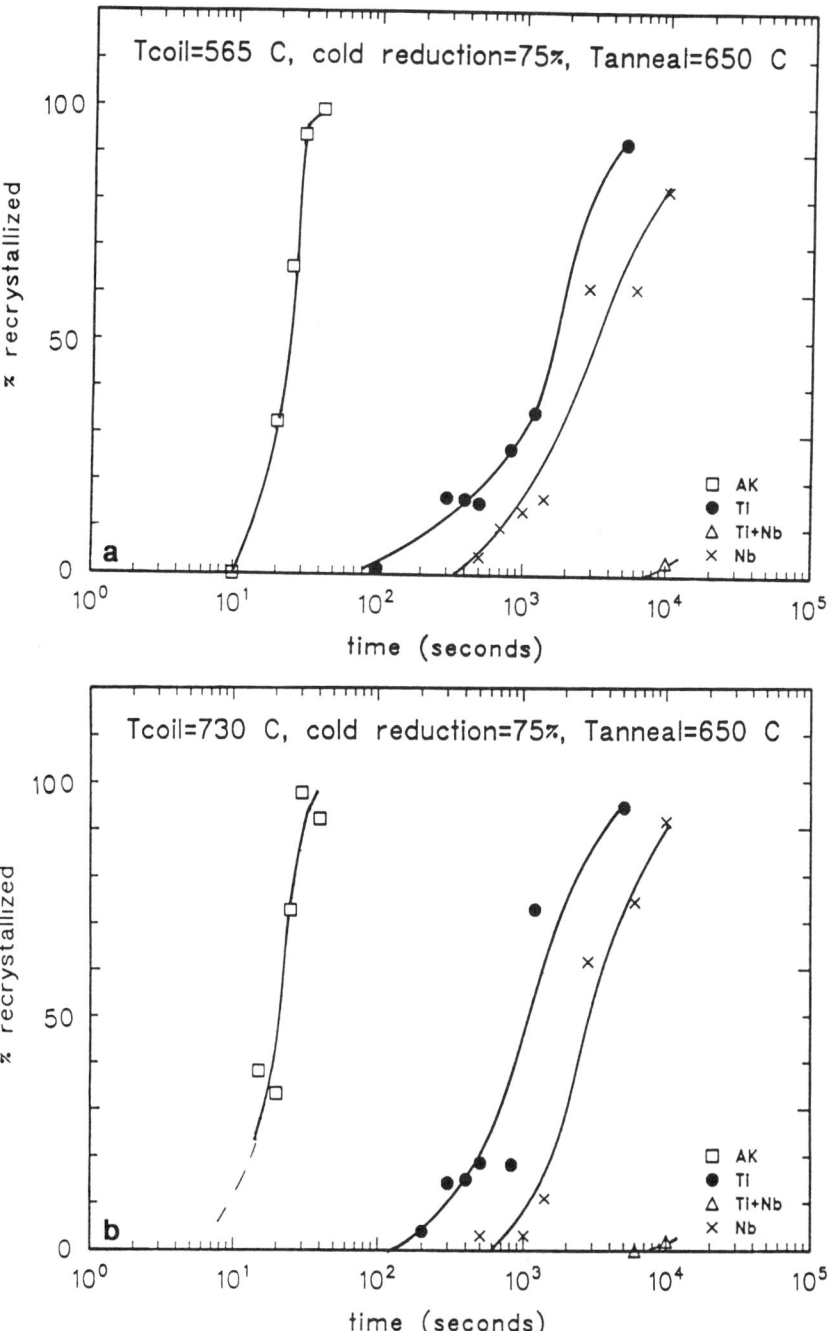

Figure 1 - Recrystallization curves for the series of IF steels, cold-rolled 75%, isothermally annealed at 650°C, a) coiling temperature = 565°C, b) coiling temperature = 730°C.

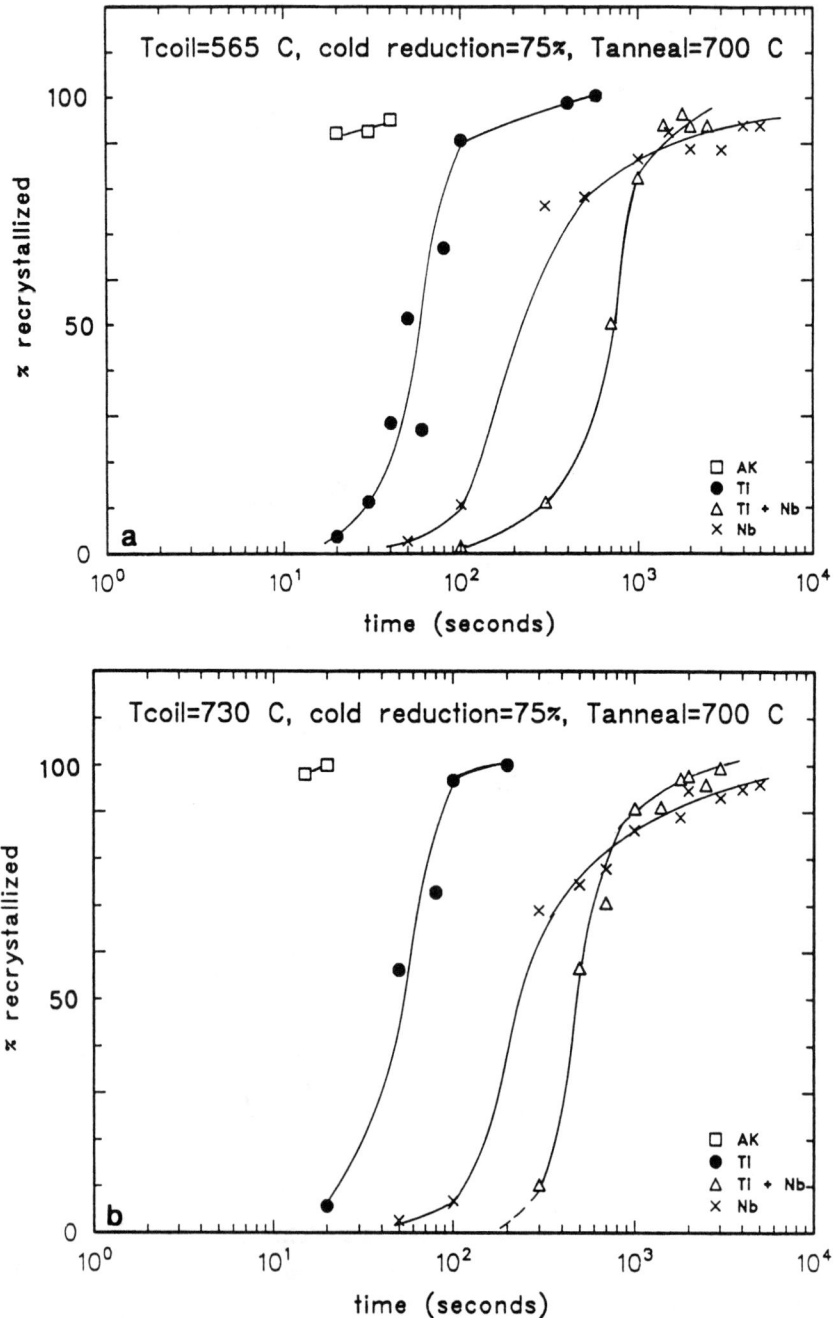

Figure 2 - Recrystallization curves for the series of IF steels, cold-rolled 75%, isothermally annealed at 700°C, a) coiling temperature = 565°C, b) coiling temperature = 730°C.

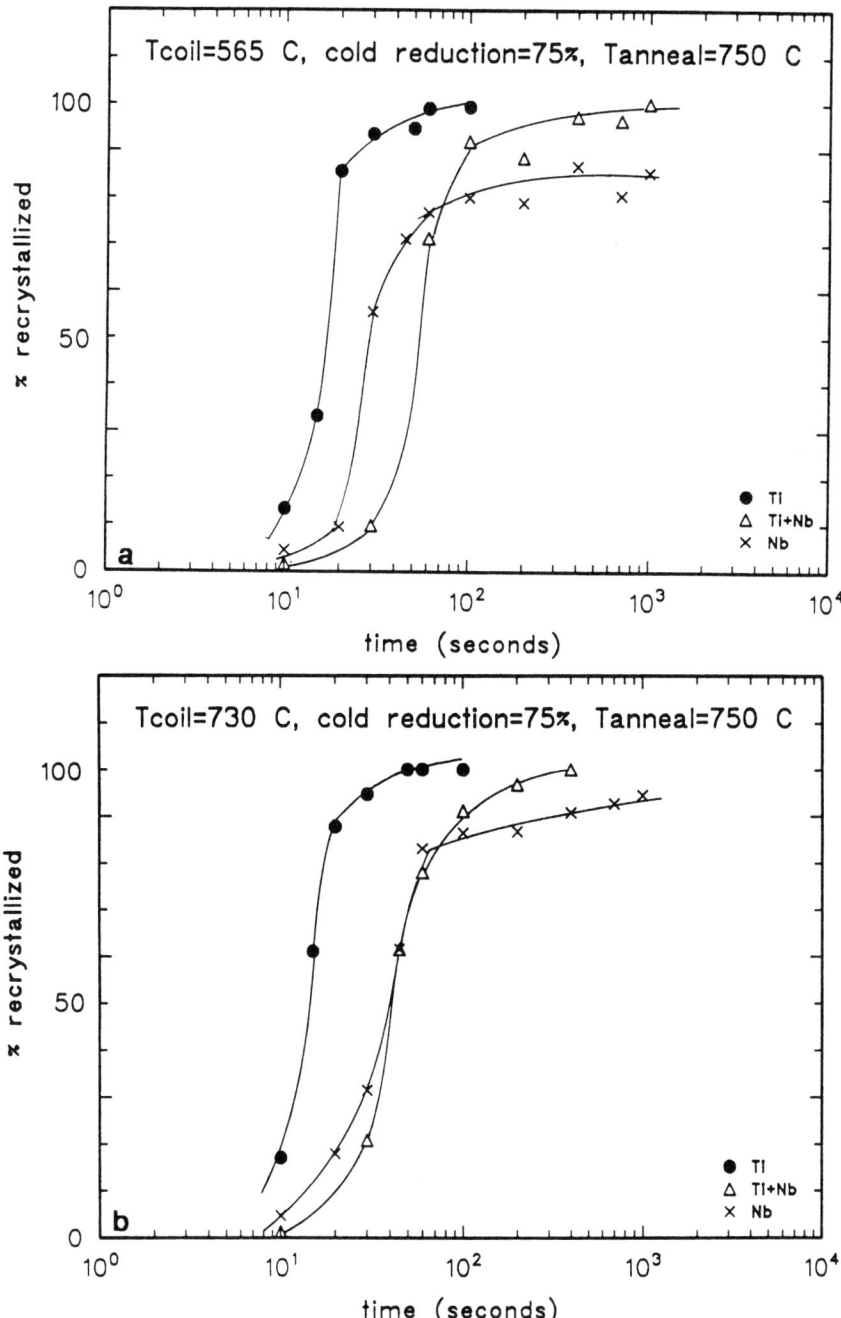

Figure 3 - Recrystallization curves for the series of IF steels, cold-rolled 75%, isothermally annealed at 750°C, a) coiling temperature = 565°C, b) coiling temperature = 730°C.

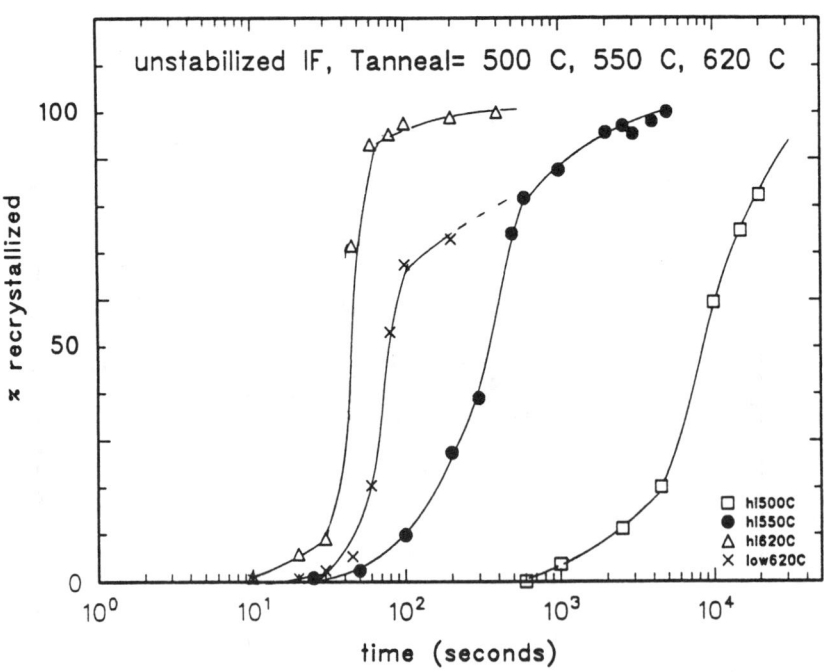

Figure 4 - Recrystallization curves for unstabilized IF steel, cold-rolled 75%, isothermally annealed at 500°C, 550°C, and 620°C.

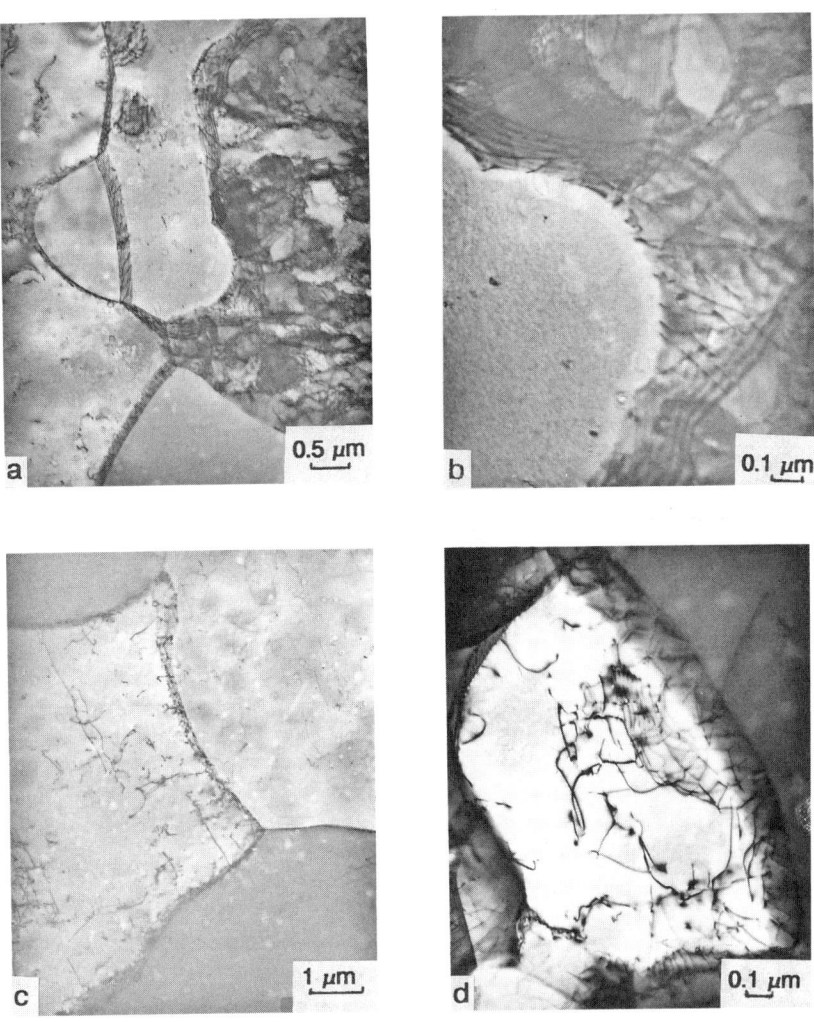

Figure 5 - Partially recrystallized unstabilized-IF steel coiled at 730°C, annealed at 650°C a) recrystallized grain with low precipitate distribution growing into cold-worked material. b) higher magnification of a, c) other region, showing low precipitate distribution d) another recrystallized grain with low precipitate distribution. Rolling plane section. TEM bright-field.

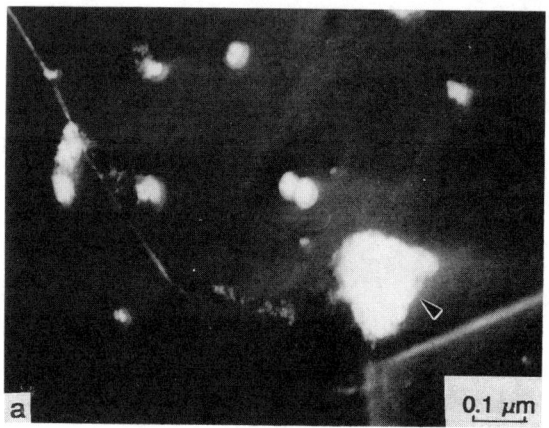
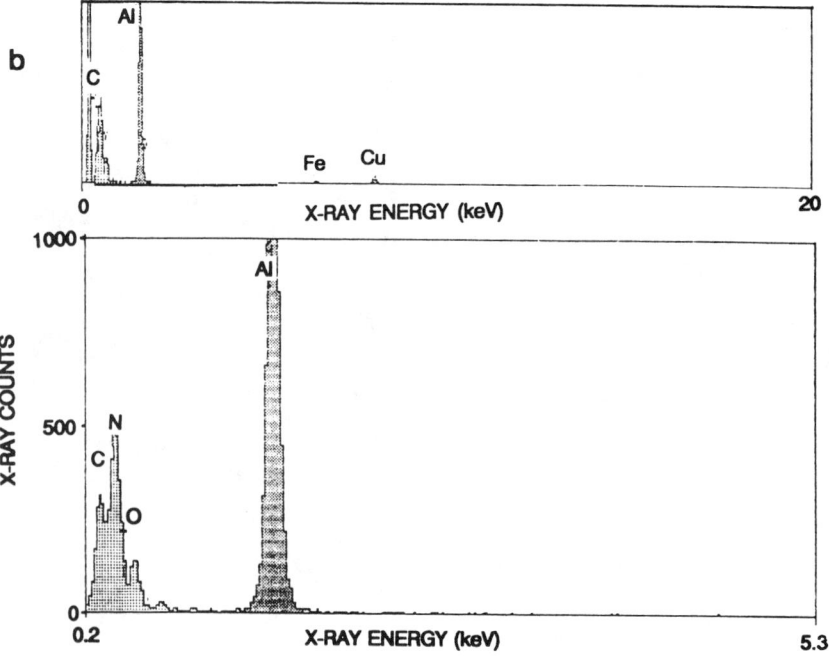

Figure 6 - a) Annular Dark-Field (ADF) image of carbon replica, b) x-ray spectrum for AlN precipitate cluster in unstabilized IF hot band coiled at 730°C.

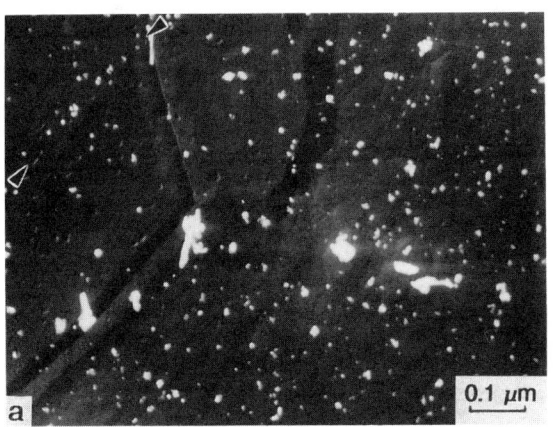

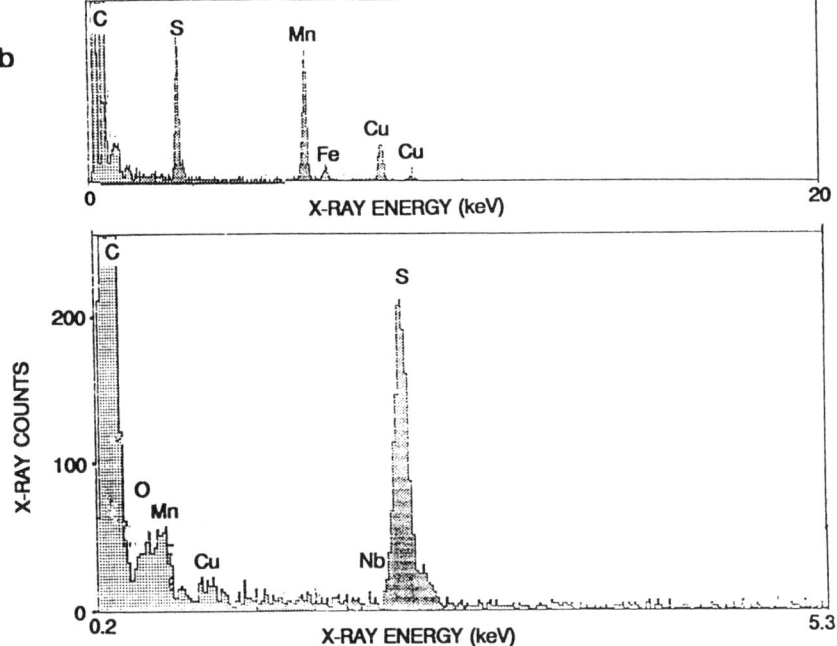

Figure 7 - a) ADF image and b) x-ray spectrum for MnS precipitates in unstabilized IF hot band coiled at 730°C.

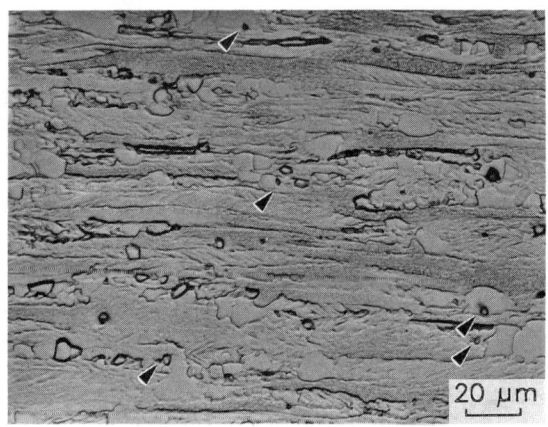

Figure 8 - Partially recrystallized at 700°C Ti-stabilized IF steel coiled at 565°C with large Ti-rich precipitates acting as preferential nucleation sites, as shown by arrows. Longitudinal section. Picral, Marshall's reagent, and 1:1 4% picral:4% nital. Light micrograph.

stabilized IF steel, Figure 11. In Figure 11a, the density of fine precipitates is low, an observation which may be related to the presence of a nearby, large cubic Ti-rich precipitate which locally depletes the matrix of Ti. In other regions, Figures 11b-d, the fine precipitate distribution appears much denser. Even though the recrystallization rate is accelerated by the presence of the large cubic Ti-rich precipitates, the fine precipitates retard the rate of recrystallization in the Ti-stabilized steel, resulting in lower recrystallization rates relative to the unstabilized IF steel. The large Ti-rich precipitates were found in all of the steels containing Ti.

Large, cubic precipitates and fine, 80-400 Å, precipitates were observed in the Ti+Nb-stabilized IF steels. A large cubic precipitate, shown in Figure 12, was identified as TiN; only a small amount of Nb was in this precipitate. In the Ti+Nb stabilized IF steel most of the precipitates were not large cubic precipitates but smaller, spherical or rod-like precipitates as shown in Figure 13a. The chemical analysis of the fine precipitates showed varying levels of Ti, Nb, S, and N; typical fine precipitate chemistries are shown in Figures 13b-c. The finest precipitates, that is, those less than 100 Å, seem to be Ti and S rich, as shown in Figure 14. The large, cubic, precipitates apparently drain the matrix of Ti, and not until the temperature is quite low, is the solubility product high enough to drive Ti to precipitation. Typical precipitate distributions for the Ti+Nb-stabilized steel are shown in bright field images, Figure 15. Figure 15a shows a typical region with high density of fine precipitates, and Figure 15b illustrates a recrystallized grain with a large, cubic Ti precipitate at the boundary, suggesting that the precipitate acted to accelerate recrystallization. Figure 15c shows both fine precipitates and a larger cubic precipitate, A higher magnification of the fine and large precipitates is shown in Figure 15d.

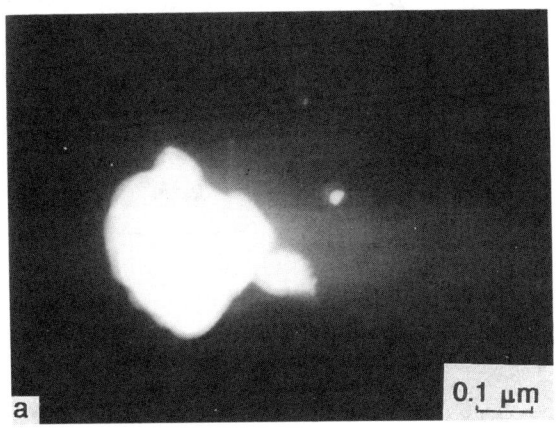

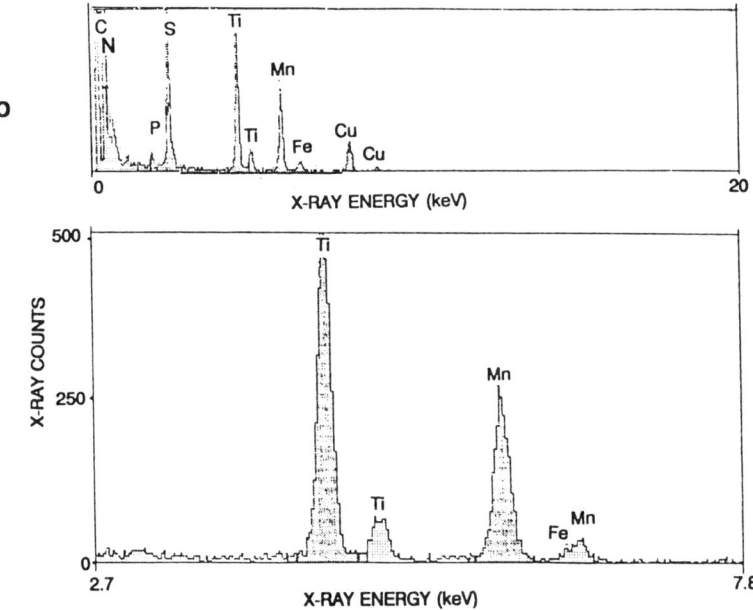

Figure 9 - Large (Ti,Mn)N precipitate in Ti-stabilized IF hot band, coiled at 730°C, a) ADF image, b) x-ray spectrum.

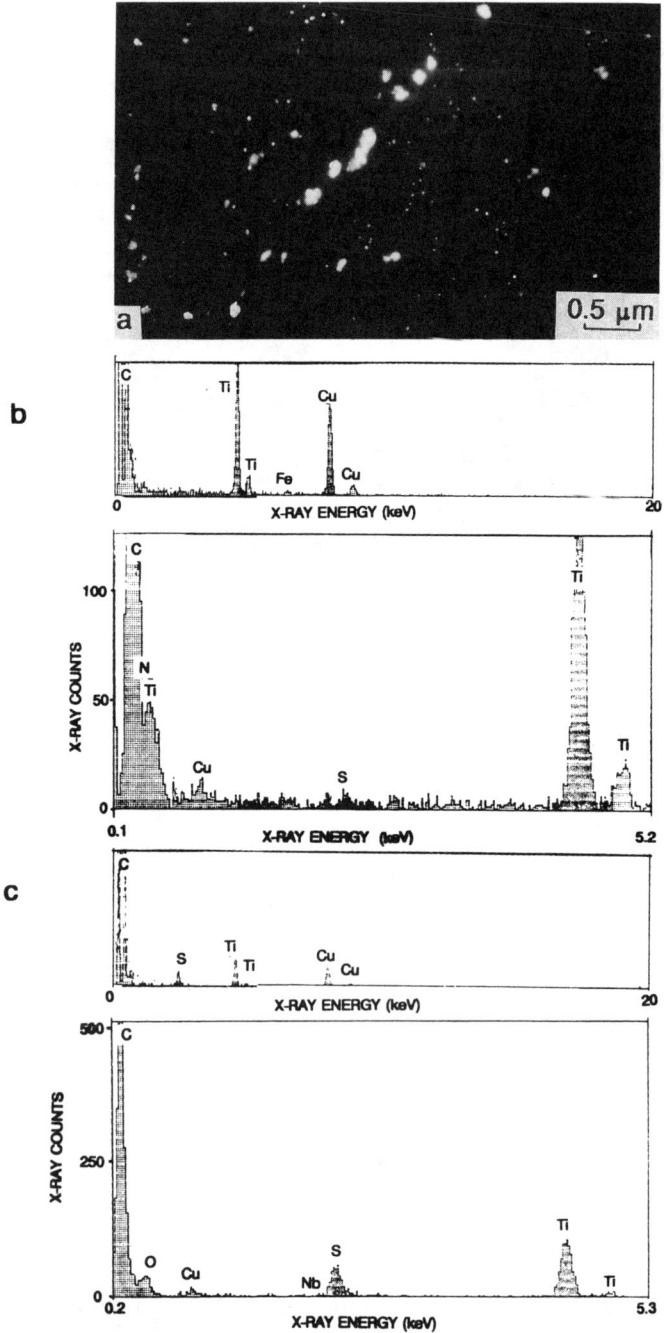

Figure 10 - Linear array of TiN and TiS precipitates in Ti-stabilized hot band coiled at 730°C, a) ADF image, b) c) x-ray spectra.

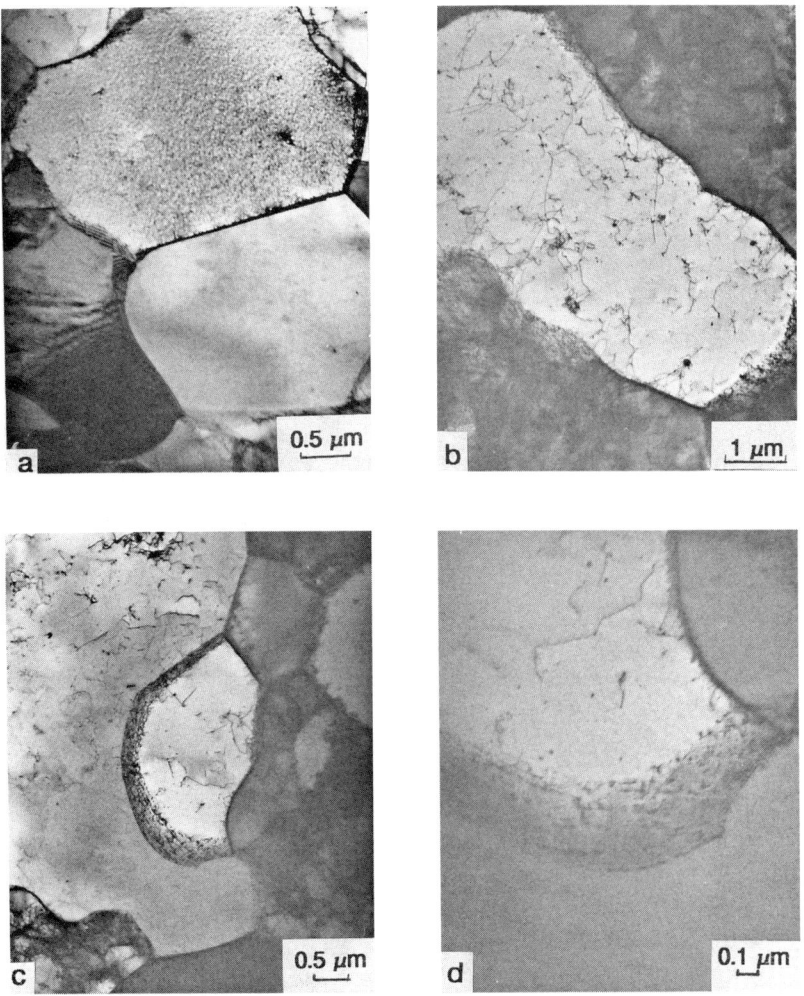

Figure 11 - Partially recrystallized at 700°C Ti-stabilized IF steel coiled at 730°C, a) two grains with very low density of fine precipitates, b) another region showing a high fine precipitate density, c) another region showing a high fine precipitate density, curvy boundary of recrystallized grain indicates presence of precipitates prior to recrystallization d) higher magnification of c.

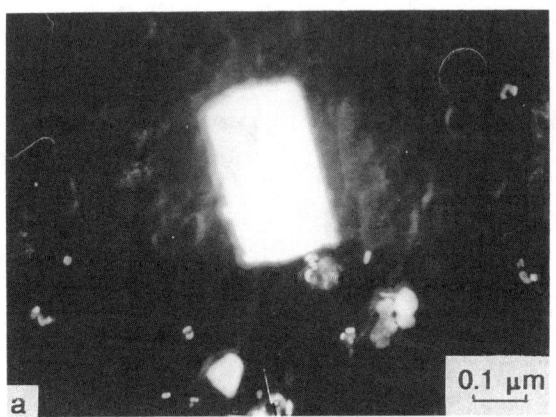
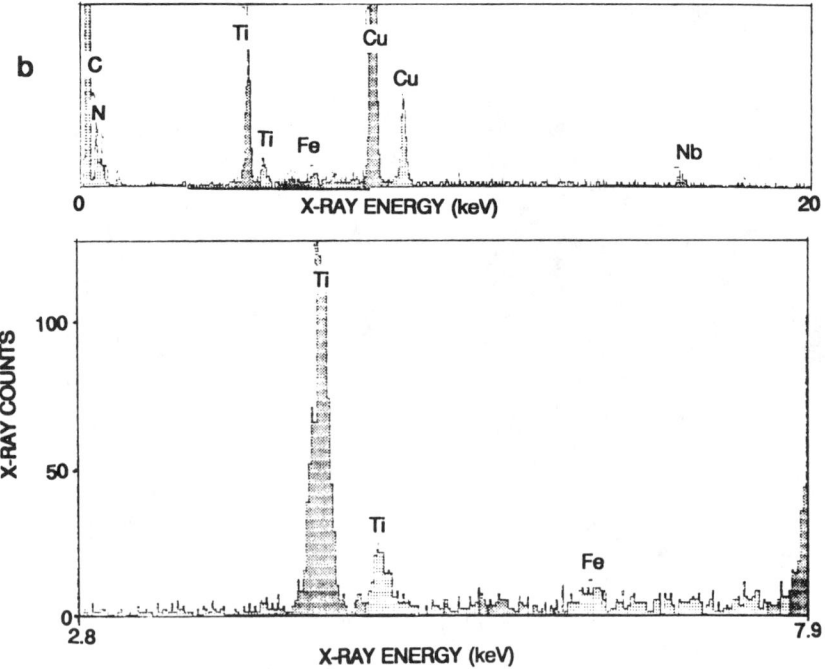

Figure 12 - Large TiN precipitate in Ti+Nb-stabilized IF hot-band coiled at 730°C, a) ADF image, b) x-ray spectrum.

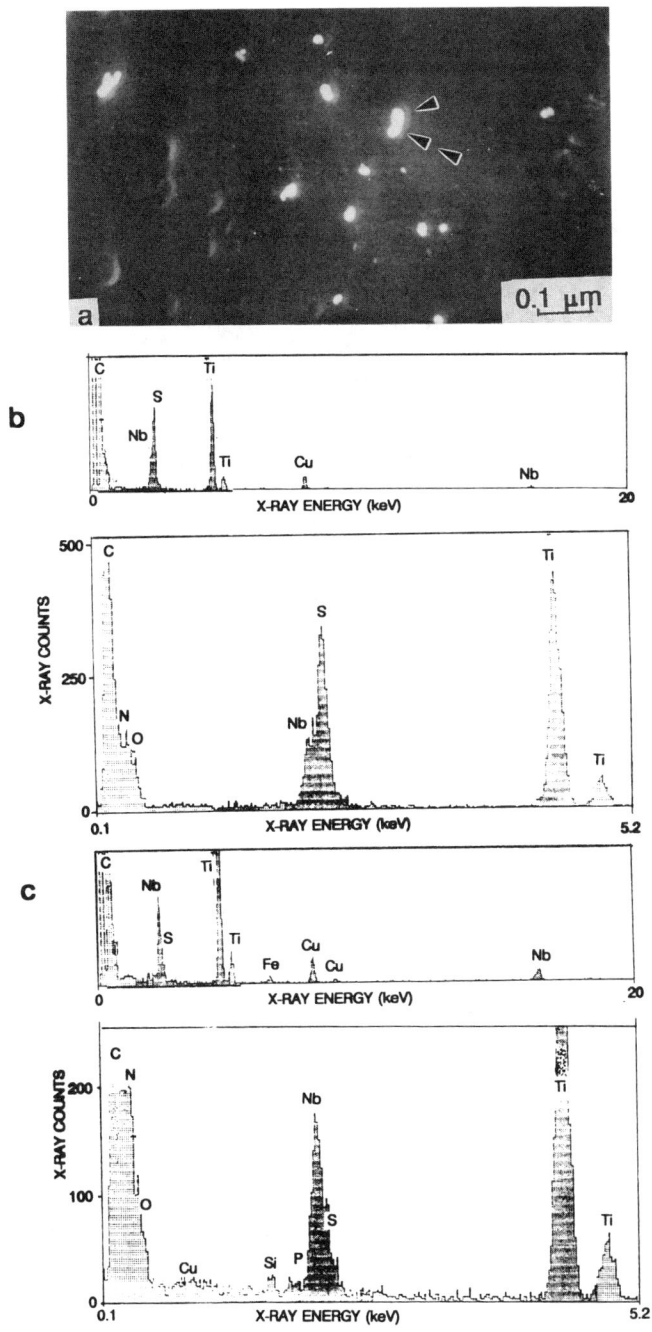

Figure 13 - Fine precipitates with varying S levels in Ti+Nb-stabilized hot band, coiled at 730°C, a) ADF image, b) c) x-ray spectra.

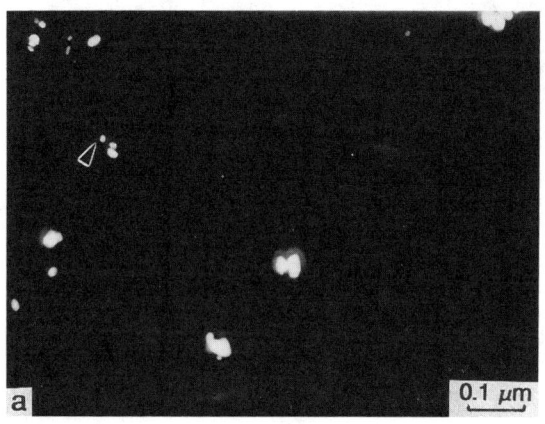
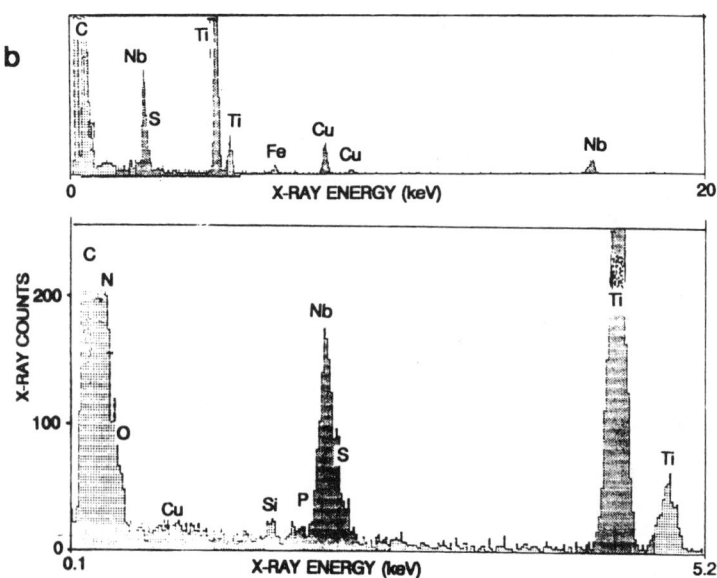

Figure 14 - Ti and S rich precipitates, smaller than 100 Å, in Ti+Nb-stabilized IF hot band coiled at 730°C, a) ADF image, b) x-ray spectrum.

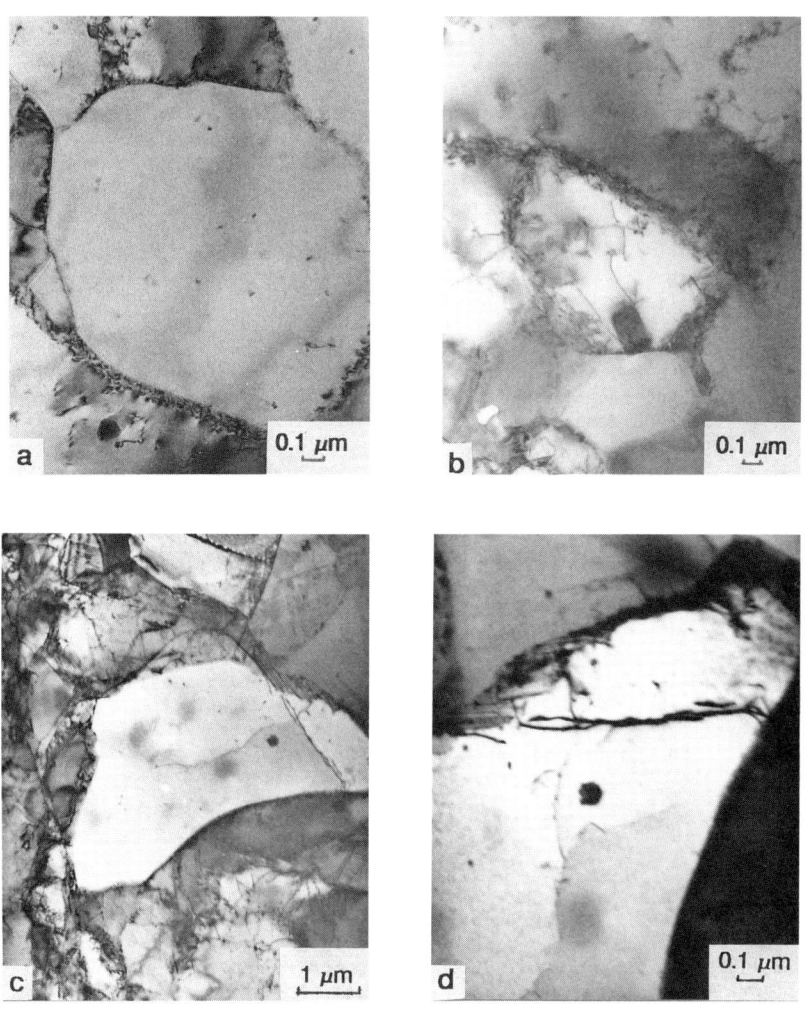

Figure 15 - Partially recrystallized at 700°C Ti+Nb-stabilized IF hot-band coiled art 730°C, a) fine precipitates distributed throughout the grain, b) recrystallized grain with a large precipitate adjacent to grain boundary c) large, Ti-rich precipitate, area surrounding the large precipitate is depleted of fine precipitates, d) higher magnification of c.

The Nb-stabilized IF steel has a high density of fine precipitates as shown in the partially recrystallized material, Figure 16. There were a few large Ti-rich precipitates, owing to the trace amount of Ti in this steel, but the majority of the precipitates were fine Nb-rich precipitates; typical chemical compositions and distribution of the fine precipitates are shown in Figure 17. The Ti was apparently depleted during the formation of large precipitates at high temperatures, and none of the fine precipitates examined contained Ti.

The relative density of fine precipitates may explain the differences in recrystallization behaviour between the Ti+Nb-stabilized and the Nb-stabilized IF steels. The incubation time for recrystallization of the Ti+Nb stabilized steel increases as the isothermal annealing temperature decreases from 750°C to 650°C. As the isothermal annealing temperature decreases, the critical radius for a stable precipitate is expected to decrease (9). As a result, at the lower temperature, the Ti that has not precipitated as large cubic precipitates, may now form fine precipitates that impede recrystallization. Also, the Ti+Nb-stabilized steel has a higher Nb level than the steel that is solely Nb-stabilized, a factor which may also contribute to the long incubation time for recrystallization for the Ti+Nb-stabilized IF steel.

In order to better delineate differences in recrystallization behaviour, apparent activation energies were calculated as follows: the data from the experimentally derived recrystallization curves were linearized with the Johnson-Mehl-Avrami-Kolomogorov (JMAK) expression for recrystallization:

$$X = 1 - \exp(-bt^k) \tag{1}$$

where X represents the fraction recrystallized, t is time in seconds, b and k are constants describing nucleation and growth. Manipulating this expression yields:

$$\log\ln(1/(1-x)) = \log b - k\log t \tag{2}$$

Plotting $\log\ln(1/(1-x))$ against $\log t$ yields linear plots, as shown in Figure 18, for the unstabilized IF steels. Values of time for a specified amount of recrystallization are then easily interpolated from the linear JMAK plots. Since recrystallization is a thermally activated phenomena, the ferrite recrystallization kinetics were assumed to follow an Arrhenius equation of the form:

$$1/t_R = A_o \exp(-Q/RT) \tag{3}$$

where t_R is the time to reach a specified fraction of recrystallization (in this study, time for 50% recrystallization was chosen) A_o is a preexponential frequency constant, Q is the apparent activation energy for recrystallization, which will subsequently be referred to as activation energy, and R and T have their usual meaning. The slope of an Arrhenius plot (rate versus the inverse of absolute temperature) yields the activation energy, the y-intercept yields the preexponential factor. Arrhenius plots for the series of IF steels coiled at 565°C and 730°C are shown in Figures 19a and 19b respectively. The values for the calculated activation energies and the prexponential factors for the steels examined in this investigation are summarized in Table II, and activation energies determined for low carbon steels in other recrystallization studies are summarized in Table III.

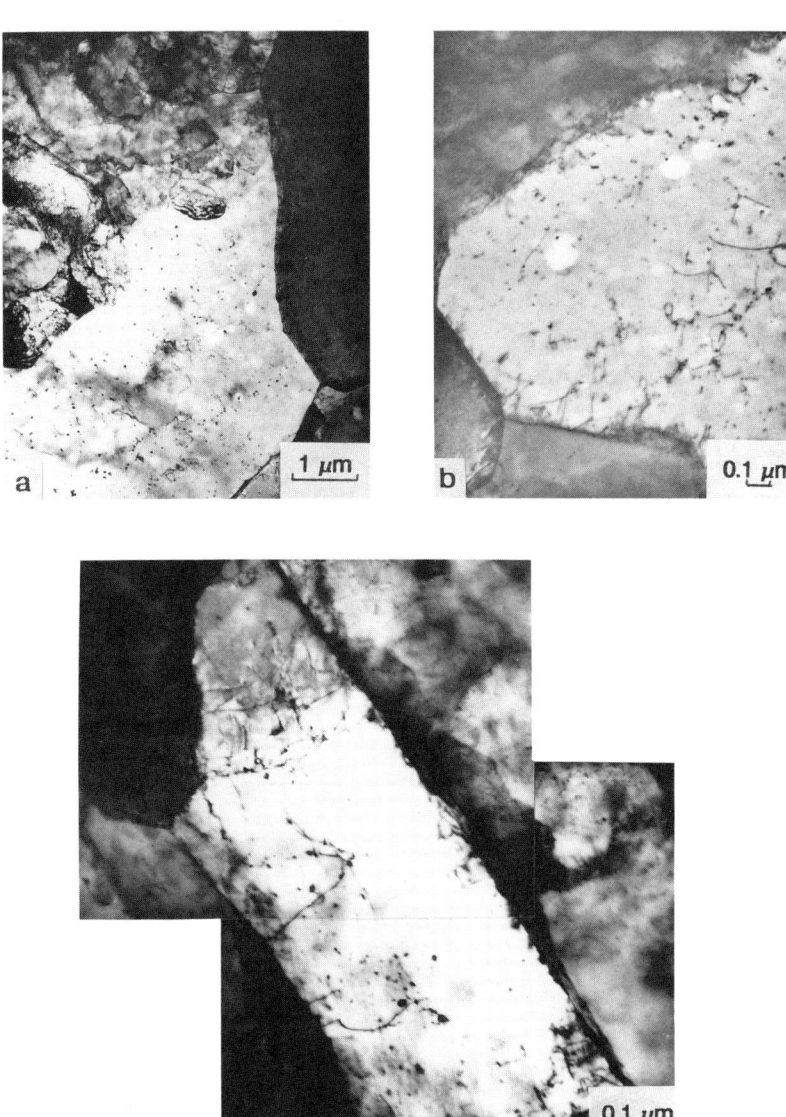

Figure 16 - Partially recrystallized Nb-stabilized IF steel coiled at 730°C, annealed at 700°C a) recrystallized grain growing into cold worked material with discontinuous boundary curvature, b) higher magnification of a, boundary is decorated with fine precipitates. c) Blocky recrystallized grain, high precipitate density. Rolling plane section. TEM bright-field.

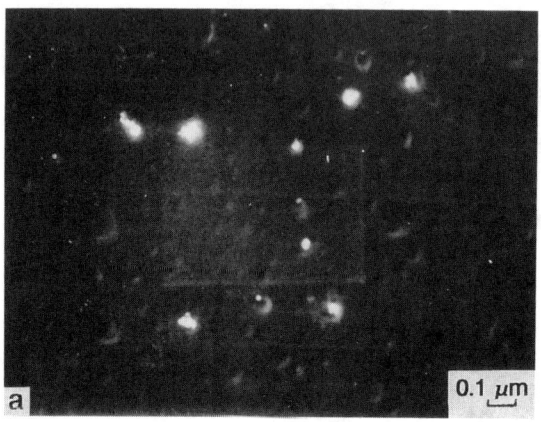

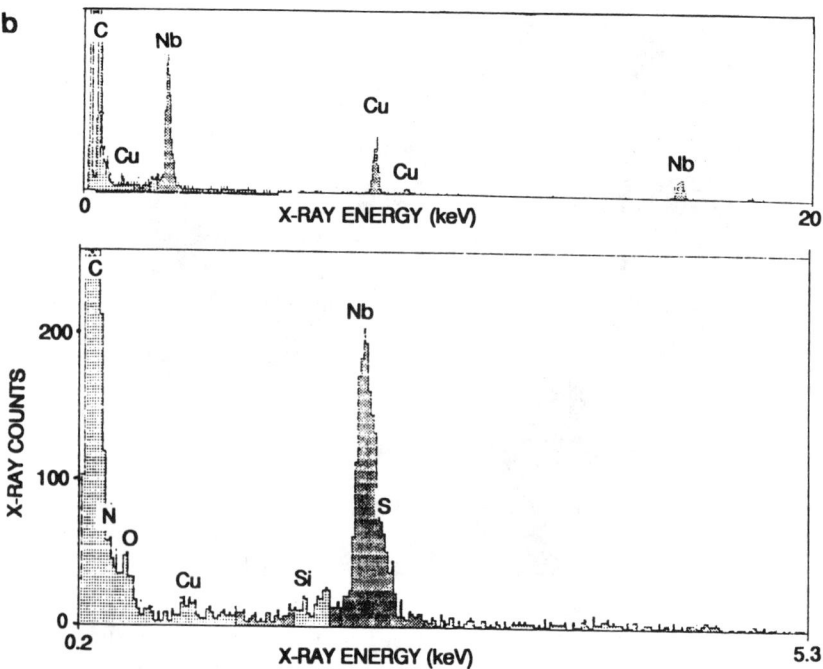

Figure 17 - Typical fine precipitate distribution in Nb-stabilized hot band coiled at 730°C, a) ADF image, b) x-ray spectrum.

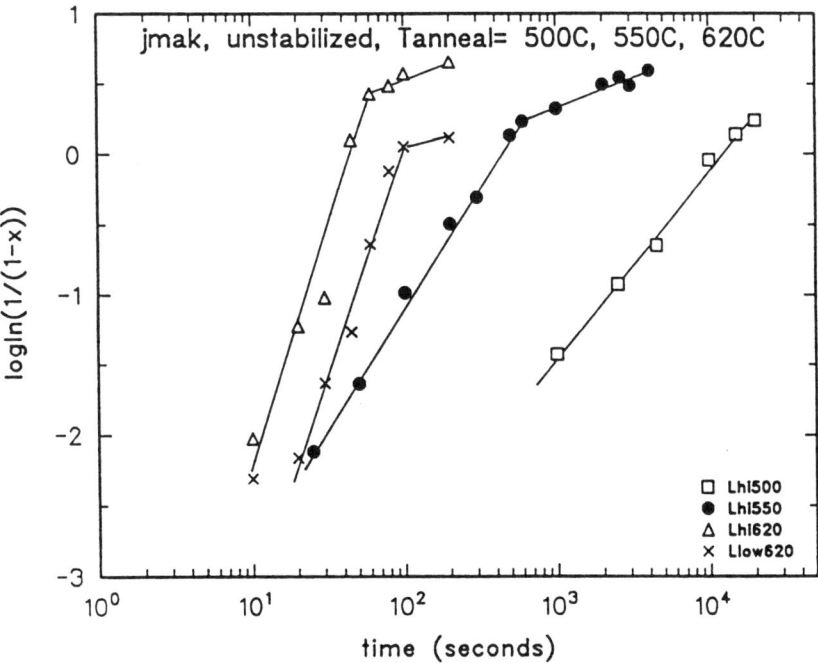

Figure 18 - Johnson-Mehl-Avrami-Kolomogorov plots for unstabilized IF steel cold-rolled 75%, and isothermally annealed at 500°C, 550°C, and 620°C.

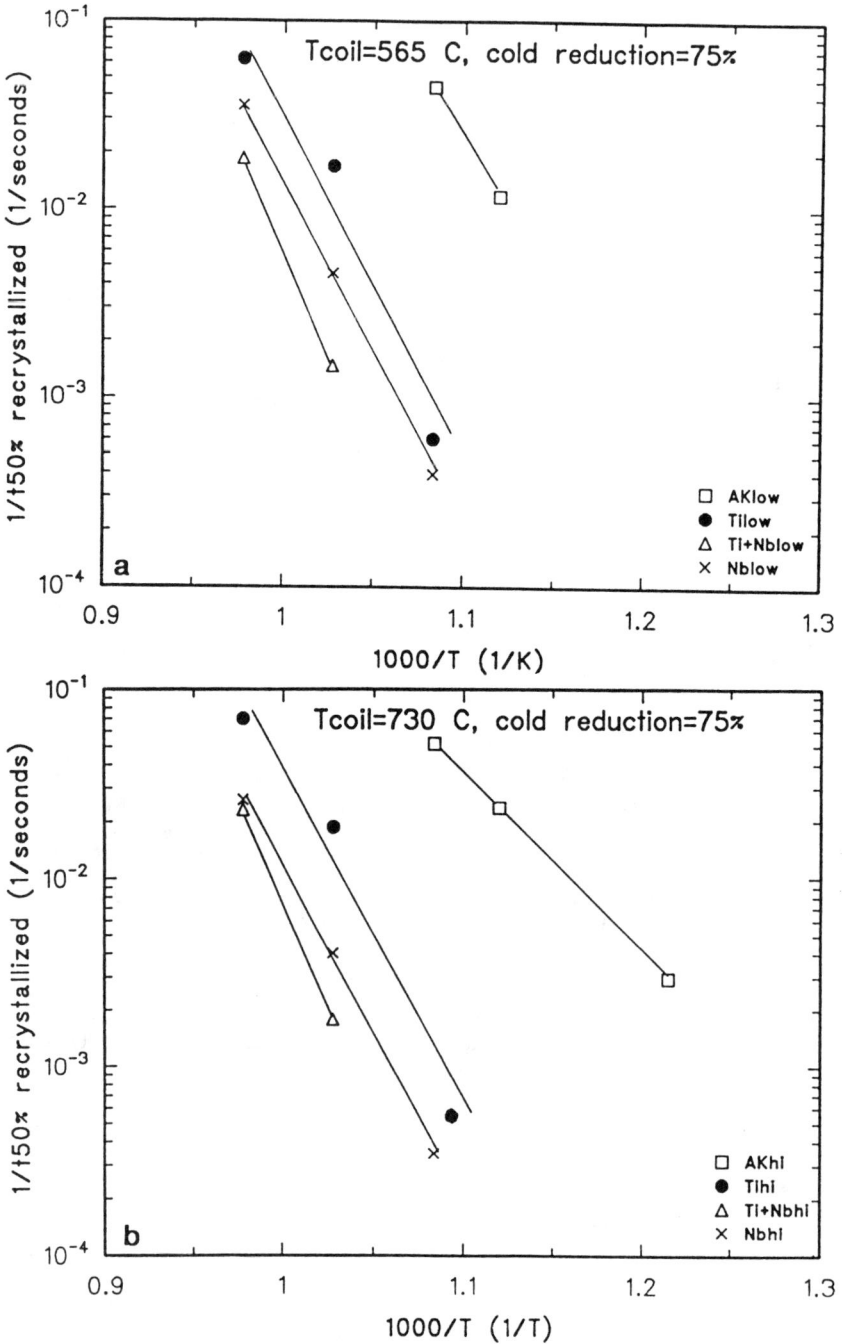

Figure 19 - Arrhenius plots for apparent activation energy for recrystallization of IF steels cold rolled 75%, a) T_{coil}=565°C, b) T_{coil}=730°C.

Table II. Apparent Activation Energies for Recrystallization and Prexponential Factor for Experimental Series of IF Steels of this Study

steel	T_{coil} (°C)	Q kJ/mol	kcal/mol	A_o
unstabilized	565	305	73	7.6×10^{15}
unstabilized	730	180	43	9.3×10^{8}
+ Ti	565	370	88	4.6×10^{17}
+ Ti	730	390	93	6.9×10^{18}
+ (Ti+Nb)	565	420	101	5.6×10^{19}
+ (Ti+Nb)	730	425	102	1.2×10^{20}
+ Nb	565	345	82	1.2×10^{16}
+ Nb	730	340	81	6.0×10^{15}

Table III. Apparent Activation Energies for Recrystallization in Low Carbon Steels

steel	Q kJ/mol	kcal/mol	reference
+ Ti IF	312	77	10
+ (Ti+Nb) IF	419	100	10
+ Nb IF	339	95	10
AK, low carbon	122	29	10
AK, C=0.08%, Mn=1.45%	272	65	11
AK, low carbon, direct cast	324	77	12
Fe + 0.6% Mn	370	88	13

Figures 19a and 19b, and Table II illustrate two points: the activation energy for recrystallization is independent of the coiling temperature for the stabilized IF steels, but is very sensitive to coiling temperature for the unstabilized IF steel. The preexponential factors, A_o, listed in Table II, are almost the same for the stabilized steels at both coiling temperatures, but vary with coiling temperature for the unstabilized IF steels.

In the unstabilized IF hot band, the precipitates were identified as AlN and MnS; however, the precipitation of AlN depends on hot mill coiling temperature. A high coiling temperature will allow AlN precipitation in the hot mill, as seen in this work, and reported elsewhere, whereas a low coiling temperature will suppress AlN precipitation (14). Therefore, in the unstabilized IF steel, the different coiling temperatures should result in a difference in fine precipitate density and, correspondingly, a difference in matrix solute level. The difference in solute level in the unstabilized IF steels coiled at 565°C and 730°C is reflected in the different values of activation energy: 73 kcal/mol at a low coiling temperature, and 43 kcal/mol at a high coiling temperature, where AlN was able to precipitate. That is, higher solute levels in unstabilized IF steels are reflected in higher values for the activation energy of recrystallization.

The stabilized IF steels all had stabilization ratios ((Ti^*+Nb)/C) greater than 1, and thus the excess stabilizing elements must then be present as solute in α-iron. The values for recrystallization activation energy for the stabilized IF steels range from 81 to 101 kcal/mol; this range compares well to the values generated by others for IF steels (10). The values for the activation energy of recrystallization for the stabilized IF steels are greater than the activation energy for self-diffusion of α-iron, 60kcal/mol (15).

The basis for the higher activation energies in the IF steels can be understood by considering the fundamentals of thermodynamics and kinetics as applied to diffusion. The activation energy for diffusion, Q_{diff}, by the vacancy mechanism consists of both the enthalpy of vacancy formation, ΔH_v, and the enthalpy of motion, ΔH_m, and is given as (16):

$$Q_{diff} = \Delta H_v + \Delta H_m \qquad (4)$$

and the diffusivity, D, is expressed as:

$$D = D_o \exp(-Q_{diff}/RT) \qquad (5)$$

It has been shown(16) that ΔH_m increases as the solute concentration increases. The range of values for the activation energy of recrystallization obtained in this investigation is therefore consistent with the activation energy of self-diffusion for α-iron corrected for solute.

The rationale for relating the activation energy for recrystallization to that for self diffusion of iron can described with the aid of Figure 20, a series of schematic drawings which illustrate the effects of both solute and precipitation on the recrystallization behaviour of cold-worked iron. In each figure a recrystallized grain, indicated by the light area, is shown growing into a heavily cold-worked region, indicated by the crosshatched area. In Figure 20a, recrystallization in pure iron is shown; this process is achieved by the short range diffusion of iron atoms from the unrecrystallized portion to the recrystallized portion. There are no solute atoms or precipitate particles to impede the progression of recrystallization, and the activation energy for recrystallization can be taken to be the same as that for the self diffusion of α-iron.

The effects of solid solution elements, without the presence of precipitates is illustrated in Figure 20b. Recrystallization occurs by the short range diffusion of iron which, however, is impeded by the presence of

a Q_{rextl}
   pure Fe

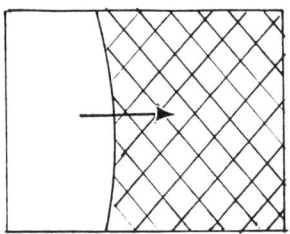

$Q_{rextl} = Q_{D\alpha}$

b Q_{rextl}
   Fe + solute

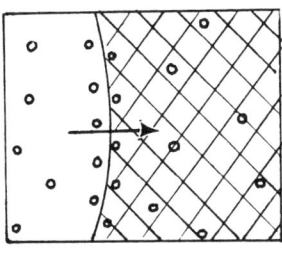

$Q_{rextl} = Q_{D\alpha}$
$Q_{D\alpha} = f(solute)$

c Q_{rextl}
   Fe + precipitate

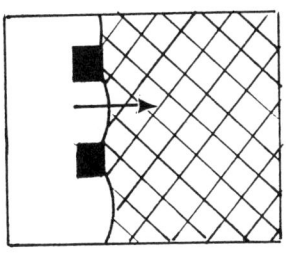

$Q_{rextl} = Q_{D\alpha} + f(ppt)$

d Q_{rextl}
   Fe + precipitate
   + solute

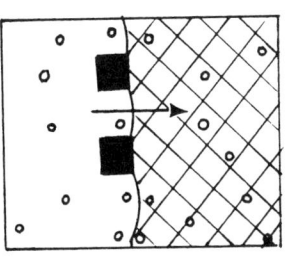

$Q_{rextl} = Q_{D\alpha} + f(ppt)$
$Q_{D\alpha} = f(solute)$

Figure 20 - Schematic representation of recrystallization in a) pure iron, b) iron containing solute, c) pure iron containing fine precipitates, and d) iron containing solute and fine precipitates.

solute atoms (17). In this instance, the activation energy for recrystallization must be corrected for the presence of the solute, which results in a value that exceeds that for the self diffusion of pure α-iron. Figure 20c schematically illustrates the recrystallization of iron containing equilibrium precipitate particles for which there is no excess solute present in the matrix. The fine precipitate distribution may be reflected in activation energies for recrystallization as follows. To consider the interaction between recrystallization and precipitation (18), one must consider the force of recrystallization, F_R, which is the force per unit area related to minimizing the strain energy of cold work, given as:

$$F_R = (\mu b^2/2)(\Delta\rho) \qquad (6)$$

where μ is the shear modulus, b is the burger's vector, and $\Delta\rho$ is the change in dislocation density between the cold-worked and recrystallized material. The restraining force of spherical particles on boundary motion can be written as (18):

$$F_p = 6\gamma f/\pi r \qquad (7)$$

where f is the volume fraction of precipitates, γ is the interfacial energy per unit area of boundary, and r is the particle radius. Boundaries will be pinned when F_p is greater than F_R, but not when the opposite is true. If F_p is subtracted from F_R, then multiplied by a characteristic length, for example b, then a net energy for recrystallization corrected for the pinning force of precipitates, U_{R-net}, term can be written as:

$$U_{R-net} = [(\mu b^2/2)(\Delta\rho) - 6\gamma f/\pi r]b \qquad (8)$$

The term U_{r-net} can be considered as the driving force for recrystallization which encompasses the amount of cold-reduction and the precipitate distribution. However, it has been shown that a variation in amount of cold-reduction results in variation in the activation energy for recrystallization (19). Therefore, it is not unreasonable to assume that the value of U_{r-net} will affect the value of the activation energy for recrystallization.

Finally, Figure 20d schematically illustrates the recrystallization of iron containing both solute atoms and fine precipitates; this most accurately reflects the systems investigated in this study. Here, the activation energy for recrystallization includes both contributions of solute atoms and fine precipitates. As a result, the activation energy for recrystallization in the stabilized IF steels should be higher than for pure α-iron, as observed in this study.

Conclusions

1. The recrystallization rates of interstitial-free sheet steels are severely retarded by the stabilizing additions; Nb appears to have the greatest effect.

2. Titanium also retards the recrystallization rate; however, a relatively high titanium content, 0.089 wt pct, leads to high temperature precipitation and the formation of coarse, greater than 1 μm, Ti-rich precipitates, which act as sites of preferential recrystallization.

3. Fine precipitates, present in the hot band prior to recrystallization, were identified as AlN and MnS in the unstabilized steel, Ti compounds, Nb compounds and Ti+Nb compounds in the Ti-stabilized, Nb-stabilized and Ti+Nb-stabilized IF steels respectively. All the Ti-bearing IF steels contained large Ti-rich precipitates.

4. The kinetics of recrystallization of stabilized IF steels during isothermal heat treatments at temperatures ranging between 650°C and 750°C, are independent of coiling temperature.

5. The activation energies of recrystallization for the stabilized interstitial-free steels exceed that for the self-diffusion of α-iron; the increase is attributed to solute elements which modify the self diffusivity of iron. The fine precipitates retard high angle boundary migration during recrystallization and thereby decrease the driving force for recrystallization.

6. The recrystallization kinetics of the unstabilized steels varies as a function of coiling temperature; this is related to the difference in AlN precipitation at the two different coiling temperatures.

Acknowledgements

The authors thank the Bethlehem Steel Company for providing the experimental series of IF steels, especially Dr. W. Furdanowicz for the chemical microanalysis of the precipitates on Bethlehem's dedicated STEM, and Ms. P. Iwamasa, formerly a student at CSM and presently at Carnegie Mellon University, for generating samples for the recrystallization curves for the unstabilized and Ti-stabilized IF steel coiled at 565°C and isothermally annealed at 700°C. This work was supported by the Advanced Steel Processing and Products Research Centre, at the Colorado School of Mines, a NSF joint university-industry cooperative research centre.

References

1. Metallurgy of Continuous-Annealed Sheet Steel, B.L. Bramfitt and P.C. Mangonon, Jr. editors, (TMS-AIME, Warrendale, PA, 1982).

2. Technology of Continuously Annealed Sheet Steel, R. Pradhan, editor, (TMS-AIME, Warrendale, PA,1985).

3. Metallurgy of Vacuum Degassed Products, R. Pradhan, editor, (TMS-AIME, Warrendale, PA, 1990).

4. Interstitial Free Steel Sheet: Processing, Fabrication and Properties, L.E. Collins and D.L. Baragar, editors, (CIM, Ottawa, Canada, 1991).

5. I. Gupta, and D. Bhattacharya, "Metallurgy of Formable Vacuum Degassed Interstitial Free Steels", in reference 3, 43-72.

6. D.O. Wilshynsky, D.K. Matlock, and G. Krauss, "Microstructure and Mechanical Properties of Interstitial-Free Sheet Steels with Different Stabilizing Additions" in reference 3, 247-261.

7. P. Shewmon, Transformation in Metals (New York, NY, McGraw-Hill Publishing Company, 1969), 69-138.

8. W.C. Leslie, Physical Metallurgy of Steels (McGraw Hill International Student Edition), 247-261.

9. D.A. Porter and K.E. Easterling, Phase Transformations in Metals and Alloys (New York, NY, Van Nostrand Reinhold Company, 1981), 263-271.

10. W. Bleck, R. Bode, F-J. Hahn, "Interstitial-Free Steels:Processing, Properties and Applications" in reference 4, pp. 73-90.

11. D.Z. Wang, E.L. Brown, D.K. Matlock, and G. Krauss, "Ferrite Recrystallization and Austenite Formation in Cold-Rolled Intercritically Annealed Steel", Met. Trans. A, 16A, (August 1985) 1385-1392.

12. J.F. Bingert, "Cold Rolling, Annealing and Texture of Direct-Cast Strip Steel" (M.S. Thesis Colorado School of Mines, September 1990), 76.

13. W.C. Leslie, F.J. Plecity, and J.T. Michalak, "Recrystallization of Iron and Iron-Manganese Alloys", Trans. Met. Soc. of AIME, 221 (August)(1961), 691-700.

14. W.B. Hutchinson, "Development and Control of Annealing Textures in Low Carbon Steels", International Metals Review, 29, (1)(1984) 25-42.

15. H. Hu, "Annealing of Silicon-Iron Single Crystals" in Recovery and Recrystallization of Metals, ed. L. Himmel, (Interscience Publishers, 1963), 311-361.

16. P. Shewmon, Diffusion in Solids (J. Williams Book Company, Oklahoma, 1983) 44-45, 57-61.

17. P. Cotterrill and P.R. Mould, Recrystallization and Grain Growth in Metals (John Wiley and Sons, Great Britain, 1976) 101.

18. S.S. Hansen, J.B. Vandersande, and M. Cohen, "Niobium Carbonitride Precipitation and Austenite Recrystallization in Hot-Rolled Microalloyed Steels", Met. Trans. A 11A, (3)(1980) 387-402.

19. R.E. Reed-Hill, Physical Metallurgy Principles, 2nd edition (Van Nostrand Company, Toronto, 1973) 289-290.

EFFECT OF PRECIPITATE SIZE AND DISPERSION ON LANKFORD VALUES OF Ti STABILIZED INTERSTITIAL-FREE STEELS

S.V. Subramanian and M. Prikryl
Dept. of Materials Science and Eng., McMaster University, Hamilton, Ontario
B.D. Gaulin
Department of Physics, McMaster University, Hamilton, Ontario
M. Koch and S. Benincasa
Dofasco Inc., Hamilton, Ontario

Abstract

The deep drawing characteristics of interstitial free steels stabilised with Ti, Nb or Ti + Nb are critically dependent upon the promotion of the favourable texture {111} in the recrystallisation process during the continuous annealing cycle. The control of the size and dispersion of the precipitates, the residual solute content in the matrix and the grain size are important factors that influence the deep drawing property of the final sheet product. The objective of the work is to develop a quantitative database on the size, dispersion and microchemistry of two hot dip galvanised Ti stabilised interstitial free steels of comparable chemistry and grain size but with a distinct difference in Lankford values. Small angle neutron scattering (SANS) technique was used complementary to electron optical techniques (TEM/STEM) to resolve the ultrafine precipitates (size <10 nm). A dense dispersion of fine precipitates decreases the deep drawing property of the steel. The effect of ultra fine precipitates on the retardation of recrystallisation growth through particle pinning effect on the grain boundary will be discussed.

Background

The deep drawing characteristics of an interstitial free sheet steel, as measured by Lankford value from a tensile test, are critically dependent upon the development of strong {111} recrystallisation texture during in-line annealing (1,2,3). Prior studies on the hot band have shown that a fine grain size and an interstitial-free or "pure" iron matrix brought about by gettering the interstitials as a sparse dispersion of coarse precipitates using Ti or Nb are important requisites to promote {111} recrystallization texture (4-9). In the process of removing interstitials using microalloying elements, excess solute concentrations of microalloying elements could be left over in the matrix that may adversely affect the recrystallization process (10). However, excess solute titanium in the matrix is far less effective than niobium in retarding the recrystallization process during in-line annealing (11). This characteristic of Ti is considered a distinct advantage in the design and processing of Ti stabilised interstitial free steel for extra deep drawing grades.

The size and dispersion of the precipitates in a Ti stabilised interstitial free steel can interact with the recrystallization and texture evolution. It is believed that a dense dispersion of fine precipitates does not impair the recovery and nucleation of recrystallised grains with {111} orientation. However, the growth of these grains is retarded by the pinning force exerted by the particles on the grain boundary(12). The pinning force is directly proportional to the volume fraction of the precipitates but is inversely related to the average precipitate size. With the retardation in the growth of grains with {111} orientation, competitive growth of grains with other orientations, notably {100} sets in, resulting in a decrease in Lankford value (6). A Ti stabilised interstitial free steel with the same base chemistry can yield different Lankford values under varying mill processing conditions because of the differences, in the size and dispersion of the precipitates as well as the grain size.

Objective

The objective of this work is to quantify the effect of precipitate size and dispersion on Lankford values of Ti stabilised interstitial free steels.

Experimental

The experimental work consisted of developing a quantitative data-base on the size, dispersion and microchemistry of precipitates in two hot dip galvanised Ti stabilised interstitial free steels of comparable chemistry and grain size but with a distinct difference in deep drawing property as measured by r bar values. The precipitates were extracted on carbon replicas and examined in Phillips CM-12 TEM/STEM. The microchemistries of the precipitates were determined in the dedicated STEM-VG HB5.

Small Angle Neutron Scattering (SANS) technique was used complementary to electron optical techniques (TEM/STEM) to resolve ultrafine precipitates (size <10 nm in diameter) in bulk specimens.

Texture Measurements

Texture measurements were carried out at Dofasco Research Laboratories using a Rigaku RU300 unit. The Zr filtered Mo radiation was used. The measurements were carried out with an alpha step size of 5 degrees. The beta step size was kept at 5 degrees over the range from 0 to 360 degrees.

Texture measurements were made at the surface and in the mid plane of the steel sheets. The samples were cut into 31.25mm dia discs, which were then machined, polished and etched in 2% Nital solution. The sample was then placed in the goniometer and (110), (200) and (211) pole figures were measured. The intensity measurements were corrected for background and normalised to units of "times random". The pole figures were rotated, smoothened and symmetrized. The pole figures were used to calculate the Orientation Distribution Function (ODF).

Results

The base chemistry of the two Ti stabilised interstitial free steels H1 and H7 used in this investigation are summarised in Table 1 and their mechanical properties are given in Table 2. The Lankford values (r) were determined after stripping the coating. The two steels exhibit comparable grain size (11.4 - 11.5 µm) but distinctly different deep drawing properties. Steel-H1 exhibits a r value of 2.12, whereas steel-H7 gave a rather low r value of 1.69.

Table I

Sample I.D.	Base Chemistry (w%)								
	C	N	Ti	Nb	S	Mn	P	Si	Al
H1	0.004	0.0022	0.078	0.003	0.010	0.16	0.010	0.008	0.043
H7	0.0042	0.0017	0.068	0.004	0.013	0.15	0.007	0.010	0.043

Table II

Sample I.D.	Mechanical Properties							
	Thickness (mm)	Y.S. (KSI)	T.S. (KSI)	El (%)	n	r	Zinc Coating (gm/m^2)	Grain Size (µm)
H1 (785430-B-L)	0.78	25.4	44.7	47.9	0.23	2.12	139	11.5
H7(784358-L)	0.99	25.9	44.9	43.8	0.22	1.69	95	11.4

Thermodynamic Analysis of Precipitation Behaviour

The equilibrium precipitation behaviour of Steel H-1 and Steel H-7 are analysed quantitatively using binary solubility product data for TiN (13), TiS (14) and TiC (15) in austenitic phase. The precipitate mole fraction is plotted as a function of temperature for Steel-H1 and Steel-H7 and the results are given in Figures 1 and 2, respectively. Using a regular solution parameter of -35 kJ/mole of Ti,

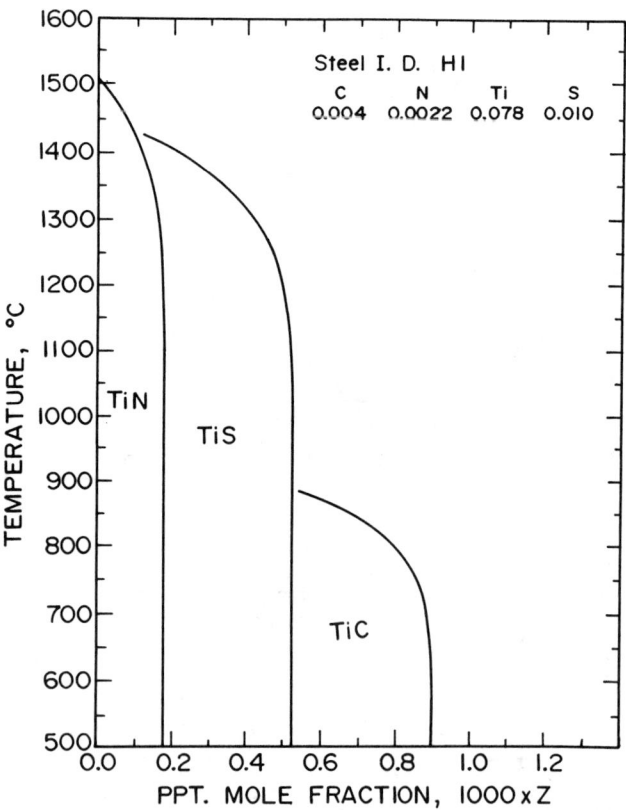

Figure 1 Equilibrium precipitate mole fraction as a function of temperature using binary solubility data for steel H1.

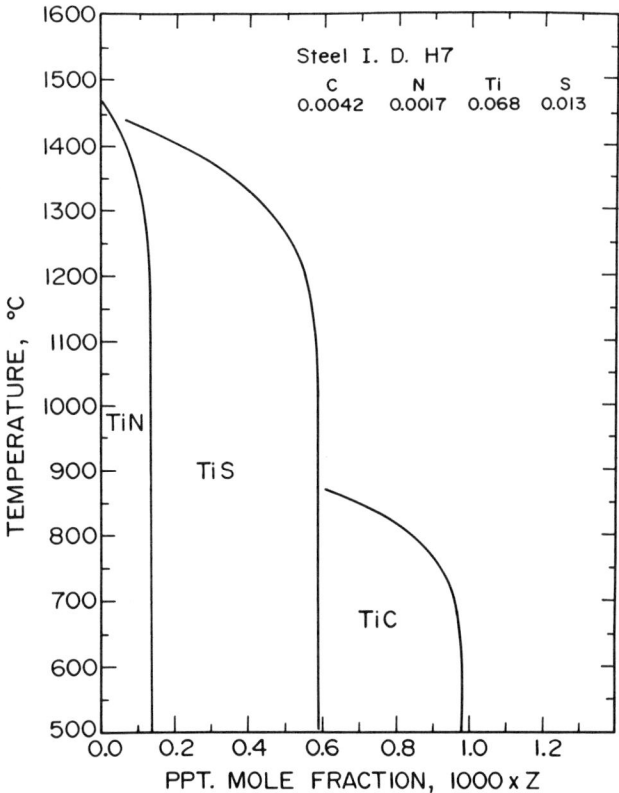

Figure 2   Equilibrium precipitate mole fraction as a function of temperature using binary solubility data for steel H7.

equilibrium precipitation of $Ti_4C_2S_2$ is calculated for each case and the results are superimposed on the precipitation of binary compounds for Steel H1 and H7 in Figures 3 and 4 respectively.

The precipitate dissolution temperatures for each steel are summarised below

Steel	TiN	TiS	$Ti_4C_2S_2$	TiC
H1	1508°C	1401°C	1280°C	887°C
H7	1470°C	1412°C	1283°C	874°C

In terms of sequence of precipitation and cumulative equilibrium mole fraction of precipitates, Steels H1 and H7 are comparable in their equilibrium precipitation behaviour. As the solubility data for TiC in ferrite is not well established, the precipitation in ferrite phase is not included in the precipitation diagram. However, it must be noted that the bulk of the precipitation of TiC occurs in ferrite.

Particle size and microchemistry:

TEM/STEM observations on precipitates extracted by carbon replicas from Steel H1 and Steel H7 showed a lot of features common to both steels. These results are summarised below:

Microchemistry of the precipitates	Size Ranges
TiN	>1 μm
TiS occasionally with MnS encapsulation	500 nm - 1 μm
$Ti_4C_2S_2$	30 - 50 nm
TiC	50 nm at subboundaries, 10-20 nm within the grain

The typical STEM results are given in Figures 5-9. Figure 5 shows a STEM image and X-ray spectra of a coarse TiS/MnS precipitate in Steel H1. Figure 6 is a STEM image of finer precipitates in Steel H1; the precipitates are inferred to be TiC from the X-ray spectrum. Figure 7 is a STEM image of typical precipitates in Steel H7; the X-ray spectra of a coarse TiS precipitate (A) and a fine TiC precipitate (B) are shown. Figure 8 shows a STEM image and the X-ray spectra of a row of precipitates in steel H7. Some of the precipitates showed an intensity ratio of Ti/S of 2, corresponding to $Ti_4C_2S_2$; some of the precipitates exhibited X-ray spectrum of Ti only, which is inferred to be TiC. Figure 9 shows two STEM images, one at low (5k) and the other at high (200k) magnification. The low magnification image

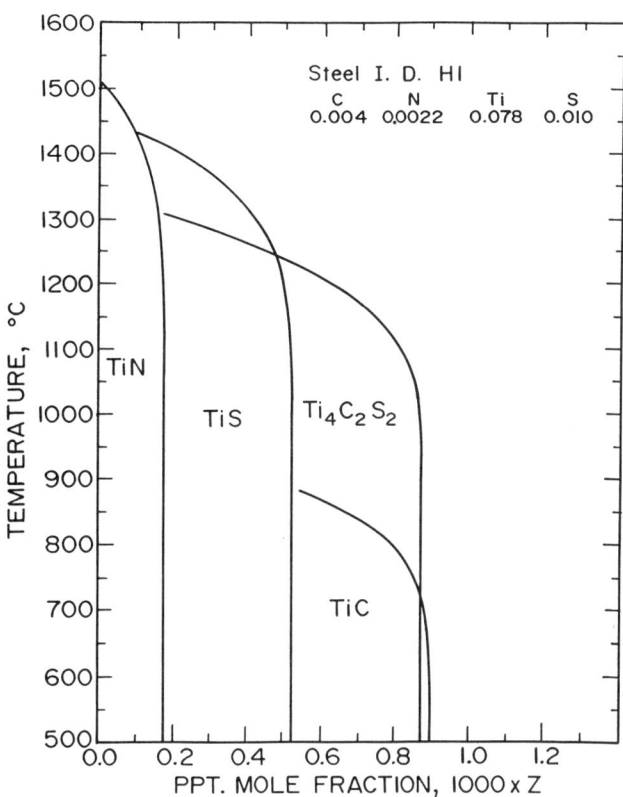

Figure 3 Equilibrium precipitate mole fraction as a function of temperature modified to include $Ti_4C_2S_2$ precipitation for steel H1.

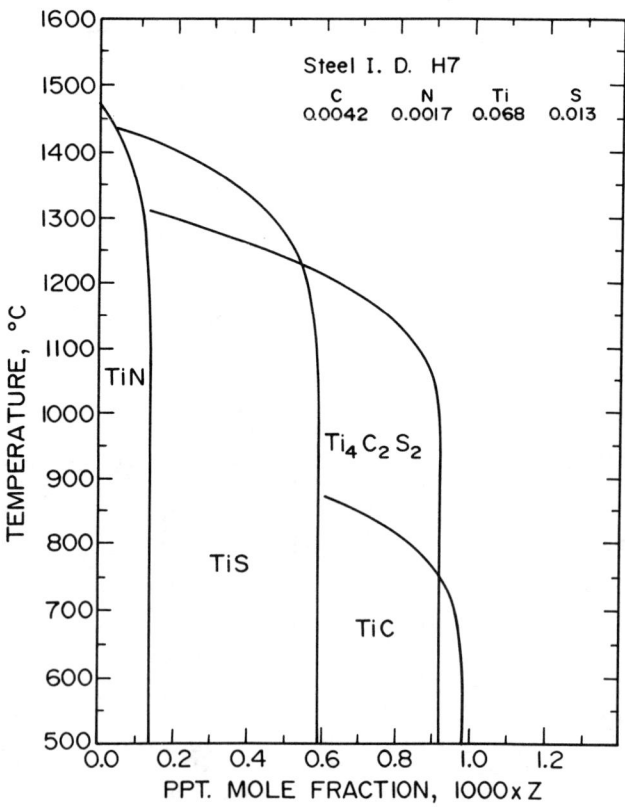

Figure 4 Equilibrium precipitate mole fraction as a function of temperature modified to include $Ti_4C_2S_2$ precipitation for steel H7.

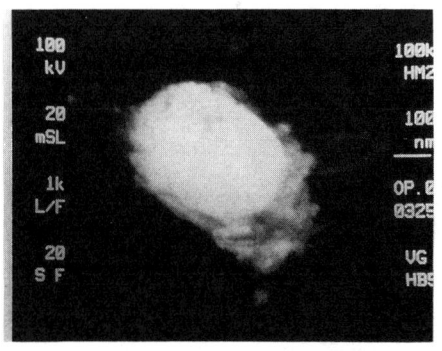

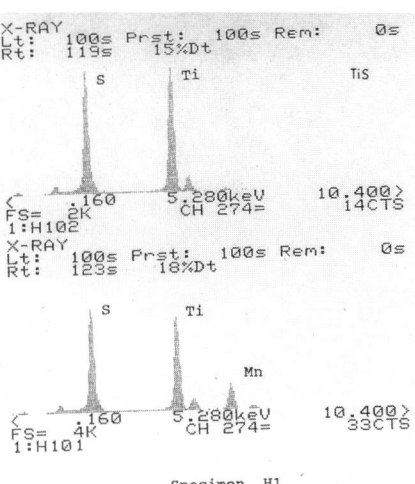

Figure 5 STEM image and X-ray spectra of a coarse TiS/MnS precipitate in steel H1. Carbon extraction replica on Al grid.

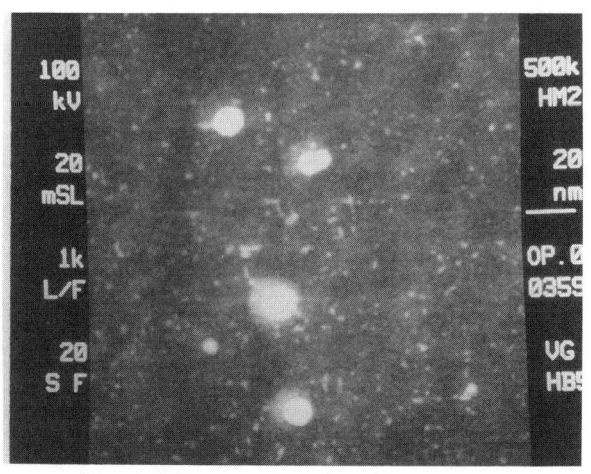

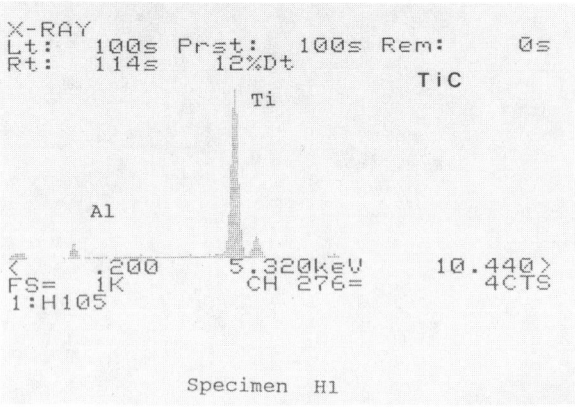

Figure 6   STEM image and X-ray spectrum of fine TiC precipitates in Steel H1. Carbon extraction replica on Al grid.

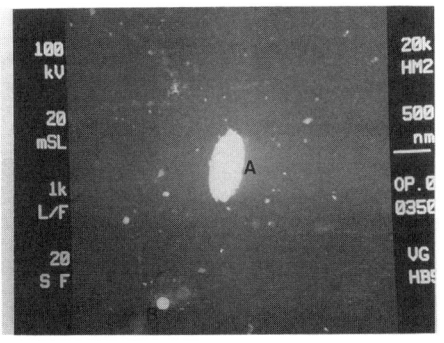
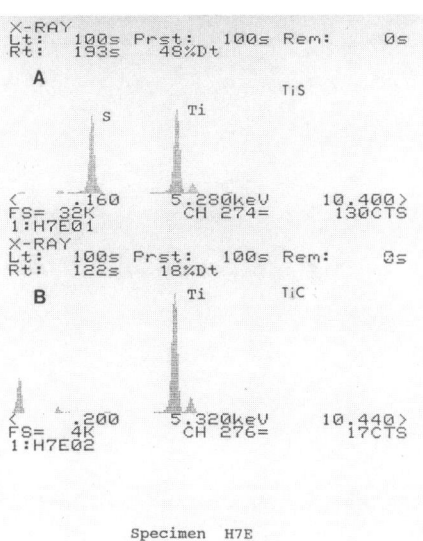

Figure 7   STEM image and X-ray spectrum of a coarse TiS precipitate (A) and a TiC precipitate (B) in steel H7. Carbon extraction replica on Al grid.

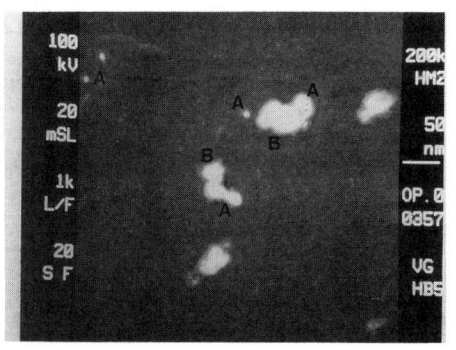
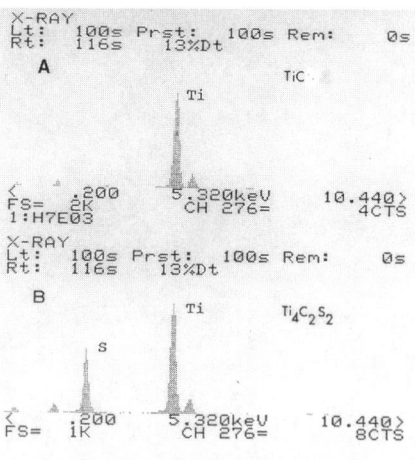

Figure 8   STEM image and X-ray spectrum of a row of precipitates in steel H7. Some of the precipitates were analyzed as TiC (A), and some as $Ti_4C_2S_2$ (B). Carbon extraction replica on Al grid.

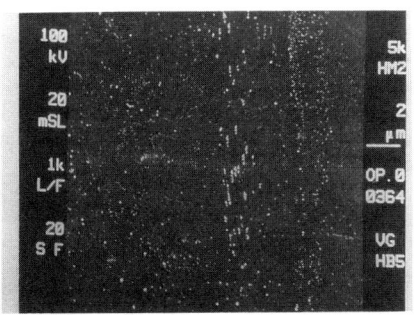

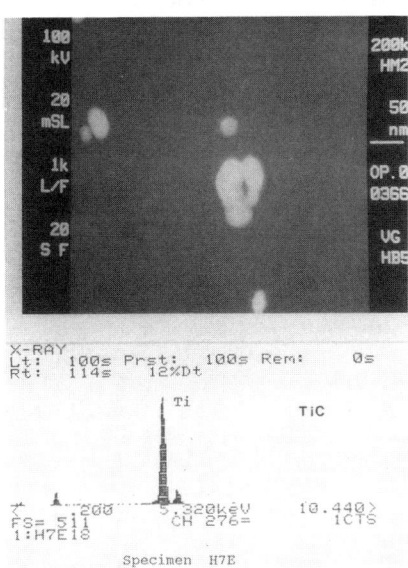

Figure 9   Two STEM images and X-ray spectrum of precipitates in steel H7. Low magnification image shows fine precipitate dispersion within the ferritic grain and a band of coarser precipitates along a grain boundary. Both fine and coarser precipitates were analyzed as TiC. Carbon extraction replica on Al grid.

shows a dense dispersion of fine precipitates within the ferritic grain and a band of coarser precipitates along a grain boundary. Both the precipitates gave X-ray signals characteristic of Ti and the precipitates are inferred to be TiC.

Size and Dispersion of the Precipitates

Figures 10A and 11A show TEM pictures of precipitates extracted on carbon replicas from the surface of Steel Sheet H1 and H7 respectively. The size and dispersion of the precipitates obtained by image analysis in each case are shown in Fig. 10B and 11B respectively. The precipitates at the surface of steel sheet H1 exhibit a sparser distribution of coarser precipitates compared to those of Steel H7. The mean idealised diameter of the precipitates of steel H1 is 7.5 nm, whereas that of steel H7 is 4.8 nm.

Figures 12 and 13 show TEM pictures of precipitates extracted on carbon replicas from the midplane sections of H1 and H7 steel sheets respectively. Clearly the midplane of Steel H1 exhibits a sparse dispersion of coarse precipitates, whereas Steel H7 midplane shows a dense dispersion of fine precipitates. A comparison of surface and midplane specimens of each steel shows that the precipitates are uniformly coarse at the surface and in the midplane of steel H1. The precipitates in Steel H7 are finer than those of H1, and the precipitates in the midplane of H7 are even finer in size and denser in dispersion than those at the surface of H7. Thus there is a variation in the size and dispersion of precipitates in Steel H7, the midsection plane exhibiting a denser dispersion of finer precipitates than the surface of the specimen.

Results From Small Angle Neutron Scattering (SANS)

Small angle neutron scattering is a diffraction technique, which measures fluctuations in the scattering density of the matrix caused by precipitates or voids of rather small size (<35 nm in diameter) from a truly bulk specimen (up to 20mm thick).

The neutron scattering intensity I is plotted as a function of scattering vector Q for the control (C) and heat-treated specimen (H) of each steel. The results for steel H1 and H7 are given in Figures 14 and 15 respectively. The neutron scattering intensity of steel H7 (with $r=1.69$) is greater than that of steel H1 (with $r=2.12$), suggesting that the volume fraction of the ultrafine precipitates of steel H7 is more than that of steel H1. In order to confirm the validity of the technique, each steel was heat treated to decrease the volume fraction of the precipitates. The scattering intensity is found to decrease with the decreased volume fraction of the precipitates in both the steels. The heat treatment consisted of holding for 1 hour at 950°C followed by quenching.

Figures 16 and 17 show plots of log I (Intensity) against Q^2 (scattering vector) for steel H1 and H7 respectively. Assuming that the scattering is from a distribution of spherical precipitates of a mean diameter, and that the volume fraction is low such that the interference between neighbouring particles is negligible, the mean particle size can be calculated. The mean precipitate diameter for steel H7 is about 80Å compared with 90Å for steel H1. As a result of heat treatment, the calculated mean precipitate size increases to 98Å in the case of steel H7 and 94Å for steel H1.

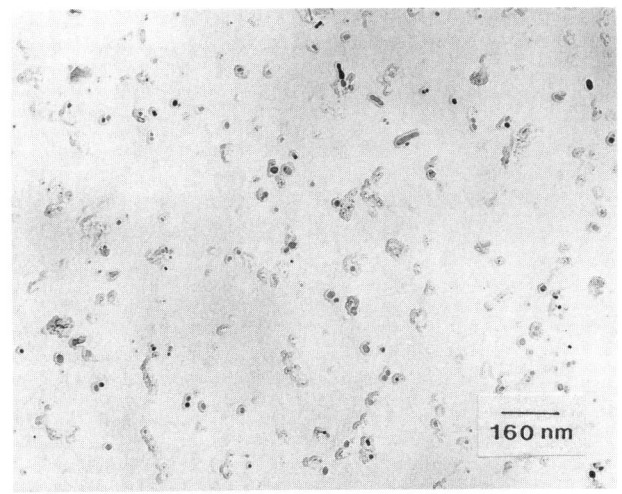

Figure 10A TEM image showing fine TiC precipitates within the ferritic matrix in steel H1 surface. Carbon extraction replica. 62000 x.

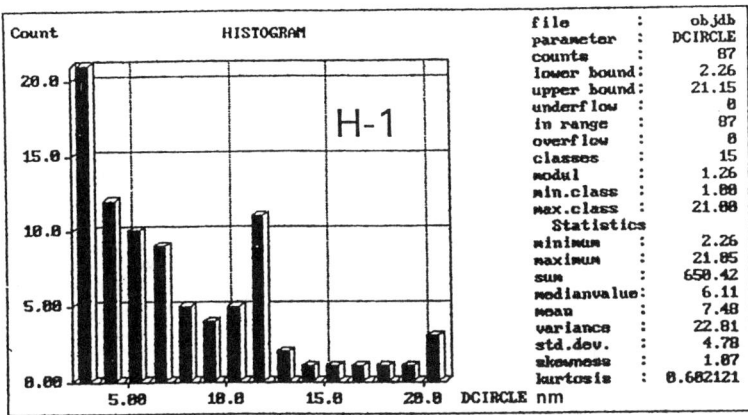

Figure 10B Idealised diameter of precipitate versus count from image analysis of Figure 10A.

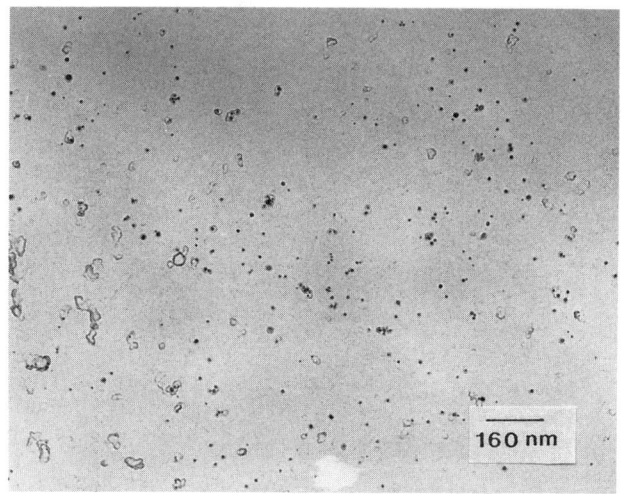

Figure 11A TEM image showing fine TiC precipitates within the ferritic matrix in steel H7 surface. Carbon extraction replica 62000 x.

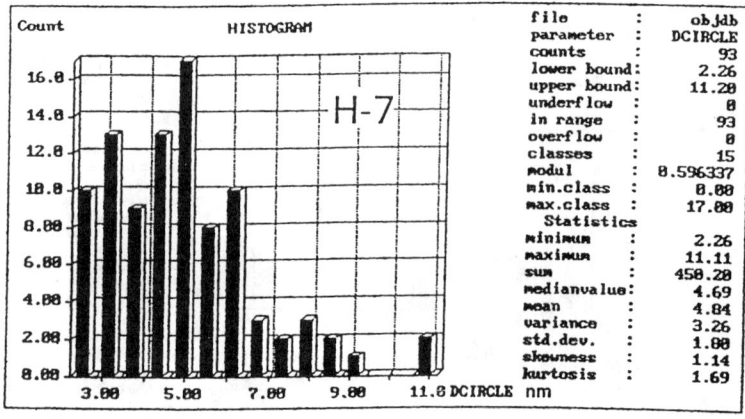

Figure 11B Idealised diameter of precipitate versus count from image analysis of Figure 11A.

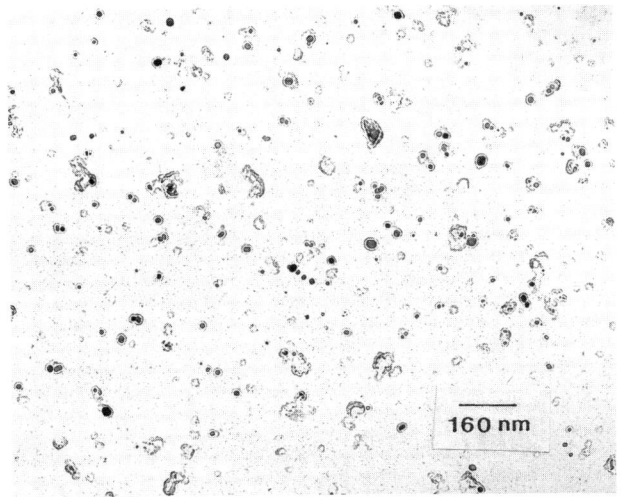

Figure 12   TEM image showing a sparse dispersion of coarse precipitates of TiC occurring in the midsection of Steel sheet H1. The precipitates are comparable in size and dispersion to those occurring on the surface of sheet H1.

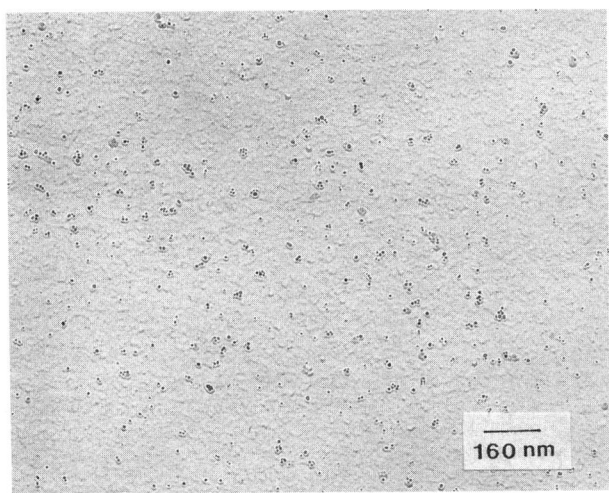

Figure 13   TEM image of ultrafine TiC precipitates occurring in the midsection of Steel sheet H7. The precipitates are denser in dispersion and finer in size than those observed at the surface of Steel H7.

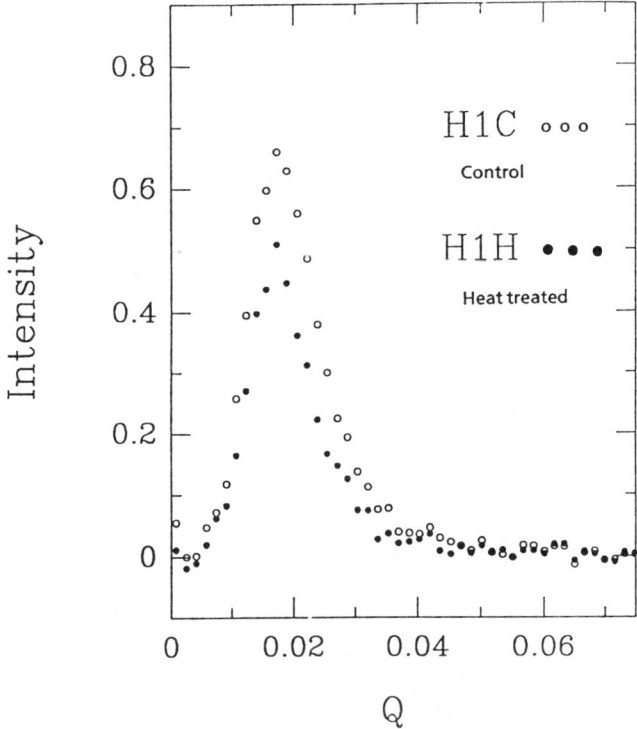

Figure 14   The neutron scattering intensity I plotted as a function of scattering vector Q for the control (H1) and heat treated (950°C 1 hr-WQ) specimen of H1.

Texture Results

The ODF results for the surface and midplane samples of Steel H1 are summarised in Figure 18 and those for Steel H7 are given in Figure 19. The texture as seen from the ODF results exhibit <111> fibre with fibre axis parallel to the sheet normal. The maximum ODF density observed at the surface was 6.0 for H1 and 5.0 for H7 respectively.

The texture at the midplane differed for the two samples. The H1 sample exhibits a strong gamma fibre (i.e. <111> parallel to the sheet plane) with a maximum times random value 0f 6.0. The H7 sample exhibits a weaker gamma fibre with a maximum times random value of 3.0.

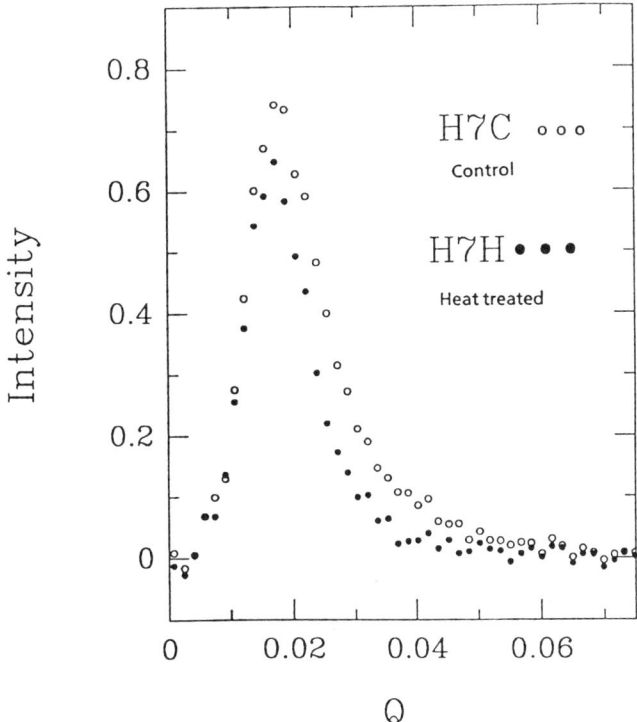

Figure 15 The neutron scattering intensity I plotted as a function of scattering vector Q for the control (H7) and heat treated (950°C 1 hr-WQ) specimen of H7.

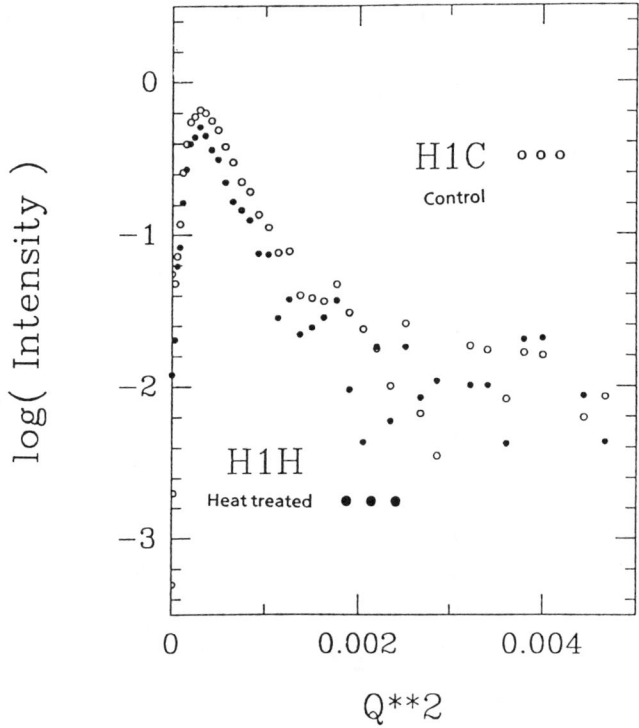

Figure 16 Log I (Intensity) versus Q^2 (scattering vector) for steel H1 control and heat treated.

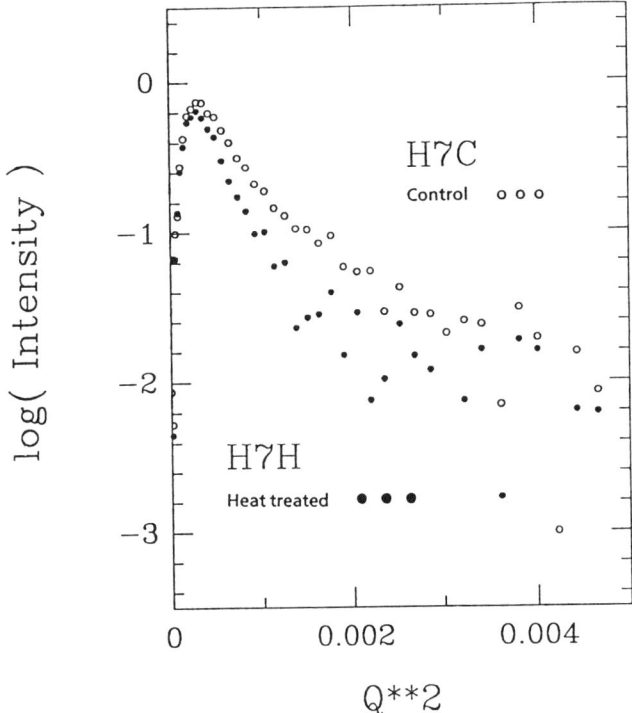

Figure 17　Log I (Intensity) versus Q^2 (scattering vector) for steel H7, control and heat treated.

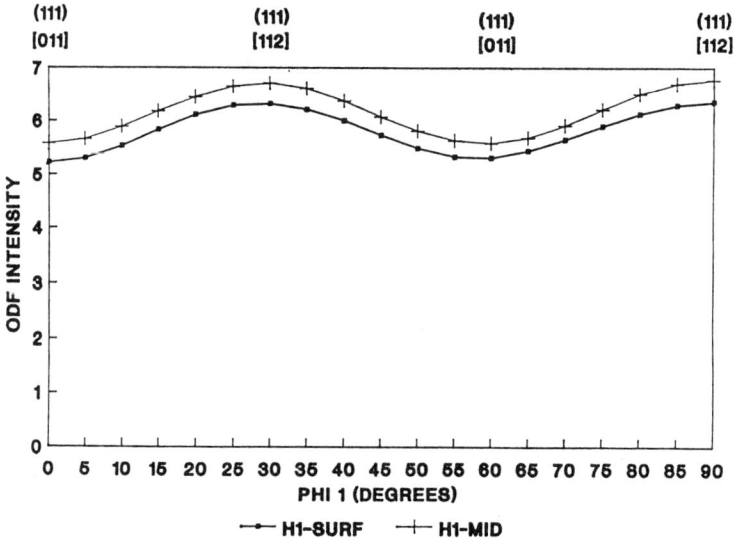

Figure 18   ODF represented in the form of gamma fibres for H1 sample taken from the surface and midplane.

Figures 20A and 20C show the development of strong gamma fibre at the surface and midplane of Steel H1. By comparison, Figures 20B and 20D show low intensity values obtained at the surface and mid plane of Steel H7. Further, a significant texture gradient through the thickness of H7 sheet was found.

Discussion

The grain size, the precipitate size and dispersion, and the residual solute content are identified as important hot band parameters that influence {111} texture evolution in interstitial free steels (3-6). The plastic anisotropy, as measured by the r-bar value correlates well with {111}/{100} texture component intensity ratio (2). In most studies, the effect of precipitate size and dispersion on the r bar values could not be isolated from that of grain size, as the grain size changes when controlling the dispersion of precipitates before cold rolling and annealing.

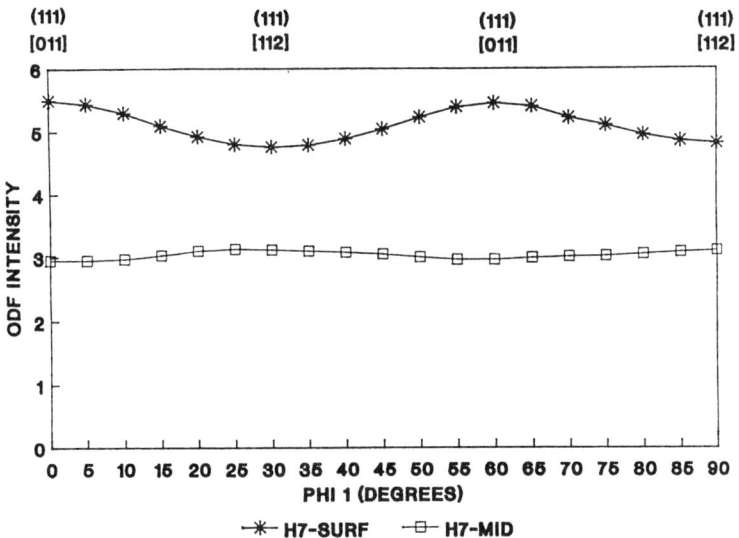

Figure 19 ODF represented in the form of gamma fibres for H7 sample taken from the surface and midplane.

The effect of precipitate dispersion on recrystallisation texture of niobium added extra low carbon cold rolled steel was investigated by Satoh, Obara and Tsunoyama (6). By varying the hot reduction, the size and dispersion of the precipitates was changed widely using the same steel, but the grain size could not be kept constant. However, Satoh et al. concluded that fine and dense precipitates strongly suppress the growth of recrystallised grains and result in a decrease in {111} texture. The pinning force exerted by the precipitates on the boundary mobility during recrystallisation is considered to be the most important factor controlling the recrystallisation texture in an interstitial free steel. In the present work, two Ti stabilised interstitial free steels of comparable base chemistry with the same final grain size but with distinct difference in r bar values were investigated. An attempt is made to relate the texture results to the precipitate size and dispersion.

Thermodynamic analysis of the precipitation behaviour of Steel H1 and H7 are comparable in terms of the sequence of precipitation and equilibrium mole fraction of precipitates. The dissolution temperature of TiN is slightly higher for Steel H1 and Steel H7 because of the high Ti content in the base chemistry but the dissolution temperatures of TiS, $Ti_4C_2S_2$ and TiC for the two steels are comparable.

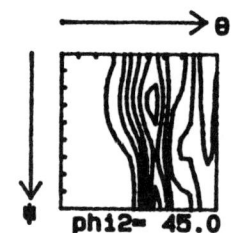

Contors at 1 2 3 4 5 6

(a) $I_{(001)<110>} = 0.08$
$I_{(554)<225>} = 4.29$
$I_{(111)<112>} = 5.23$
$I_{(111)<110>} = 6.36$

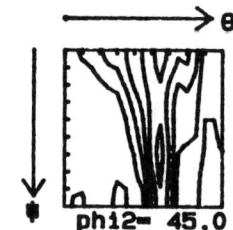

Contors at 1 2 3 4 5

(b) $I_{(111)<110>} = 0.76$
$I_{(554)<225>} = 4.83$
$I_{(111)<112>} = 5.50$
$I_{(111)<110>} = 4.82$

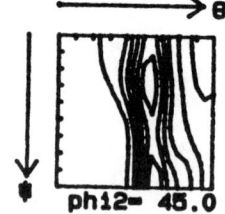

Contors at 1 2 3 4 5 6

(c) $I_{(001)<110>} = 0.16$
$I_{(554)<225>} = 4.33$
$I_{(111)<112>} = 5.58$
$I_{(111)<110>} = 6.77$

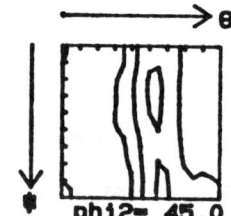

Contors at 1 2 3

(d) $I_{(111)<110>} = 1.04$
$I_{(554)<225>} = 2.73$
$I_{(111)<112>} = 2.95$
$I_{(111)<110>} = 3.10$

Figure 20   PHI2 = 45 Degrees
          (a) ODF section for H1 surface
          (b) ODF section for H7 surface
          (c) ODF section for H1 midplane
          (d) ODF section for H7 midplane

TEM/STEM characterisation of precipitates correlate well with the sequence of precipitation, the coarsest one, i.e., TiN being the first one to form, the next one in size being TiS for both steel. There was no distinguishable difference in size and dispersion of coarse precipitates (size >30nm) between steel H1 and H7. However, there was distinguishable difference in the ultrafine precipitates between Steel H1 and H7. The mean diameter of the precipitates in Steel H7 is finer than that of Steel H1. Since the replicas were taken from the surface of the steel samples H1 and H7, the observed difference in particle size and dispersion at the surface of the sheet can be related to the ODF density differences. Even though both steels exhibit strong gamma fibre at the surface, the maximum ODF density observed at the surface was 5 for H7 whereas that for steel H1 is 6.

The texture at the midplane differed for the two steels, H1 sample exhibiting a strong gamma fibre with a maximum times random value of 6.0, whereas H7 showing a weaker gamma fibre with a maximum times random value of 3.0. The r bar values of Steel H1 is 2.12, correlating well with the strong gamma fibre, whereas r bar value of Steel H7 is 1.69 corresponding to the weak gamma fibre.

Small angle neutron scattering results are representative of bulk specimens. The neutron scattering intensity of steel H7 (with $r=1.69$) is found to be greater than that of Steel H1 (with $r=2.12$) confirming that the volume fraction of ultrafine precipitates of Steel H7 is more than that of Steel H1. From the slope of log I (Intensity) versus Q^2 (scattering vector), the average particle size of steel can be calculated. The average precipitate size of Steel H7 is finer than that of Steel H1.

The limited precipitate size and dispersion data obtained from TEM results correlate well with the texture results. Strong {111} recrystallisation texture is obtained at the surface and in the centre of the sheet Steel H1. A sparse dispersion of coarse precipitates is observed both at the surface and in the centre of sheet Steel H1. The neutron scattering intensity of the bulk specimens of Steel H1 is lower than that of Steel H7 as well, suggesting that the volume fraction of the precipitates (f) in steel H1 is lower than that of Steel H7. Thus the Zener drag force, f/r (where f is the volume fraction of the precipitates and r the average radius of the precipitates) of Steel H1 is lower than that of Steel H7. By comparison the precipitates of Steel H7 are finer than those observed in Steel H1. The midplane specimens of Steel H7 are ultrafine, even finer than those observed at the surface of H7. Thus a variation in the Zener drag force is expected from the surface to the centre of the sheet in H7. There is a gradient in the {111} texture from the surface to the centre of the steel in H7. The weak intensity of {111} texture noted in the midsection of H7 correlates well the high Zener drag force associated with a high volume fraction of ultra fine precipitates in H7. Therefore, it is concluded that the evolution of {111} texture is suppressed by a strong pinning force resulting from a dense dispersion of ultrafine precipitates.

Kudobera and Imagaki (12) have proposed that fine precipitates result in a very strong pinning effect when a recrystallised grain grows in a recovered matrix. Satoh, Obara and Tsunoyama (6) have reported that irrespective of precipitate size and dispersion, {111} recrystallised grains recover faster than {100} oriented grains. {111} recrystallised grains nucleate in situ in the deformed matrices. Therefore, a dense dispersion of fine precipitates strongly suppresses the growth of recrystallised grains with {111} orientations through pinning force exerted by the particles on the boundary mobility in recovered matrices. Hu (16,17) has pointed out that the driving force for the mobility of boundary during primary recrystallisation is residual strain energy in the matrix. This driving force is increased by increasing the prior strain, as is the case with a large cold reduction in an interstitial free steel, and by decreasing the grain size. The driving force for the subsequent growth of the newly formed strain free grains is the grain boundary surface free energy, which is substantially smaller in magnitude than the driving force for the primary recrystallisation. Thus, the Zener drag force on the boundary mobility is expected to be even more dominating during the secondary growth than during primary recrystallisation in recovered matrix, as printed out by Satoh et al. (6).

Technological Implications

A dense dispersion of fine precipitates suppresses the growth of recrystallised grains with {111} texture, thereby impairing the deep drawing property of the steel. The removal of interstitials as a sparse dispersion of coarse precipitates is a key factor for promoting high r-bar values.

In a Ti stabilised interstitial free steel, a low slab reheat temperature is beneficial to promote the formation of $Ti_4C_2S_2$ at elevated temperature (13). This facilitates the removal of carbon as coarse $Ti_4C_2S_2$ precipitates that form at elevated temperatures. The ultrafine precipitates that form in ferrite can be coarsened by selecting a high coiling temperature.

Conclusions

1. Small Angle Neutron Scattering studies and TEM examination of Ti stabilised interstitial free steels confirm that steels with a sparse dispersion of coarse precipitates exhibit a strong {111} texture and high r bar values.
2. The pinning effect of the particles on the boundary mobility during recrystallization growth is considered the most important factor for controlling the recrystallization texture in an interstitial free steel.
3. A low slab reheat temperature is a distinct advantage in a Ti stabilised interstitial free steel to remove carbon as a sparse dispersion of coarse precipitates, thereby promote a strong {111 } texture to achieve high r bar values.

Acknowledgements

The financial support of this research by Ontario Centre for Materials Research (OCMR) and Manufacturing Research Corporation of Ontario (MRCO) is gratefully acknowledged. Grateful thanks are expressed to Dr. J. Berlinsky Director, Institute for Materials Research and Dr. G.C. Weatherly and Dr. D.A.R. Kay, Department of Materials Science and Engineering for their active interest and contributions to this research. The authors wish to thank Dofasco Inc., Hamilton, Ontario for technical support of this project. Special thanks are expressed to Dr. B. Strathdee and Mr. F. Goetz of Dofasco Research Department.

References

1. R.L. Whiteley and D.E. Wise in, "Flat Rolled Products III", 47-63, 1962, New York, Interscience.

2. J.F. Held in, "Mechanical Working and Steel Processing IV", Paper 3, 1965, New York, The Metallurgical Society AIME.

3. W.B. Hutchinson, "Development and Control of Annealing Textures in Low Carbon Steels", Int. Met. Rev., 1984, vol. 29, No. 1, p. 25-43.

4. M. Matsuo, S. Hayami and S. Nagashima, Proc. ICSTIS II, Suppl. to Trans. ISIJ, II, 1971, 867.

5. W.B. Hutchinson and K. Ushioda, "Texture Development in Continuous Annealing", Scan. J. of Met., 1984, 13, p. 269-275.

6. S. Satoh, T. Obara and K. Tsunoyama, "Effect of Precipitate Dispersion on Recrystallisation Texture of Nb Added Extra Low Carbon Cold Rolled Steel Sheet", Trans. ISIJ, 1986, Vol. 26, p. 737-744.

7. O. Hashimoto, S. Satoh and T. Tanaka, Trans., ISIJ, Vol. 27, 1987, p. 746-754.

8. T. Obara, S. Satoh, S. Okada, S. Masii and K. Tsunoyama, Proc. Thermec 88, ISIJ, Vol. 2, p. 676-683.

9. T. Tsunoyama, K. Sakata, T. Obara, S. Satoh, K. Hashiguchi and T. Irie, Proc. Symp. on "Hot and Cold Rolled Sheet Steels", Ed. R. Pradhan and G. Ludkovsky, The Metallurgical Society, 1988, p. 155-164.

10. H. Takechi, "Development and Production of Interstitial Free Steel", Paper presented at the Symp. on Metallurgy and Application of modern IF steel grades, VDEh, Dusseldorf, May 1990.

11. H. Inagaki, "Effect of Ti on the Development of Rolling Textures in High Purity Iron, Textures and Microstructures", 1988, Vol. 8 and 9, p. 173-179, published by Gordon and Breach Science Publishers Inc. U.K.

12. H. Kubodera and H. Inagaki, Bull. Japan Inst. Metals, 7, 1968, 383.

13. K. Balasubramanian, "Thermodynamics of Microalloyed Austenites and Non-Stoichiometric Carbides and Nitrides", Ph.D. Thesis, McMaster University, 1988, p. 170.

14. S.V. Subramanian, M. Prikryl, A. Ulabhaje and K. Balasubramanian, "Thermokinetic Analysis of Precipitation Behaviour of Ti Stabilised Interstitial Free Steel", Proc. Int. Symp. on Interstitial Free Steel Sheet: Processing, Fabrication and Properties, Aug. 1991, C.I.M. Conf., Ottawa, ed. L.E. Collins and D.L. Baragar, p. 15-38.

15. H. Ohtani, T. Tanaka, M. Haseba and T. Nishizawa, "Solubility of NaCl-Type Carbides (NbC, VC and TiC) in Austenite", Proc. Japan - Canada Seminar on Secondary Steelmaking, Dec. 3/4, 1985, Joint publication by The Canadian Steel Industry Research Association and The Iron and Steel Institute of Japan, J-7-1.

16. H. Hu, Proceedings of the 5th Int. Conf. in Textures of Materials, vol. II, G. Gottstein and K. Lucke, Ed., Springer-Verlag, 1978, p. 3.

17. H. Hu, ASM Metals Handbook vol. 9, p. 697.

EFFECTS OF MANGANESE CONTENT AND HOT-BAND COILING TEMPERATURE
ON DEEP DRAWABILITY OF CONTINUOUS-ANNEALED AL-KILLED SHEET STEELS

Naomitsu Mizui* and Atsuki Okamoto**

Sumitomo Metal Industries Ltd.
* Iron and Steel Research Laboratories.
1-8 Fuso-cho, Amagasaki, 660 Japan
** Personal Division, Osaka Head Office.
5-15 Kitahama, Higashi-ku, Osaka, 541 Japan

Abstract

In a previous study on the effect of C content on mechanical properties of continuous-annealed Al-killed sheet steels, 0.08%Mn steel exhibited a maximal mean r-value at C content of 0.012%, when the coiling after hot rolling was conducted at 700℃. For further improvement of the deep drawability in continuous-annealed 0.012%C Al-killed sheet steel, the effect of Mn content on the r-value has been studied in a range between 0.02 and 0.14% Mn for the case of several hot-band coiling temperatures. As the result, it has been clarified that the r-value in the rolling direction exhibits a maximal value at a medium Mn content. Furthermore this optimum Mn content increases with the increase in the coiling temperature. It is concluded that the above effect of Mn content on the r-value is due to the changes in the distribution of MnS, AlN and cementite.

Introduction

Continuous annealing process has been widely adopted for the cold rolled sheet steel production in place of conventional batch annealing process because of the high efficiency of production. Such an alternation of annealing process stimulated the development of materials for cold rolled sheet steels. For example, the total production of ultra-low carbon Ti, Nb, and Ti+Nb added sheet steels has been increasing. Nevertheless, low-carbon Al-killed sheet steels are still produced in a large quantity.

When low-carbon Al-killed sheet steels with a high deep drawability are produced through continuous-annealing process, the following two important metallurgical factors should be taken into account.

(1) Rapid heating in continuous annealing process is incapable of controlling the recrystallization texture formation by means of the precipitation of AlN or AlN-cluster.

(2) A significantly large amount of C is retained in solution after annealing after rapid cooling in continuous-annealing process. This causes strain aging at room temperature or the stretcher strain problem.

Therefore the recent research work has been mostly focused on the improvement of deep drawability and the suppression of the strain aging at room temperature.

It has been clarified that the coiling of the hot bands at a high temperature is effective for the improvement of the deep drawability of the continuous-annealed sheet steels (1-4), because of the reduction in the solute C and N through the coarsening of cementite particles and the precipitation of AlN. The deep drawability is also considerably improved by decreasing the N content. When the N content is lowered less than 10 ppm, the coiling temperature can be lowered (5).

It has also been clarified that by the adjustment of the cooling pattern and the over-aging condition, the amount of remained solute C in the steel can be lowered enough to suppress the stretcher strain (6,7). When the steel is cooled at a rate faster than 50°C/s before OA, intragranular cementite nucleates at the beginning of over-aging, and the density of precipitation sites increases. As a result, after a short-time over-aging the amount of remained solute C becomes sufficiently small.

For the improvement of mechanical properties and aging properties of continuous-annealed Al-killed steels, the authors examined the effect of C content by using steels with 0.08% and 0.24%Mn (8). As shown in Fig.1, 0.08%Mn steels coiled at 700°C exhibit a maximal mean r-value at C content of about 0.01wt%. However in case of 0.08%Mn steels coiled at 500°C and 0.24%Mn steels coiled at 700°C, the mean r-value decreases with the increase in C content. The result suggests that Mn content also influences strongly on the r-value of continuous-annealed sheet steels. All the steels exhibited a maximal bake-hardenability at 0.005%C. In the range of C content more than 0.005%, the bake-hardenability increases with an increase in Mn content and a decrease in coiling temperature. The optimum Al-killed steel suitable for continuous annealing is a steel with C content ranging from 0.01 to 0.02% and Mn content of 0.08%. However the optimum Mn content has not been clarified by this previous experiment.

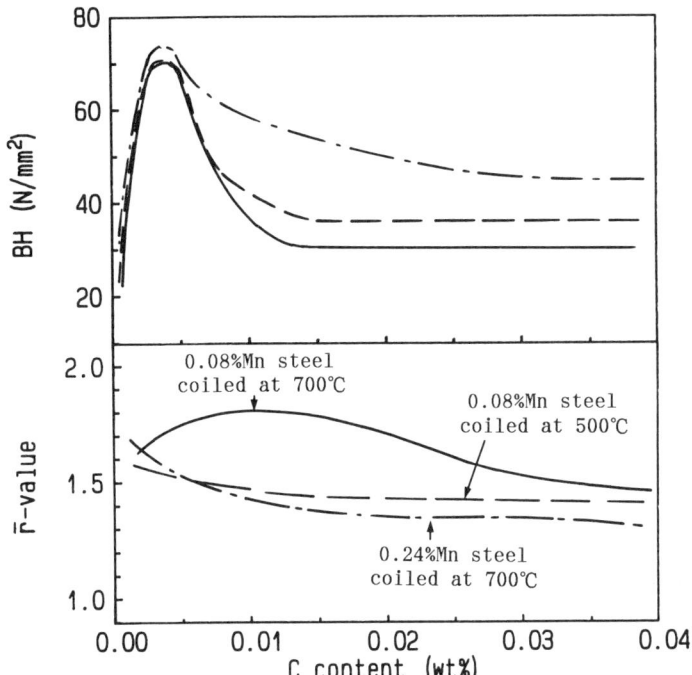

Fig.1 Effect of C content on mean r-value and bake-hardenability of steels with 0.08 or 0.24 wt% Mn, coiled at 500 or 700°C in hot rolling process, cooled at 100°C/s before over-aging stage in continuous annealing process(8).

Hu and Goodman (9) reported that the r-value of rimmed steels decreased with an increase in Mn content. However they reported that a rimmed steel containing 0.005% Mn exhibited a very small r-value. Matsudo et al (1). examined the effect of Mn content ranging from 0.03 to 0.8 wt% on the r-value of continuous-annealed 0.04%C Al-killed steels with pearlite or coarse cementite structure in the hot bands. They reported that the r-value decreased with the increase in Mn content independent of the carbide morphology. However Osawa and Kurihara conducted AlN precipitation treatment of hot bands before changing the cementite morphology by furnace-cooling of hot bands from 300 or 750°C. In both reports the effect of the AlN precipitation in the coiling stage was not clarified.

In the present study the effects of Mn content and coiling temperature on the r-value was examined with 0.012%C Al-killed steels which exhibited a maximal r-value in the previous study (8).

Materials

Al-killed steels with Mn contents ranging from 0.022 to 0.14 wt% were vacuum-melted in a laboratory induction furnace. The results of chemical analysis are given in Table I

Table I Chemical composition of steels (wt%).

Steel	C	Si	Mn	P	S	sol.Al	N
A	0.013	0.003	0.022	0.014	0.006	0.029	0.0060
B	0.011	0.003	0.041	0.014	0.006	0.049	0.0037
C	0.010	0.003	0.071	0.013	0.006	0.054	0.0033
D	0.011	0.003	0.140	0.013	0.006	0.058	0.0032

Experimental procedure

Figure 2 shows the schematic illustration of experimental procedure. The 20 mm thick slabs hot forged from 17 kg ingots were soaked at 1150℃ for 1h and hot rolled into 5 mm thick plates. After hot rolling, the coiling simulation was carried out as follows. The hot bands were cooled either by air-jet or water-spray to a temperature between 500 and 700℃. After keeping for 30 min at each temperature, the hot bands were furnace-cooled to room temperature at a rate of 20℃/h.

After removing the surface layers by machining, the 3.5 mm thick hot bands were cold rolled to 0.8 mm thick with 77% reduction. The cold rolled sheets were annealed in an infrared image furnace. The annealing process consists of 5 stages ; heating to 820℃ at 10℃/s, soaking at 820℃ for 40s, cooling to 450℃ at 50℃/s, over aging for 2.5 min by very slow cooling from 450℃ to 350℃ at 40℃/min and final cooling at 10℃/s to room temperature.

Microscopic observation was made for the hot bands and annealed sheets. Tensile test to measure r-values and x-ray pole intensity measurement were conducted for the annealed sheets. Furthermore the electron microscopic observation was carried out for several samples of hot bands.

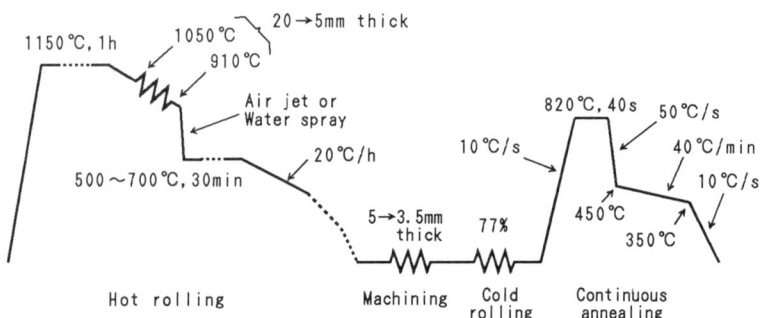

Fig.2 Schematic illustration of experimental procedure.

Experimental results

Microscopic observation

Microstructure of hot bands: The changes in the hot band microstructure of steel C with the coiling temperature are shown in Fig.3. The change in the grain size with the coiling temperature was very small. The average grain diameter was between 10 and 15 μm. The morphology of cementite depended strongly on coiling temperature, as well known. When the hot bands were coiled at 500 or 600°C, flaky cementite particles were observed along grain boundaries. On the other hand, when the hot bands were coiled at 650 or 700°C, a very few coarse cementite particles were observed. No pearlite was observed. As shown in Fig.4 interparticle spacing changes abruptly by raising the coiling temperature from 600 to 650°C. The change in the interparticle spacing with the Mn content was small.

Microstructure of annealed Steel: Figure 5 shows the changes in the average grain diameter of continuous-annealed sheet steels. The average grain diameter of steels coiled at 500 and 600°C became maximum at 0.04 and 0.07% Mn respectively. However those of the steels coiled at 650 and 700°C increased monotonously with the increase in Mn content.

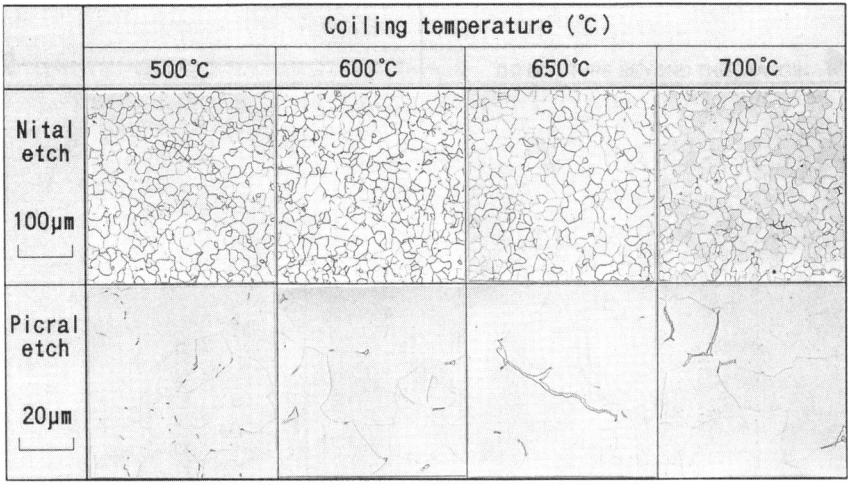

Fig.3　Microstructures of hot bands of 0.07%Mn steel (Steel C), coiled at various temperatures.

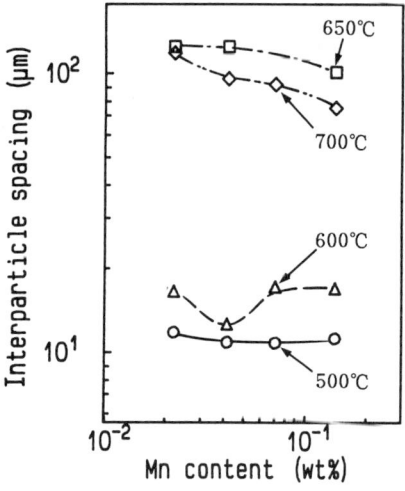

Fig.4 Effect of Mn on interparticle spacing in hot bands of 0.012%C Al-killed steels coiled at various temperatures.

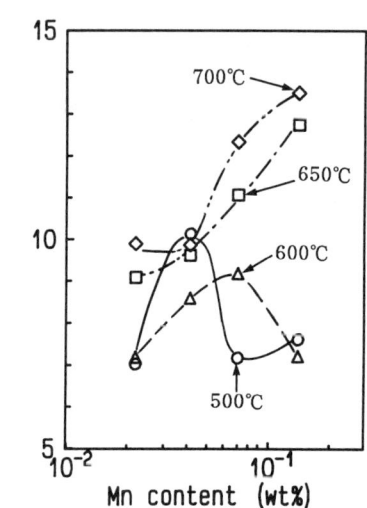

Fig.5 Effect of Mn on grain size of 0.012%C Al-killed continuous-annealed steels coiled at various temperatures.

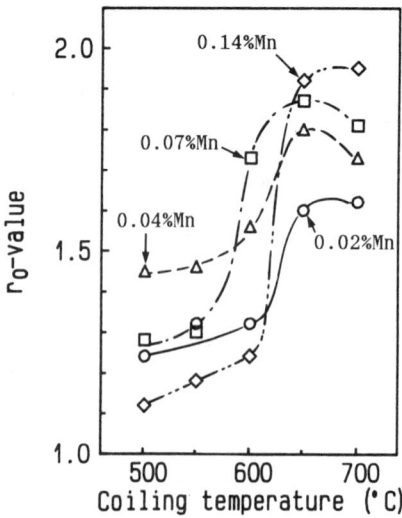

Fig.6 Effect of coiling temperature on r_0-value of 0.012%C Al-killed continuous-annealed steels with different Mn contents.

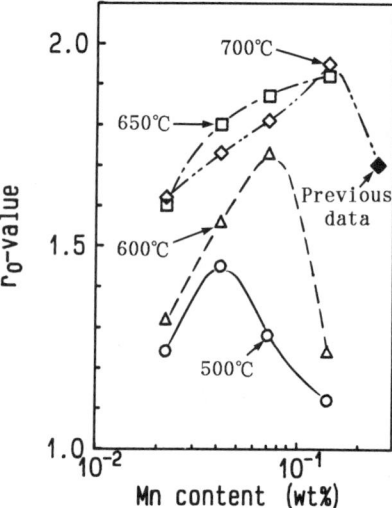

Fig.7 Effect of Mn on r_0-value of 0.012%C Al-killed continuous-annealed steels, coiled at various temperatures.

R-value measurement

Figure 6 shows the changes in the r-value in the rolling direction (r_0-value) with the coiling temperature. With an increase in coiling temperature, the r_0-value is increased. The difference in r_0-values between the steels coiled at 600 °C and 650 °C was large. This is due to the change in cementite morphology.

Figure 7 shows the changes in the r_0-value with Mn content. The steels coiled at 500 and 600 °C exhibited maximal r_0-value at 0.04%Mn and 0.07%Mn respectively. On the other hand the r_0-values of the steels coiled at 650 and 700 °C increased monotonously with coiling temperature. However according to our previous data, 0.012%C-0.24%Mn Al-killed continuous-annealed sheet steel, which was coiled at 700 °C and continuous-annealed in almost the same condition, exhibited r_0-value of 1.7. Therefore the steels coiled at 700 °C exhibit a maximal r_0-value at a Mn content of about 0.14%. The optimum Mn content for a high deep drawability depends on the coiling temperature.

X-ray measurement

Figure 8 shows the changes in pole intensity with Mn content. The {110} pole intensity of the steels coiled at 500, 600 and 650 °C exhibited a minimum at 0.04, 0.07, and 0.07%Mn respectively. However the {110} pole intensity of the steels coiled at 700 °C decreased monotonously with the increase in Mn content. The {200} pole intensity changes with the increase in Mn content in almost the same manner.

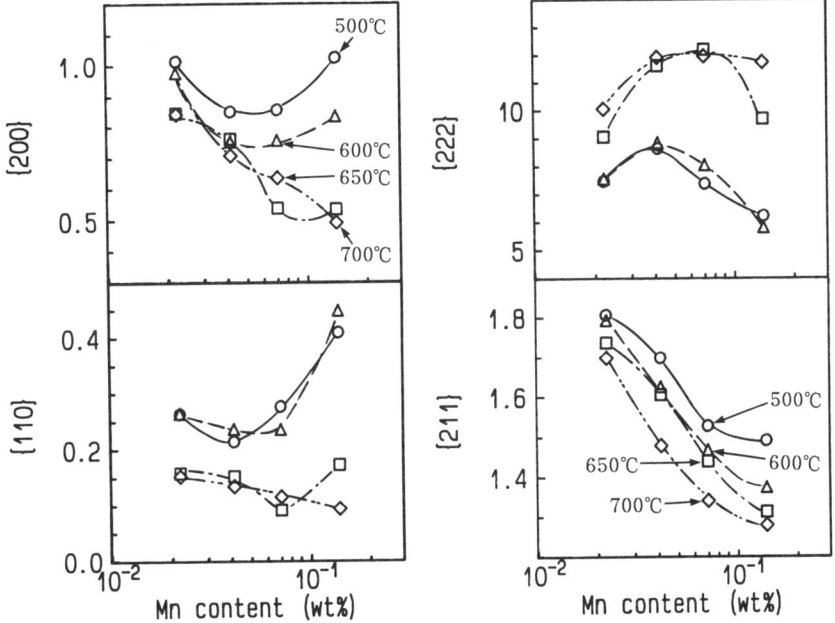

Fig.8 Effect of Mn on pole intensity of continuous-annealed 0.012%C Al-killed sheet steels, coiled at various temperatures.

However Mn contents for a minimal {200} pole intensity differ from those of {110} pole intensity. The {222} pole intensity changes in the reverse manner. The {211} pole intensity decreases monotonously with the increase in Mn content.

Discussion

In the present study, it was clarified that there is an optimum Mn content for each coiling temperature. It is noticeable that the optimum Mn content increases with an increase in the coiling temperature as shown in Fig.9. The result seems to be related to the changes in MnS distribution, AlN precipitation and cementite morphology. In relation to cementite morphology, the recrystallization texture formation has been discussed by many researchers (8,11). In the following section the effect of Mn on the recrystallization texture formation will be discussed after the effects of Mn on MnS distribution and AlN precipitation are discussed.

Fine precipitates in hot bands

Figure 10 shows the examples of microstructure of hot bands with various Mn contents coiled at 600°C. In the steel with 0.02%Mn many fine particles are distributed homogeneously. Most of them were MnS. With the increase in Mn content these particles are decreased in number as shown in Fig.11 and a re distributed along the grain boundaries. In case of 0.04%Mn steel, fine AlN particles could also be counted together. Therefore the effective MnS density for suppression of the nucleation and grain growth in the recrystallization process is considered to be decreased monotonously with the increase in Mn content.

In the steels with Mn content higher than 0.04%, feather-like AlN was observed. The AlN seems to nucleate at MnS. Figure 12 shows the change in AlN precipitation ratio with the increase in Mn content. There may be an object

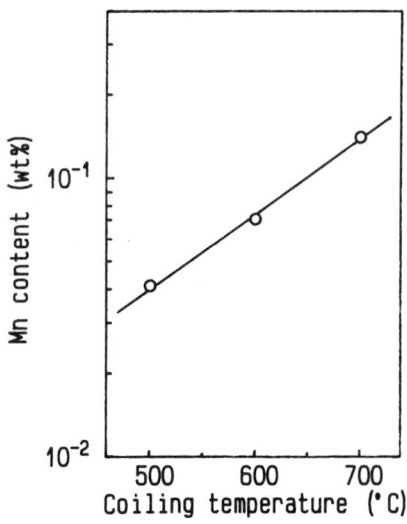

Fig.9 Relationship between optimum Mn content for a maximal r_0-value and coiling temperature.

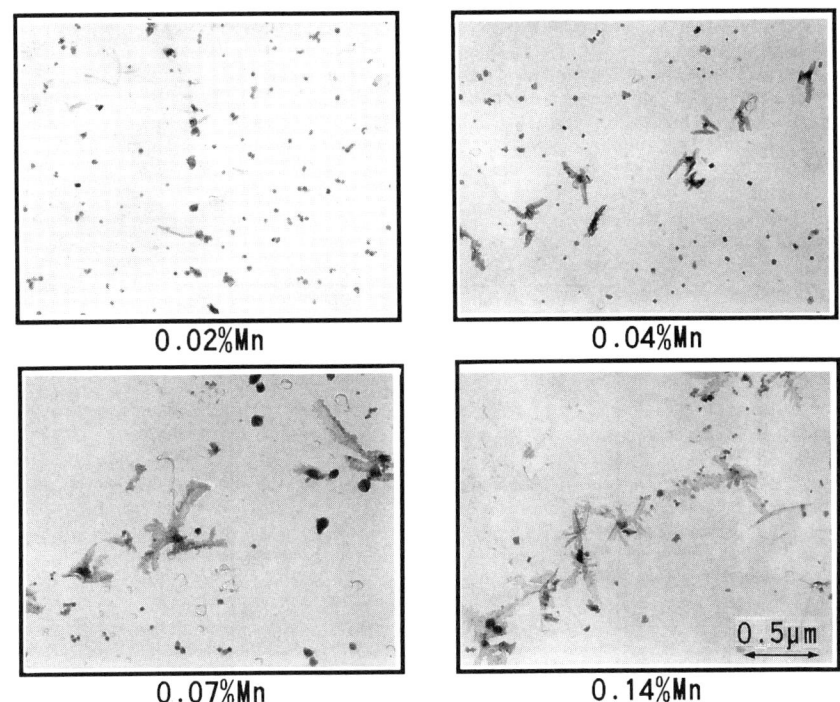

Fig.10 Microstructures of steels with various Mn content and coiled at 600°C.

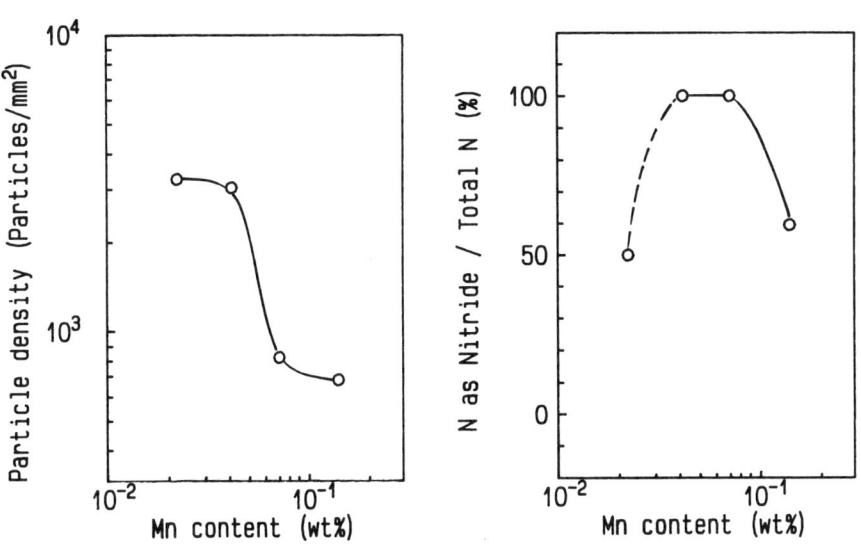

Fig.11 Effect of Mn on density of fine precipitates in steels coiled at 600°C.

Fig.12 Effect of Mn on AlN precipitation ratio in steels coiled at 600°C.

that the Al and N contents of the 0.02%Mn steel differ from the other three steels(Table I). Nevertheless the curve in Fig.12 can be explained as shown in Fig.13. AlN precipitation is accelerated with the increase in Mn content, because the activity of N in ferrite is increased by an addition of Mn. However with the increase in Mn content, MnS precipitates become coarser. Therefore the number of nucleation sites of AlN will decreases. Then AlN precipitation is suppressed with the increase in Mn content. These two opposite effects of Mn yields the maximal precipitation ratio with at medium Mn content.

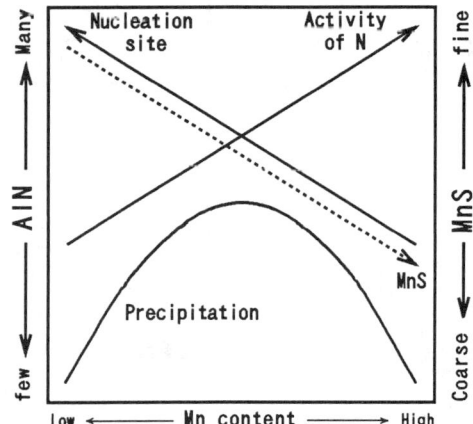

Fig.13 Schematic illustration of effect of Mn on AlN precipitation ratio.

Effect of Mn on r-value

The changes in pole intensity of annealed steels did not correspond perfectly with those in r_0-values. It is suggested that the effect of Mn on recrystallization texture formation differs from one orientation component to another, because each orientation component exhibits a different recrystallization process in both nucleation rate and growth rate. As the result of the effect of Mn on the recrystallization process of each orientation, the r-value is considered to exhibit maximum at a medium Mn content.

Figure 14 shows a schematic illustration of the effect of Mn on the r-value. As mentioned above, the increase in Mn content yields the following changes.

(1) MnS becomes coarser. It leads to a higher r-value. This effect of Mn does not depend on the coiling temperature.

(2) AlN precipitation ratio becomes maximum at a medium Mn content, when the hot bands are coiled at temperatures lower than 600℃. However the increase in Mn content does not suppress AlN precipitation any more, when the hot bands are coiled at temperatures higher than 650℃, because the interaction between Mn and N is small.

Furthermore the change in the amount of C in solution at the recrystallization temperature should be taken into account. According to optical micro-

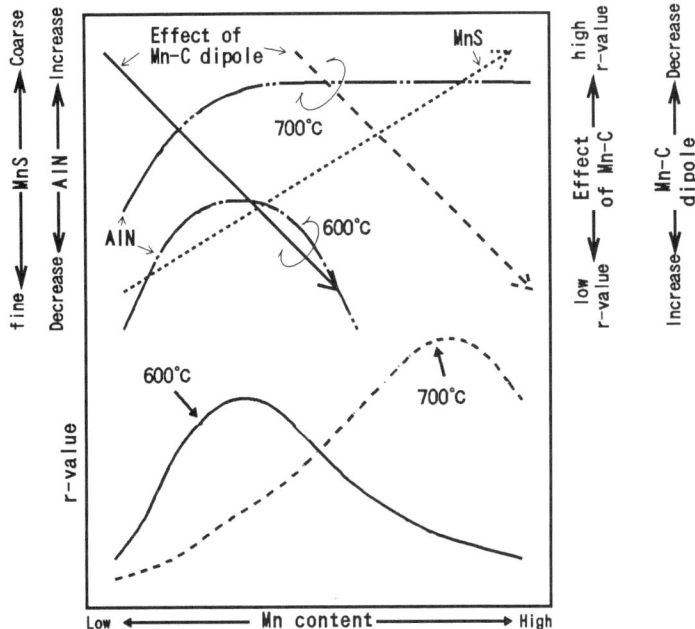

Fig.14 Schematic illustration of effect of Mn on r-value

scopic observation there was a small change in cementite dispersion with Mn content. Therefore the change in the amount of C in solution with Mn content will be negligibly small. However part of solute C will combine with Mn atoms to form Mn-C dipoles. Then not only single C atoms but also Mn-C dipoles suppress the formation of recrystallization texture favorable for deep drawability. The amounts of single C atoms and Mn-C dipoles increase with the decrease in coiling temperature, because of the change in cementite morphology. With the increase in Mn content, the amount of Mn-C dipole increases. Then the formation of the recrystallization texture favorable for deep drawability is suppressed stronger. Therefore as shown in Fig.13, the effect of Mn-C dipoles on the r-value can be expressed by different lines for two coiling temperatures.

By the coexistence of the three effects of Mn on the r-value, the r-value has a maximum at a medium Mn content. In case of coiling at a temperature as low as 600℃, MnS and AlN seem to dominate the texture formation. On the other hand in case of coiling at a temperature as high as 700℃, MnS and Mn-C dipole seem to dominate the texture formation, because of a high precipitation ratio of AlN.

To eliminate the effects of both MnS and AlN, the change in r_0-value of 0.02%Ti-added 0.012%C steel with the increase in Mn content was examined in case of coiling at 600℃. As shown in Fig.15 the r_0-value of 0.02%Ti-added steels decreased monotonously with the increase in Mn content. The average

r_0 value of the 0.02%Ti-added steels is lower than that of Ti-free steels, because of TiC precipitation.

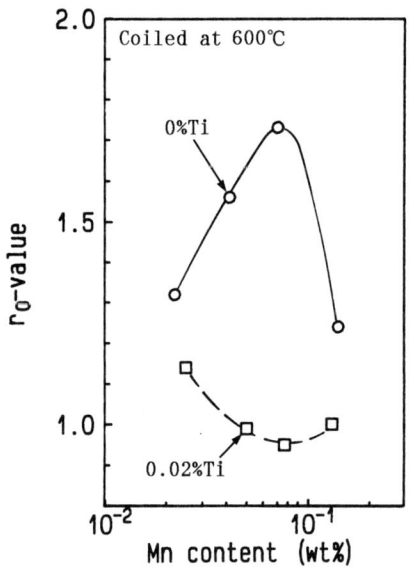

Fig.15 Effect of Mn on r_0-value of Ti-free and 0.02%Ti added 0.012%C steels coiled at 600 ℃.

Conclusions

For the improvement of deep drawability of continuous-annealed 0.012%C Al-killed sheet steel, the effect of Mn content has been studied for the steels with Mn contents ranging from 0.02 to 0.14% and for several coiling temperatures in the hot rolling process. The following results have been obtained.

(1) The r-value in the rolling direction exhibited a maximum value at a medium Mn content which increases with the coiling temperature.

(2) It is due to the changes in the distribution of MnS, and the precipitation ratio of AlN.

(3) In case of coiling at temperatures between 500 and 600 ℃, the effects of MnS and AlN seems to be dominant.

(4) In case of coiling at temperatures higher than 600 ℃, the effects of MnS and Cementite seemed to be dominant.

References

1. K.Matsudo, T.Shimomura and O.Nozoe, "Effect of Carbide Size, Cold Reduction and Heating Rate in Annealing on Deep-drawability of Low-carbon, Capped Cold-rolled Steel Sheet, "Texture of Crystalline Solid, 3 (1979), 53-72.

2. S.Ono, O.Nozoe, T.Shimomura and K.Matsudo, "Effect of Carbon on the Mechanical Properties of Continuous-Annealed Drawing-Quality Steels,"Metallugy of Continuous-Annealed Sheet Steel ed. B.L.Bramfitt and P.L.Mangonorv,Jr., (TMS-AIME,New York, 1982), 99-115.

3. K.Toda, E.Gondo, H.Takechi, M.Abe, N.Uehara and K.Komiya, "Metallurgical Investigations on Continuous Annealing of Low-Carbon Capped-Steel Sheet," Tetsu-to-Hagane, 61 (1975), 2363-2374.

4. A.Okamoto, and N.Mizui, "Effect of Cooling Condition of Hot Coil on the Recrystallization Texture Development during Rapid Annealing," Technology of Continuously Annealed Cold-Rolled Sheet Steel, ed. R.Pradhan (TMS-AIME, New York, 1984), 139-150.

5. N.Mizui and A.Okamoto, "Effect of Nitrogen Content on the Deep Drawability of Continuous-annealed Low Carbon Al-killed Sheet Steel," Sumitomo Kinzoku, 41 (1989), 493-502.

6. T.Obara, T.Sakata, M.Nishida and T.Irie, "Effect of Heat Cycle and Carbon Content on the Mechanical Properties of Continuous Annealed Low Carbon Steel Sheet," Kawasaki Steel Technical Report, 16 (1984), 264-272.

7. K.Koyama, H.Kato and M.Nagumo, "A Kinetics Model for Carbide Precipitation during Over-aging in Continuous Annealing of Low-carbon, Cold-rolled Sheet Steels," Tetsu-to-Hagane, 72 (1986), 823-830.

8. N.Mizui and A.Okamoto, "The effect of Carbon Content on the Mechanical Properties of Continuous-annealed Al-killed Sheet Steels,"Sumitomo Search, No.44, (1990),113-119.

9. H.Hu and S.R.Goodman, "Effect of Manganese on the Annealing Texture and Strain Ratio of Low-Carbon Steels,"Metallurgical Transactions., 1 (1970), 3057-3064.

10. K.Matsudo, K.Osawa, and K.Kurihara, "Metallurgical Aspect of the Development of Continuous Annealing Technology at Nippon Kokan,"Technology of Continuously Annealed Cold-Rolled Sheet Steel, ed. R.Pradhan (TMS-AIME, New York, 1984), 3-36.

11. K.Ushioda, W.B.Hutchinson, J.Agren, and U.von Schlipenbach, "Investigation of structure and texture development during annealing of low-carbon steel,"Materials Science and Technology, 2 (1986), 807-815

METALLURGICAL INVESTIGATION FOR PRODUCING NON-AGING DEEP-DRAWABLE

LOW-CARBON AL-KILLED STEEL SHEETS BY CONTINUOUS ANNEALING

K. Ushioda*, O. Akisue*, K. Koyama** and T. Hayashida***

* Sheet & Coil, Steel Research Lab., Technical Development Bureau, Nippon Steel Corp., 20-1 Shintomi, Futtsu, 299-12, Japan
** Kimitsu R & D Division, Technical Development Bureau, Nippon Steel Corp., 1 Kimitsu, Kimitsu, 299-11, Japan
*** Hirohata R & D Division, Technical Development Bureau, Nippon Steel Corp., 1 Fujicho, Hirohata, Himeji, 671-11, Japan

Abstract

Production of non-aging deep-drawing quality (DDQ) cold-rolled sheets from low-carbon aluminum-killed steel by the continuous annealing process called for the development of two techniques: (1) technique for ensuring the non-aging property of sheet products; and (2) technique for coiling hot-rolled bands at a low-temperature. A metallurgical study was carried out to this end and accomplished new progress in continuous annealing technology. The non-aging property was achieved by optimizing heat cycle in the overaging process, namely by slight supercooling, reheating and ramp overaging (R-OA). In this process, effective use was made of the phenomenon that manganese sulfide (MnS) precipitates act as preferential nucleation sites for transgranular cementite. The conditions under which cementites coarsen and aluminum nitride (AlN) precipitation are accelerated even in the hot-rolled band coiled at a low-temperature were attained by optimizing the carbon, manganese, and aluminum contents of the steel. The validity of these findings was verified by mill tests.

Introduction

The continuous annealing of cold-rolled sheet steels is a major trend of the times because it brings about many advantages, such as process step continuation, labor savings, and product quality consistency. The switch to the continuous annealing process was slow, however. The principal reason was that deep-drawing quality (DDQ) high-grade cold-rolled sheets for automotive panels could not be produced from low-carbon aluminum-killed steel (Fig. 1). Despite 30 years of research on continuous annealing, it was not practical to produce continuously annealed sheets using low-carbon aluminum-killed steel with a non-aging property comparable to that of batch annealed sheets. The DDQ grades had to be manufactured from low-carbon aluminum-killed steel by the batch annealing process, or from ultralow-carbon steel microalloyed with titanium or niobium by the continuous annealing process.

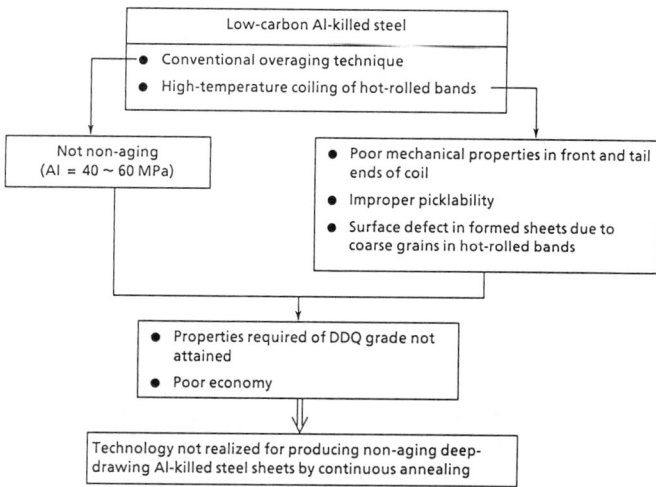

Figure 1 - Problems with conventional continuous annealing technology for producing non-aging deep-drawing sheets using low-carbon Al-killed steel.

It is a well-known fact that the batch annealing process is inferior to the continuous annealing process in quality consistency and production process rationality. Titanium or niobium-microalloyed ultralow-carbon steel is costly because it requires intense vacuum degassing and addition of expensive alloying elements. Therefore development of technology for manufacturing non-aging DDQ sheets from low-carbon aluminum-killed steel by continuous annealing had long been awaited.

Moreover the conventional continuous annealing technology for low-carbon aluminum-killed steel had the following problems (Fig. 1): Hot-rolled bands had to be coiled at a high temperature of ≥ 730°C to ensure deep drawability. High-temperature coiling involved such problems as: (1) sharp drop of quality at the front and tail ends of the coil, which decreases the product yield; (2) increase in oxide scale thickness to the detriment of pickling perform and (3) formation of coarse grains in the surface layer of the hot-rolled band, resulting in surface defects of the cold-rolled sheet product after press-forming. These problems led to the growth of demand for a low-temperature coiling techniques.

This study tackled the development of techniques to manufacture non-aging deep-drawing quality sheets from low-carbon aluminum-killed steel by the continuous annealing process. Basic research resulted in finding solutions to the two basic technical problems - assurance of the non-aging property and low-temperature coiling of hot-rolled bands. Optimum composition and process conditions were proposed, and verified by mill tests. These research and development activities are reported here.

Non-aging techniques

Conventional non-aging techniques:

The aging phenomenon that results from solute carbon is the most basic problem in the manufacture of cold-rolled sheets from low-carbon aluminum-killed steel by continuous annealing. Many studies have been made on the overaging treatment for the purpose of developing non-aging techniques. For example, the aging index (AI) that can be achieved by 3 to 5 min of overaging treatment after gas cooling in the heat cycle of Fig. 2a) is 50 to 60 MPa. This is not the non-aging level as described in the following section. As compared with the heat cycle of Fig. 2a), that of Fig. 2c) uses air-water mist cooling or roll cooling as the cooling method after recrystallization annealing. As the cooling rate is high, an aging index of 45 to 55 MPa can be obtained by about 2 min of overaging treatment. This aging index is still not the non-aging level described in the next section. Cementite that precipitates during overaging is located for the most parts at grain boundaries in the heat cycles of Figs. 2a) and c). When the steel is water quenched to room temperature after recrystallization annealing and overaged by reheating as shown in Fig. 2b), supersaturated carbon precipitates very densely as cementite within the matrix grains. In this case, the overaging time can be shortened further, but the aging index obtained is about 40 to 60 MPa for about 1 min of overaging. This value does not represent the non-aging property, however.

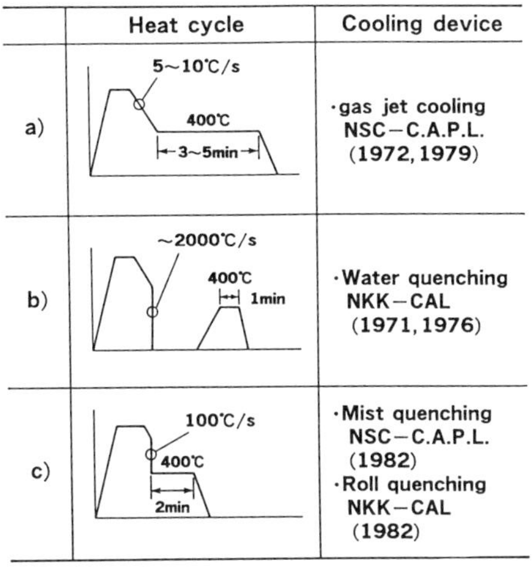

	Heat cycle	Cooling device
a)	5~10°C/s, 400°C, 3~5min	·gas jet cooling NSC−C.A.P.L. (1972, 1979)
b)	~2000°C/s, 400°C, 1min	·Water quenching NKK−CAL (1971, 1976)
c)	100°C/s, 400°C, 2min	·Mist quenching NSC−C.A.P.L. (1982) ·Roll quenching NKK−CAL (1982)

Figure 2 - Overaging technologies in conventional continuous annealing.

Furthermore the extremely dense transgranular cementite raised yield strength and degraded ductility. It was difficult to achieve both non-aging property and good formability at the same time. In terms of energy use, this procedure is not favorable because large-scale reheating is indispensable.

As noted above, it was difficult to manufacture non-aging steel sheets by the conventional overaging techniques. The aging index to be aimed at for non-aging sheets was not clear. Further, the phenonmenon that transgranular cementite preferentially precipitates at MnS as described later was still unknown.

This study first set the target value of the aging index that the non-aging sheets should have. Taking advantage of the phenomenon that MnS precipitates act as preferential nucleation sites for transgranular cementite, a new overaging heat cycle was devised and employed to manufacture non-aging sheet steels.

Strain aging rate and target value of aging index:

The stretcher-strain that occurs when the steel sheet is press-formed is correlated with the yield-point elongation of the sheet determined by the tensile test. It is a well-known fact that if its yield-point elongation is 0.2% or less, the sheet does not develop stretcher-strains when press formed(1). A slab of the 0.024% C - 0.15% Mn - 0.008% P - 0.006% S - 0.045% Al - 0.0021% N - 0.0004% B composition in mass % was hot-rolled, cold-rolled and continuously annealed on production equipment, changed in the aging index over a range of 10 to 30 MPa by adjusting the solute carbon content by heat treatment in the laboratory, and temper-rolled by 0.8%. The material was then strain-aged by various temperature-time combinations and tensile tested. Figure 3 shows the relationship between the aging index and the aging conditions required for the yield-point elongation to recover to 0.2%. If the aging index is 30 MPa or less, the yield-point elongation falls below 0.2% under normal aging conditions (for example, 30°C for 3 months, or 100°C for 1 h). The aging index of 30 MPa was thus taken as the substantial non-aging target value. To secure such total elongation and yield strength after strain aging, and stretchability under plane strain conditions (LDH_0: the minimum value of limiting dome height)(2) that are comparable to those of batch annealed steel (Fig. 4), it was found necessary to aim at an aging index of 20 MPa. The aging index of 20 MPa was thus taken as target for the complete non-aging property.

MnS as preferential nucleation sites of cementites during overaging:

Solute carbon present in supersaturation prior to the overaging treatment precipitates as cementite during the overaging treatment. The solute carbon content decreases in the overaging process. Particularly when its supersaturation degree is high, solute carbon preferentially precipitates at MnS preexisting within matrix grains(1), and on dislocation(3). In the case of commercial steel, MnS is the major precipitation site. When cementite is transgranularly nucleated in this way, its growth proceeds as controlled by the short-distance diffusion of carbon. Effective utilization of MnS leads to an efficient overaging treatment. Slabs of low-carbon aluminum-killed steels of different sulfur contents as listed in Table I were reheated at four different temperatures as shown in Fig. 5 to change the distribution density of MnS in the hot-rolled band (Fig. 6). Cementite preferentially precipitated on MnS after overaging (Fig. 6). Figure 7 shows the relationship between the aging index and MnS distribution density, and the relationship between the transgranular cementite density and MnS distribution density after the overaging treatment. As evident from Fig. 7, the transgranular cementite density increases and the aging index decreases with increasing MnS distribution density. Control of the MnS distribution density is extremely important for the non-aging technique. Factors that govern the MnS distribution density, sulfur content, and slab reheating temperature in particular, were investigated. The results are shown in Fig. 8. Steels A, B, and C listed in Table I were experimentally processed as schematized in Fig. 5. The number of MnS precipitate particles that measure 0.05 μm or more in size and effectively act as transgranular cementite precipitation sites(1) is maximal in the vicinity of the 1050°C slab reheating temperature when the sulfur content is about 0.015%, as shown in Fig. 8. The MnS distribution density is about 1 particle/μm^3. Figure 9 shows the change in the MnS distribution density when the sulfur content is held constant at 0.007% and the manganese content is changed. The slab reheating temperature is 1100°C. The MnS distribution density increases with decreasing manganese content.

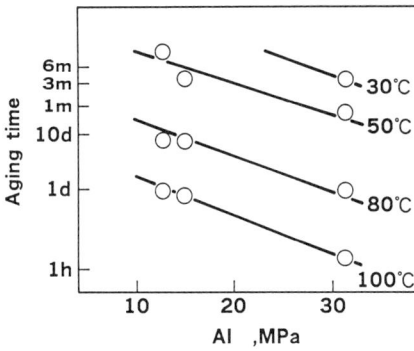

Figure 3 - Relationship between AI (aging index) and time to return of 0.2% YP-El.

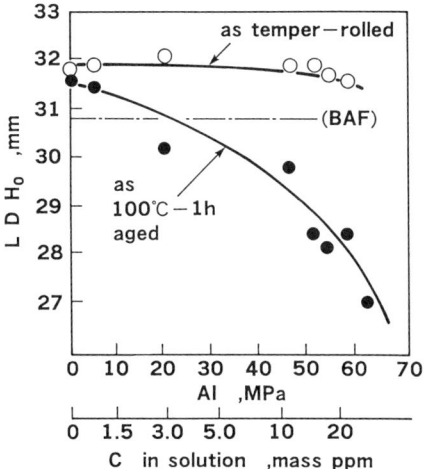

Figure 4 - Relationship between AI (aging index), content of C in solution and LDHo (2). Specimen has 0.022% C - 0.18% Mn - 0.006% P - 0.004% S - 0.060% Al - 0.0018% N (mass %) and 0.8mm in thickness. BAF means the average LDHo of DDQ steel sheets produced by batch annealing process.

Table I Chemical composition of steels used (mass %).

steel	C	Si	Mn	P	S	Al	N
A	0.020	0.012	0.15	0.003	0.005	0.042	0.0017
B	0.020	0.012	0.15	0.004	0.010	0.045	0.0019
C	0.020	0.012	0.15	0.003	0.015	0.046	0.0019

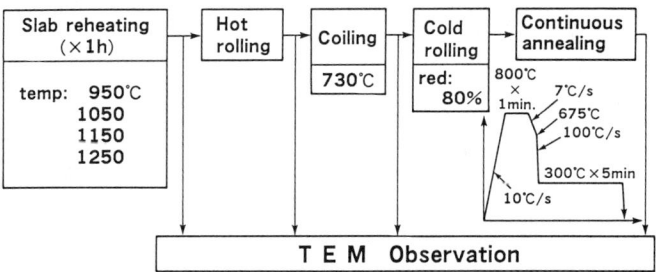

Figure 5 - Experimental procedure to investigate the precipitation behaviors of both MnS and Fe_3C.

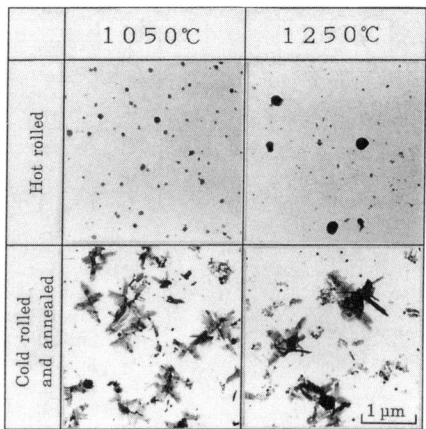

Figure 6 - TEM observation of extraction replica showing the distribution of MnS and Fe_3C (steel B in Table I).

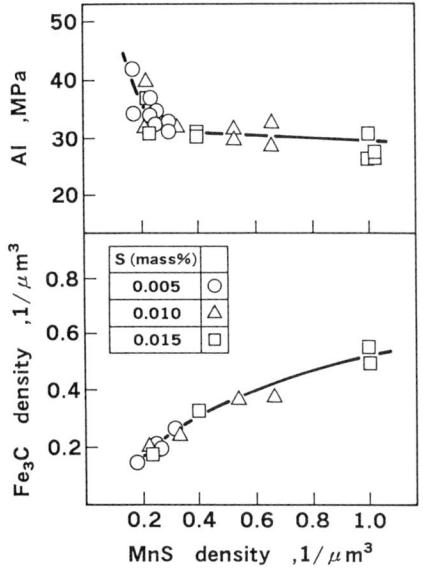

Figure 7 - Influence of the density of MnS (1/μm³) in hot-rolled bands larger than 0.05μm on Fe₃C density and Al of the annealed sheets.

Figure 8 - Influences of S content and slab reheating temperature on the density of MnS (1/μm³) in hot-rolled bands larger than 0.05μm.

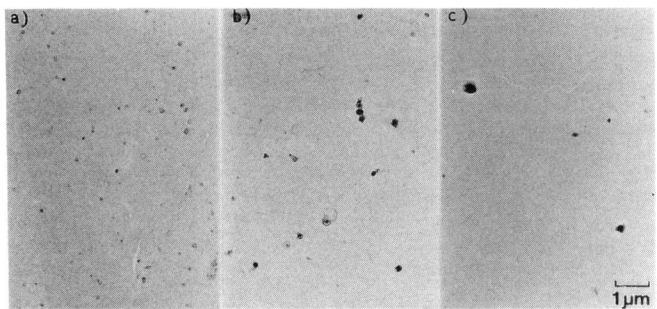

Figure 9 - TEM observation of extraction replica showing the influence of Mn content on precipitates in hot-rolled bands coiled at 650°C (Slab reheating temperature: 1100°C, Steel a) 0.05% Mn (M2), b) 0.11% Mn (M3), c) 0.28% Mn (M5) in Table III).

Kinetic model for cementite precipitation:

As clarified in the preceding section, cementite transgranularly precipitates when the superstauration degree of carbon is high. Since grain boundaries are the most stable precipitation sites for carbon, intergranular precipitation actually proceeds at the same time as transgranular precipitation. To understand this competition between intergranular and transgranular precipitation of carbon, it is most effective to utilize the kinetic model reported by Katoh et al.(4). This section briefly describes the precipitation model and presents the results of research conducted to build a new overaging heat cycle to ensure the non-aging property.

Table II summarizes the basic equations of the cementite precipitation model. The model equations calculate the number and size of transgranular cementite precipitates nucleated at specific time intervals as well as the growth of cementite nucleated beforehand, and intergranular cementite precipitation. The particle size and distribution density of cementite can thus be determined by the kinetic model. The model has only one experimental constant K_2 to be adjusted. As material parameter, it only needs to give the ferrite grain size, the parameter α for transgranular cementite nucleation (ease of heterogeneous cementite nucleation on MnS), and the parameter N_S for the number of precipitation sites that is equal to the number of MnS. The model can completely describe the precipitation of carbon, including cases in which temperature changes from time to time. The application of the model to the cementite precipitation during continuous annealing has been described in detail in previous reports(1, 4, 5, 6).

Table II Summary of cementite precipitation model.

	Fundamental Equations
1) Balance of carbon	$C_0 = C + C_G + C_T$
2) Grain boundary cementite precipitation (C_G)	$\dfrac{dC_G}{dt} = (\dfrac{\pi}{a})^2 \cdot D \cdot (C-S)$
3) Transgranular cementite precipitation (C_T)	$C_T = \dfrac{4\pi}{3} R^3 \cdot C_\theta \cdot m$
a) nucleation	$\dfrac{dm}{dt} = K_2 \cdot N_S \cdot (1 - \dfrac{m}{N_S}) \cdot D \cdot (C-S) \cdot \exp\left\{-\dfrac{\alpha\beta}{T^3 (\ln C/S)^2}\right\}$
	$\beta = (256\pi\sigma^3 \Omega_\theta^2)/(3kR_g)^2 = 1.29 \times 10^{12} (K^{-3})$
	R_{cr} (critical radius) $= 8 \cdot \sigma \cdot \Omega_\theta \cdot \alpha^{1/3}/\{R_g \cdot T \cdot (\ln C/S)\}$
b) growth	$\dfrac{d(R^2)}{dt} = 2 \cdot D \cdot (C-S)/C_\theta$
4) Note: C_0 = total C content, C = content of C in solution, C_G = content of C as Fe_3C in grain boundary, C_T = content of C as Fe_3C within grain, t = time, a = radius of ferrite grain, D = diffusion constant of C, S = solubility limit of C, R = radius of Fe_3C, C_θ = C content in Fe_3C, m: density of Fe_3C, K_2 = experimental const., N_S = density of MnS (number of nucleation site within grain), α = coefficient of reduction in activation energy, T = temperature, σ = interfacial energy between Fe_3C and matrix, Ω_θ = molar volume of Fe_3C, k = Boltzman const., R_g = gas const.	

This paper presents the results of basic research leading to the new overaging heat cycle described in the next section. Computer simulation was performed on the precipitation of carbon during isothermal annealing, using grain size ASTM No.8 and the values for MnS observed in commercial steels (α = 0.02, N_s = $10^9 mm^{-3}$). The computer simulation put C_0, the initial supersaturated carbon content, at 100 ppm. Figure 10 shows the change with time in the number of cementite nuclei (m) at different temperatures. It is apparent that the most efficient cementite nucleation is achieved at temperatures of 200 to 300°C.

When the temperature is further raised slightly, the diffusion of carbon is accelerated to ensure the smooth growth of cementite nuclei.

Slight supercooling, reheating and ramp overaging:

A slight supercooling, reheating and ramp overaging (R-OA) as shown in Fig. 11 has already been proposed as a new overaging heat cycle for rationally reducing solute carbon to the non-aging level (4). The heat cycle is characteristic in: (1) that cementite is transgranularly nucleated as densely as possible; and (2) that the cementite nuclei are grown in the shortest possible time to efficiently reduce the amount of solute carbon remaining in the product sheet. To attain objective (1), (a) the grain size is increased by high-temperature annealing, to minimize the cementite precipitation at grain boundaries; (b) the steel is slowly cooled to the temperature at which the amount of solute carbon is maximized, and is then rapidly cooled to maximize the amount of supersaturated solute carbon; and (c) with the maximized supersaturated solute carbon as driving force, cementite is nucleated on MnS precipitates at temperatures of 200 to 300°C. To attain objective (2), (a) the steel is reheated to a small degree to accelerate the diffusion of carbon and ensure the efficient growth of cementite, and at the same time, nucleation and growth of fresh transgranular cementite are promoted during the heat elevation; and (b) the steel is then slowly cooled to further reduce the amount of remaining solute carbon.

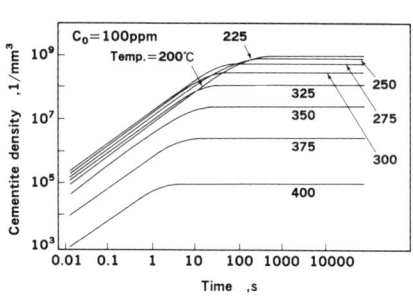
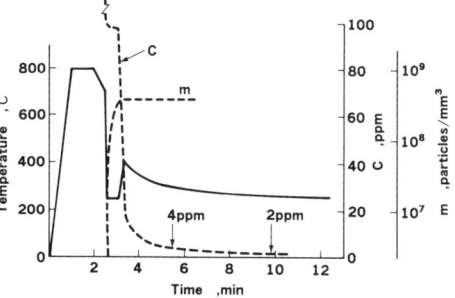

Figure 10 - Variation of calculated density of Fe$_3$C with time isothermally annealed at various temperatures.

Figure 11 - One of the typical annealing cycles for the production of commercially non-aging steels and the variations of carbon in solution (C) and density of cementites (m) during R-OA cycle (rapid cooling 100°C/s)(4).

The steel sheets manufactured by R-OA is sufficiently low in the amount of solute carbon, as shown in Fig. 12, which attests to the effectiveness of the new heat cycle. A mill cold-rolled sheet steel of the 0.024% C - 0.15% Mn - 0.011% P - 0.007% S - 0.052% Al - 0.0013% N composition was continuously annealed in the laboratory by an overaging heat cycle with supercooling (a) and by an overaging heat cycle without supercooling (b). The overaging time required to obtain the aging index of ≦ 20 MPa, a complete non-aging condition, is 5 min with supercooling and is 10 min without supercooling. The overaging time required to accomplish an aging index of ≦ 30 MPa, a substantial non-aging condition, is 1.5 min with supercooling and 5 min without supercooling. The effectiveness of R-OA is also clear in Fig. 13. A mill cold-rolled sheet steel of the 0.024% C - 0.15% Mn - 0.011% P - 0.007% S - 0.052% Al - 0.0013% N composition was continuously annealed in the laboratory by the 3-min overaging treatment as illustrated in Fig. 13. As evident from Fig. 13, a substantially non-aging level can be achieved over the wide temperature range of T_1 and T_2. Moreover, the solute carbon value calculated by the model closely agrees with the observed amount of solute carbon (Fig. 13). This also proves the validity of the cementite precipitation model. In either case, R-OA can sharply reduce the overaging time.

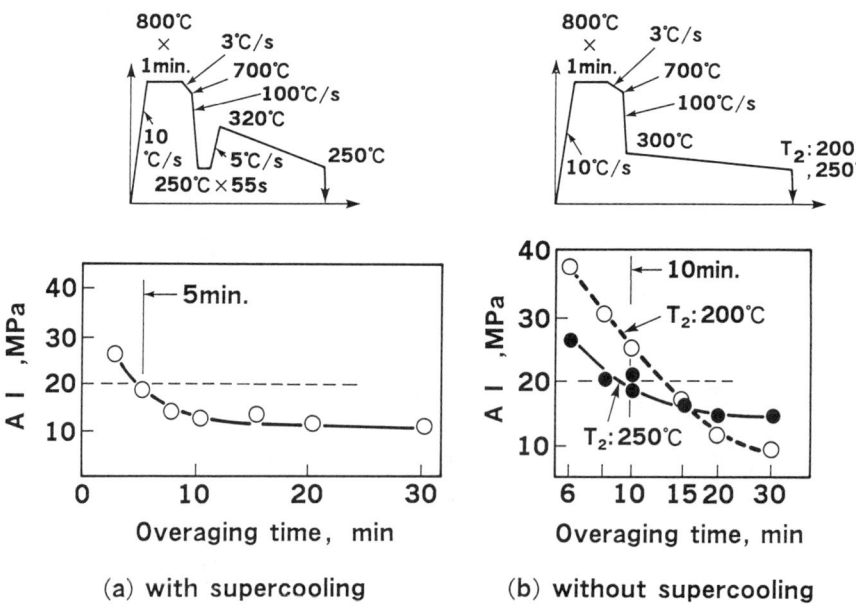

Figure 12 - Variation of AI (aging index) of annealed sheets with overaging time.
(a) with supercooling, (b) without supercooling

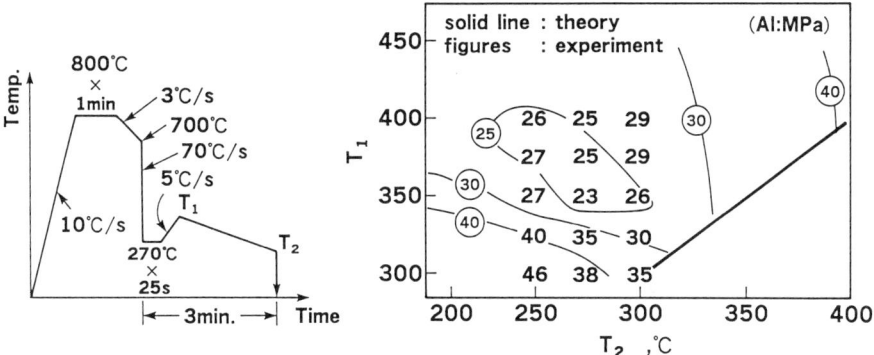

Figure 13 - Influence of T_1 and T_2 on Al (aging index) of annealed sheets (overaging time = 3 min.).

Figure 14 summarizes the effects of the cooling rate β, cooling stop temperature T_E, holding time t_E, and reheating rate α, all of which are important variables for the new overaging treatment. The test material is a 0.8mm thick cold-rolled sheet steel of the 0.022% C - 0.01% Si - 0.18% Mn - 0.006% P - 0.004% S - 0.06% Al - 0.0018% N composition. It was annealed with the heat cycle shown in Fig. 14. As evident from the figure, the minimum cooling rate β and maximum cooling end temperature T_E must be 70°C/s and 300°C, respectively, in order to achieve an aging index of $\leq$ 30 MPa by 150 s of overaging treatment. The effect of holding at the cooling end temperature is not so large. A slow reheating rate is favorable, but its effect was insignificant within the experimental range. Conversely, a rapid reheating rate, for instance, 80°C/s is feasible, which should facilitate the shortening of the furnace length. The non-aging property can be achieved if the cooling rate after slight supercooling and reheating falls within the cooling range shown in Fig. 13. Moreover, Hayashida et al. proposed the optimum cooling rate to reduce the solute carbon effectively, based on the experiments on the influence of cooling rate after reheating(7). It is most preferred to cool the steel along the optimum cooling curve theoretically derived by Kurihara et al(8).

Effect of carbon content on aging property:

Figure 15 shows the effect of carbon content on the aging property of low-carbon aluminum-killed steels containing 0.008 to 0.022% C, 0.008% Si, 0.13% Mn, 0.007% P, 0.008% S, 0.063% Al and 0.0017% N, and given the R-OA. The slab reheating temperature was 1050°C, the coiling temperature was 720°C, the reduction of cold-rolling was 80%, the sheet thickness was 0.8mm, and the overaging time was 150 s. The minimum carbon content should be preferably set at 0.012% in order to obtain the substantial non-aging property. This suggests that a supersaturated carbon degree of 100ppm or over is necessary prior to the overaging treatment.

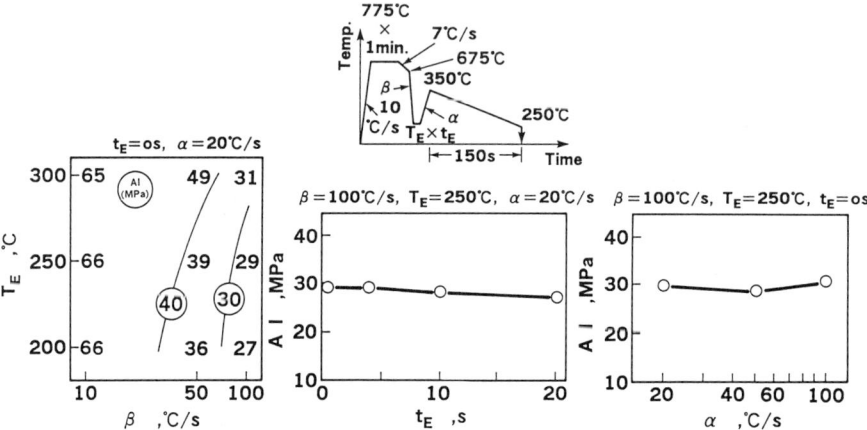

Figure 14 - Influences of the combination of β and T_E, t_E and α on AI (aging index) after annealing with R-OA cycle shown in the figure.

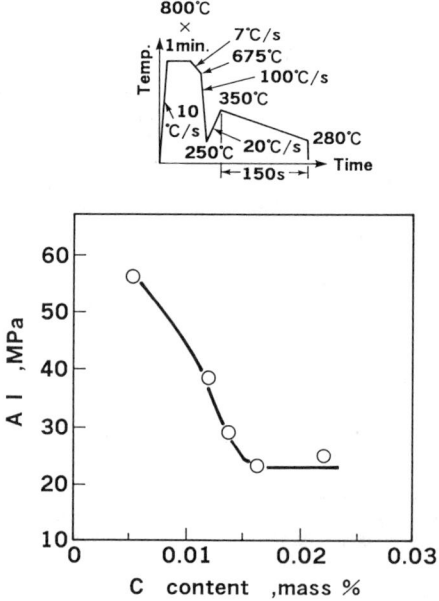

Figure 15 - Influence of C content on AI (aging index) after annealing with R-OA cycle shown in the figure.

Technique for coiling hot-rolled band at low-temperature

Conventional technique and position of present study:

The technique of coiling hot-rolled bands at high-temperatures has been conventionally used to obtain excellent formability from low-carbon aluminum-killed steel even by such short-time annealing involving rapid heating as continuous annealing(9). The metallurgical significance of this high-temperature coiling technique lies in the coarsening of cementite and full precipitation of aluminum nitride (AℓN) in the hot-rolled band. The coarsening of cementite reduces the frequency of nucleation of the recrystallized grains with random orientations nucleating from the vicinity of cementite, lowers the amount of dissolved carbon from cementite during annealing, and thus promotes the evolution of the {111} recrystallization texture. The full precipitation of AℓN reduces the amount of nitrogen dissolved in the hot-rolled band, which in turn inhibits the precipitation of fine AℓN and encourages grain growth during annealing. The high-temperature coiling technique poses the problems already discussed however. This study therefore was aimed at developing a low-temperature coiling technique that can overcome such problems. A basic study was made in order to obtain the aforementioned metallurgical benefits even if low-temperature coiling is employed. As it was known that the problems with high-temperature coiling can be solved if the coiling temperature is 650°C or lower, 650°C was made the target temperature for low-temperature coiling.

Technique for cementite coarsening in hot-rolled bands

Conditions for coarse cementite:

The influence of carbon on the morphology of cementite in hot-rolled bands was examined using specimens C1 to C6 in Table III, which were reheated to 1050°C and hot-rolled followed by coiling at 650°C (Fig. 16). Figure 17 shows the relationship between the inter-particle spacing of the cementite and the carbon content. The coarse dispersion of cementite particles in hot-rolled bands was achieved by lowering the carbon content below 0.02%. The inter-particle spacing of the specimen with 0.017% C, for an example, was about 60μm, which is comparable to the value about 70μm in the conventionally processed specimen (0.03% C) coiled at 730°C(10). However, from the point of view of non-aging property, a lower limit of the carbon content is considered to be about 0.012% (Fig. 15).

Table III Chemical composition of steels used (mass %).

Steel	C	Si	Mn	P	S	Al	N
C1	0.005	0.011	0.110	0.007	0.0070	0.071	0.0020
C2	0.009	0.015	0.120	0.009	0.0050	0.063	0.0020
C3	0.015	0.015	0.120	0.005	0.0060	0.061	0.0021
C4	0.017	0.014	0.130	0.005	0.0070	0.056	0.0029
C5	0.022	0.014	0.110	0.007	0.0060	0.075	0.0017
C6	0.032	0.013	0.110	0.007	0.0060	0.073	0.0019
A1	0.013	0.013	0.110	0.005	0.0070	0.023	0.0023
A2	0.013	0.013	0.110	0.005	0.0070	0.043	0.0025
A3	0.014	0.013	0.110	0.005	0.0070	0.068	0.0026
A4	0.013	0.014	0.110	0.005	0.0070	0.095	0.0028
M1	0.013	0.011	0.025	0.008	0.0066	0.054	0.0011
M2	0.013	0.009	0.050	0.008	0.0069	0.057	0.0019
M3	0.011	0.010	0.110	0.008	0.0063	0.059	0.0012
M4	0.014	0.009	0.150	0.005	0.0057	0.080	0.0014
M5	0.011	0.009	0.280	0.005	0.0051	0.070	0.0013
M6	0.014	0.010	0.460	0.005	0.0056	0.078	0.0013

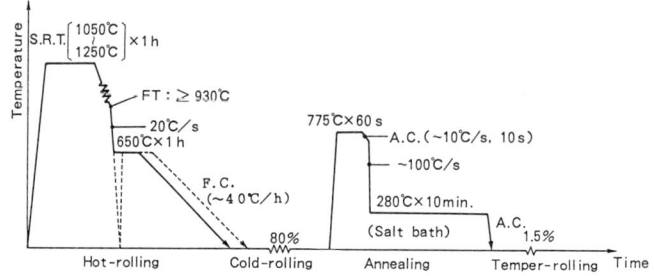

Figure 16 - Diagram showing experimental procedures.

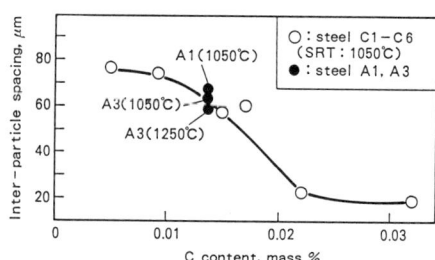

Figure 17 - Changes in inter-particle spacing of cementite with C content in hot-rolled bands coiled at 650°C (steels C1 ~ C6 in Table III).
Data showing the influence of Al content and slab reheating temperature are also plotted (steels A1 (0.023% Al), A3 (0.068% Al) in Table III).

The aluminium content was confirmed as not influencing the coarsening of cementite in the hot bands coiled at 650°C within the range examined (A1~A4 in Table III and Fig. 17).

The influences of manganese content and slab reheating temperature on the coarsening of cementite in hot-rolled bands were represented in Figs. 18 and 19 (M1 ~ M6 in Table III). The decrease in manganese content, especially below 0.1%, is distinctively effective in the coarsening of cementite, whilst the slab reheating temperature has almost no effect on it.

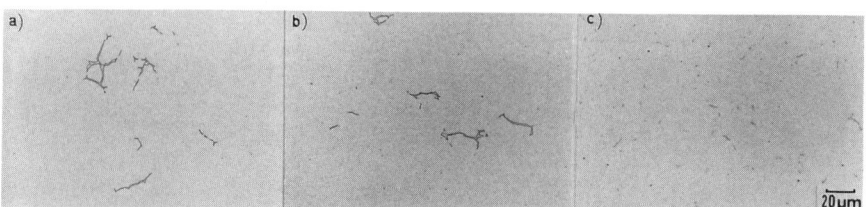

Figure 18 - Optical micrographs showing the influence of Mn content on cementite coarsening in hot-rolled bands coiled at 650°C (Slab reheating temperature: 1150°C; steels a) 0.025% Mn (M1), b) 0.11% Mn (M3), c) 0.28% Mn (M5) in Table III).

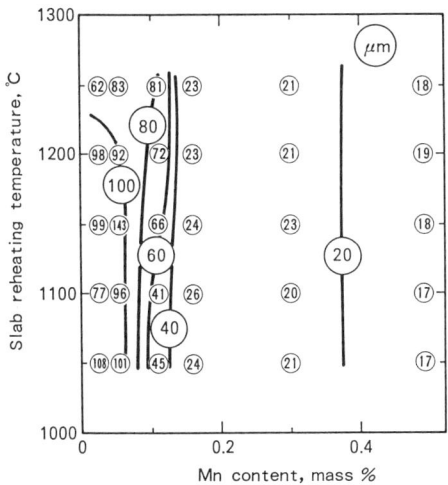

Figure 19 - Influences of Mn content and slab reheating temperature on inter-particle spacing of cementite in hot-rolled bands coiled at 650°C (steels M1 ~ M6 in Table III).

Mechanism of cementite coarsening

Influence of carbon content:

Cementite in hot-rolled bands forms during rapid cooling on the run-out-table after hot-rolling and/or during slow cooling after coiling.

Firstly the phase transformation and the cementite precipitation behaviors during continuous cooling from a single austenite region were experimentally investigated, taking into account the fact that the cooling rate in the practical run-out-table is about 20°C/s. The specimens used were low carbon and low manganese aluminum-killed steels with two different carbon contents, i.e. 0.012% C (LC) and 0.038% C (MC), whose chemical compositions are listed in Table IV. The specimens were reheated once to 930°C in the austenite region and held for 20 min followed by continuous cooling at a rate of 20°C/s. The microstructual changes during cooling were examined using optical microscope for the specimens quenched from various temperatures during cooling.

Table IV Chemical composition of steels used (mass %).

steel	C	Si	Mn	P	S	Al	N
LC	0.012	<0.010	0.07	0.007	0.0081	0.073	0.0024
MC	0.038	<0.010	0.08	0.008	0.0083	0.074	0.0017

In the specimen with 0.012% C, cementite commences to precipitate in the ferrite grain boundaries at around 560°C after the completion of phase transformation from γ to α. In the specimen with 0.038% C, on the contrary, pearlite transformation starts around 690°C followed by its growth during cooling. Based upon these experimental data, it is possible to draw schematic CCT curves (Fig. 20). As Seter et al.(11) has pointed out, steels with more than 0.02% C have pearlite transformation on the run-out-table resulting in the fine pearlite nodules in hot-rolled bands coiled at low temperature (Fig. 20 c)②). The steel with less than 0.02% C is thought conversely to form coarse cementite particles in hot-rolled bands even when they are coiled at 650°C, since cementite starts to precipitate in ferrite grain boundaries during slow cooling after coiling (Fig. 20 b)①). Therefore it is necessary to reduce the carbon content to less than 0.02% in order to obtain coarse cementite in hot-rolled bands coiled at 650°C.

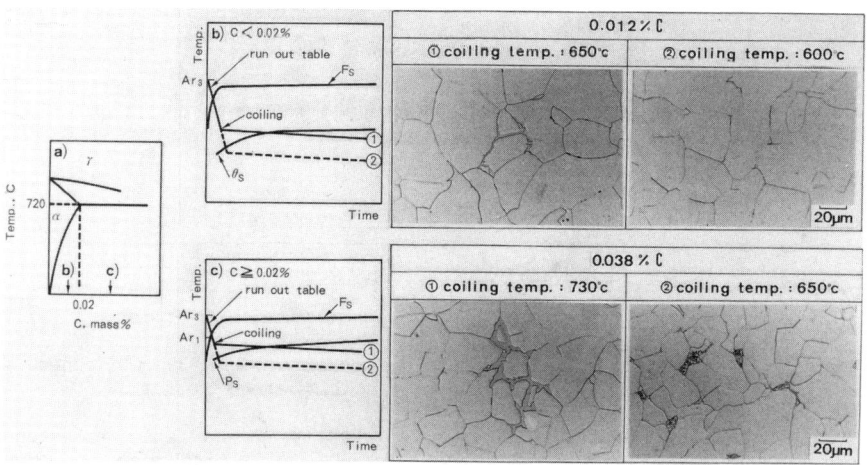

Figure 20 - Schematic illustration of CCT curves for two low-C steels showing the influences of C content and coiling temperature on cementite coarsening.

Influence of manganese content:

Cementite in hot-rolled bands coiled at 650°C tends to aggregate with decreasing manganese content, the mechanism of which is discussed here. Since the preferred steels contain less than 0.02% C, cementite is supposed to precipitate in ferrite grain boundaries of hot-rolled bands after the completion of the phase transformation from γ to α. Cementite is assumed to form by the nucleation and growth mechanism.

The influence of manganese on the solubility of carbon in ferrite which is in equilibrium with cementite is of most importance, since the solubility is closely related to the temperature at which cementite commences to precipitate. One of the present authors reported that manganese hardly affects the solubility(12). Therefore, the commencing temperature of cementite precipitation is assumed almost independent of manganese content, as long as the carbon content is identical.

Taking into consideration the fact that the nucleation site of cementite is the ferrite grain boundary, the influence of manganese on the following two factors with respect to the nucleation process becomes important: a) the ferrite grain size, b) the distribution density of precipitates such as MnS in the ferrite grain boundaries of hot-rolled bands. Firstly, it was confirmed that the grain size of hot-rolled bands is scarcely affected by the presence of manganese, namely, about 20 to 25μm for all the specimens. This negates the possibility of condition a) mentioned above. Secondly, the variation in the distribution density of precipitates with manganese content in hot-rolled bands is demonstrated in Fig. 9. The slab reheating temperature was identical at 1100°C. It is evident that the distribution density of precipitates supposed as MnS increases with deceasing manganese content. Supposing that the precipitates in grain boundaries act as the preferential nucleation sites of cementite, coarse cementite particles are expected to form by increasing the manganese content. This expectation is in complete disagreement with the experimental observations shown in Figs. 18 and 19. Therefore, the factor b) as a mechanism can also be disregarded. This conclusion is further supported by the fact that cementite coarsening is not affected by the slab reheating temperature despite the decrease in the distribution density of precipitates by reducing the slab reheating temperature.

Next, the influence of manganese on cementite coarsening is discussed from the aspect of the growth of cementite. The factors affecting the growth rate are classified as follows: a) influence of manganese on carbon diffusion in ferrite, b) influence of manganese on the growth rate of cementite due to the partitioning of the slower diffuser manganese between ferrite and cementites. Regarding factor a), no reliable experimental data are thought to be available. Nishizawa(13) investigated the influence of a third element on carbon diffusion in austenite thermodynamically. In the present study, his analysis was extended to the case of ferrite. The presence of the third element (M) modifies the diffusivity of carbon in ferrite as described in eq. (1),

$$D_c^M = D_c \left(1 - \frac{W_{MC}}{RT} \cdot x_M \cdot x_c\right), \quad (1)$$

where D_c and D_c^M are the diffusion coefficients of carbon in Fe-C and in Fe-C-M systems, respectively, W_{MC} is the interaction parameter between M and C atoms, and x_M and x_c are the atomic fraction of M and C atoms, respectively. Substituting $x_M = 3.0 \times 10^{-3}$ (0.3% Mn), $x_c = 9.3 \times 10^{-4}$ (0.02% C), $W_{MC} = 0.26eV(14)$ and T=923K into eq. (1), D_c^M was found to be almost the same as D_c. Therefore, factor a) mentioned above is concluded not to be significant.

The mechanism of the influence of manganese on the growth rate of cementite is discussed with respect to factor b). The line chemical analyses of cementite in hot-rolled bands coiled at 650°C by means of EPMA reveal that manganese enriched to cementite in the specimen M5 in Table III containing 0.28% Mn (Fig. 21). The enrichment of manganese was observed in a number of cementite particles. Although EPMA allows us only semi-quantitative evaluation in terms of the amount of the enrichment, the content of manganese in cementite was roughly evaluated to be 1% for the specimen with 0.28% Mn. According to Hillert et al.(15), the equilibrium content of manganese in cementite at 650°C is expected to be 4.5% using the partitioning coefficient with the value of about 15. The partitioning of manganese between ferrite and cementite has the possibility of retarding the growth rate of cementite. In the present study, the growth rate is evaluated as follows: Figure 22 shows the isothermal compositional section at 650°C. Supposing that i) the steels considered here contain 0% and 0.3%

Mn, as described A and B in Fig. 22, respectively, ii) cementite grows with the common supersaturation of carbon with the amount of 20ppm. Firstly, in steel A, the growth rate of cementite, v (cm· $s^{-1/2}$), can be evaluated using eq. (2) which was proposed by Coates(16), based on the mass conservation of carbon at the growing interface,

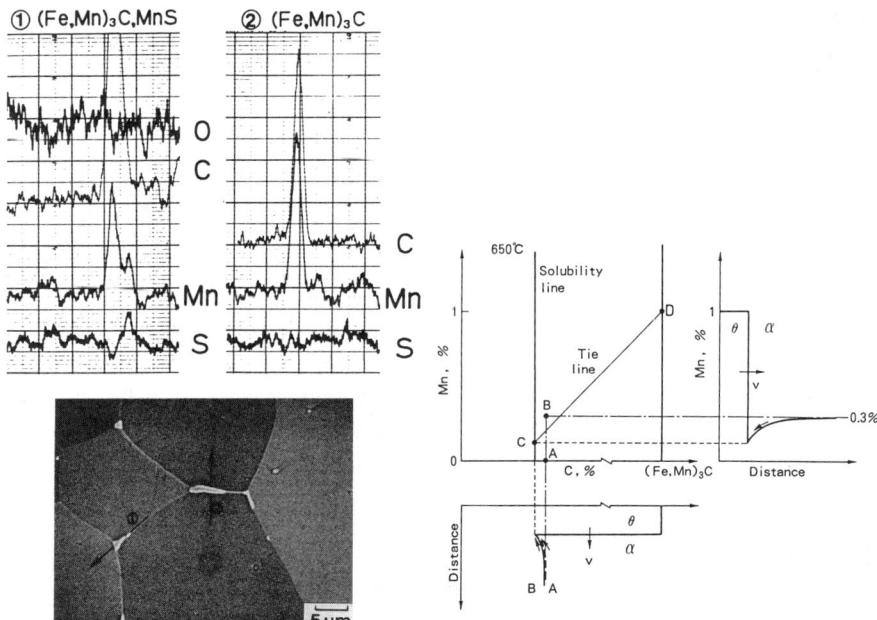

Figure 21 - EPMA analyses showing the enrichment of Mn in cementite in hot-rolled band coiled at 650°C (steel M5 (0.28% Mn) in Table III).

Figure 22 - Schematic illustration of isothermal compositional section showing the influence of Mn on the growth rate of cementite.

$$\frac{\bar{C} - C^{\alpha\theta}}{C^{\theta} - C^{\alpha\theta}} = \sqrt{\frac{\pi}{4D_c}} \cdot v \cdot \left[1 - \mathrm{erf}\left(-\frac{v}{\sqrt{4D_c}}\right)\right] \cdot \exp\left(\frac{v^2}{4D_c}\right), \qquad (2)$$

where $\bar{C}$ is the average carbon content in matrix, $C^{\alpha\theta}$ is the equilibrium C content in ferrite at the α/θ interface, C^{θ} is the equilibrium C content in cementite, and D_c is the diffusion coefficient of carbon in ferrite. In the equation (2), i) one dimensional parabolic growth rate, and ii) no soft impingement during growth were assumed. The growth rate of cementite (v) in steel A was evaluated to be 3.8×10^{-7} cm·$s^{-1/2}$. Furthermore, Coates(16) reported a similar approach in ternary alloys (Fe-C-M) like the steel B. Assuming local equilibrium, the growth rate of cementite was determined so as to satisfy the mass balance at the α/θ interface with respect to both carbon and manganese atoms. The growth rate (v) in steel B was calculated to be 4.1×10^{-8} cm·$s^{-1/2}$. The tie line was determined to be the line CD as described in Fig. 22. The absolute values of the growth rates were found out to be lower than the actual values, since the bulk diffusivities of carbon and manganese atoms were used. However, the relative comparison is considered to be acceptable. It was clarified that the addition of manganese up to 0.3% may retard the growth rate of cementite by the order of one because of the enrichment of manganese atoms in the cementite. Therefore, it is natural to speculate that in the case of steels with the addition of manganese, cementite particles in hot-rolled bands tend to be finely distributed due to the retardation of the growth of cementite, even if the nucleation of cementite particles themselves are not affected.

Precipitation of AℓN in hot-rolled bands

The influence of aluminium content on the precipitation ratio of AℓN (N as AℓN/Total N) in hot-rolled bands was investigated using specimens A1 to A4 in Table III. As Fig. 23 shows, the increase in aluminium content, more than 0.05%, makes it possible to fix more than 70% of total nitrogen as AℓN even if hot-rolled bands are coiled at 650°C. This is rather close to that coiled at a high temperature (e.g. 730°C). Furthermore lower slab reheating temperatures tend to fix more nitrogen as AℓN.

The extraction replica images of the precipitates observed in hot-rolled bands are shown in Fig. 24. The globular precipitates observed in the specimen A1, namely 0.02% Aℓ, was confirmed to be MnS (Fig. 24 a)), whereas, the precipitates observed in specimen A3, namely 0.068% Aℓ, are similar to those observed in specimen A1 with respect to their size and distribution (Fig. 5 b)). However, they have a rather angular shape and were confirmed to be the compound precipitates of (MnS + AℓN) (1) based on the energy dispersive X-ray analyses and electron diffraction patterns. The frequency of observing the compound precipitates of (MnS + AℓN) increases with increasing aluminium content. Moreover higher slab reheating of the specimen A3 up to 1250°C resulted in a large number of fine precipitates in hot-rolled bands (Fig. 24 C)). They were also confirmed to be the compound precipitates of (MnS + AℓN).

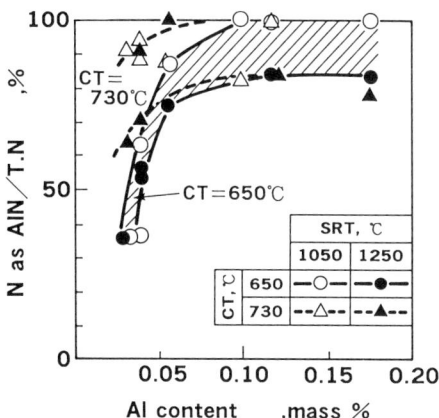

Figure 23 - Influence of Al content, coiling temperature, and slab reheating temperature on AlN precipitation in hot-rolled bands (Steels A1 ~ A4 in Table III).

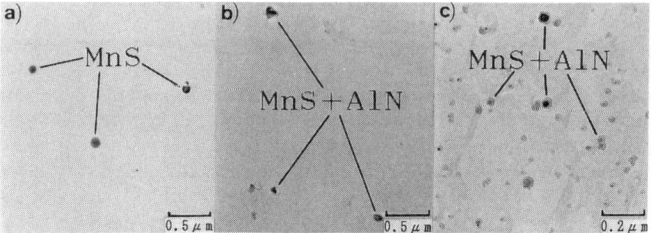

Figure 24 - TEM observation of exaction replica of precipitates in hot-rolled bands coiled at 650°C.
a) steel A1 (0.023% Al), SRT: 1050°C,
b) steel A3 (0.068% Al), SRT: 1050°C,
c) steel A3 (0.068% Al), SRT: 1250°C, SRT: Slab reheating temperature

According to Turkdogan et al.(17), the solution temperature of MnS in the specimen A1 ~ A4 is 1153°C, whilst after Leslie et al.(18) those of AℓN in specimen A1 (0.024% Aℓ) and A3 (0.068% Aℓ) are calculated to be 1002°C and 1142°C, respectively. Therefore, it is considered that the larger compound precipitates observed in the specimens with high aluminium content (e.g. A3) are formed during the slab reheating stage, when low slab reheating temperatures of 1050°C are used. In the case of high slab reheating temperatures of 1250°C, it is speculated that the dissolved MnS precipitates again very finely during hot-rolled followed by the precipitation of AℓN on them. When the aluminium content is low such as in the specimen A1, aluminium can hardly fix nitrogen as AℓN in hot-rolled bands coiled at low temperature, which results in the single precipitates as MnS. Therefore the addition of amounts of aluminium of more than 0.05% is considered to be very effective in fixing nitrogen as AℓN.

Tensile properties of annealed sheet

In general, the decrease in coiling temperature tends to impair formability, for instance, the deterioration of r̄ - value, the rise in yield strength and the decrease in elongation. However, Fig. 25 demonstrates that the decrease in manganese content of the specimens with C ≦ 0.02% and Al ≧ 0.05% and the reduction in the slab reheating temperature have a marked effect on improving formability with regard to

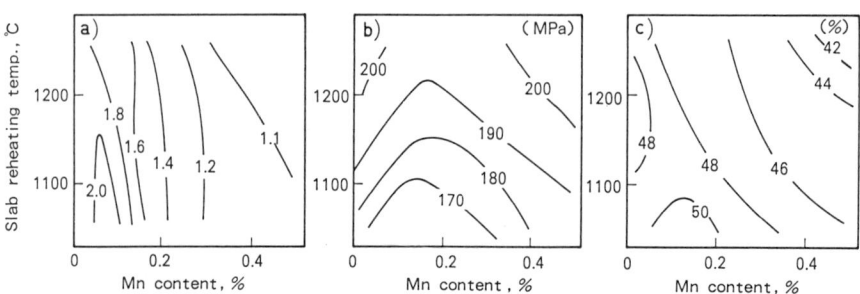

Figure 25 - Influences of Mn content and slab reheating temperature on tensile properties of annealed sheets (coiling temperature: 650°C; steels: M1 ~ M6 in Table III).
a) r̄, b) YP, c) El

the r̄ - value, yield strength and elongation (steel M1 - M6 in Table III). These results are inferred to be closely related to the coarsening of cementite and the precipitation of AℓN in hot-rolled bands as well as the decrease in the content of Mn-C complex(10) during annealing.

Prevention of hot-shortness by low slab reheating temperature

The reduction in manganese content has been shown to play a significant part in improving the mechanical properties of the final products. On the other hand, it has a shortcoming with regard to the hot-shortness. Figure 26 shows the influences of manganese content and slab reheating temperature on the hot-shortness. It is evident that the reduction in the slab reheating temperature has a significant effect in preventing the embrittlement caused by lowering the manganese content. For example, in the case of the specimen with 0.1% Mn and 0.007% S, no hot-shortness was observed when reducing the slab reheating temperature below 1150°C.

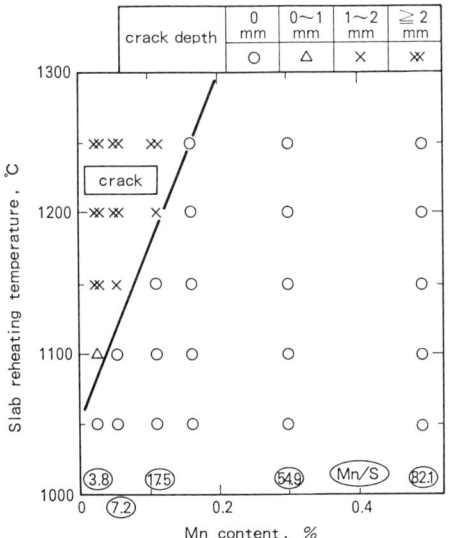

Figure 26 - Influences of Mn content and slab reheating temperature on the hot-shortness in hot-rolled bands (steel: M1 ~ M6 in Table III).

The observation of the fracture surface by means of SEM allowed us to speculate on the mechanism as the liquid film embrittlement of austenite grain boundaries, the reason for which is as follows. In the case where the manganese content decreases too much and/or the slab reheating temperature increases high enough to dissolve MnS, the content of sulfur unfixed by manganese as MnS increases. Since sulfur has a tendency to segregate into grain boundaries, the content of sulfur in the austenite grain boundaries is supposed to become significantly high. Taking into consideration the fact that the molten steel with the high content of sulfur has a eutectic reaction at 988°C, the existence of the liquid film in the austenite grain boundaries is expected above the eutectic temperature which caused hot-shortness(19).

Results of mill tests

Non-aging test:

If the overaging treatment during continuous annealing is the R-OA described before, solute carbon can be efficiently reduced and the non-aging property can be obtained. For instance, a 1000-kW experimental induction heater was installed in the upward path section after the air-water mist cooling zone of a commercial continuous annealing furnace to provide small-scale, reheating as shown in Fig. 27. Steel strip was actually processed through the continuous annealing line to evaluate its aging and mechanical properties. The test materials were conventional low-carbon aluminum-killed steels containing 0.016 to 0.022% C, 0.10 to 0.19% Mn, 0.003 to 0.012% P, 0.004 to 0.014% S, 0.034 to 0.053% Aℓ and 0.0015 to 0.0025% N and coiled at high-temperatures of 721 to 741°C, and a newly developed low-carbon aluminum-killed steel having the 0.016% C - 0.09% Mn - 0.007% P - 0.006% S - 0.063% Aℓ - 0.0012% N composition and coiled at the low-temperature of 650°C. The annealing temperature was about 780°C. The quenching end temperature T_E was about 250 to 300°C for most of the specimens. Some specimens were cooled to 200 to 250°C to increase the degree of supercooling. The reheating temperature T_R was about 350°C at a heating rate of 80°C and the ramp overaging end temperature T_{OAE} was about 250°C. A complete overaging property of AI $\leq$ 20 MPa is stably obtained by over 5 min of overaging, as shown in Fig. 28. If the T_E is low enough and the supercooling degree is large enough, complete non-aging property can be secured even by 3 min of overaging. A substantial non-aging level of AI $\leq$ 30 MPa can be stably attained by about 2 min of overaging. The aging index of the new low-carbon aluminum-killed steel developed for low-temperature coiling falls within the aging index range of conventional steels as shown in Fig. 28 and is the non-aging level in accord with what has been discussed above. The tensile properties of the new steel after an accelerated aging treatment at 100°C for 1 h are a yield strength of 182 MPa, elongation of 46.5%, and r̄ value of 1.80. These values are higher than the mechanical properties required for non-aging deep-drawing quality cold-rolled sheet steels.

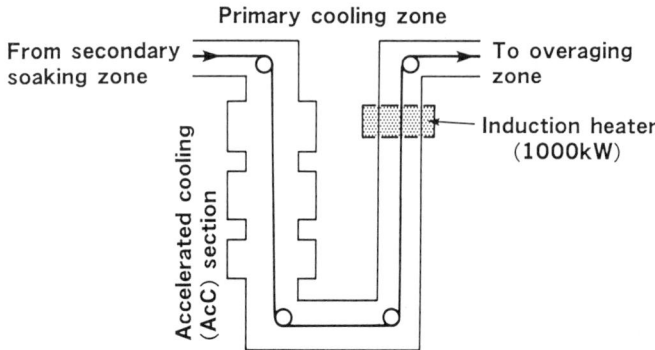

Figure 27 - Outline of experimental reheating facility installed in practical continuous annealing line.

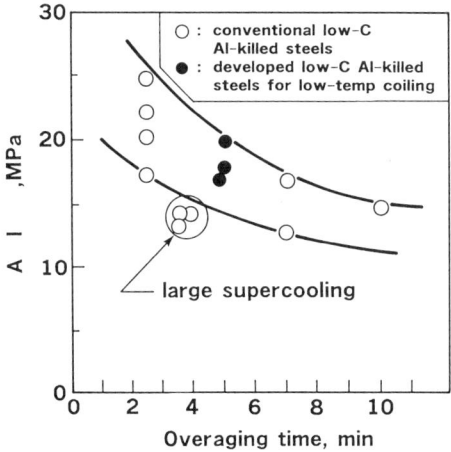

Figure 28 - Relationship between ΔI (aging index) and overaging time of the continuously-annealed sheets with practical R-OA cycle.

Low-temperature coiling test:

Based upon the findings described above, practical mill test for low-temperature coiling was also carried out. Table V shows the chemical composition and the production conditions from hot-rolling to temper-rolling of the steel tested. The coiling temperature was 650°C. Figure 29 shows the along-the-length distribution of the $\bar{r}$ - value of the final product, which is very significant from the industrial point of view. For the purpose of comparison, a representative result in the case of the conventional steel with about 0.03% C, 0.2% Mn, and 0.03% Aℓ coiled at 750°C was also plotted in the same figure. It is evident that by employing low temperature coiling based on the present proposal, not only is the level of $\bar{r}$ - value obtained sufficiently high, but also is the along-the-length distribution significantly improved.

The effectiveness of the present technology was also confirmed also by another practical mill test, as reported by Matsuzu et al.(20). Therefore the improvement of production yield is expected by the present low-temperature coiling technology.

Table V Chemical composition and production parameters of steels mill-tested for low-temperature coiling.

Chemical composition (mass %)						Hot-rolling (°C)			Cold-rolling	Annealing	Temper-rolling
C	Mn	P	S	Al	N	SRT	FT	CT	Reduction (%)	(°C - s)	Reduction (%)
0.016	0.09	0.004	0.006	0.068	0.0015	1070	915	650	80	780 - 60	1.0

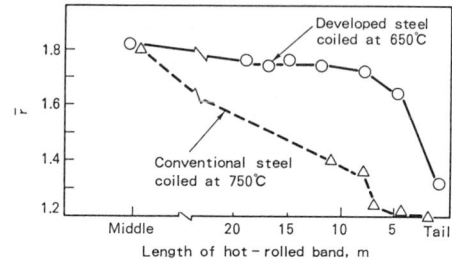

Figure 29 - Through-the-length distribution of $\bar{r}$-value of annealed sheets made on a production scale for developed steel coiled at 650°C, in comparison with that resulted from conventional steel with 0.03% C, 0.2% Mn and 0.03% Al coiled at 750°C.

Conclusions

Basic technology was developed for manufacturing non-aging deep-drawing quality cold-rolled sheets (DDQ) from low-carbon aluminum-killed steel by continuous annealing.

(1) Non-aging technique

(a) A target value of the aging index was established as the non-aging property. An aging index of ≦ 30 MPa is considered a substantial non-aging level, while an aging index of ≦ 20 MPa is considered a complete non-aging level.

(b) A new heat cycle, called the slight supercooling, reheating and ramp overaging (R-OA), was proposed. This treatment method can accomplish the aging indexes of ≦ 30 MPa and ≦ 20 MPa by about 120 and 300 s of overaging time, respectively.

(c) The R-OA utilizes MnS precipitates as preferential nucleation sites for transgranular cementite, with the result that the amount of solute carbon can be sufficiently reduced by the short-distance diffusion of carbon. A cementite precipitation model was developed and used to optimize the heat cycle.

(2) Technique for low-temperature coiling of hot-rolled band:

Cold-rolled sheets with excellent formability can be produced from hot-rolled bands coiled at a low-temperature of 650°C by optimizing the chemical composition (0.012 ≦ C ≦ 0.02%, Mn ≦ 0.15%, Aℓ ≧ 0.05%) and reheating the slab at a low temperature of under 1150°C.

(3) According to the findings noted in (1) and (2) above, an experimental apparatus for applying the R-OA was installed in a commercial continuous annealing furnace and was used in mill tests. As a result, it was confirmed that cold-rolled sheets with non-aging characteristics and superior formability can be produced at high yield.

Acknowledgements

The authors express their thanks to Drs. H. Takechi, and H. Katoh in Nippon Steel Corp. for their valuable discussion and suggestions. Thanks are also extended to Mr. M. Takahashi in Sheet and Coil, Steel Research Lab., Technical Development Bureau of Nippon Steel Corp. for the calculation of cementite growth rate.

References

1. H. Katoh, H. Takechi, N. Takahashi and M. Abe, "Cold-rolled steel sheets produced by continuous annealing" Technology of Continuously Annealed Cold-Rolled Sheet Steel, ed. R. Pradhan, TMS-AIME, (Detroit, Michigan, 1984), 37-58

2. T. Katayama, K. Ushioda and M. Takita, "Effect of carbon on stretchability under plane strain state," Proceedings of Advanced Technology of Plasticity, vol. 2 (Kyoto, 1990), 1361-1368

3. M. Abe and K. Ushioda, "Effect of carbon on mechanical properties in mild steel sheet," Scand J. of Metallurgy, 13 (1984), 276-282

4. H. Katoh, K. Koyama and K. Kawasaki, "The kinetics of precipitation of cementites during over-aging in continuous annealing," Technology of Continuously Annealed Cold-Rolled Sheet Steel, ed. R. Pradhan, TMS-AIME, (Detroit, Michigan, 1984), 79-94

5. K. Koyama, S. Kuroda, H. Katoh and M. Nagumo, "Derivation of kinetics model for solute carbon during rapid-cooling in continuous annealing process for low-carbon, cold-rolled sheet steels," Tetsu-to-Haganè, 71 (1985), 1497-1503

6. K. Koyama, H. Katoh and M. Nagumo, "A kinetics model for carbide precipitation during over-aging in continuous annealing of low-carbon, cold-rolled sheet steels," Tetsu-to-Haganè, 72 (1986), 823-830

7. T. Hayashida, M. Oda, T. Yamada and T. Ukena, "Effect of thermal histories of over-aging on the rate of precipitation of Fe_3C in continuous annealed low-carbon $A\ell$-killed steel sheets," CAMP-ISIJ, 3 (1990), 1822

8. K. Kurihara and K. Nakaoka, "Metallurgical analysis of the over-aging process in continuous annealing," Metallurgy of Continuous-Annealed Sheet Steel, ed. B.L. Bramfitt and P.L. Mangonon, Jr., TMS-AIME (Dallas, Texas., 1982), 117-132

9. H. Kubodera, K. Nakaoka, K. Araki, K. Watanabe and K. Iwase, "Effect of the high-temperature coiling on the properties of continuously annealed product, Tetsu-to-Haganè, 62 (1976), 846-855

10. K. Ushioda, W.B. Hutchinson, J. Ågren and U. von Schlippenbach, "Investigation of structure and texture development during annealing of low-carbon steel," Mater. Sci. Technol., 2 (1986), 807-815

11. B. Seter, U. Bergström and W.B. Hutchinson, "Extra deep-drawing quality steels by continuous annealing", Scand. J. Metallurgy, 13 (1984), 214-219

12. H. Saitoh and K.Ushioda, "Influences of manganese on internal friction and carbon solubility determined by combination of infrared absorption in ferrite of low-carbon steels, ISIJ International, 29 (1989), 960-965

13. T. Nishizawa, "Thermodynamics of ferro-alloy", Bull. Jpn. Inst. Met., 12 (1973), 401-417

14. H. Abe, "Mn-C dipole formed in ferrite of low-carbon steel", Memoirs of the Low-carbon Sheet Steels Research Committee, ISIJ, Tokyo, (1987), 203-213

15. M. Hillert, T. Wada and H. Wada, "The α-γ equilibrium in Fe-Mn, Fe-Mo, Fe-Ni, Fe-Sb, Fe-Su and Fe-W systems", J. Iron Steel Inst., 205 (1967), 539-546

16. D.E. Coates, "Diffusion-controlled precipitate growth in Ternary Systems I", Metall. Trans., 3 (1972), 1203-1212

17. E.T. Turkdogan, S. Ignatowicz and J. Pearson, "The solubility of sulphur in iron and iron-manganese alloys", J. Iron Steel Inst., 180 (1955), 349-354

18. W.C. Leslie, R.L. Rickett, C.L. Dotson and C.S. Walton "Solution and precipitation of aluminum nitride in relation to the structure of low carbon steels", Trans. Am. Soc. Met., 46 (1954), 1470-1499

19. H. Ohtani and T. Nishizawa, "Calculation of Fe-C-S Ternary Phase Diagram", Tetsu-to-Hagané, 73 (1987), 152-159

20. N. Matsuzu, K. Koyama, N. Uehara, K. Ushioda and T. Yamada, "Low temperature coiling of hot-rolled bands for producing deep-drawable Aℓ-killed steel sheets by continuous annealing", CAMP-ISIJ, 3 (1990), 788

CONTINUOUS ANNEALING OF ULC-Ti FERRITIC HOT-ROLLED STRIPS

P. MESSIEN, J.C. HERMAN, V. LEROY, C.R.M., Abbaye du Val Benoît
B-4000 LIEGE, Belgium.
Ph. HARLET, F. BECO, L. RENARD, R.D.C.S., Bld de Colonster, B57
Sart Tilman, B-4000 LIEGE, Belgium.

Abstract

Ferritic rolling of IF steels is a new economical rolling schedule to produce a thin hot-strip with a strained ferrite microstructure either for direct recrystallization in a continuous annealing line without intermediate cold-rolling, or for subsequent cold-rolling and continuous annealing. From laboratory studies at CRM and industrial trials at COCKERILL-SAMBRE, it is concluded that direct annealing of the strained hot-strips allows to produce a soft DQ quality in gauge ranging from 0.8 to 1.6 mm. Cold rolling and annealing of the ferritic hot-strips produce a DDQ quality in thinner gauges (0.4 mm), using limited cold reduction rates in the tandem mill. Results show that lubricated rolling improves the drawability reducing the superficial shear texture <110>//ND of the annealed strips. Comparative results are obtained on ELC steels.

Introduction

Thin hot strips (< 2,25 mm) cannot be easily produced by rolling in the homogeneous austenite on actual finishing mills because of the important temperature drop in the last stands of the mill. This difficulty increases due to higher transformation temperature when rolling the new steel grades with reduced carbon and manganese contents as in the low ELC and ultra low ULC-Ti steels. Partial rolling in the austenite-ferrite two phases region in the last stands leads to reduced quality[1] of the mill products : poorer flatness, thickness inhomogeneities, microstructural defects, reduced drawability.

A solution to the problem would be to reduce the reheating and entry temperatures in the finishing mill and roll the thin strips entirely in the ferrite[2,3,4]. Many advantages can be derived from such a ferritic hot rolling process : energy saving for slabs being reheated at lower temperatures, saving in costly additions, conditions for reduced oxidation of the strips and rolls, increase of the overall yield, reduction of the water consumption on the runout table. Hot rolled strips in reduced thicknesses (1.5 - 2.5 mm) are actually produced on a finishing mill at Cockerill-Sambre by rolling ELC steels in the ferrite[5]. Such ELC ferritic strips are softer, more ductile than the classical ones rolled in the austenite and particularly well adapted for stretch-forming. Most of the users requirements can be met using the new ferritic ELC quality with an energy saving up to 20% at the user's shop.

Moreover as rolling forces and torques are significantly reduced in the cold mill when rolling ferritic hot strips, the rolling limits can be enlarged for the production of steel sheets in CQ and DQ qualities.

For the production of soft, fully recrystallized, ferritic hot strips specified rolling conditions have to be met on the mill and a constant high coiling temperature must be guarantee on the entire length of the coils[5]. This observation is particularly important when processing IF steels. The requirement of a sufficient high coiling temperature are not so easily fullfilled when rolling strips in thickness lower than 1.5 mm. Partially strained microstructures are then observed on the strips with reduced ductility so that a subsequent annealing treatment is needed.

The aim of the present research work at Cockerill-Sambre and CRM was to provide the guidelines for a production of thin hot-rolled strips (≤ 1.0 mm) in DQ quality after direct annealing of the strained hot strips. The production of such a DQ quality requires that the plastic anisotropy ($\bar{r}$-value) and therefore the texture can be improved when compared to the one of classical coiled strips. An important straining of the microstructure on the hot strips is thought to be a solution to develop an important <111>//ND texture during direct continuous annealing of the ferritic strained microstructure obtained after ferritic rolling. At first sight, the ULC-Ti steel for which the recrystallisation temperature is higher than for the ELC steel, appears to be better adapted for the production of such a DQ quality. A second product considered in the present work is a cold-rolled and continuously annealed sheet from the strained thin hot strip, the objective being to produce thin sheets (≤ 0.5 mm) in drawing qualities using reduced cold rolling drafts in the tandem mill.

Experimental Details

Chemical compositions of the continuous-cast ULC-Ti and ELC steels are shown in Table I. The compressions specimens (20 x 20) mm and the 5 mm thick slabs for laboratory ferritic hot-rolling were machined from the industrials continuously cast slabs.

TABLE I - CHEMICAL COMPOSITIONS OF THE STEELS (10^{-3} % wt)

Grade	C	Mn	S	P	Si	Al	Ti	N
ULC-Ti	3	213	11	9	8	38	81	3.0
ELC	50	227	19	18	5	35	-	3.9
ELC Low Mn	18	130	9	11	3	31	-	1.3

A simulator TMT-S for compression tests was used to determine the stress-strain curves in the ferrite up to $\epsilon = 0.45$; strain rate was 2 or 10 sec^{-1}. This simulation equipment was previously described elsewhere[6]. Records were corrected to take into account the important increase of the temperature during the straining. Interpolated stress-strain curves at constant temperature were established - each 50°C in the range 50-800°C. Compression specimens were heated to 1000°C (20 min) then cooled at 1°C sec^{-1} mean rate down to the testing temperature. Graphite was used as a lubricant.

Laboratory rolling conditions in the ferrite are shown in Table II : reheating was at 1050/950°C (20 min), rolling intervals for strained microstructures were 700-600°C, 600-500°C, 550-450°C, rolling schedule was in 3 passes from 5 mm to the final thickness (0.7 - 1.6 mm), rolling speed was 40m/min and beef tallow was used on rolls as a lubricant. Coiling was at 400 or 300°C. The rolling loads were measured to get the mean resistance to hot deformation in the ferrite and comparative results were obtained for austenitic and intercritical rolling.

TABLE II - HOT ROLLING CONDITIONS

REHEATING : 1050°C/950°C - 20 min.
ROLLING INTERVALS : 700 - 600°C, 600 - 500°C, 550 - 450°C
COILING : 400°C/300°C
ROLLING : 3 passes from 5 mm to 1.6 - 0.7 mm
ROLLING SPEED : 40 m/min.
LUBRICANT : beef tallow.

Direct annealing in laboratory for the hot strips was at 825°C (ULC-Ti steel), at 720°C (ELC steels) with 60 sec. soaking time. Other continuous stop-quench annealing conditions were : 12-18°C sec^{-1} heating rate, 25-55°C sec^{-1} intermediate cooling down to 400°C overaging temperature (3 min) for ELC steels. Full quench annealing conditions were used for ULC-Ti steel.

Cold rolling was to 0.4 mm final thickness for all the strained hot strips. Continuous annealing was at 825°C (ULC-Ti) and at 720°C (ELC) for 60 sec soaking time.

Deformation cells were evaluated by TEM on the strained compression (ϵ= 0.1 and 0.4) specimens. Cellular aspects were compared in function of the straining temperature for both steels. Intercellular and intergranular fine precipitation of the cementite at high temperature in the strained ferrite was highlighted in ELC steel as a consequence of the dynamic strain aging.

The crystalline textures were determined by inverse poles figures in the rolling plane of the hot-strained and annealed materials. Texture gradients through the thickness were established from surface, 1/4 thickness and midsection measurements of the poles densities ; $\bar{r}$-values were evaluated from the textures measurements. Mechanical properties and $\bar{r}$-values were measured on triplicate specimens by using a strain rate of 0.4 mm/min ; tensile specimens with a 50 mm by 12.5 mm gauge section were machined in the rolling, transverse and diagonal directions.

Results and Discussion

Hardening and softening behaviours in the hot-strained ferrite

The appearent mean resistance to hot deformation (Km) from mean pressure measurements of the ULC-Ti and ELC steels is compared at figure 1 in function of the rolling temperature. Rolling in the ferrite at low temperatures was either on lubricated or on unlubricated rolls. As seen at figure 1, the Km values increase more rapidly for the ELC than for the ULC-Ti steels when the rolling temperature decreases in the ferrite. Mechanical damages in the finishing mill cannot be fully prevented therefore when rolling the ELC grades at low temperatures unless use can be made of a lubricant to reduced the appearent resistance in the ferrite (fig. 1). As it is assessed from the compression results, the higher Km values in ELC steel are obviously the consequence of the sensitivity to dynamic strain aging[7,8] of these steels with high solute atoms contents at the straining temperature. The compressive stresses have been plotted (fig. 2) in function of the temperature for both steels strained ($\dot{\epsilon}$ = 10 sec^{-1}) at ϵ = 0.1 and ϵ = 0.4. The monotonic increase of the stress of the ULC-Ti steel when deformation temperature decreases in the ferrite is no longer observed in the ELC steel for which the compressive stress shows a maximum at intermediate temperatures (400-500°C). As shown in figure 3, quite different morphological aspects of the deformation cells are observed in both steels strained in compression (ϵ = 0.4). Dynamic strain aging in ELC steel deformed at 450°C appears to be the consequence of solute carbon, leading to finer and irregularly shaped cells. A fine intracellular - intergranular in the recrystallized grains - cementite precipitation is clearly appearent in ELC steel quenched after straining at a higher temperature (750°C) in the ferrite.

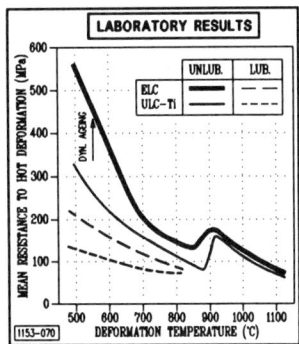

Figure 1 - Mean resistance to hot deformation of ELC and ULC-Ti steels (from rolling tests)

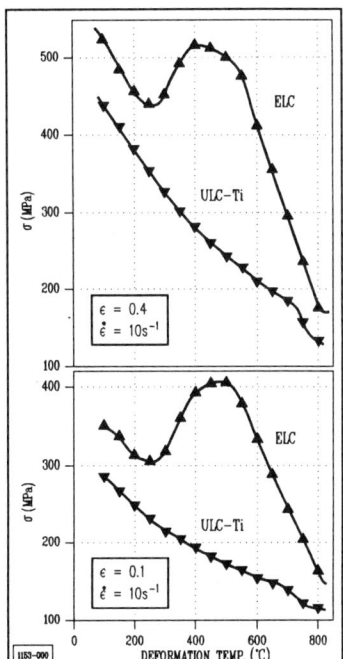

Figure 2 - Evidence of dynamic strain aging in ELC steel from TMT-S compression results.

Dynamic strain aging in ELC steel is thought therefore to be the result of a pinning effect on the moving dislocations by the segregating carbon atoms. Stress, strain and average dislocations density have to be related to get an understanding of the hardening and softening behaviours of both steels which are to be quantified as functions of the straining temperature in the ferrite.

In absence of dynamic strain aging in ULC-Ti steels, it is usual to postulate that the flow stress has two additive contributions: $\sigma = \sigma_i + \sigma_e$ where σ_i is the "internal stress" related to the micro-structure (dislocation density) and σ_e the "effective stress" or "friction stress". When deformation proceeds at high temperature, the second term is small and consequently neglected in the present study.

At the first step, dislocation density and strain have been related as in Bergström approach[8], using a modified relationship taking into account that the mean free path $s(\epsilon)$ of the mobile dislocations generated in the ferrite decreases as the straining proceeds.

It is assumed that $s(\epsilon)$ is linearly related to the average dislocations cells size $l(\epsilon)$.

$$s(\epsilon) = C.l(\epsilon) \tag{1}$$

The average size of the cells $l(\epsilon)$ is assumed to decrease during the straining according to :

$$l(\epsilon) = l(o).H(\epsilon)/H(o) \tag{2}$$

where $l(o)$ is related to the initial mean free path and $H(o)$ and $H(\epsilon)$ the initial and momentary heights of the compression specimen ; hence for constant volume :

$$s(\epsilon) = C.l(o).e^{-\epsilon} \tag{3}$$

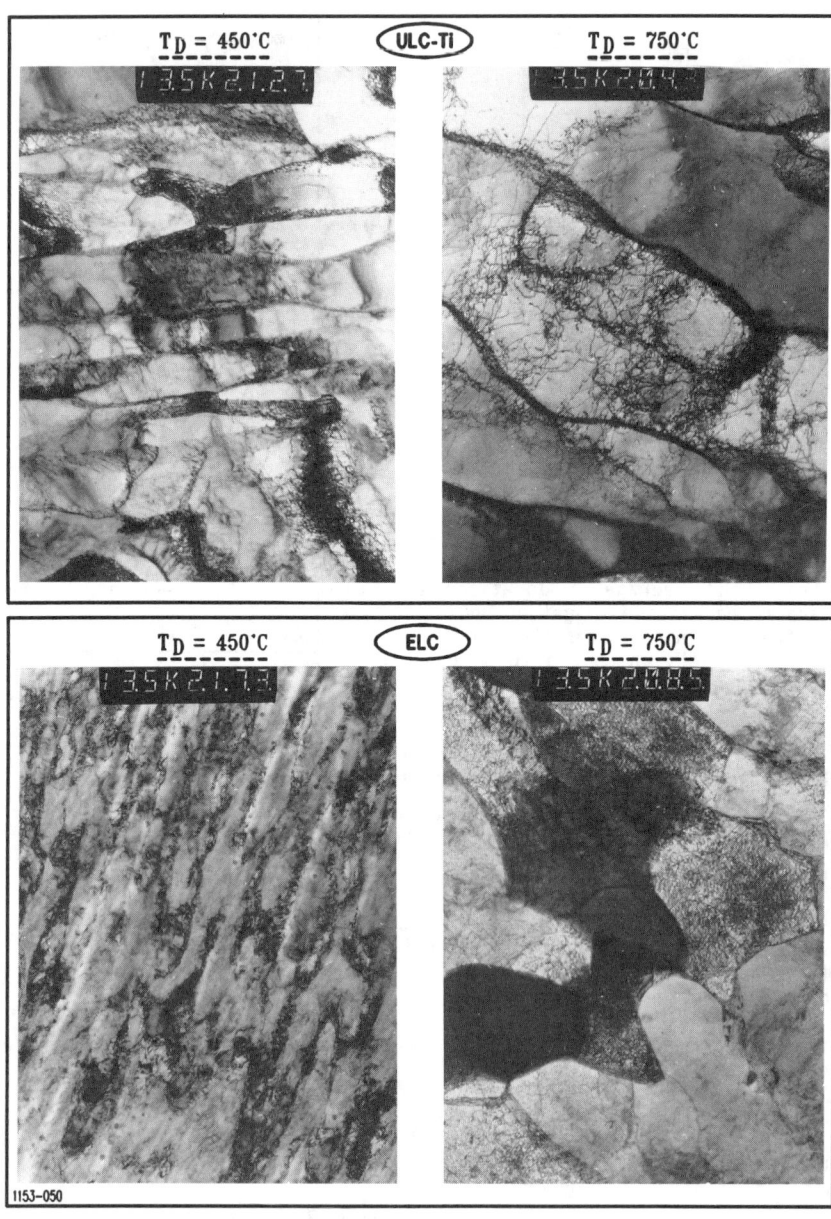

Figure 3 - Typical aspects of the deformation cells of ULC-Ti and ELC steels strained (TMT-S) at 450°C and 750°C ($\epsilon = 0.4$, $\dot{\epsilon} = 10$ s^{-1})

Taking into account this dependence of the dislocations mean free path on the strain, the Bergström's hardening parameter h which is a measure of the rate of dislocation immobilization is given by :

$$h = h_o \cdot e^\varepsilon \qquad (4)$$

The change in the dislocations density is then formulated according to relationship 5 where h_o and r are respectively the strain independant hardening parameter and the softening parameter.

$$d\rho = (h_o \cdot e^\varepsilon - r\rho) \, d\varepsilon \qquad (5)$$

The dislocations density is obtained by integration of relationship 5.

$$\rho = \frac{h_o}{1+r} (e^\varepsilon - e^{-r\varepsilon}) + \rho_o \, e^{-r\varepsilon} \quad (m^{-2}) \qquad (6)$$

The flow stress σ has been related experimentally and theoretically[9] to the dislocations density as

$$\sigma = \alpha G b \sqrt{\rho} \quad (MPa) \qquad (7)$$

where α a constant ($\alpha \sim 1$ for c.c metals), G the shear modulus[10] and b the dislocations Burgers vector.

For the present fully recrystallized and coarse grained microstructures ($\bar{d}$(ELC) = 40-50µm, $\bar{d}$(ULC-Ti) > 100 µm), the initial dislocations density ρ_o can be neglected in relationship 6. The flow stress is thus related to the strain according to relationship 8.

$$\sigma = \alpha G b \sqrt{\frac{h_o}{1+r} (e^\varepsilon - e^{-r\varepsilon})} \quad (MPa) \qquad (8)$$

Experimental stress-strain curves are correctly described by relationship 8. A good fit of experimental values has been obtained for ULC-Ti steel strained up to ε = 0.45 in the temperature range 50 - 800°C in the ferrite. The strain independant hardening parameter h_o and the softening parameter r have been expressed in function of the straining temperature. The hardening parameter h_o in ULC-Ti steel (Fig. 4a), in abscence of any dynamic strain aging, is slightly reduced as the temperature increases in the ferrite according to relationship 9.

$$H_o = 0.9610^{15} \exp(755/RT_k) \quad (m^{-2}) \quad (\dot{\varepsilon} = 10\text{sec}^{-1}) \qquad (9)$$

As seen in figure 4.b, no important softening occurs in ULC-Ti steel strained at temperatures lower than 700°C. Nevertheless softening increases rapidly at temperatures in excess of 700°C, being described by the second term in relationship 10. This high temperature softening which depends on the strain rate is reduced at high strain rates as seen in figure 5 computed from the second term of relationship 10.

$$r = 20.0 \exp(-1100/RT_k) + 1.1 \, 10.^7 \, \dot{\varepsilon}^{-0.46} \cdot \exp(-28400/RT_k) \qquad (10)$$

As a consequence interest can be on the ULC-Ti steel to get DQ or DDQ qualities by direct annealing of the strained hot strips.

On the other hand, in ELC steels, an additional stress contribution is observed due to dynamic strain aging, as seen at figure 2. This contribution, which doesn't greatly affect the microstructure cannot be expressed as a function of the dislocation density.

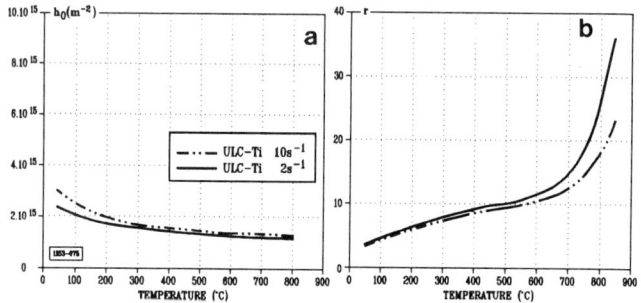

Figure 4.a - Hardening parameter in function of the straining temperature in the ferrite.
b - Softening parameter in function of the straining temperature in the ferrite

It is more related to a pinning effect of the solute carbon content. Consequently this additional contribution consumes energy during rolling, but doesn't contribute to an increase of the dislocation density in the ferritic matrix or an increase of the retained strain in the hot-rolled product, which is important for subsequent annealing.

Lubrication to reduce the shearing of the surfaces

Due to the high friction between strip and rolls an important shear strain is generated in the ferrite which tends to accumulate near the surfaces to form severely sheared zones[11,12].

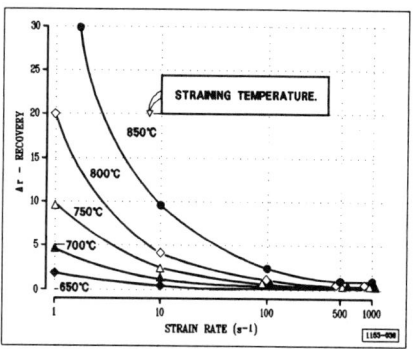

Figure 5 - Computed softening dependence temperature and strain rate in ULC-Ti steel.

As shown in figure 6a, fine recrystallized grains can appear in these severely sheared bands when rolling at low temperature (< 600°C) in the ferrite with a sufficient high strain ($\epsilon > 0.4$). Goss grains (110) [001] are found (fig. 6b) to originate from the shear deformation of the ferrite crystals. The shear texture relative intensity and its penetration depth increase with increasing the rolling strain (fig. 7a). As shown in figure 7b, the penetration depth of this texture (110) [001] also extends the more, the thinner is the ferritic hot strip. Recovery as in ELC steels rolled at intermediate temperatures in the ferrite or recrystallization after coiling or annealing at a sufficient high temperature reduce the penetration depth of the shear texture as it is observed for ULC-Ti steels in figure 7b. Nevertheless the effects of the shearing cannot be eliminated in full recrystallization; an important shear texture when compared to the one on strips rolled in the austenite, still remains extending to the 1/4 thickness of the strips. Moreover, as a consequence of the surfaces material sticking to the rolls, the mean resistance to hot deformation (Km) increases when high friction conditions prevail during ferritic rolling. As seen in figure 8 the more reduced is the entry thickness of the strip in the last stand of the mill, the more important is the Km value for the same reduction.

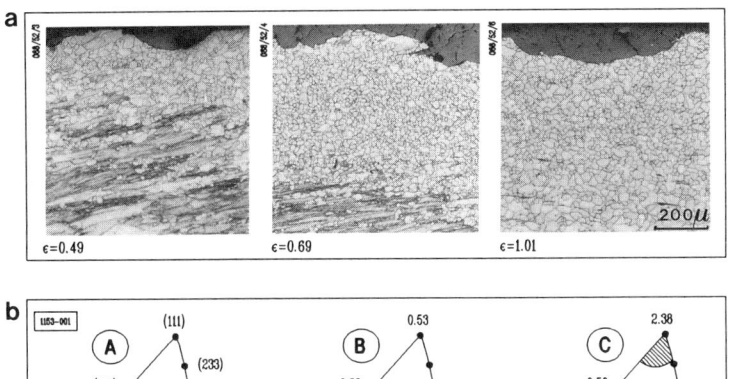

Figure 6.a - Shearing at the surfaces induces dynamic recrystallisation; ELC steel rolled at 650°C without lubricant. b - Typical textures. A : orientation, B : surface, C : center

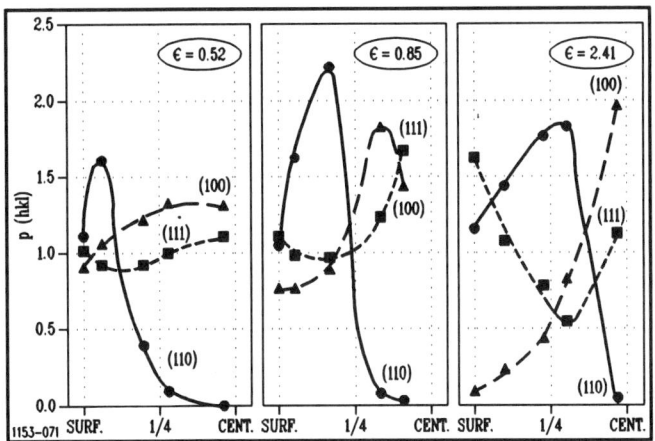

Figure 7.a - Influence of the strain (ϵ) on the texture gradient across the thickness; ELC steel rolled in the ferrite (680-640°C); reheating : 950°C; air cooled.

Lubricated rolling (beef tallow) reduces the sticking of the surfaces and the Km values decrease markedly becoming almost independent on the thickness of the strips, as shown on the same figure 8.

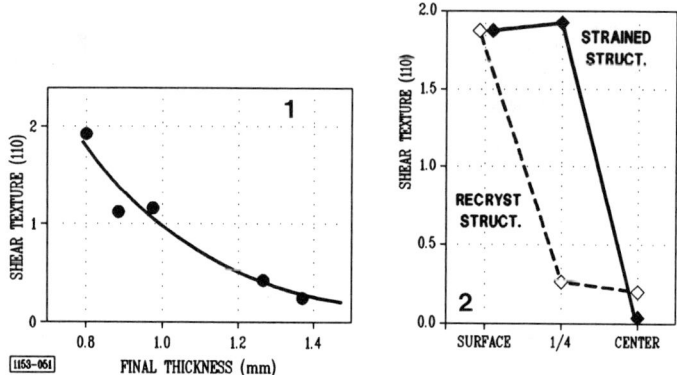

Figure 7.b - ULC-Ti steel hot-rolled in the ferrite without lubricant.
1. effect of final thickness, 1/4 strip thickness; 2. texture gradient, t = 0.8 mm.

The shear texture intensity on the strips surfaces can be reduced in lubricated ferritic rolling conditions. ULC-Ti and ELC steel strips have been rolled at decreasing temperatures in the ferrite (Fig. 9) on a laboratory mill and coiled at low temperatures to get strained ferritic microstructures. Estimated straining was more than 95% for ULC-Ti strips, about 70% for ELC strips from the hardness measurements. The texture gradients through the thickness are compared in the Figure 9 for lubricated and unlubricated rolling conditions. The Goss orientation (110) [001] is no longer the main component of the surface texture for lubricated hot-rolled strips; compression textures <111>//ND and <100>//ND can develop in the surfaces layers of the strips.

The shear texture is totally removed on the intermediate layers so that <111>//ND becomes the main orientation at 1/4 thickness of lubricated hot-rolled strips. No modification of the midthickness texture is observed due to the lubrication.

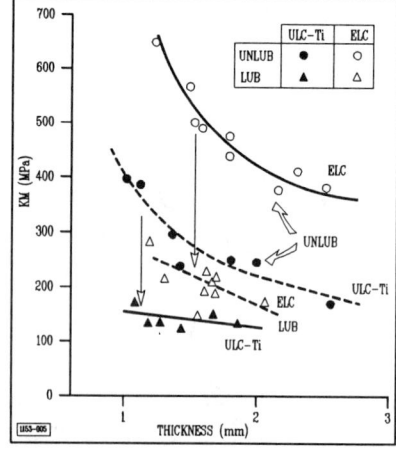

Figure 8 - Mean resistance to hot deformation in function of the entry thickness of the strip; rolling : 550-450°C. Strain : 20-40%.

In absence of any shearing, the compression textures <111>//ND and <100>//ND develop at midthickness. As the final objective is to improve the drawability of the directly annealed strips, an important <111>//ND texture has to be developped as a first step thourought the thickness of the strained strips before annealing. Lubricated rolling is thought therefore to be a prerequisite to increase the deep drawability of the directly annealed thin strips.

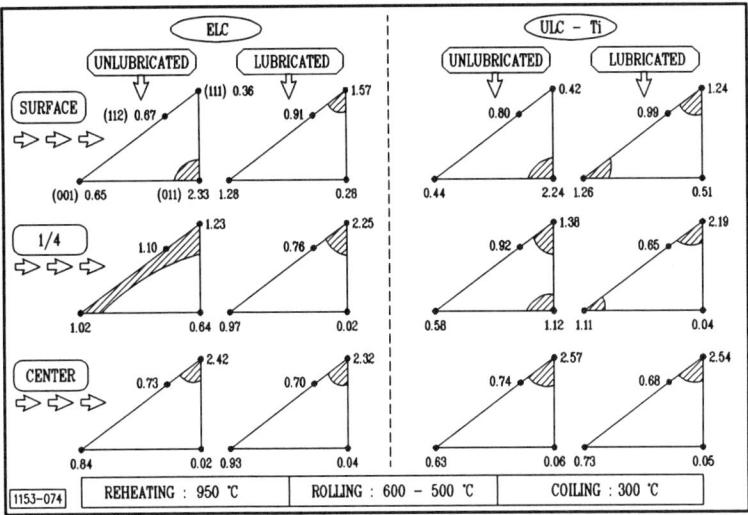

Figure 9 - Influence of a lubrication on the texture gradient through the thickness of ferritic hot-rolled strips.

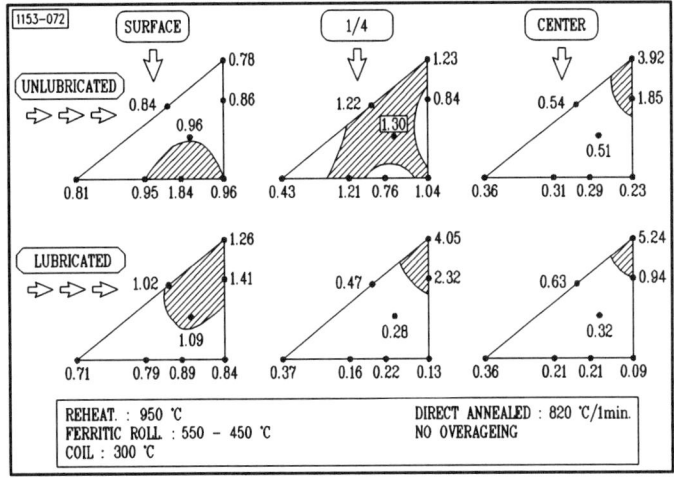

Figure 10 - ULC-Ti steel; influence of a lubrication on the texture gradient through the thickness of directly annealed ferritic hot-rolled strips.

Direct continuously annealed hot strips. The texture gradients of the recrystallized hot strips in ULC-Ti steel are compared in Figure 10 for the influence of the lubrication of the hot rolls. The main component of the surface texture is (210), which is near the (110) [001] shear texture, in annealed strips hot-rolled in unlubricated conditions; the <111>//ND texture cannot develop in these severely sheared regions.

For strips hot-rolled with a lubricant, the surface shear texture decreases with recrystallization and some increase of the <111>//ND texture is allowed.

A complex texture is formed at 1/4 thickness of the recrystallized hot strips initially hot-rolled without a lubricant ; the shear orientation cannot be removed by annealing and the <111>//ND axis density remains well below the one required for a DQ quality strip.

The effects of the lubrication on the texture at 1/4 thickness of the recrystallized hot strips are to reduce the shear orientation and to enhance <111>//ND at the expense of <100>//ND so that (111)/(100) texture ratio corresponding to DQ-DDQ qualities are reached at 1/4 strip thickness. The positive effect of the lubrication for an increase of the (111)/(100) texture ratio is also observed at midthickness of the recrystallized hot strips.

The same positive influence of the lubrication for a reduction of the shear orientation is observed (fig. 11) in ELC steels hot-rolled in the ferrite and directly annealed. Nevertheless an important <111>//ND orientation cannot be developped in the recrystallized ELC steel in presence of the <100>//ND texture.

From changes in r-value through the thickness calculated from the texture[13] in Table III it clearly appears that the effect of the lubrication for an increase of the drawability is more important at 1/4 thickness of the recrystallized strips. Mechanical properties and drawability of the continuously annealed ELC and ULC-Ti steel strips hot-rolled in the ferrite are compared in Table IV for lubricated and unlubricated rolling.

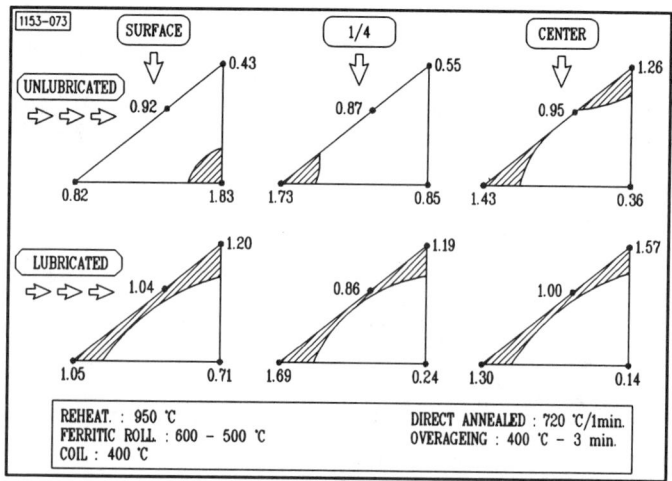

Figure 11 - ELC steel; influence of a lubrication on the texture gradient through the thickness of directly annealed ferritic hot-rolled strips.

The DQ quality requirements are met only for directly annealed strips in ULC-Ti steel hot-rolled with a lubricant. An isotropic CQ quality ($\bar{r} = 1$, $\Delta r \leq 0.1$) is achieved for the ELC steel. Recrystallization in presence of solute atoms (carbon) is thought[3,14] to be the main reason for the reduction of the (111)/(100) texture ratio which is observed, hence for the lower drawability of the annealed strips in ELC steel (fig.12). On the contrary for IF steels, recrystallization may improve significantly the (111) texture.

TABLE III - CHANGES IN THE r-VALUES (from texture) THROUGH THE THICKNESS OF ULC-Ti AND ELC STEEL STRIPS HOT-ROLLED IN THE FERRITE ON LUBRICATED AND UNLUBRICATED ROLLS AND DIRECTLY ANNEALED

Steel	reheat/roll/coil °C	Surface		1/4		Center	
		lub	unlub	lub	unlub	lub	unlub
		$r =$					
ULC-Ti	1050/700-600/400	1.34	-	1.49	-	1.43	-
	950/600-500/400	1.08	-	1.68	-	1.58	-
	950/550-450/300	1.15	0.99	1.65	1.28	1.72	1.64
ELC	1050/700-600/400	1.01	0.79	0.96	0.73	0.89	0.90
	950/600-500/400	1.07	-	0.91	-	1.04	-
	950/550-450/300	1.04	0.83	0.91	0.69	1.05	0.97

TABLE IV - MECHANICAL PROPERTIES AND DRAWABILITY OF ULC-Ti AND ELC STEELS HOT-ROLLED IN THE FERRITE AND DIRECTLY ANNEALED ; LUBRICATED (L) AND UNLUBRICATED (U) ROLLING

Steel	Conditions reheat/roll./coil. C°		YS MPa	TS MPa	El %	$\bar{r}$	Δr
ULC-Ti	1050/700-600/400	L	218	278	37.0	1.40	0.72
		U	191	282	38.1	-	-
	950/600-500/400	L	208	267	42.4	1.33	0.70
		U	210	291	37.6	-	-
	950/550-450/300	L	196	277	43.2	1.37	0.80
		U	190	288	-	0.79	0.38
ELC	1050/700-600/400	L	235	319	39.4	1.02	0.07
		U	250	356	38.0	0.88	-0.15
	950/600-500/400	L	228	325	38.0	0.97	0.00
		U	-	-	-	-	-
	950/550-450/300	L	236	335	34.7	1.09	-0.03
		U	283	357	32.0	0.94	0.54

Cold rolled sheets from strained ferritic hot-rolled strips. The textures of the thin cold-rolled (0.4 mm), continuously annealed sheets from the strained hot strips in ELC and ULC-Ti steels are compared in Figure 13. Hot rolling was on lubricated or on un-lubricated rolls. As a consequence of the reduced thicknesses of the strained hot strips, cold rolling drafts were well below the usual 70%-75% reduction rates.

The initial texture of the strained hot strips i.e. the reduction of the shear structure (110) [001] and the increase of the <111>//ND axis density on the strained lubricated hot-rolled strips has a determinent influence on the final texture of the cold-rolled and continuously annealed sheets.

In addition to the <111>//ND orientation, important <100>//ND and <112>//ND texture components are observed on the cold-rolled and annealed sheets from the strained strips hot-rolled without a lubricant.

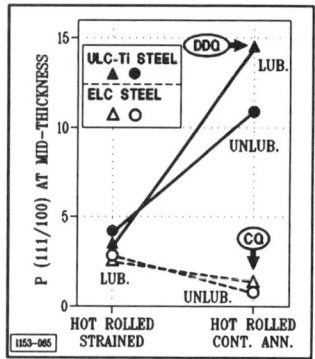

Figure 12 - (111)/(100) texture ratio in the center of hot strips hot-rolled in the ferrite; influence of soluble carbon.

High (111)/(100) texture ratio and consequently drawing qualities cannot be produced as thin sheets from the unlubricated hot-rolled strips. The texture change, as seen in figure 13 is towards an increase of <111>//ND in the annealed thin sheets for the initial lubricated hot rolling conditions. The increase of <111>//ND is particularly important in ULC-Ti steel.

As seen in Table V, soft thin sheets with a high ductility can be produced in ULC-Ti steel. Higher strength levels are obtained for the ELC steel sheets of which the ductility (total elongation, n-value) is somewhat lower. A positive effect of a reduction of the carbon and manganese contents is observed on the mechanical properties of the ELC steel.

TABLE V - MECHANICAL PROPERTIES AND DRAWABILITY OF COLD-ROLLED AND CONTINUOUSLY ANNEALED THIN SHEETS FROM LUBRICATED HOT-ROLLED STRIPS IN ELC AND ULC-Ti STEELS

Steel	Ferrit. roll. cond. reheat/roll./coil. °C	YS MPa	TS MPa	El %	r	n
ULC-Ti	1050/700-600/400	122	249	40.4	2.11	0.241
	950/600-500/400	124	255	39.1	1.73	0.243
ELC	1050/700-600/400	214	341	34.0	1.03	0.201
	950/600-500/400	203	339	33.8	0.97	0.195
ELC Low Mn	1050/700-600/400	166	290	36.1	1.18	0.208
	950/600-500/400	167	297	35.0	1.17	0.216

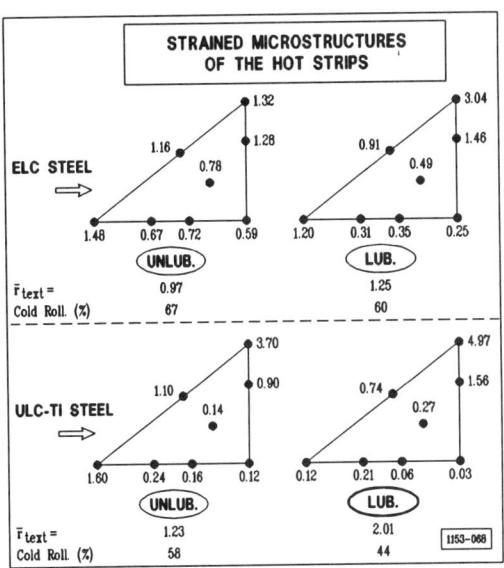

Figure 13 - Influence of lubrication on the crystalline texture of C.R. and continuously annealed ELC and ULC-Ti steels hot-rolled in the ferrite.

The distributions of the individual r-values in the rolling plane are shown in figure 14 for the thin annealed sheets in ELC steel. A comparison is made of sheets from lubricated and unlubricated hot strips. As seen on the Figure 14 a CQ quality is produced with a low planar anisotropy Δr for reduced cold-rolling drafts ($\leq$ 60 %).

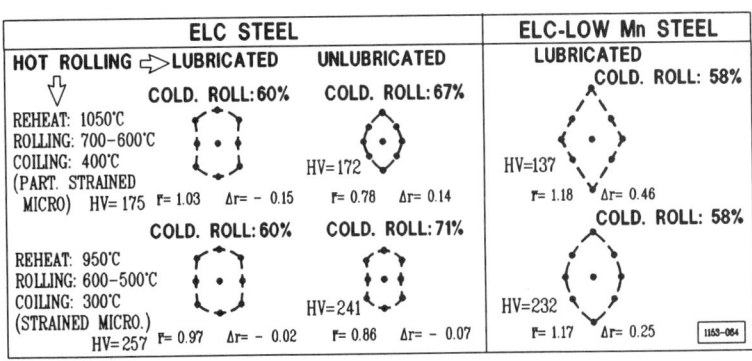

Figure 14 - Plastic anisotropy of C.A. (720°C/60 s) sheets (0.4 mm) from ferritic hot-rolled strips; influence of the lubrication and steel chemistry.

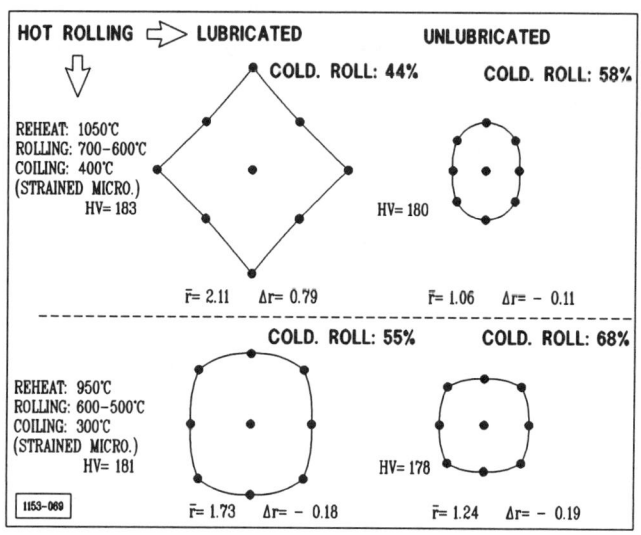

Figure 15 - Plastic anisotropy of C.A. (820°C/60 s) sheets (0.4 mm) from ferritic hot-rolled strips; influence of the lubrication and steel chemistry.

Lubrication in the finishing mill is needed to get such a quality which can be increased in ELC steel with reduced carbon and manganese contents. As shown in Figure 15 lubrication of the hot rolls in the ferrite is a prerequesite to get thin cold-rolled and annealed ULC-Ti steel sheets in DDQ quality, such a quality being obtained for reduced ($\leq$ 60 %) cold rolling drafts.

Conclusions

Conditions have been investigated for the production on the finishing mill of thin strips hot-rolled in the ferrite in strained microstructures for direct recrystallization in a continuous line or for subsequent cold rolling and annealing as thin sheets. Dynamic strain hardening and softening behaviours have been quantified and compared for ELC and ULC-Ti steels. Moreover as an improved drawability is to be provided to such materials with reduced thicknesses, conditions were assessed for the development of a strong <111>//ND texture thourought the thickness of strained strips and recrystallized materials.
- ULC-Ti steels only appear to be well adapted for the production of drawing and deep-drawing qualities ; a more important strain can be cumulated in ULC-Ti steel even at rather high (700-600°C) rolling temperatures in the ferrite and the higher is the strain rate ($\dot{\varepsilon}$) the more important will be the cumulated strain ; a strong <111>//ND texture develops inside the strained strips ; moreover, as ULC-Ti steel is insensitive to dynamical strain aging, the resistance to hot deformation in the ferrite is lower thus more compatible with the capacity of the mill.
- Softening of ELC steels is still important at low temperatures (500-°C); rolling at very low temperatures ($\leq$ 450°C) is thus necessary to cumulate strain and develop a strong <111>//ND texture in those steels which harden markedly in this temperature range by dynamic strain aging; reduced drawing qualities will be produced therefore using rather high rolling loads.

- High friction between strip and rolls generates a shear strain in the ferrite ; the intensity and the penetration depth of the associated Goss texture increase with increasing the cumulated strain and reducing the thickness of the strip ; appearent resistance to hot deformation in the ferrite is markedly increase by sticking, especially for very thin gages.
- Lubrication of the rolls is imperative to reduce the texture gradient thorought the thickness of the ferritic hot-rolled strips and to supress the shear texture developpement at 1/4 thickness ; <111>//ND texture can develop in these conditions. Resistance to the hot deformation in the ferrite is also reduced in lubricated rolling conditions;
- Direct continuous annealing of the strained ULC-Ti steel strips hot-rolled with a lubricant allows to produce a soft DQ quality in reduced thickness ($\leq$ 1.5 mm) ; an isotropic CQ quality is obtained in directly annealed ELC steel strips. Annealing in presence of solute carbon is thought to be the main reason for the reduction of <111>//ND texture by annealing ELC steel strips.
- Cold-rolled and continuously annealed sheets (0.4 mm) have been produced from the strained hot strips lubricated rolled in the ferrite in DDQ quality (ULC-Ti steel), in CQ quality (ELC steel) for reduced (60 %) cold rolling drafts.

Acknowledgements

The authors are grateful for the financial assistance from IRSIA (Belgian Institute for the Encouragement of Scientific Research in Industry and Agriculture) and ECSC (European Coal and Steel Community).

REFERENCES

1. G. Kim and O. Kwon, "Formation of abnormally coaarse grain structure in hot-rolled strips", (Proceedings of Thermec-88, the Iron and Steel Institute of Japan, (1988), pp. 668-675.

2. European patent 0196788, Kawasaki Steel Corp.

3. S. Hashimoto, (Kobe Steel Eng. Reports, 39, n°3 (1989), pp. 73-76).

4. T. Senuma, H. Yada, Y. Matsumara and K. Yamada, Tetsu-to-Haganè, 73 (1987), pp. 1956-1963.

5. P. Messien, J.C. Herman, V. Leroy, J.M. Detry and P. Cantinieaux, "Laminage ferritique de bandes minces an acier doux (ferritic rolling of mild steels in thin strips)", (Paper presented at ATS Steelmaking Conf. Paris 4-6 Dec. 1990).

6. J.C. Herman and V. Leroy, "Incoherent and Coherent Precipitations Induced by Hot Deformation and Accelerated Cooling in HSLA Steels", Proceedings of THERMEC-88, the Iron and Steel Institute of Japan (1988) pp. 283-290.

7. P. Dadras, Trans. JIM, 19 (1978), pp. 230-232.

8. Y. Bergstrom, The Plastic Deformation of Metals, Metallography, K.T.H., Stockholm (1982).

9. J.E. Bailey and P.B. Hirsch, <u>Phil. Mag.</u> 5 (1960), pp. 485-497.

10. <u>Werkstoff Handbuch Stahl und Eisen</u>, 3 Auflag, B 25-4 (1965)

11. M. Matsuo, T. Sakai ans Y. Suga, <u>Metall. Trans.</u>, 17A, Aug. (1986), pp. 1313-1322.

12. T. Sakai, Y. Saito and K. Kato, <u>Trans. ISIJ</u>, 28 (1988), pp.1036-1042.

13. P. Messien, J.C. Baret and A. de Leval, <u>CRM Reports</u>, 31 (1972), pp. 39-45.

14. T. Senuma et al., <u>J. Japan Inst. Metals</u>, 52, n°12 (1988), pp.1212-1220.

INFLUENCE OF THERMAL HISTORY PRIOR TO HOT ROLLING ON MECHANICAL PROPERTIES OF CONTINUOUSLY ANNEALED HIGH STRENGTH SHEET STEELS CONTAINING TITANIUM, MANGANESE AND PHOSPHORUS

K. G. Chin, H. J. Kang and S. K. Chang

Research Institute of Industrial Science and Technology
699 Kumho-dong, Dongkwangyang 544-090, Korea

Abstract

Hot charge rolling (HCR), which is the process that continuously cast slabs are hot-charged into reheating furnace for energy saving and then supplied to hot strip mill, has been applied to produce cold rolled sheet steels.

In this paper, the influence of thermal history of slabs before hot rolling in HCR process on mechanical properties of interstitial-free high strength steel containing titanium, manganese and phosphorus was studied. It was found that the plastic strain ratio, $\bar{r}$-value, was dependent on the hot charge temperature but yield and tensile strengths were not. High $\bar{r}$-values more than 1.9 were obtained in the steels which were hot-charged at 750°C and coiled at 600°C. In these steels, recovery and recrystallization were delayed to a higher temperature and $\{111\}$ texture increased rapidly at the later stages of continuous annealing. These results are attributed to an increase in the amount of fine precipitates of TiC and (Ti,Fe)P formed in hot band.

Introduction

Recently, high strength cold rolled sheet steels for automotive applications have been widely used to reduce the weight of automobile for fuel economy and to improve the dent-resistance of whitebody. For outer panels, various types of deep drawing high strength sheet steels with a high $\bar{r}$-value have been developed. Rephosphorized steel and interstitial-free high strength steels (IF-HSS) are well known as typical examples of the steels. IF-HSS containing titanium and/or niobium is strengthened by adding such elements as phosphorus,silicon and manganese. IF-HSS has a high $\bar{r}$-value and a non-aging property, however, if the steel contains phosphorus as a strengthening element, interstitial atoms are removed from the grain boundaries and phosphorus atom is segregated in the grain boundaries, resulting in increasing the frequency of the grain boundary fracture during secondary working. To improve the secondary workability of IF-HSS, It is necessary to strengthen the grain boundary by addition of boron (1) or reduce impurities such as phosphorus and sulfur in grain boundary (2). In the IF-HSS containing titanium, manganese and phosphorus, Brun et al.(3) found that the retardation of recrystallization was caused by the precipitation of (Ti,Fe)P,resulting in a higher recrystallization temperature, and suggested that the r-value could be improved by combinations of low temperature coiling after hot rolling and continuous-annealing or high temperature coiling and batch-annealing. Okamoto et al.(4) showed that in the same steel, the annealing texture was influenced by the heating rate. That is,the faster the rate,the stronger the {111} texture after recrystallization and a higher $\bar{r}$-value could be attained through continuous-annealing. They proposed that it was accounted for FeTiP precipitation after annealing.

On the other hand, since energy crises in 1974 and 1979, hot charge rolling process was adopted in several steel works for energy saving(5). This is the system in which the hot slab is charged into reheating furnace and then supplied to the hot strip mill. Recently, the production of cold rolled sheet steels by hot charge rolling process has been gradually increased.

In the present paper, the influence of hot charge temperature on mechanical properties of IF-HSS containing titanium, manganese and phosphorus was studied for the development of deep drawing high strength cold rolled sheet steels with tensile strength of 35-40kgf/mm².

Experimental Procedures

Materials

Three interstitial-free steels were prepared by vacuum induction melting. All the steels have the same base composition of approximately 0.003 wt pct carbon, 0.07 wt pct phosphorus, 0.07 wt pct sulfur, 0.03 wt pct soluble aluminum and 0.002 wt pct nitrogen. The major difference in composition is the amount of manganese and titanium and the chemical analyses of these steels are listed in Table 1. As listed in Table 1, the atomic ratio of (Ti*/C) as a parameter ranges from 2.2 to 8.3, where Ti* = Ti(wt%) - (48/32)S(wt%) - (48/14)N(wt%). Also,in order to study the

Table 1 Chemical composition of steels, vacuum melted.(wt pct)

Steels	C	Mn	P	S	sol.Al	N	Ti	Ti*/C	Ti**/P
UT1	0.0020	0.60	0.073	0.007	0.035	0.0023	0.085	8.3	0.80
UT2	0.0021	0.57	0.083	0.007	0.028	0.0028	0.059	4.6	0.36
UT3	0.0030	0.25	0.064	0.009	0.035	0.0022	0.048	2.2	0.23

Ti* = Ti(wt%)-(48/32)S(wt%)-(48/14)N(wt%)
Ti** = Ti*-(48/12)C(wt%)

effect of the titanium content on the precipitation in hot band, another parameter of (Ti**/P) is used, where Ti** = Ti*-(48/12) C(wt%), and it ranges from 0.23 to 0.8. These ratios indicate the degree of stabilization of carbon and phosphorus; a ratio of 1 would mean that the exact stoichiometric amount of stabilizing element of titanium is present to form TiC with carbon and TiP with phosphorus, repectively. Each of 30kg heats was cast into four 7.5kg ingots with dimension of 30mm X 70mm X 200mm and charged into reheating furnace at four temperatures of room tepemperature, 600°C, 750°C and 900°C for HCR simulation. The ingots were reheated to 1200°C for one hour and hot rolled to 3.2mm thickness in 5 passes. The finishing temperature for all steels was 910°C. The steels then cooled down to the coiling temperature of 600°C by water spray (15-30°C/s). The hot bands were kept at that temperature for one hour in order to simulate coiling and were then slowly cooled in a furnace. A schematic diagram showing various thermal histories of slabs and hot rolling condition was illustrated in Figure 1.

After the hot bands were descaled by sand-blasting, they were cold rolled to 0.8mm in thickness. For the simulation of continuous annealing, steels were heated at a rate of 7°C/s up to the ranges between 740°C and 840°C in an infrared image furnace for 30 sec, rapidly cooled and then overaged at 400°C. In order to

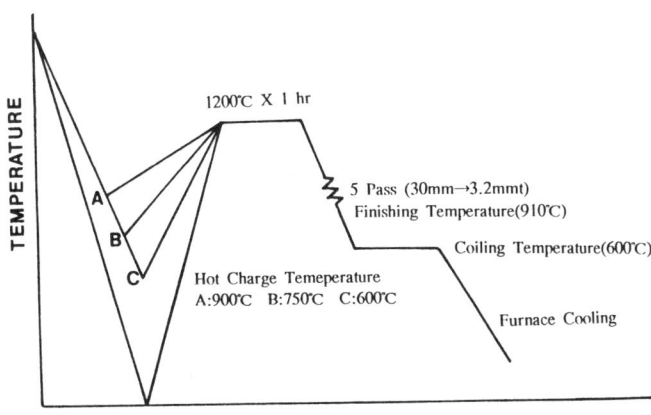

Figure 1 - Schematic diagram showing the thermal history and hot rolling conditions for HCR and CCR simulations.

investigate the recrystallization behavior, specimens were isochronal- annealed at various temperatures within the range between 600℃ and 825℃ for 30 seconds and then water-quenched down to room temperature. For the study of the phosphorus segregation with the coiling temperatures after hot rolling, hot bands were rapidly heated to 1000℃ for one hour and water-quenched down to room temperature and reheated to the ranges of 600- 775℃ for one hour and then slowly cooled in a furnace.

Experimental Methods

Optical microscopy was performed to examine the microstructure of the hot bands and annealed sheets. Grain size was determined by linear intercept measurements. Thin foils and extraction replicas for transmission electron microscopy were prepared and then examined in JEOL JEM 200-CX operated at 200KV to determine the size and distribution of precipitates. Crystalline textures were evaluated by the X-ray integrated reflection intensity. The tensile tests were carried out at a cross head speed of 2mm/min using a tensile testing machine of 20 tonnage capacity . ASTM 370A standard test pieces (gauge length: 50mm) were used to determine the yield strength, tensile strength and elongation. $\bar{r}$-values were estimated for angles of 0° ,45° and 90° to the rolling direction after 15% prestraining. Hardness was measured to investigate the softening during the recrystallization annealing. The susceptibility of grain boundary fracture due to the segregation of phosphorus with coiling temperature was measured from the Charpy-V absorbed energy test. The segregation of phosphorus in hot bands was analyzed by Auger electron spectroscopic observation on the fracture surface. For the Auger analysis a cylindrical specimen of 3.7mm diameter and 30mm length was mounted in an ultra-high vacuum chamber operating at a base pressure of 4×10^{-10} torr. The specimens were fractured by impact with a hammer. Auger spectra were taken at least three different spots on the fracture surface using a primary electron beam with 1000Å diameter. The peak heights of phosphorus, carbon and titanium were normalized with respect to the iron peak at 650 eV and called as a peak height ratio $A_{[P,C,Ti]}/A_{[Fe]}$. The internal friction measurement was carried out on the hot bands by means of an inverse torsion pendulum($\pm$1Hz) in order to evaluate the solute C content before annealing as a function of the applied thermal history from casting to hot-rolling.

Results and Discussion

Microstructure and Precipitation in Hot Bands

The microstructural changes of hot bands with various hot charge temperatures are shown in Figures 2 and 3. The microstructures of hot-rolled bands were affected by the titanium content and the hot charge temperature. In the steel UT1 containing 0.085 wt pct titanium, the ferrite grain size was minimized at 750℃ and then increased with increasing the temperature, but it was increased monotonously in the steel UT2 containing 0.059 wt pct titanium. When hot charging was conducted at 900°C in the steel UT1, a mixed microstructure of fine and coarse grains was obtained. The influence of the hot charge temperature on the microstructure of hot band seems to be due to the occurence of transformation

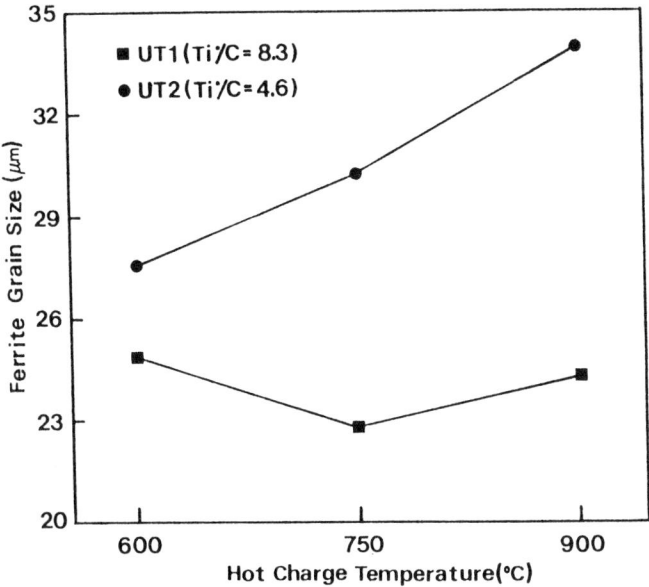

Figure 2 - Grain sizes in hot bands of the steels UT1 and UT2.

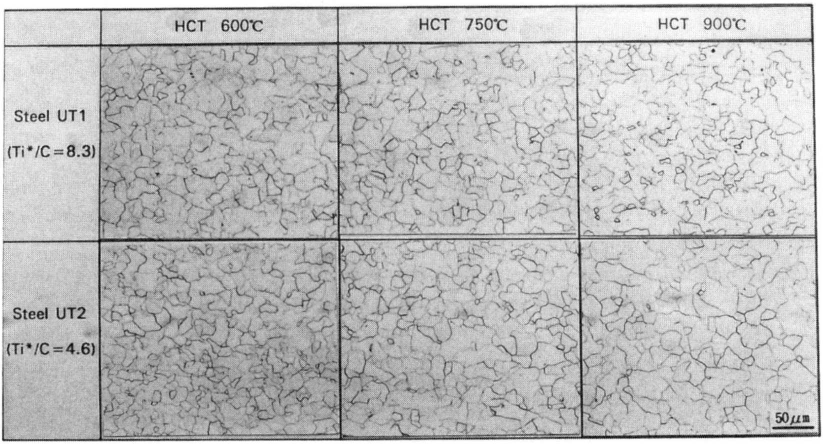

Figure 3 - Microstructural changes in hot bands with various hot charge temperatures in the steels UT1 and UT2.

from austenite to ferrite before charging into reheating furnace. As it were, if the transformation does not occur, a large grain size will be obtained in hot band, except for the steel with a large amount of precipitates. In the steel, the grain size is affected by rather the amount of precipitates than the hot charge temperature.

The transmission electron micrographs and EDX analysis of the particles in hot bands of the steel UT1 depending on the hot charge temperatures are shown in Figure 4. When the hot charge temperature increased, the large particles of 0.1 to 0.3μm began to precipitate along the grain boundaries and the fine particles over the matrix increased. These particles were classified into two type of

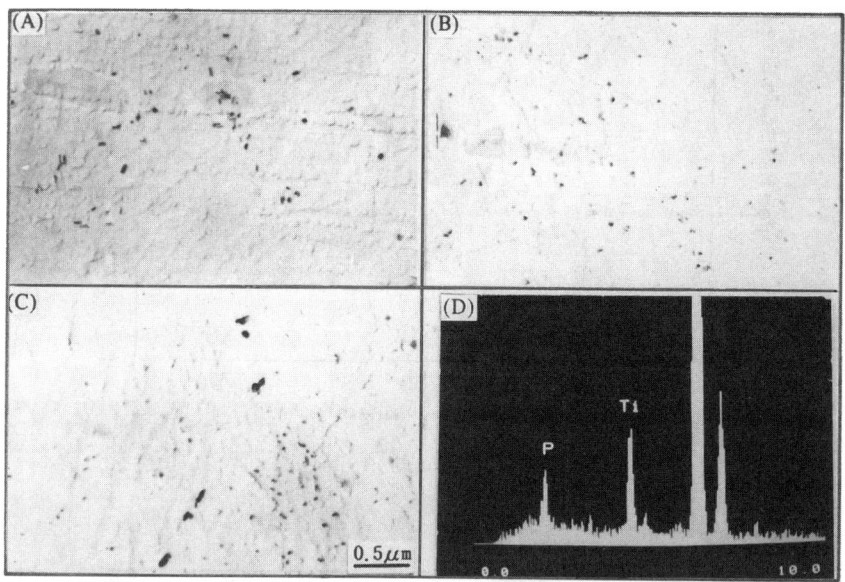

Figure 4 - Precipitates in hot bands of the steel UT1 hot-charged at various temperatures.(A)600℃, (B)750℃, (C)900℃ and (D)EDS of (Ti,Fe)P

precipitates. One of the particles consists of titanium, iron and phosphorus. The other was observed only in matrix and identified as TiC. These results can be explained as follows; the precipitates do not form before charging into reheating furnace in the steels hot-charged at the higher temperatures, the dense and fine precipitates will be subsequently obtained in the hot band. For the absence of TiC at the grain boundaries, Brun et al.(3) repoted that the large TiC particles lying at the grain boundaries were replaced by (Ti,Fe)P and however did not clearly propose the mechanism of replacement from TiC to (Ti,Fe)P. Also, Brun et al(3) and Okamoto et al.(4) reported that an increase of manganese retarded the formation of TiC or $Ti_4C_2S_2$. Kurosawa et al.(6) also reported that,in 0.01%C-0.14%Mn-0.014%P-0.16%Ti steel, the maximum amount of $Fe_{0.4}Ti_{1.2}P$ was precipitated at the temperatures between 700℃ and 800℃ when the steel was soaked for one hour. Furthermore, it is well known that TiC is formed at a higher temperature than (Ti,Fe)P during hot rolling. According to Kaneko et al.(7), titanium is more strongly interacted with carbon than phosphorus. From the present study, there was neither an occurence of yield point elongation nor an evidence of carbon in solution at the internal friction test for hot band of the steel UT1. It can be assumed from this results that all carbon is precipitated as TiC with in steel during hot rolling. A typical scanning electron micrograph of the fracture

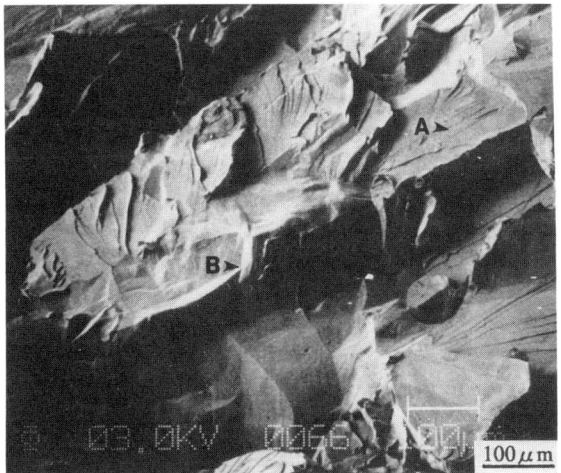

Figure 5 - Scanning electron micrograph of the fracture surface of the steel UT1 hot-charged at 900°C.
A:Transgranular fractured facet, B:Intergranular fractured facet

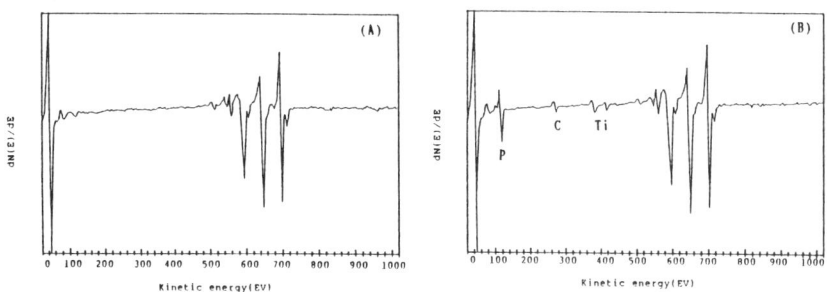

Figure 6 - Auger spectra obtained from the fracture surface in the steel UT1 hot-charged at 900°C.
(A)Transgranular fractured facet, (B)Intergranular fractured facet

surface of hot bands is shown in Figure 5 and Auger spectra taken from the surface are presented in Figure 6. All examined specimens showed partially the mixed type of transgranular and intergranular fractures. The spectrum which was obtained from the cleavage facet showed only the Auger peak of iron, whereas the peaks of phosphorus, carbon, titanium and iron were observed in the spectra of the grain boundaries. Considering these results, it can be suggested that TiC at the grain boundaries was not replaced by (Ti,Fe)P but might be combined with titanium phosphide or iron phosphide. Erhart et al.(8) also proposed that the phosphorus atoms could be trapped at the interface of TiC/ferrite matrix. Figure 7 shows the size and distribution of precipitates with the changes in titanium and manganese contents. The particle size decreased with an increase of titanium content and the precipitation of (Ti,Fe)P appeared to be enhanced by an increase of manganese content. In the steel UT3, which contains 0.25% Mn, most of the precipitates were $Ti_4C_2S_2$ but (Ti,Fe)P was not observed. This is because of the retardation of (Ti,Fe)P precipitation which is attributed to the decrease of manganese content and low temperature coiling.

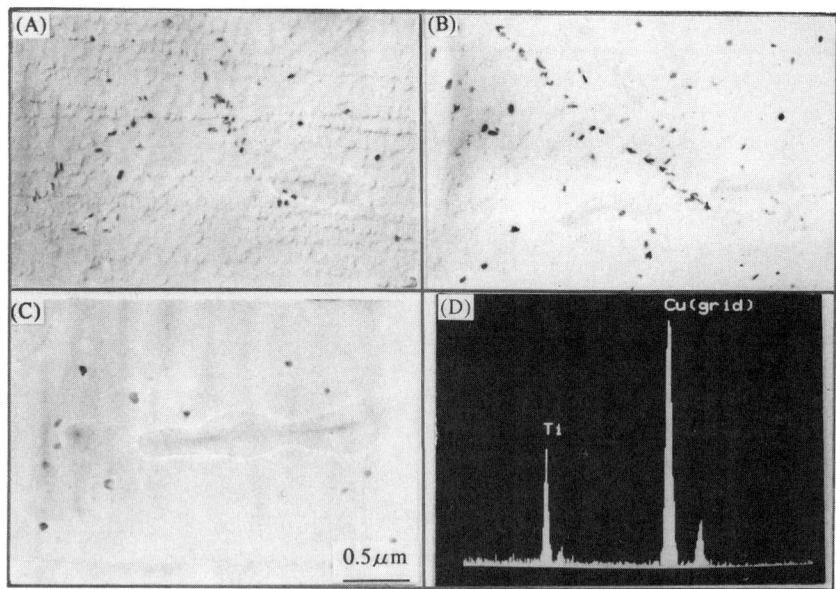

Figure 7 - Precipitates in hot band of the steels with various compositions. (HCT 600°C, CT 600°C) (A) Steel UT1, (B) Steel UT2, (C) Steel UT3 and (D) EDS of precipitate in the steel UT3

Segregation of Phosphorus in Hot Bands

In the steel UT3, the phosphorus segregation behavior at the grain boundaries and the Charpy-V absorbed energy of hot bands with the coiling temperature are shown in Figures 8 and 9, respectively. The Auger peak height ratio, $A_{[P]}/A_{[Fe]}$, increased exponentially at 650°C. This increase seems to be related to the precipitation of (Ti,Fe)P at the grain boundaries. In the steels containing titanium and phosphorus, the phosphorus segregation and the precipitation of (Ti,Fe)P have a strong effect on development of {222} texture in the annealed sheets. Therefore, it is assumed that the r̄-value of the steel UT3 will be affected by the coiling temperature. On the other hand, the phosphorus segregation at the grain boundaries can cause embrittlement of steels. In Figure 9, the fracture mode changed from ductile fracture to brittle fracture at above 650°C. This corresponds

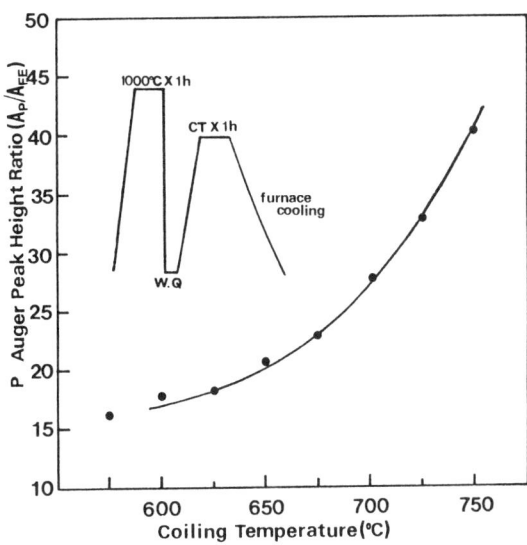

Figure 8 - Effect of the coiling temperature on Auger peak height ratio, $A_{[P]}/A_{[Fe]}$, in hot bands of the steel UT3.

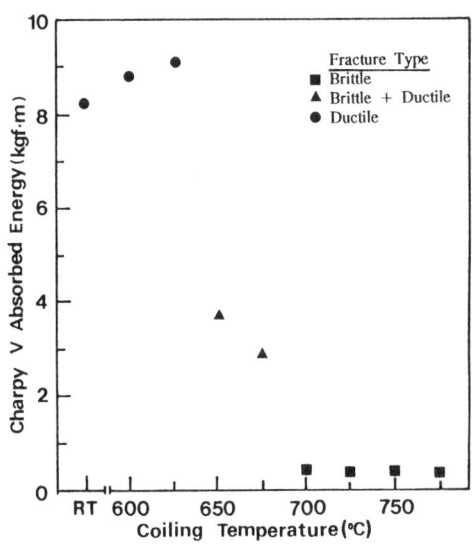

Figure 9 - Effect of the coiling temperature on the Charpy absorbed energy in hot bands of the steel UT3.

with the results in Figure 8. In the production of high strength sheet steels containing titanium and phosphorus, planar cracking was somewhat observed when the bending stress was applied to the hot strips during uncoiling or tension levelling. This separation crack is a result of phosphorus segregation to ferrite grain boundaries during the slow cooling of the hot coils. From the results in Figures 8 and 9, the cracking could be suppressed by the coiling below 650°C.

Characteristics of Annealed Sheets

Recrystallization Behavior

In the steel UT1, the hardness as a function of annealing temperature with different hot charge temperatures is shown in Figure 10. It was found that the recrystallization was delayed by an increase in the hot charge temperature. This retardation of recrystallization appears to be attributed to a large amount of precipitates formed in hot band. However, in the steel hot charged at the temperature of 900°C, it is assumed that a coarse grain size in hot band lowers the recrystallization temperature, contrary to the effect of precipitates on the recrystallization.

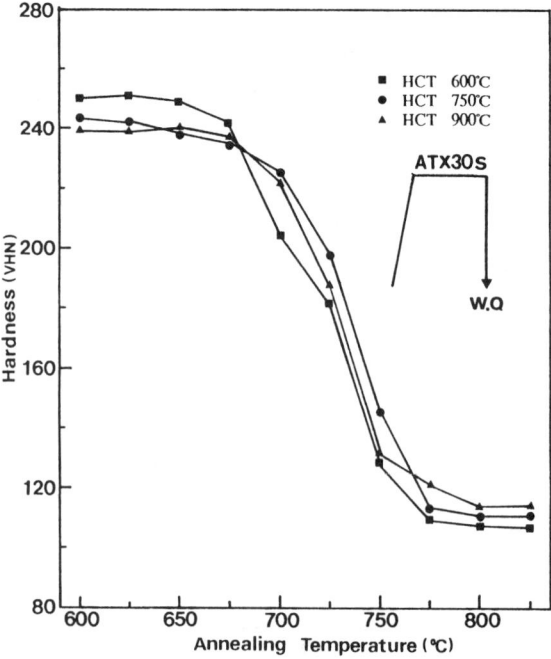

Figure 10 - Changes in hardness with the annealing temperature in the steel UT1 hot-charged at various temperatures.(CT 600°C)

The effect of annealing temperature on the pole intensities for various hot charge temperatures is also shown in Figure 11. In all the steels, {222} intensity increased and {200} and {110} intensities drastically decreased as the annealing temperature increased to 725°C.

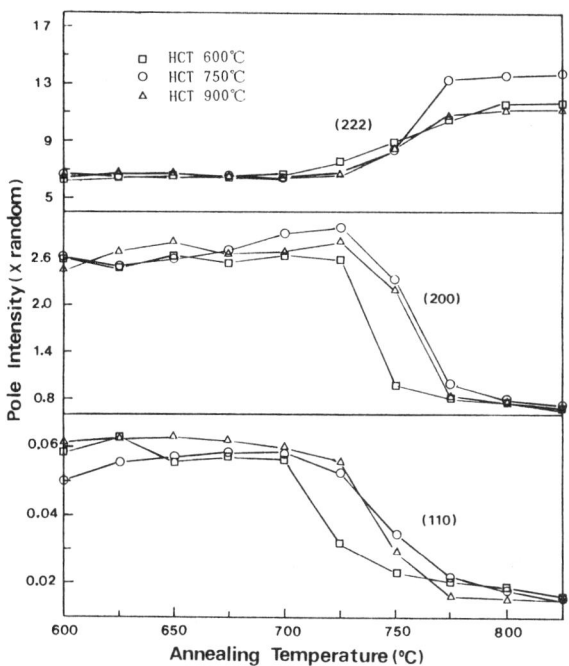

Figure 11 - Effect of the annealing temperature on the pole intensities in the steel UT1 hot-charged at various temperatures.(CT 600°C)

The hot charge temperature provided a strong effect on the texture development at the later stages of recrystallization. The high temperature charging at 750°C enhanced especially the development of {222} at the higher annealing temperature. The development of {222} texture in the steels seems to be attributed to the distribution of precipitates such as TiC and (Ti,Fe)P and phosphorus segregation in hot bands. As seen in Figure 4, a lower charge temperature increases the amount of precipitates at the grain boundaries, whereas reduces the a mount of fine precipitates in matrix. Ohashi et al.(9) and Brun et al.(3) reported that the coarse (Ti,Fe)P particles at the grain boundaries give a bad influence on {222} texture formation. From the present results, it was noticed that the precipitates formed at the grain boundary, regardless of precipitate size, suppressed the development of {222} and {110} intensities in recrystallization texture of the steels. This tendency was not affected by the hot charge temperature.

Microstructures and Mechanical Properties

The effect of the hot charge temperature on the $\bar{r}$-values of continuously annealed sheets in the steels UT1, UT2 and UT3 is shown in Figure 12. These steels were annealed at 830°C for 30 seconds. In the steel UT1, the $\bar{r}$-value was maximized at the hot charge temperature of 750°C ; thereafter, it decreased as the hot charge temperature increased. However, the $\bar{r}$-value was not changed by the hot charge temperatures of 600°C and 750°C in the steels UT2 and UT3.

The changes in ferrite grain size with hot charge and annealing temperatures for the steel UT1 are shown in Figure 13. At the annealing temperatures below

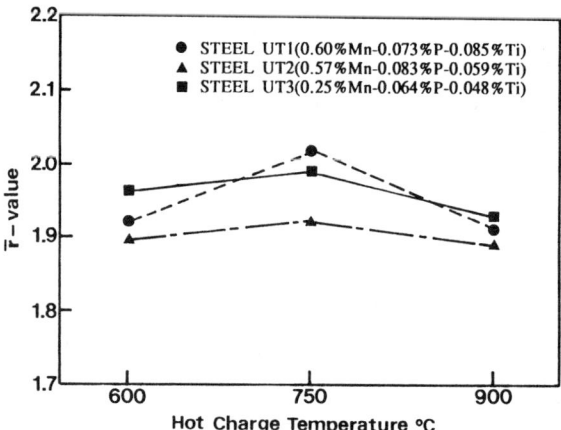

Figure 12 - Effect of the hot charge temperature on the $\bar{r}$-value of the continuously annealed sheets in the steels UT1, UT2 and UT3. (CT 600°C)

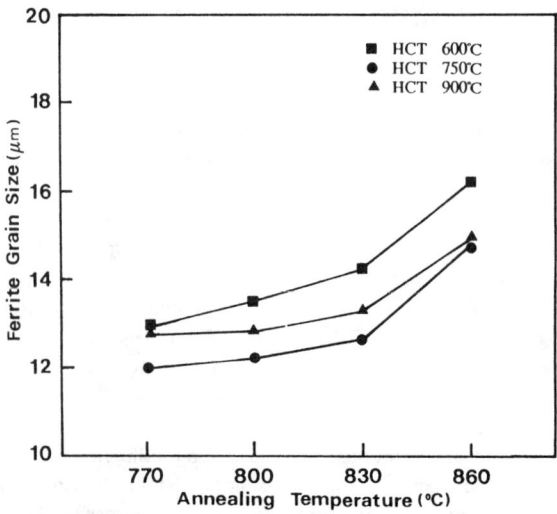

Figure 13 - Grain sizes in the annealed sheets of the steel UT1 hot-charged at various temperatures. (CT 600°C)

316

830°C, a grain size increases only in the steel hot-charged at 600°C but it is not changed in steels hot-charged at 750°C and 900°C. On the other hand, the ferrite grain size was increased by high temperature annealing for all the steels. This is due to the precipitation during annealing. The precipitates in annealed sheets of the steel UT1 are presented in Figure 14. In Figure 14, the dense and fine precipitates in the matrix increased as the hot charge temperature and the amount of grain boundary precipitates increased. The increase of $\bar{r}$-values in the steel UT1 is likely to be related to the fine precipitates in matrix. However, if the fine precipitate disperses excessively in matrix as in the case of hot charging of 900°C, the grain growth during annealing will be retarded and the $\bar{r}$-value of the sheet will be decreased.

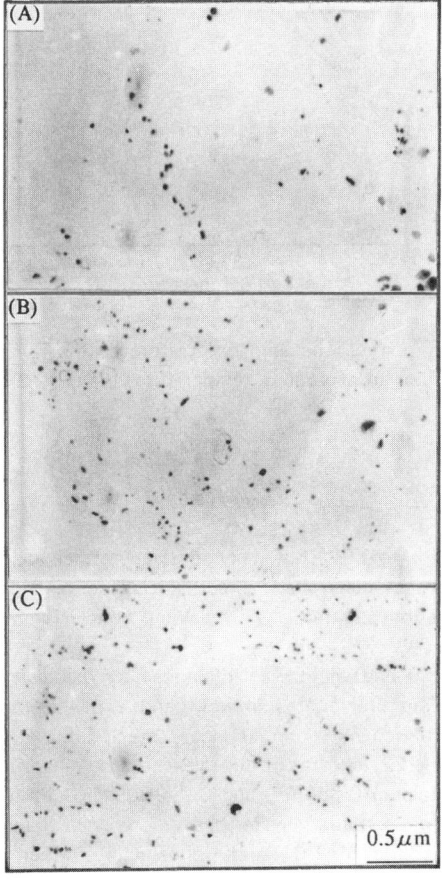

Figure 14 - Precipitates in the continuously annealed sheets of the steel UT1.
(annealed at 830°C, CT 600°C)
(A) HCT 600°C, (B) HCT 750°C, (C) HCT 900°C

Figure 15 shows the changes in the r̄-value with the annealing temperature in the steel UT1 hot charged at various temperatures. The r̄-values increase significantly for all the steels as the annealing temperature increases. The steel charged at 750°C reveals higher r̄-values. These results correspond to the variation of recrystallization texture.

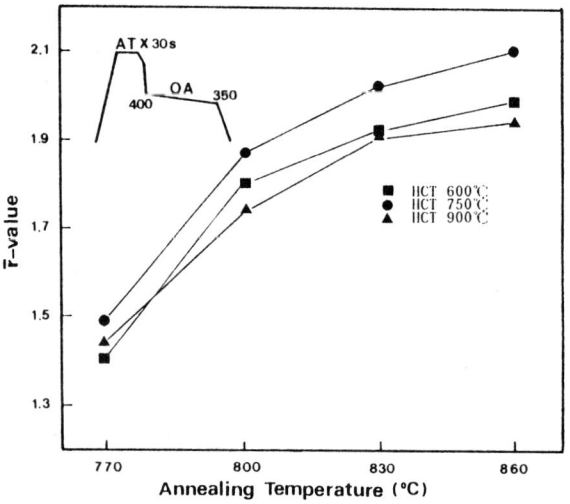

Figure 15 - Effect of the annealing temperature on the r̄-value in the steel UT1 hot-charged at various temperatures.(CT 600°C)

Conclusions

The influence of the hot charge temperatures on microstructures and mechanical properties of cold-rolled sheet steels containing titanium, manganese and phosphorus was investigated. The following conclusions were obtained;

1) The r̄-value of annealed sheet was influenced by the distribution of precipitates in hot band. The fine precipitates in hot band enhance the development {111} textures, whereas the coarse grain boundary precipitates in hot band suppresses the development of {222} texture of annealed sheets.

2) In the steel containing 0.086%Ti and 0.073%P, TiC and a particle consisting of titanium, iron and phosphorus were observed in hot bands. The particles are assumed as a compound of TiC and phosphide rather than (Ti,Fe)P. The amount of dense and fine precipitates increased as the hot charge temperature increased from 600°C to 900°C, as a result, the maximum r̄-value was obtained when the steel was hot- charged at 750°C. In the steel hot-charged at 900°C, r̄-value was lower than in the steel hot-charged at 750°C and this is accounted to the effect of a microstructure of mixed grain size in hot band.

3) In the steel containing 0.048%Ti and 0.064%P, the phosphorus peak height ratio, $A_{[P]}/A_{[Fe]}$, increased exponentially in the coiling temperature ranges above 650℃, resulting in a decrease of Charpy-V absorbed energy in hot band. A brittle fracture mode appeared at the fracture surface of the steel and it can be controlled by lowering temperature coiling.

4) It was found that the increase of Ti**/P made the precipitates finer and manganese accelerated the precipitation of (Ti,Fe)P during hot rolling process.

5) High strength cold rolled sheet steels having tensile strengths of 35-40kg/mm^2 and higher $\bar{r}$-values above 1.9 were attainable from the hot-charge rolling process.

Reference

(1) M.Yamada, Y.Tokunaga and M.Yamamoto : "Effect of Nb and Ti on resistance to cold work embrittlement of extra-low-carbon high strength cold-rolled steel sheet containing phosphorus", Tetsu-to-Hagane, 73(1987),p.1049-1056.

(2) K.Sakata, K.Hashikuchi, S.Okano, T.Higashino and M.Inoue : "Cold work brittleness of extra-low carbon hot rolled steel sheet", Tetsu-to-Hagane, 73(1987), S549.

(3) C.Brun, P.Patou and P.Parnierenie : "Influence of Phosphorus and Manganese on the Recrystallization Texture Development during Continuous Annealing in Ti-IFS Steels", p.173-195 in Metallurgy of continuous annealed sheet steel, B.L.Bramfitt and P.L.Mangonon,ed., TMS-AIME, New York, 1982

(4) A.Okamoto and N.Mizui : " Texture Formation in Ultra-Low-Carbon Ti-added Cold-Rolled Sheet Steels Containing Mn and P", p.161-180 in Metallurgy of Vacuum-Degassed Steel Products, R.Pradhan,ed.,TMS, Warrendale, Pa, 1990

(5) M.Takahashi and Y.Masui : "Hot Direct Rolling", Tetsu-to-Hagane, 64(1978), p.2012-2019.

(6) F.Kurosawa, I.Taguchi and R.Matsumoto : "Observation and Analysis of Phosphides in Steels using a Non-aqueous Electrolyte-Potentiostatic Etching Method", J.Jpn.Inst.Met., 44(1980), p.539-548.

(7) H.Kaneko,T.Nishizawa and K.Tamaoki :"Phosphide-Phases in Ternary Alloys of Iron, Phosphorus and Other Elements", J.Jpn.Inst.Met., 29(1965), p.159-165.

(8) H.Erhart,H.J.Grabke and R.Möller : "Effect of titanium on the grain boundary segregation of phosphorus in Fe-Ti-P and Fe-Ti-C-P alloys and on the embrittlement of steels", Arch.Eisenhüttenwes, 54(1983), p.285-289.

(9) N.Ohashi, M.Konishi, A.Yasuda, S.Sato and T.Irie : " Effects of Phosphorus and Solute Carbon on (111) Texture Formation in Extra-Low Carbon,Titanium- or Niobium-Added Cold Rolled Steel Sheets", p.195-208 in Proceedings for 6th Int.Conf.on Textures of Materials, ISIJ, Tokyo, 1981.

COLD-ROLLED, INTERCRITICALLY ANNEALED, AND ISOTHERMALLY

TRANSFORMED SHEET STEELS

Yasuharu Sakuma, D.K. Matlock, and G. Krauss

Department of Metallurgical and Materials Engineering
Colorado School of Mines
Golden, Colorado 80401

Abstract

Improved combinations of strength and ductility, relative to those obtainable by other processing approaches for cold-rooled and annealed sheet steels, can be produced by intercritical annealing and subsequent isothermal transformation. This paper presents the results of a study of steel sheets containing 0.1 and 0.14 pct C, 1.5 pct Mn and 1.2 pct Si, intercritically annealed at 770 or 760°C and isothermally transformed at 400 and 450°C. Microstructures consisted of ferrite, bainite, martensite and up to 12 pct retained austenite. Tensile testing was performed at temperatures between -80 and 120°C. The maximum benefit of strain-induced austenite transformation was observed between 20 and 70°C, and total elongations of 30 and 37 pct at strength levels of 750 and 620 MPa, respectively, were measured. Ductility was limited at low testing temperatures by low-strain austenite transformation and at high temperatures by strain aging and increased austenite stability. A Mo addition raised the benefit of retained austenite to higher testing temperatures.

Introduction

Cold-rolled sheet steels may be subcritically or intercritically annealed. When the steels are intercritically annealed, depending on cooling rate and alloy content, the intercritically-formed austenite transforms on cooling to various microstructures, including martensite. Steels in the latter condition are referred to as dual-phase steels and were extensively studied in the period 1975 to 1985 (1-3]). Dual-phase steels, by virtue of their martensite content, demonstrated continuous yielding and higher strain-hardening rates than conventionally treated steel sheets, and extended the envelope of ultimate tensile strength versus total elongation for sheet steels to high combinations of strength and ductility.
This paper describes a modification of dual-phase steel technology for cold-rolled and annealed sheets (4-13). Instead of direct quenching from the intercritical annealing temperature, the steel is isothermally held at a temperature where austenite transforms to bainite. Figure 1 shows a schematic of the thermomechanical processing. Intercritical annealing

temperatures are maintained just above the A_{C1}, in order to produce a matrix of recrystallized ferrite with a small amount of austenite. The isothermal transformation temperatures are varied between 300 and 500°C in order to search for conditions which cause the intercritically-formed austenite to transform to combinations of microstructural constituents which provide good combinations of strength and ductility in the processed steel.

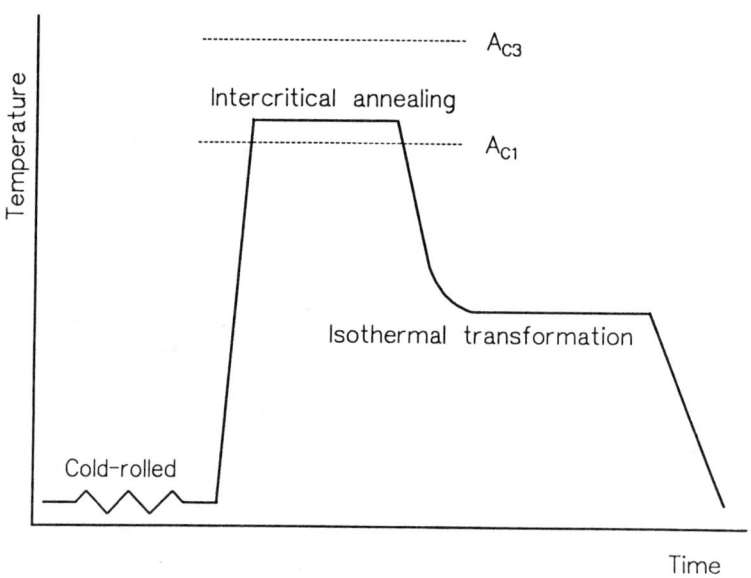

Figure 1.   Schematic diagram of thermomechanical processing.

Isothermal transformation of the intercritically-formed austenite results in various amounts of bainite and untransformed austenite. The latter may transform partly to martensite on cooling to room temperature. Thus, the microstructures of cold-rolled intercritically annealed and isothermally transformed (IA/IT) steels consist of various mixtures of retained ferrite, bainite, retained austenite, and martensite. The selection of steels with Si, Mn, and Mo retards the formation of cementite in association with bainite and therefore promotes the retention of austenite in the bainitic structures. Bainite formation introduces much lower densities of dislocations into retained ferrite than does martensite formation. Thus, the bainite, together with significant amounts of retained austenite, contribute to produce higher ductility in IA/IT steels compared to direct-quenched intercritically annealed steels.

This paper describes the effect of IA/IT processing on the microstructure and properties of a series of cold-rolled sheet steels containing Si, Mn, and Mo at two carbon levels, 0.1 and 0.14 pct. In order to vary the stability of austenite to deformation-induced transformation, the IA/IT processed specimens were tensile tested above and below room temperature.

Experimental Procedure

Table 1 presents the chemical compositions of the three vacuum-melted steels used in this study, together with A_{C1} and A_{C3} transformation temperatures estimated from Andrews empirical formula (14). The steels permit the evaluation of a base composition consisting of 1.2 pct Si and 1.5 pct Mn at two levels of carbon and with and without Mo.

TABLE 1 - CHEMICAL COMPOSITION (WT PCT) AND ESTIMATED TRANSFORMATION TEMPERATURE (°C) OF STEELS INVESTIGATED

C	Si	Mn	P	S	Mo	Sol. Al	Total N	A_{C1}	A_{C3}
0.138	1.206	1.567	0.006	0.0060	<0.005	0.027	0.0035	741	846
0.096	1.198	1.495	0.004	0.0056	<0.005	0.028	0.0056	742	859
0.099	1.212	1.507	0.005	0.0057	0.140	0.032	0.0051	742	863

Slabs 30 mm thick were heated to 1250°C, and hot-rolled to 4.0 mm thick strip with finishing temperatures between 830 and 890°C. After surface grinding to 3.0 mm thickness, the hot-rolled strips were cold-rolled 67 pct to 1.0 mm thickness. According to ASTM E8, tensile specimens measuring 101.6 mm long and 12.7 mm wide were prepared with tensile axis oriented transverse to the rolling direction.

The tensile specimens were heat-treated following cycles illustrated in Fig. 1. First they were intercritically annealed for 5 min by immersion in a neutral salt bath. The temperature of the bath was kept at 760°C for annealing of the 0.1 pct C steels and at 770°C for annealing of the 0.14 pct C steel. After annealing, the specimens were isothermally transformed for 1 to 60 min in another salt bath kept at 400 or 450°C. The isothermal heating was followed by oil-quenching to room temperature.

Tensile testing and microstructural evaluation, including light microscopy, scanning electron microscopy (SEM), transmission electron microscopy (TEM), and x-ray diffraction, were performed. Tensile testing was done at a strain rate of 8.8×10^{-4} sec^{-1} on a screw driven test machine with a liquid-filled chamber mounted to the movable crosshead. In addition to testing at 20°C, selected specimens were strained at temperatures between -80 and 120°C. The temperature was maintained by completely submerging the specimens in a methanol bath at -80 to 20°C or in an oil bath at 70 to 120°C. A submersible extensometer with initial length of 25.4 mm was directly attached to the sample, and load-displacement data were recorded on a computer-based data acquisition system.

Light and scanning electron microscopic measurements were done on longitudinal sections which were cut from unstrained grip ends of the tensile specimens, polished, and etched with 2 pct nital. TEM foils parallel to the rolling plane were made by thinning to 0.07 mm, punching out 3 mm diameter discs and perforating in a twin jet electropolisher with an electrolyte of 5 vol pct perchloric and 95 pct acetic acid. A Philips 400 EM was used in the examination with an accelerating voltage of 120 kV. For x-ray diffraction measurements, chemically polished surfaces, about 0.27 mm below the original surface, were prepared not only from unstrained grip ends but also from the gage sections of unloaded specimens strained at various temperatures. The amount of retained austenite was determined from the integrated intensities of $(200)\alpha$, $(211)\alpha$, $(220)\gamma$, and $(311)\gamma$ diffraction peaks produced by monochromatic Cu Kα radiation (15,16).

Results

Microstructure

Figure 2 shows a series of SEM micrographs for the isothermally transformed 0.14 C-1.2 Si-1.5 Mn steel. Microstructures of the steel transformed at 450°C for 1 and 4 min are presented in Fig. 2(a) and (b), respectively, and those of the steel transformed at 400°C for 1 and 4 min are presented in Fig. 2(c) and (d), respectively. About 75 pct of the microstructures are occupied by somewhat elongated ferrite grains and the remaining 25 pct corresponds to transformation products of the intercritically-formed austenite. The 450°C-1 min microstructure contains the most retained austenite, about 12 pct. This austenite is present as smooth, featureless regions, about 1 to 2 μm in diameter, Fig. 2(a). After longer transformation times at 450°C, the austenite transforms to a microstructure sensitive to etching, but with few identifiable features revealed by SEM, Fig. 2(b). This microstructure may represent a non-classical form of bainite which consists largely of ferrite with a high dislocation substructure and some retained austenite.

The microstructure of a specimen transformed at 400°C for 1 min contains large amounts of martensite formed by cooling to room temperature. The martensite consists of laths with bright intervening features, presumably retained austenite, Fig. 2(c). With increasing time at 400°C, the austenite isothermally transforms to bainite and less martensite is formed. As a result, there is more austenite present in the 4 min specimen than the 1 min specimen, compare Figs. 2(c) and 2(d).

Characteristic features associated with retained austenite are shown in Fig. 3, which presents (a) SEM and (b) TEM micrographs of the 0.14 C-1.2 Si-1.5 Mn steel transformed at 400°C for 15 min. The high magnification SEM micrograph shows that most of the retained austenite is present at the ferrite-austenite interfaces formed by intercritical annealing. The TEM micrograph shows very low dislocation densities in a crystal of retained austenite and in the ferrite surrounding the austenite. As pointed out by the letter T in Fig. 3(b), annealing twins are imaged in some austenite particles.

Figure 4 summarizes the amounts of retained austenite measured by x-ray diffraction of the 0.14 C-1.2 Si-1.5 Mn steel as a function of transformation time and temperature. Data from specimens transformed between at 300 and 500°C are included,[11] and contours are drawn for various austenite contents. Two types of microstructural change affect the retained austenite content measured at room temperature. Isothermal transformation of austenite to bainite is one type, and bainite formation causes the amount of austenite to decrease as a function of time at a given temperature. A second type is martensite transformation during cooling to room temperature. Although isothermal transformation at 450°C for 1 min maximizes the retained austenite content in this alloy, austenite completely transforms to bainite within 15 min at 450°C and 500°C. In the specimens transformed between 300°C and 400°C, the amount of retained austenite goes through a maximum with transformation time. Smaller amounts of austenite present in short-time transformed specimens are related to more athermal transformation to martensite on cooling, which is consistent with the SEM micrograph shown in Fig. 2(c). The transformation time associated with maxima in austenite shifts to longer times, and the maximum amounts of retained austenite decreases, with decreasing transformation temperature.

Changes of retained austenite content with the isothermal transformation time at 400 and 450°C are given in Fig. 5 for the three steels. With increasing C content, larger amounts of austenite are retained after isothermal transformation. However, retained austenite content in the 0.14 pct C Mo-free steel significantly depends on the isothermal

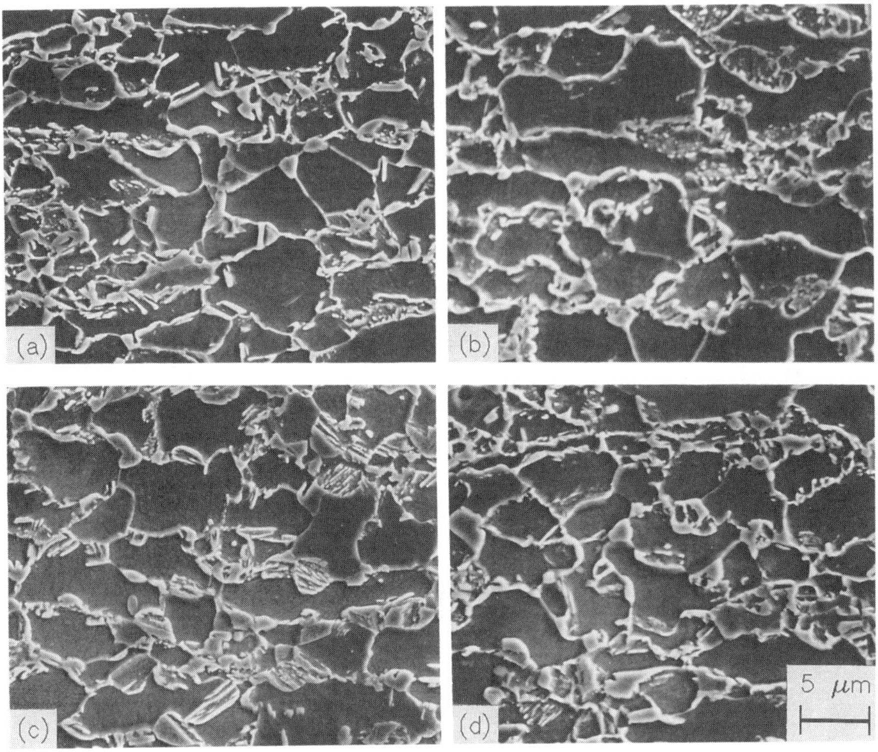

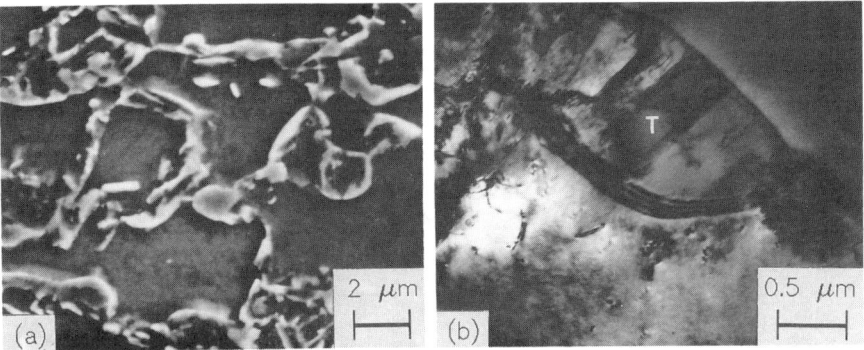

Figure 2. Microstructures of specimens of the 0.14 C-1.2 Si-1.5 Mn steel transformed at (a) 450 °C for 1 min, (b) 450 °C for 4 min, (c) 400 °C for 1 min, and (d) 400 °C for 4 min. Nital etched; SEM micrographs.

Figure 3. Microstructures of specimens of the 0.14 C-1.2 Si-1.5 Mn steel transformed at 400 °C for 15 min. (a) Nital etched; SEM micrograph and (b) TEM micrograph. Letter T in (b) indicates annealing twin in retained austenite.

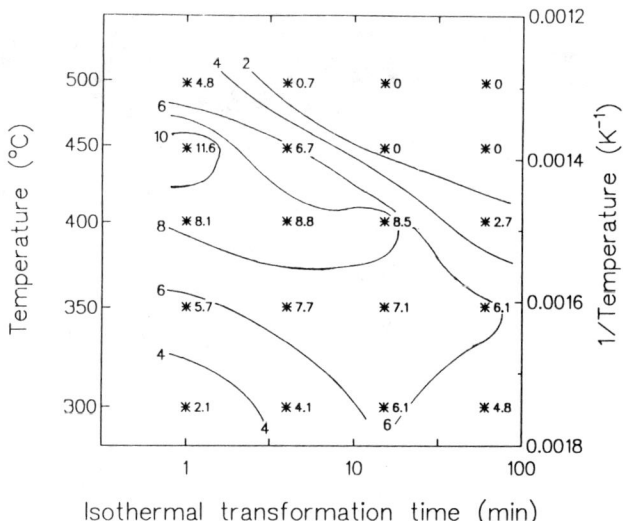

Figure 4. Changes in retained austenite content versus transformation temperature and time for isothermally transformed specimens of the 0.14 C-1.2 Si-1.5 Mn steel. Contours are drawn for various retained austenite contents.

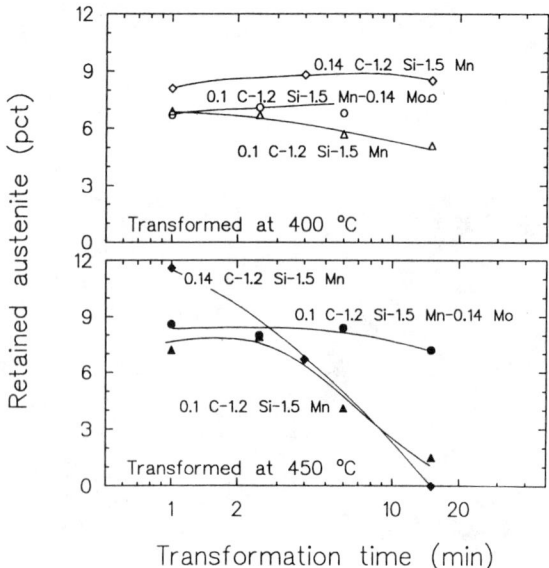

Figure 5. Changes in retained austenite content for the three steels as a function of isothermal transformation time at 400 °C and 450 °C.

transformation time at 450°C and no austenite is retained after isothermal transformation at 450°C for 15 min. In contrast, molybdenum does not affect the amounts of austenite retained after short time transformation, but increases austenite retained after prolonged transformation time, especially at 450°C. Therefore relatively large amounts of retained austenite are incorporated into the intercritically annealed microstructures of the 0.14 pct Mo steel regardless of isothermal transformation condition.

Room Temperature Mechanical Properties

The effect of transformation time on the room temperature engineering stress-strain curves of 400°C-transformed specimens of the 0.1 C-1.2 Si-1.5 Mn steel is given in Fig. 6. Deformation behavior of the 1 min transformed specimen resembles that of dual-phase steels, i.e. deformation is characterized by continuous yielding and low yield to ultimate tensile strength ratios. In contrast, curves for the specimens isothermally transformed more than 2.5 min show discontinuous yielding. With increasing isothermal transformation time from 1 min to 6 min, yield strength increases from 310 MPa to 430 MPa and ultimate tensile strength decreases slightly from 670 MPa to 620 MPa. At the same time total elongation increases from 33 pct to 39 pct. Increasing transformation time to 15 min does not change strength, but decreases ductility.

The Jaoul-Crussard analyses of the curves from the four specimens shown in Fig. 6 are given in Fig. 7. The Jauol-Crussard plots characterize strain hardening as a function of true plastic strain, and the dependency of strain hardening on strain is monitored from yielding to necking. Curves are presented after 0.2 pct strain for the continuous yielding specimen or after Lüders strain for the discontinuous yielding specimens. The strain hardening rates of the 1 min specimen are higher than those of the other specimens at small strains, but decrease rapidly with increasing strain. In contrast, the strain hardening rate of the 6 min specimen decreases gradually with increasing strain, and exceeds those of the 1 min specimen after 5 pct true plastic strain. These high strain hardening rates of the 6 min specimen at large strains deter the onset of plastic instability, and account for the large uniform elongation presented in Fig. 6. The strain hardening rates of the 15 min specimen change with strain similarly to those of the 6 min specimen, but are somewhat lower.

Figure 8 summarizes the room temperature mechanical properties of the 400°C-transformed specimens of the three steels as a function of isothermal transformation time. Although values of the yield and ultimate tensile strength (YS, UTS) and uniform and total elongations (e_U, e_T) are dependent on the steel chemistry, changes of mechanical properties as a function of isothermal transformation time follow the same patterns for all three steels. With increasing C content from 0.1 pct to 0.14 pct, the yield and ultimate tensile strengths increase 30 to 70 MPa and 100 to 130 MPa, respectively. The ultimate strengths are the highest and the yield strengths are the lowest for specimens transformed for the shortest times. The measures of strength converge with increasing transformation time. Uniform and total elongations decrease about 5 and 7 pct, respectively, with increasing C content for all conditions. The different values of the decrease in uniform and total elongations indicate that post-uniform elongation also decreases with increasing C content.

The Mo addition also increases the yield and ultimate tensile strengths, but does not affect uniform and total elongations except for the 6 min specimen. Elongations of the 0.1 pct C Mo-free steel are quite sensitive to the isothermal transformation time and peak after the transformation at 400°C for 6 min, whereas elongations of the 0.1 pct C-0.14 pct Mo steel are little affected by the isothermal transformation time at 400°C between 2.5 and 15 min.

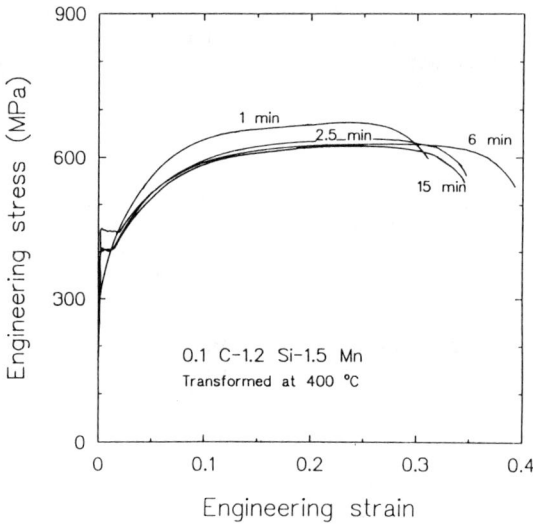

Figure 6. Room temperature engineering stress-strain curves of specimens of the 0.1 C-1.2 Si-1.5 Mn steel transformed at 400 °C. Curves plotted to failure.

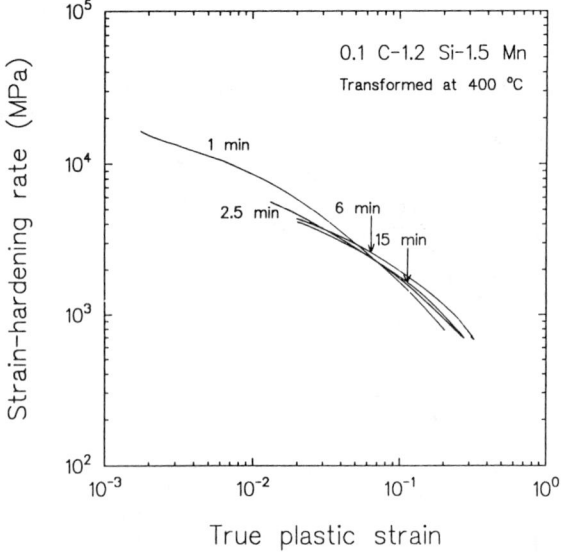

Figure 7. Jaoul-Crussard plots of room-temperature-tested specimens of the 0.1 C-1.2 Si-1.5 Mn steel transformed at 400 °C.

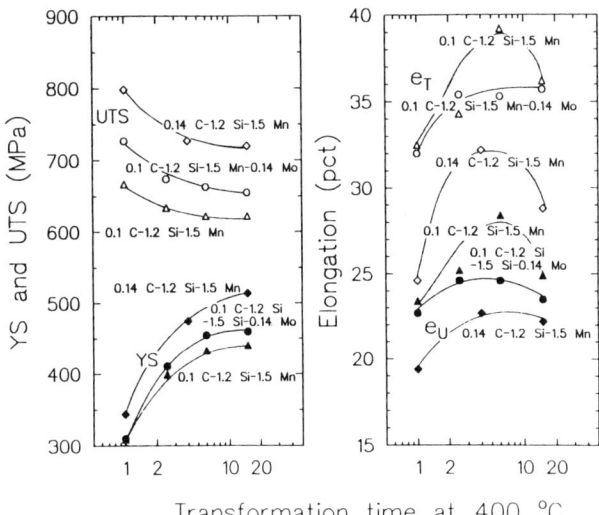

Figure 8. Changes in room temperature mechanical properties as a function of transformation time for isothermally transformed specimens of the three steels.

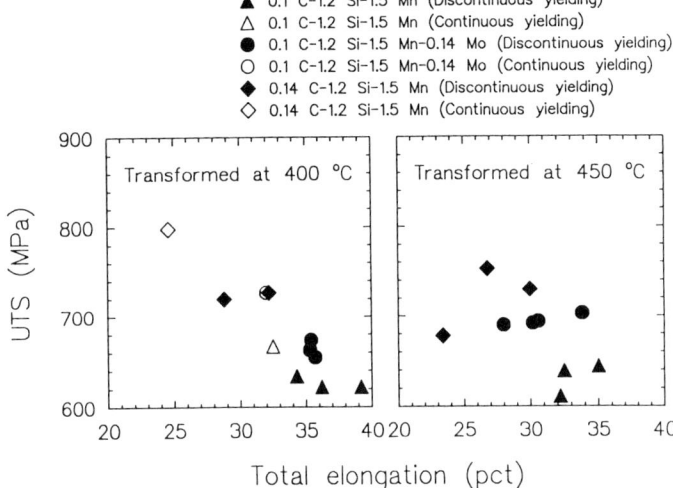

Figure 9. Ultimate tensile strength vs. total elongation for isothermally transformed specimens of the three steels.

Ultimate tensile strength versus total elongation is presented in Fig. 9 for the three steels isothermally transformed at 400 and 450°C. Maximum ductility is obtained after isothermal transformation at 400°C for the three steels. For the Mo-free steels, the maximum ductility obtained after 450°C transformation is 2 to 4 pct smaller, at the same strengths, than that obtained after 400°C transformation. In contrast, the maximum ductility decreases slightly and the ultimate tensile strength increases about 40 MPa for the 0.14 pct Mo steel with increasing isothermal transformation temperature from 400°C to 450°C. Therefore transformation at 450°C is preferred for the 0.14 pct Mo steel for the best combination of ductility and high strength.

Mechanical Properties Above and Below Room Temperature

Engineering stress-strain curves from the 0.1 C-1.2 Si-1.5 Mn steel isothermally transformed at 400°C for 6 min, obtained at -80°C to 120°C, are presented in Fig. 10. Serrated stress-strain curve and surface relief on deformed specimens, typical indications of dynamic strain aging, are observed after testing at 120°C. At 70°C, flow stresses are slightly lower than those at 120°C, and the effects of dynamic strain aging are not observed. With decreasing testing temperature from 70°C to -80°C, flow stresses increase and ductility goes through a maximum.

The Jaoul-Crussard analysis of the temperature-dependent plastic strain behavior of the 0.1 C-1.2 Si-1.5 Mn steel isothermally transformed at 400°C for 6 min is given in Fig. 11. The strain-hardening rates at low strains increase with decreasing testing temperature. At lower testing temperatures, the slopes of the strain-hardening curves decrease steeply with accumulating strain. The change is significant at strains over 8 pct for the curves obtained at -80°C. With increasing testing temperature, the steep changes of slope shift to larger strain, and at higher testing temperatures, the slopes tend to remain the same even at larger strains. Therefore, at high strains the highest strain-hardening rates are observed at 20°C and 70°C. The highest strain-hardening rates at 20°C testing correlate with the highest uniform and total elongations of this specimen, Fig. 10.

The effects of testing temperature on the mechanical properties are summarized in Figs. 12 and 13 for all three steels transformed at 400 and 450°C, respectively. All specimens were transformed for 6 min except for the 0.14 pct C Mo-free steel which was transformed for 4 min at 400°C or 450°C. All specimens have a similar temperature dependency of the yield and ultimate tensile strengths. The yield strengths are insensitive to testing temperature, but the ultimate tensile strengths increase with decreasing testing temperature below 70°C. Changes of the ultimate tensile strength with testing temperature are most dominant in the 0.1 pct C-0.14 pct Mo steel transformed at 450°C. Compared to the Mo-free steel of the same C content, the addition of 0.14 pct Mo increases the ultimate tensile strength 70 MPa at -80°C, but only 25 MPa at 120°C. Therefore, at -80°C, the strength of the 450°C-transformed 0.1 pct C-0.14 pct Mo steel is almost the same as that of the 450°C-transformed specimen of the higher C Mo-free steel.

The differences between the total and uniform elongations are insensitive to testing temperature, and the larger post-uniform elongation for the 0.1 pct C steels measured at room temperature is maintained above and below room temperature. Elongations of 400°C-transformed specimens peak at 20°C for all the alloys, but the peak of the 0.1 pct C-0.14 pct Mo steel is more diffuse than that of the Mo-free steels. After transformation at 450°C instead of 400°C, similar testing temperature sensitivity of ductility is observed for the 0.14 pct C Mo-free steel. However, elongations of the 0.1 pct C steels become less sensitive to testing temperature after

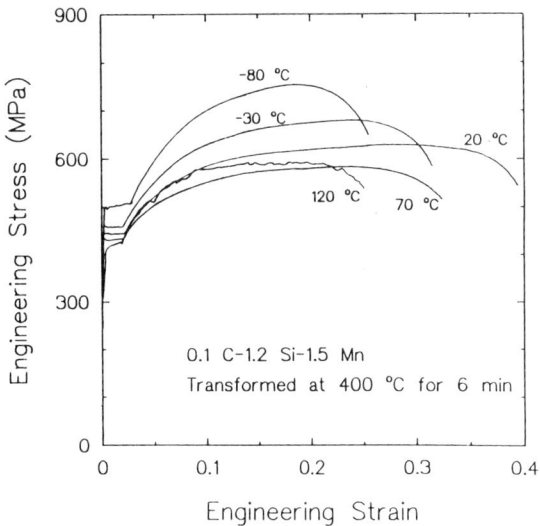

Figure 10. Engineering stress-strain curves of the 0.1 C-1.2 Si-1.5 Mn steel transformed at 400 °C for 6 min at various testing temperatures. Curves plotted to failure.

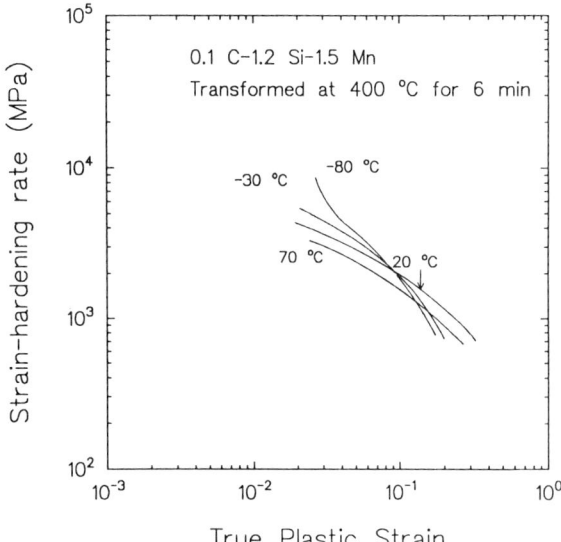

Figure 11. Jaoul-Crussard plots of specimens of the 0.1 C-1.2 Si-1.5 Mn steel transformed at 400 °C for 6 min tested at various temperatures.

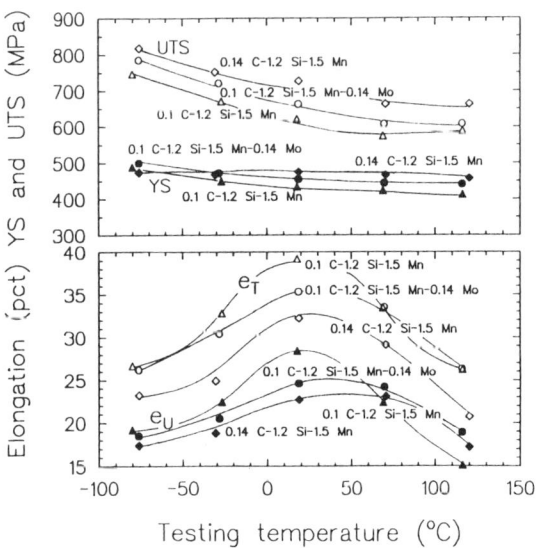

Figure 12. Mechanical properties of 400 °C-transformed specimens as a function of testing temperature. Specimens of the 0.1 pct C steels are isothermally transformed at 400 °C for 6 min and specimens of the 0.14 pct C steel are isothermally transformed at 400 °C for 4 min.

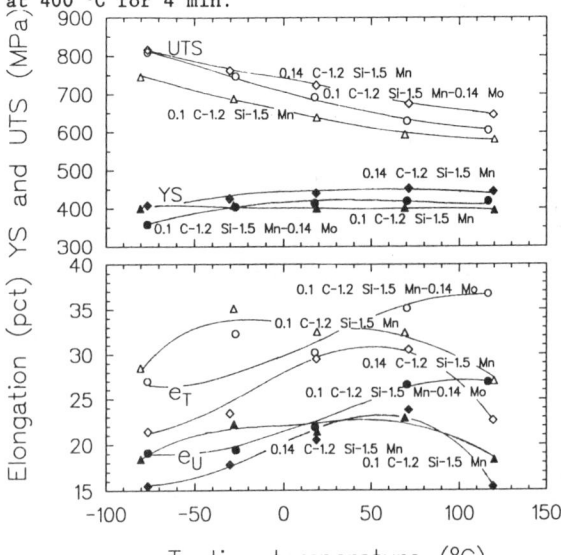

Figure 13. Mechanical properties of 450 °C-transformed specimens as a function of testing temperature. Specimens of the 0.1 pct C steels are isothermally transformed at 450 °C for 6 min and specimens of the 0.14 pct C steel are isothermally transformed at 450 °C for 4 min.

transformation at 450°C. The 0.1 pct C Mo-free steel has plateaus of
elongations between -30 and 70°C. The most noticeable difference in the
temperature sensitivities of elongations is found in the 450°C-transformed
0.1 pct C-0.14 pct Mo steel. The enhanced ductility is maintained even at
120°C, whereas ductilities of the other specimens are significantly reduced
with increasing testing temperature from 70°C to 120°C.

Decrease of Retained Austenite on Deformation

Figure 14 shows amounts of retained austenite as a function of strain
at 20°C for the most ductile specimens of the three steels isothermally
transformed at 400°C. Retained austenite decreases steadily up to the onset
of necking, and therefore the enhanced ductilities of these specimens are
attributed to the increased strain-hardening rates associated with the
strain-induced transformation of austenite.

Figure 15 presents the decrease of retained austenite with increasing
strain at 20 and 120°C for the 0.1 pct C steels with and without Mo in
specimens isothermally transformed at 450°C for 6 min. Molybdenum not only
increases the amounts of austenite retained after isothermal transformation,
but also influences significantly the mechanical stability of austenite. In
the 0.1 pct C Mo-free steel, strain-induced transformation continues with
increasing strain at 20°C, but occurs only slightly at 120°C. In contrast,
most of the retained austenite in the 0.1 pct C-0.14 pct Mo steel transforms
at small strains at 20°C. Therefore the amounts of austenite left for
transformation after 4 pct strain in the 0.14 pct Mo steel are almost
equivalent to that available in the Mo-free steel at the start of testing.
As a result, the deformation behavior of the two steels during testing at
20°C is almost the same. At 120°C, rapid transformation at small strains
does not occur, and the austenite gradually decreases with accumulation of
strain. The large ductility of the 450°C-transformed 0.14 pct Mo steel,
shown in Fig. 13, therefore correlates to the sustained transformation of
austenite to martensite.

Discussion

Figure 16 compares ultimate tensile strength-total elongation values
for similar steels either directly quenched or isothermally transformed
after intercritical annealing. All properties were obtained by room
temperature tensile testing. Although there is overlap between the two sets
of data, the IA/IT specimens have significantly better ductility than do the
direct-quenched specimens. When the data of Fig. 16 are compared to other
sets of data for sheet steels strengthened by various alloying and
processing techniques (17), the IA/IT steels also show relatively higher
combinations of strength and ductility.

The strengths of the IA/IT steels are lower than those of dual-phase
steels with martensite dispersed in retained ferrite, but much higher than
conventionally processed cold-rolled and annealed steels. The strength of
IA/IT steels is dependent on dispersion strengthening by the isothermally
transformed austenite. Short isothermal transformation times limit bainite
formation and much of the untransformed austenite transforms to martensite
on cooling. These microstructures have high strength but low ductility,
Fig. 8. The dislocations introduced into the retained ferrite cause high
initial rates of strain hardening which drop rapidly with increasing strain,
Fig. 7. With increasing isothermal transformation time, more austenite
transforms to bainite and carbon is rejected into the remaining austenite,
especially in Si-containing steels. Thus the retained austenite resists
transformation to martensite on cooling to room temperature. The bainite

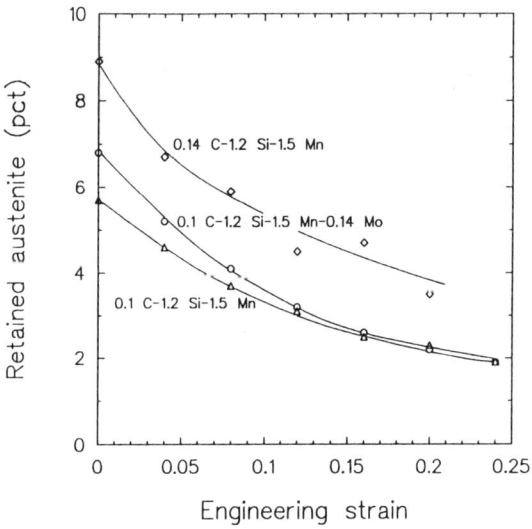

Figure 14. Decease of retained austenite as a function of engineering strain during room temperature testing. Heat treatment conditions are the same as described in Figure 12.

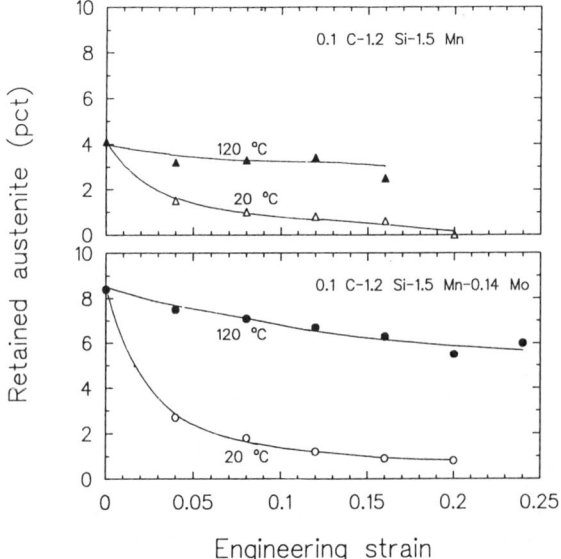

Figure 15. Decrease of retained austenite as a function of engineering strain during testing at 20 °C and 120 °C for the steels marked after transformation at 450 °C for 6 min.

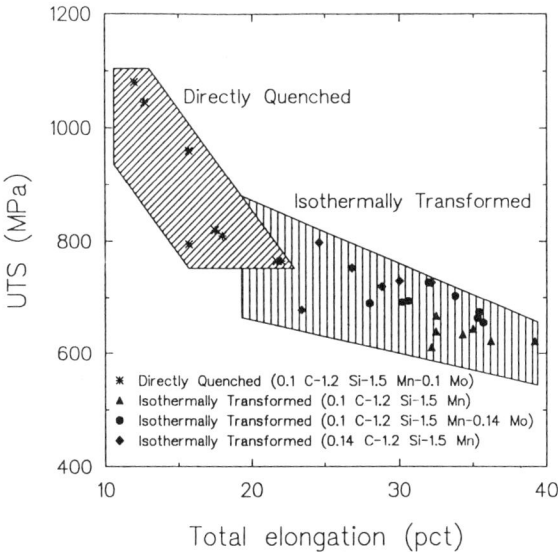

Figure 16. Comparison of ultimate tensile strength versus total elongation for various steels either directly quenched or isothermally transformed following intercritical annealing.

which forms is largely ferritic and does not substantially increase the dislocation density of the retained ferrite matrix. The combination of bainite and austenite therefore are associated with discontinuous yielding and low strain hardening rates at low strains. However, with increasing strain, the retained austenite transforms to martensite by strain-induced mechanisms. The martensite formation increases strain hardening at high strains and defers necking to higher uniform elongations. The ferrite-bainite-austenite microstructures therefore show higher ductilities than the microstructures with large amounts of martensite.

The ductilities of IA/IT steels containing austenite typically go through a maximum with testing temperature. At low testing temperatures, strengths are higher but ductilities are low because the retained austenite has low mechanical stability and transforms at low strains. During high temperature testing, the retained austenite may be too mechanically stable and therefore reduced benefits from strain-induced transformation develop at strains around instability. In the extreme, the austenite may be deformed above its M_D temperature and no strain-induced transformation may occur. Depending on alloying and isothermal transformation conditions, the peak in ductility may occur at room temperature, as shown in this investigation for the Mo-free steels, or the peak in ductility may occur above room temperature, as shown for the 0.14 pct Mo steel isothermally transformed at 450°C in this investigation. In the latter condition, molybdenum increases hardenability and causes relatively large volume fractions of austenite to be retained. The stability of this austenite is sufficient to sustain strain-induced transformation throughout testing at 120°C, Fig. 15, and produces maximum ductility in specimens tested at this temperature, Fig. 13.

Summary

The above results and discussion show that the properties of low-carbon cold-rolled and annealed steels could be pushed to a higher window of strengths and ductilities by IA/IT than by other methods of processing. The results of this and other investigations (10-13) show that mechanical properties of intercritically annealed steels within this window can be tailored by various combinations of alloying and isothermal transformation conditions. The mechanical stability of austenite is critical to obtaining good ductility, and the peak ductility may be obtained by straining at or above room temperature. Tests for the best combinations of strength and ductility should be performed at temperatures somewhat below the M_D temperature of the retained austenite where strain-induced martensite formation occurs at high plastic strains.

Acknowledgements

We acknowledge the assistance of Dr. D.L.Williamson, Physics Department, Colorado School of Mines, for providing laboratory facilities of the x-ray diffraction measurements. One of the authors, Y.Sakuma, thanks the Nippon Steel Corp. for the support of research at the Colorado School of Mines. The steels were provided by the Nippon Steel Corp. This research was part of a program supported by the Metallurgy Program, Division of Materials Research, National Science Foundation of the United Steel.

References

1. S.Hayami and T.Furukawa: in Microalloying 75, Union Carbide Corp., New York, 1975, pp.78-87.

2. W.S.Owen: Metals Technology, 1980, vol.7, pp.1-13.

3. D.K.Matlock, E.Zia-Ebrahimi, and G.Krauss: in *Deformation, Processing and Structure* ed. by. G.Krauss, ASM, Ohio, 1984, pp.47-87.

4. O.Matsumura, Y.Sakuma, and H.Takechi: Trans. Iron and Steel Inst. Japan, 1987, vol.27, pp.570-579.

5. O.Matsumura, Y.Sakuma, and H.Takechi: Scripta Metallurgica, 1987, vol.21, pp.1301-1306.

6. Y.Sakuma, O.Matsumura, and H.Takechi: Met. Trans. A, 1991, vol. 22A, pp.

7. C.Chen, H.Era, and M.Shimizu: Met. Trans. A, 1989, vol.20A, pp.437-445.

8. K.Sugimoto, M.Misu, M.Kobayashi, and H.Shirasawa: Tetsu-to-Hagane, 1990, vol.76, pp.1356-1363.

9. K.Sugimoto, M.Kobayashi, and S.Hashimoto: J. Japan Inst. Metals, 1990, vol.54, pp.657-663.

10. Y.Sakuma, D.K.Matlock, and G.Krauss: J. Heat Treat., 1990, vol.8, pp.109-120.

11. Y.Sakuma, D.K.Matlock, and G.Krauss: Submitted to Met. Trans. A in 1991.

12. Y.Sakuma, D.K.Matlock, and G.Krauss: Submitted to Met. Trans. A in 1991.

13. Y.Sakuma, D.K.Matlock, and G.Krauss: Submitted to Mater. Sci. Technol. in 1991.

14. K.W.Andrews: J. Iron and Steel Inst., 1965, vol.203, pp.721-727

15. R.L.Miller: Trans. ASM, 1964, vol.57, pp.892-899.

16. R.L.Miller: Trans. ASM, 1968, vol.61, pp.592-597.

17. G.R.Speich: in *Fundamentals of Dual-Phase Steels*, ed. by R.A.Kot and B.L.Bramfitt, AIME, New York, 1981, pp.3-45.

18. S.J.Matas and R.F.Hehemann: Trans. AIME, 1961, vol.221, pp.179-185.

19. R.LeHoullier, G.Begin, and A.Dube: Met. Trans., 1971, vol.2, pp.2645-2653.

20. E.D.Dorazil and J.Svejcar: Arch Eisenhüttenwes, 1979, vol.50, pp.293-298.

21. R.F.Hehemann, K.R.Kinsman, and H.H.Aaronson: Met. Trans., 1972, vol.3, pp.1077-1094.

22. H.K.D.H.Bhadeshia and D.V.Edmonds: Met. Trans. A, 1979, vol.10A, pp.895-907.

22. H.K.D.H.Bhadeshia and D.V.Edmonds: Met. Sci., 1983, vol.17, pp.411-419.

23. H.K.D.H.Bhadeshia and D.V.Edmonds: Met. Sci., 1983, vol.17, pp.420-425.

24. H.K.D.H.Bhadeshia and D.V.Edmonds: Acta Metallurgica, 1980, vol.28, pp.1265-1273.

25. V.T.T.Miihkinen and D.V.Edmonds: Mater. Sci. Technol., 1987, vol.3, pp.422-431

26. C.Nakao, K.Tsuzaki, and T.Maki: Current Adv. in Mat. and Processes (Report of the Iron and Steel Inst. of Japan Meeting), 1990, p.1799.

ANNEALING CONDITIONS FOR HOT-ROLLED STRIP OF 13Cr STAINLESS STEEL

K. Miura, S. Satoh, K. Yoshioka, O. Hashimoto

Technical Research Division
Kawasaki Steel Corporation
Chiba 260, Japan

Abstract

Batch annealing of hot-rolled strip of 13Cr stainless steel promotes the formation of a Cr-depleted zone on the strip surface. This Cr-depleted zone degrades oxidation resistance in finish-annealing after cold-rolling, leading to poor descalability in the following pickling process. Shortening the annealing time for hot-rolled strip, increasing the O_2 concentration and reducing the dew point of the finish-annealing atmosphere improve oxidation resistance. The combined reduction of C content, increase of N content, and addition of Al improve formability and ridging characteristics in continuously annealed hot-rolled strip.

Introduction

Martensitic stainless steels represented by type 410 have excellent formability and ridging characteristics and moderate corrosion resistance, and offer an economical alternative to stainless steels with higher chromium contents. Therefore, they have been applied widely as table and houseware materials for stainless flatware, trays, bowls and similar items.
In the conventional manufacturing process for 13Cr stainless steel sheet, the batch annealing of hot-rolled strip is indispensable for improving the formability and ridging characteristics of the strip after cold-rolling, but promotes the formation of a Cr-depleted zone on the strip surface. This Cr-depleted zone degrades the oxidation resistance of the cold-rolled strip in finish annealing after cold-rolling and leads to poor descalability in the following pickling process. In this study, the effects of annealing conditions for hot-rolled strip and the finish-annealing atmosphere on oxidation resistance were investigated.
On the other hand, the annealing conditions for hot-rolled strip affect not only the formation of the Cr-depleted zone, but also the formability of the cold-rolled strip. Shortening the annealing time applied to hot-rolled sheet to minimize the development of the Cr-depleted zone, however, leads to deterioration of the formability of cold-rolled sheets. The effects of the chemistry (C, N, Al) and annealing conditions of the hot-rolled sheets on the formability of sheet after cold-rolling and annealing were also examined.

Experimental Procedure

The chemical composition of the steels used are shown in Table I. Steel I was produced by a typical mass-production process. Steels II, III, IV and V were vacuum-melted as 100kg ingots with the various C, N and Al contents.

Table I Chemical compositions (mass %)

No.	C	Si	Mn	P	S	Cr	Al	N
I	0.058	0.39	0.39	0.032	0.005	13.30	0.011	0.0162
II	0.057	0.40	0.40	0.031	0.005	13.27	0.010	0.0153
III	0.054	0.37	0.37	0.030	0.006	13.28	0.110	0.0147
IV	0.059	0.40	0.41	0.032	0.005	13.40	0.098	0.0325
V	0.028	0.38	0.40	0.032	0.005	13.21	0.104	0.0316

Steel I was hot-rolled to a 3.5mm thickness in the laboratory and subjected to annealing at 1073K for 8hr, 1073K for 60s, 1123K for 60s, and 973K for 10hr. The Cr content profiles were measured by the atomic absorption method. The scale on the surfaces of the hot-rolled and annealed sheet was analyzed by EPMA and X-ray diffraction.
The hot-rolled and annealed sheets of steel I were then pickled in 20.0 mass% H_2SO_4 followed by 12.0 mass% HNO_3 and cold-rolled to 1.8mm. The relation between the Cr-depleted zone in the surface layer before cold-

rolling and the oxidation resistance of the cold-rolled sheets was investigated by changing the O_2 concentration and dew point (d.p.) in a finish-annealing atmosphere of 12.0 volume% CO_2-O_2-H_2O-bal.N_2 with various O_2 and H_2O contents at 1073K for 60s. The scale on the surfaces of the cold-rolled and annealed sheets was observed by SEM and analyzed by SIMS.

Steel II, III, IV, and V were also hot-rolled to a 3.5mm thickness in the laboratory and subjected to annealing at 1073K for 8hr, 1073K for 60s, and 1123K for 60s. These materials were cold-rolled to 0.8mm after descaling and subjected to finish-annealing at 1073K for 60s. Total elongation and ridging characteristics were examined by tensile testing using the JIS No.13B specimens (12.5mm width, 50mm gauge length). Ridging characteristics were evaluated by the standard reference used in Kawasaki Steel Corporation for surface appearance after 20% stretching in the longitudinal direction. Vickers hardness tests and microstructure observation of the longitudinal cross-section were also conducted.

Experimental results and discussions

Formation behavior of Cr-depleted zone

The effect of annealing conditions for hot-rolled sheet on the formation of the Cr-depleted zone is shown in Fig.1. A 6μm thick Cr-depleted zone already exists on the surface in the as hot-rolled condition. Annealing at 1073K for 8hr, simulating conventional annealing, increased its thickness to 20μm and sharply decreased the Cr content at the surface. On the other hand, annealing at 1073K or 1123K for 60s, simulating continuous annealing, increased the Cr-depleted zone by only 1~2 μm.

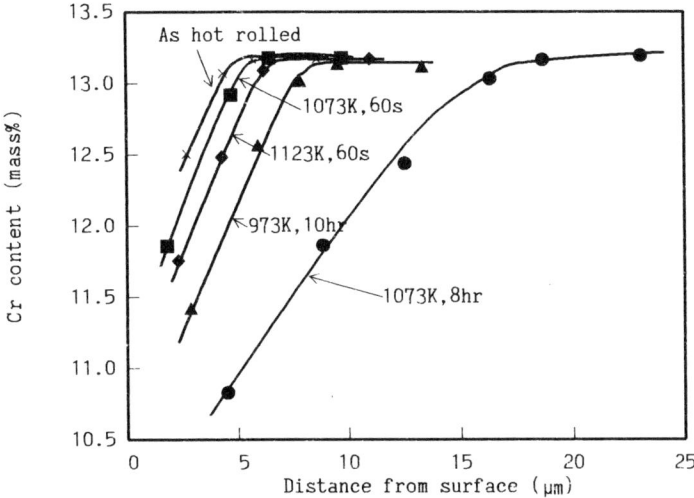

Fig.1 Effects of annealing conditions on Cr content profiles below surface of hot-rolled sheets

Cross-sectional analyses of the surface of hot-rolled sheet by EPMA are shown in Fig.2. A Cr-rich oxide was observed in the most interior oxide layer. Annealing increased the Cr content of the Cr-rich oxide layer and promoted development of the Cr-depleted zone. This behavior is more pronounced in prolonged annealing.

The effects of annealing conditions on the composition of scale are shown in Table II. While Fe_3O_4 and $FeCr_2O_4$ cannot be distinguished by diffraction because of their similar lattice constant, the major components of the scale can be estimated as Fe_3O_4 and Fe_2O_3 for hot-rolled sheet, and as $FeCr_2O_4$ and FeO for sheet annealed at 1073K for 8hr.

In general, the partial pressure of oxygen (P_{O_2}) gradually decreases toward the interior of the scale layer. Fe_3O_4 and Fe_2O_3, which are unstable at lower P_{O_2} values, are reduced in proportion to lowering of P_{O_2}; $FeCr_2O_4$, which is stable even at lower P_{O_2} values, is formed in the interior of the scale. The equilibrium dissociation pressures of Cr and Fe oxides (1) are shown in Fig.3. Therefore, the formation of the Cr-depleted zone is considered to be caused by the formation of $FeCr_2O_4$ due to the consumption of Cr in the metal underneath the scale during the annealing of hot-rolled sheet.

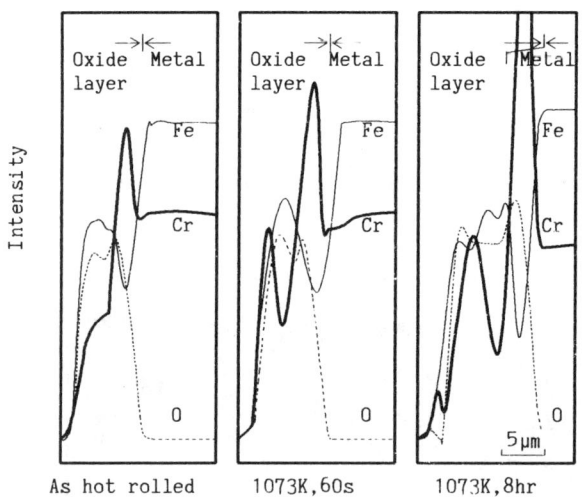

Fig.2 Cross-sectional analyses of surface of hot-rolled sheet by EPMA.

Table II Effects of annealing conditions for hot-rolled sheet on composition of scale as determined by X-ray diffraction.(mass%)

Annealing conditions	FeO	Fe_3O_4 $FeCr_2O_4$	Fe_2O_3
As hot rolled	—	56.0	44.0
1073K,60s	—	31.4	68.6
1073K,8hr	49.6	50.4	—

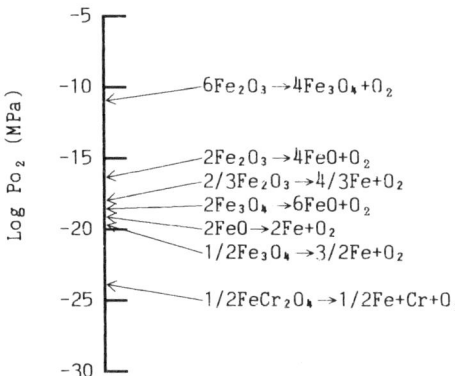

Fig.3 Equilibrium dissociation pressures of Fe and Cr oxides at 1073K

The mean diffusion distance of the Cr atoms in 13Cr steel is calculated as follows;

$$L = (2D_0 (-E/RT) t)^{1/2} \quad (1)$$

where,
- L : Mean diffusion distance of Cr atoms (mm)
- D_0: Frequency factor = 1.29   (2)
- E : Activation energy = 230318(J/mol)   (2)
- T : Annealing temperature(K)
- t : Annealing time(s)

The relation between the thickness of the Cr-depleted zone and the mean diffusion distance of Cr atoms as calculated is shown in Fig.4. A linear relationship was observed. Therefore, the formation of the Cr-depleted zone is considered to be controlled by the diffusion of Cr atoms in the metal.

Effect of annealing conditions for hot-rolled sheet on oxidation behavior of cold-rolled sheet

The Cr-depleted zone retained on the surface of cold-rolled sheet degrades oxidation resistance in finish-annealing. A weight gain of more than $0.05 \times 10^{-2} kg/m^2$ by oxidation leads to poor descalability in the following pickling process. The effect of annealing and pickling conditions for hot-rolled sheet on oxidation behavior is shown in Fig.5. Shortening the annealing time from 8hr to 60s at 1073K improved oxidation resistance, resulting in good descalability of hot-rolled sheets with only small amounts of pickling loss. Annealing for 8hr as a conventional batch annealing condition for hot-rolled sheet requires the dissolution of more than 20μm of metal by pickling before cold-rolling in order to obtain excellent descalability in the finishing process, while shortening annealing from 8hr to 60s in continuous annealing decreases the amount of dissoslution required to below 5μm.

The annealing atmosphere also affects oxidation behavior. The effect of the finish-annealing atmosphere on the oxidation behavior of cold-rolled sheet is shown in Fig.6. Increased Po_2 or reduced d.p. of the atmosphere decreased the weight gain caused by oxidation. Therefore, descalability can be improved not only by minimizing of the Cr-depleted zone, but also by controlling the annealing atmosphere.

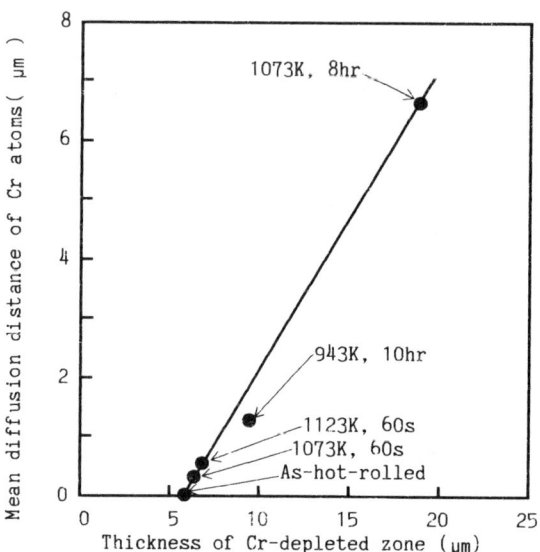

Fig. 4 Relation between thickness of Cr-depleted zone and diffusion distance of Cr atoms as calculated

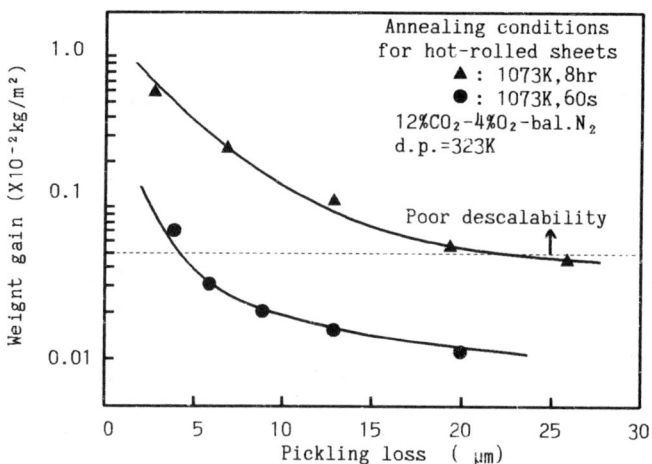

Fig. 5 Effects of annealing and pickling conditions for hot-rolled sheets on oxidation behavior of cold-rolled sheets in finish-annealing at 1073K for 60s

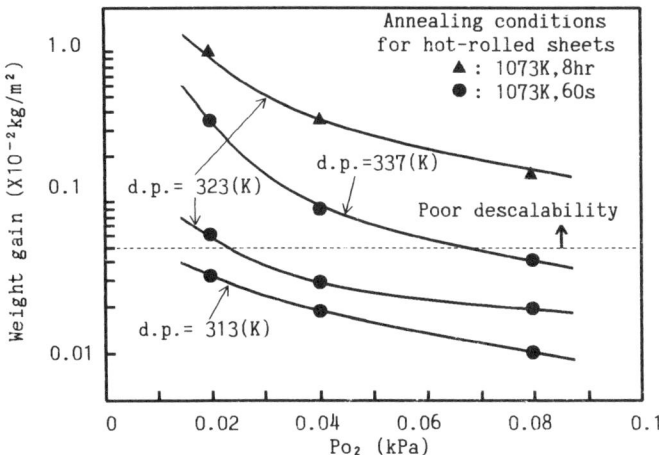

Fig.6 Effect of finish-annealing atmosphere on oxidation behavior of cold-rolled sheet (Annealing conditions for hot-rolled sheet: 1073K, 60s, 12%CO_2-O_2-bal.N_2. Pickling loss befor cold-rolling : 6μm)

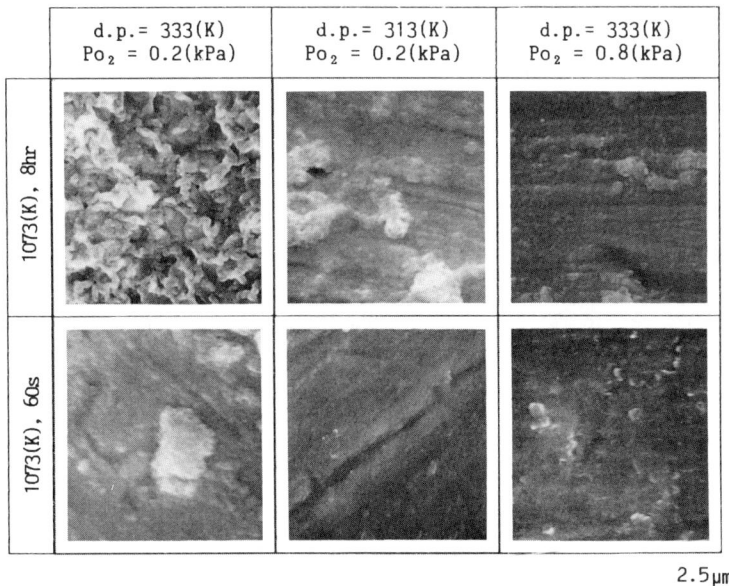

Fig.7 External surface of cold-rolled sheets after oxidation (Pickling loss before cold-rolling: 6 μm)

The surfaces of cold-rolled sheets after oxidation are shown in Fig.7. Scale with a porous structure tended to form when Po_2 was decreased and/or d.p. increased, and was markedly present in hot-rolled sheet annealed for 8hr. Cross-sectional analyses of the surface by SIMS are shown in Fig.8. In hot-rolled sheet annealed for 8hr, the preferential oxidation of Fe was great at d.p.= 333(K) and Po_2=0.2(kPa). Increasing the Po_2 or reducing the d.p. of the atmosphere, however, increased the formation of Cr-oxides protective against oxidation. It is well-known that oxidation in a moist atmosphere promotes cracking of the oxide layer, which leads to rapid oxidation of Fe accompanied by interior oxidation (3). In hot-rolled sheet annealed for 60s, while the effect of the atmosphere on oxidation was less obvious, increasing the Po_2 or reducing the d.p. of the atmosphere also suppressed the oxidation of Fe.

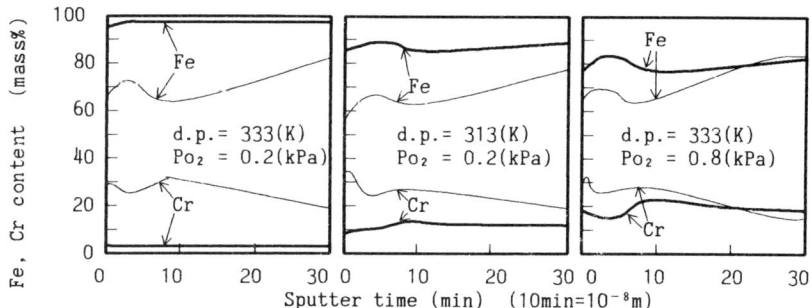

Fig.8 Cross-sectional analyses of surface of cold-rolled sheets after oxidation by SIMS.
(Annealing conditions for hot-rolled sheet:—1073K,8hr —1073K,60s)

Effects of chemistry and annealing conditions for hot-rolled sheet on formability of cold-rolled and annealed sheet

In general, hot-rolled sheets is annealed with the twin aims of softening the material by transforming the martensite formed during hot-rolling to ferrite, and of recrystallizing the hot-rolling texture. Reducing the formation of the Cr-depleted zone improves the descalability of cold-rolled and annealed sheet, but shorter annealing of hot-rolled sheet leads to insufficient transfomation and recrystallization, resulting in the degradation of the formability of cold-rolled and annealed sheet produced from this material.

The effect of C, Al and N on the A_1 transformation temperature was investigated using 13% Cr steels with various contents of C (from 0.020% to 0.063 mass%), Al (from 0.002% to 0.110 mass%), and N (from 0.012% to 0.038 mass%). The following equation was obtained by multi-regression analysis:

$$A_1(K) = -225.4(C\%) + 712.3(Al\%) - 418.5(N\%) + 1091.6 \qquad (2)$$

The metallurgical concept of the continuous (i.e. short duration) annealing process for hot-rolled sheet used in this investigation is shown in Fig.9. The addition of Al(0.1%) raises the A_1 transformation temperature and makes possible annealing at higher temperatures in order to accelerate recrystallization and transformation and achieve the desired results in a relatively short time. However, the addition of Al suppresses the formation of the austenitic phase during hot-rolling and promotes the development of the (100) hot-rolling texture, causing degradation of the ridging characteristics of cold-rolled and annealed sheet. The increase of N content (to 0.03 mass%) enlarges the (ferrite + austenite) loop over a higher region of Cr contents without significantly lowering the A_1 transformation temperature, and improves ridging characteristics.

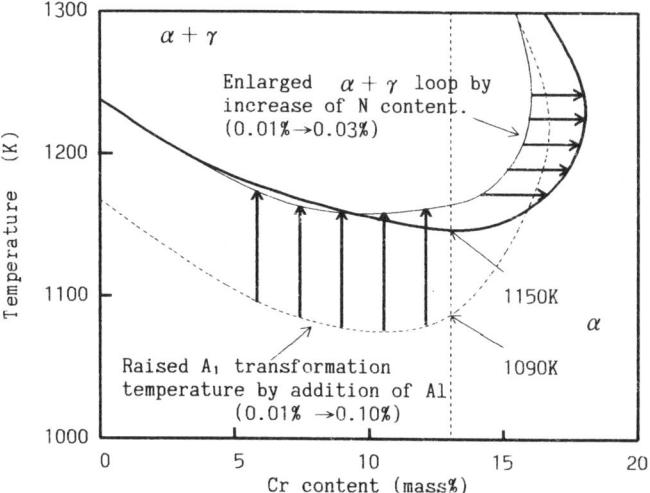

Fig.9 Metallurgical concept of the continuous annealing process for hot-rolled 13% Cr stainless steel strip

The effects of chemistry and annealing conditions for hot-rolled sheets on the formability of cold-rolled and annealed sheet are shown in Fig.10. The application of continuous annealing (60s soaking) to conventional steel (0.05 mass% C, 0.01 mass% Al, 0.01 mass% N) deteriorated its ductility and ridging characteristics. The addition of Al (0.1 mass%) made it possible to raise the annealing temperature from 1073K to 1123K and increased the ductility, however, Al addition caused deterioration of the ridging characteristics. The combined addition of Al (0.1 mass%) and increase of N (0.03 mass%) produced both good ridging characteristics and favorable ductility. Moreover decreased C content (0.03 mass%) resulted in excellent ductility and formability, and in ridging characteristics equivalent to those of conventional batch annealed steel.

The effects of the chemistry and the annealing conditions for hot-rolled sheets on longitudinal cross-sectional microstructures are shown in Fig.11. In the hot-rolled sheet with the conventional chemistry annealed at 1073K for 60s, insufficient recrystallization and a retained hard phase of martensite (transformed from austenite during hot-rolling) were observed. This structure led to an irregular grain structure in the cold-rolled and annealed sheet and to deterioration of ductility. In the hot-rolled sheet with Al addition annealed at 1073K for 60s, a sufficiently recrystallized and homogenized structure was observed, and an equiaxed grain structure was also observed in the cold-rolled and annealed sheet.

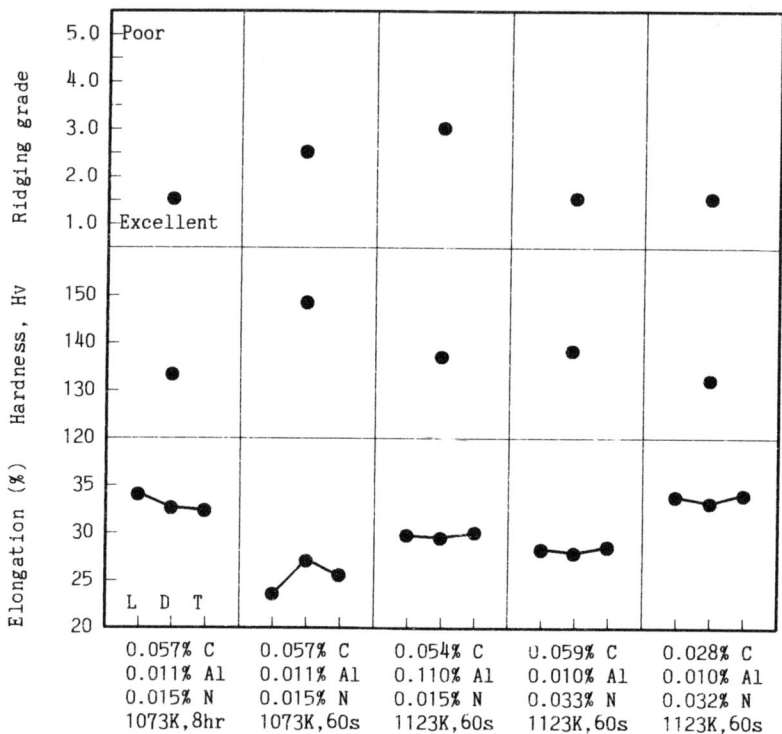

Fig.10 Effects of chemistry and annealing conditions for hot-rolled sheets on the formability of cold-rolled and annealed sheet.

Conclusions

The conventional batch annealing process for hot-rolled 13Cr stainless steel is useful in obtaining good formability and ridging characteristics, but promotes the formation of a Cr-depleted zone at the strip surface which is disadvantageous in terms of the descalability of cold-rolled and annealed strip produced from this material. Moreover, the productivity of batch annealing is also extremely poor in comparison with that of continuous annealing. Thus, in order to establish conditions for continuous annealing, the effect of chemistry, annealing conditions for hot-rolled material, and the finish-annealing atmosphere on the properties of cold-rolled and annealed strip were investigated.

(1) The formation behavior of the Cr-depleted zone at the surface of hot-rolled strip was found to be determined by the diffusion of Cr in the metal during annealing.

(2) Shortening the duration of the annealing applied to the hot-rolled strip material improved the oxidation resistance of the cold-rolled product and improved descalability in the following process.

(3) Increasing the O_2 concentration and reducing the dew point of the finish-annealing atmosphere also improved the oxidation resistance of the cold-rolled strip and descalability.

(4) A combination of reduced C content, increased of N content, and addition of Al improved the formability and the ridging characteristics of cold-rolled and annealed strip produced by the continuous annealing of hot-rolled strip.

References

1) K. Niwa, T. Yokogawa, Kinzoku Netsu Kagaku (Samgyoh Tosho, 1968)

2) R. A. Wolf, H. W. Raxton, "Diffusion in BCC Metals," Trans. Met. Soc. AIME, 230(1964), p.1426.

3) Y. Ikeda, K. Nii, "The Mechanism of Accelerated Oxidation of Fe-Cr Alloys in Water Vapor Containing Atmosphere," Bohshoku Gijutsu, 31(1982), p.156.

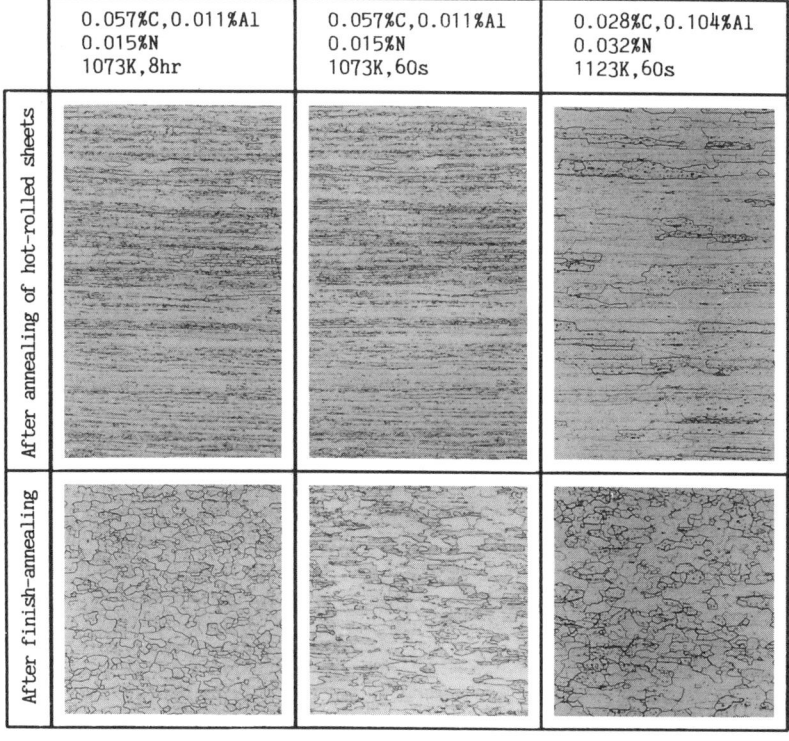

Fig.11 Effects of chemistry and annealing conditions for hot-rolled sheets on longitudinal cross-sectional microstructures

Influence of Steel Type and Surface Condition on Galvanizing Reaction

Kazumi Nishimura, Koji Kishida & Hisao Odashima

Nippon Steel Corporation
Hirohata R & D Lab.
1.Fuji-Cho Hirohata-Ku
Himeji, 671-11
Japan

Abstract

The hot dip galvanizing and galvannealing mechanisms of some types of steel with various surface conditions were investigated. In the case of hot rolled Al-killed and Si-added steel sheets, it was found that various factors of surface conditions-[the amount of residual scale and subscale layer, surface roughness and activation by mechanical pretreatment]- affected the galvanizing and galvannealing reactions. The appearance and adhesion of Zn coating on Al-killed steel sheet was improved in adequate pickling time. This pickling time affected residual scale and surface roughness. Grinding after pickling improved the wettability of molten zinc on steels more than just pickling, because surface roughness became small and the residual scale layer was removed completely. Reactivity of substrate to Zn depended on pre-treated methods. Mechanical grinding and skinpass increased Zn-Fe reactions of hot rolled steel sheet during hot dip more than electropolishing. These mechanical pretreatments increased the plastic strains on the surface. Increase of residual stress by mechanical pretreatments was observed. It seemed that active sites such as the lattice deffects of substrate increased by mechanical pretreatment, and therefore Zn-Fe reaction increased. Effect of grinding on the improvement of galvanizing and galvannealing reactions was remarkable in high Si-added hot rolled steel sheets. Homogeneous nucleation of the Zn-Fe alloyed layer occured on such mechanical treated hot rolled Si-added steel surfaces in the alloying process. According to the results of AES and ESCA analysis,in high Si-added steel sheet, it seemed that the residual subscale layer(inner oxidation layer) after pickling was removed completely by grinding. On the other hand, as cold-rolled steel sheets had relatively smoother and more active surfaces compared to hot rolled steel sheets, molten Zn-substrate reaction occured easily. Especially in low carbon steel sheets, the grain boundaries' condition after pre-annealing was an important factor in the galvanizing and galvannealing process. The initial reactive points of galvanizing process were grain boundaries in low carbon steels. Zn diffused through the grain boundaries. As Al-killed cold rolled steel sheets had relative inactive grain boundaries which contained more amount of C than Ti added low carbon steel, the Zn-Fe reaction could not occur easily. Zn-Fe reaction on Si-added cold rolled steels was restricted due to the concentration of specific steel elements on the surface during the annealing process. Mechanical grinding was also effective in increasing Zn-Fe reactions of Si-added cold rolled steels, because the Si concentrated layer was removed.

Introduction

In recent years, hot dip galvanized and galvannealed steel sheets have been widely used for various parts of construction, household electric appliances and automobiles. It is supposed that superior appearance and superior coating adhesion on various type of steels which have higher qualities(strength, workability etc.) will be required in the future. Under these circumstances, it is very important to clarify the galvanizing and galvannealing mechanism of steels with various composition and surface conditions.

Some works have already been carried out, studying the effect of steel composition on the galvanizing and galvannealing reaction.[1-5] Nevertheless, there is more open questions to clarify because the galvanizing and galvannealing process is complicated and the factors which affect these reaction are not only steel compositions but also many kinds of surface conditions, for instance – surface roughness, residual scale and subscale(inner oxidation layer), grain boundaries, concentration and oxidation of specific elements etc. Contribution of these factors on the galvanizing and galvannealing reaction will change with the types of steel, the steel making process such as hot rolled or cold rolled, the pretreatment methods of steel such as pickling, grinding and the preheating process. The purpose of this preheating process is the prereduction in the case of hot rolled steel sheets and the prereduction and annealing in the case of cold rolled steel sheets.

Hot rolled steel sheets are supplied to the hot dip galvanizing process after pickling. So surface roughness and condition of residual scales and subscales were changed with pickling time and grinding methods after pickling. Especially, in the case of Si-added hot rolled steels, the subscale layer(inner oxidation layer) easily remains after pickling. Furthermore, it is known that grinding or mechanical treatment of a steel surface change the activity of the surface by increasing the surface strains, lattice deffects etc. [6] It is supposed that the galvanizing and galvannealing process will be widely changed with the surface conditions after pickling and grinding.

The first subject of this paper is to clarify the relationship between galvanizing or galvannealing reactions and surface conditions of some types of hot rolled steel such as surface roughness, residual scale, subscale(inner oxidation layer), grain boundaries and residual strain(stress).

On the other hand, having smoother and cleanner surface than hot rolled steel sheets, cold rolled steel sheeets have relatively active surface. Their surface conditions are also changed with pre-annealing conditions, because either concentration of specific elements such as C,Si and P on the surface or oxidation occur during annealing. Secondly, galvanizing and galvannealing behavior of some types of cold rolled steel with various surface conditions were also discussed and compared to that of hot rolled steel sheets in this paper.

Experimental Procedure

Materials for investigation

Commercial hot rolled Al-killed steel, Si-added steels with various Si contents, and cold rolled Al-killed steel, Si-added steel, ultra low carbon Ti-added or Ti-P added steels were used for substrates. Si-added hot rolled steel sheets were also made in a vacuum induction furnace under laboratory conditions in order to study the effect of silicon more in detail. Their chemical composition are given in Table 1.

Galvanizing and galvannealing methods

Hot rolled steel sheets were ground in various degrees by mechanical methods-emery paper, grinder, brush, electropolishing after pickling in 8 vol% hydrochloric acid at 90 °C in order to examine the effect of surface conditions. Cold rolled steels were only degreased. Then, the steels were placed in a experimental galvanizing apparatus which is a direct resistance heating type and can be carried out from pre-heating to galvanizing or galvannealing treatment automatically. The samples sheets were reduced 30—60 s in H_2 15%-N_2 mixed gas at 550-600 ℃ and hot-dipped for 3 to 30s in a molten Zn bath containing 0.15-0.2%Al kept at 450 ℃. The Coating weight was about 135g/m^2. For the purpose of alloying experiments, sheets were passed directly to a heating section in which sheets were heated at various temperatures and times immediately after hot dipping. The coating weight in alloying experiment was about 60g/m^2. Furthermore, diffusion tests using Mo marker were also carried out.

Analysis of Zn coating layer and substrate

The Fe content in the Zn coating and Zn-Fe alloyed layer was examined by chemical analysis. Cross section and surface structure of Zn coating and Zn-Fe alloyed layer formed during hot dipping or alloying process were observed by an optical microscope, scanning electron microscope and EPMA. X-ray diffraction method and electrostripping method were used to observe the Zn-Fe alloyed layer and the interface between the coating layer and substrate. The Surface roughness and the grinding amount of substrates were examined by surface roughness analyser. Depth profiles of elements in various pretreated substrate were measured by GDS, AES and ESCA. An X-ray stress analysis method was also used to examine the residual stresses of substrates.

Adhesion tests of Zn coating and Zn-Fe alloyed coating

In order to check the formability of coated sheets, a ball impact test(Cup forming at diameter of 5 mm) for galvanizing sheets and a 60 ° bending test for galvannealing sheets were carried out.

Measuring of residual stress of substrates

The residual stress of substrates after various mechanical treatments was also examined by X-ray diffraction method using the Cr target. The measured crystal plane of substrates was Fe(211).

Table 1 Chemical composition of steels used   Mass %

	Type of steels	C	Si	Mn	P	S	sol.Al	Ti
1	Hot rolled Al-k	0.05	0.01	0.23	0.015	0.014	0.02	
2	Hot rolled Si-Al-k	0.14	0.15	0.4	0.01	0.01	0.025	
3	Hot rolled Si-added	0.14	0.26	1.36	0.02	0.001	0.03	
4	Hot rolled Si-added	0.2	1.5	1.57	0.008	0.003	0.029	
5	Cold rolled Al-k	0.04	0.01	0.15	0.015	0.015	0.03	
6	Cold rolled Ti-added	.001	0.01	0.09	0.008	0.003	0.04	0.04
7	Cold rolled P-Ti-add	.002	0.01	0.41	0.065	0.015	0.05	0.05
8	Cold rolled Si-added	0.11	1.22	1.55	0.009	0.001	0.0029	

Results and Discussions

Effect of surface condition of hot rolled steel sheets on galvanizing reaction

Surface condition after pickling. Figure 1 shows the appearance and adhesion of Zn coatings obtained from 0.2%Al, 450 ℃ bath after various pickling time of hot rolled Al-killed steel sheets. The appearance and adhesion of Zn coating were improved in adequate pickling time range of 30 to 120s. Non coating section appeared partially and coating adhesion became poor except during this range of pickling time. It seemed that coating appearance was affected by surface conditions after pickling because the surface roughness and the amount of residual scale and smudge were changed. As the residual scale and smudge were not almost observed in the pickling time range of 30 to 150s in this Al-killed steel, surface roughness seemed to the main factor on the galvanizing reaction in this range of pickling time. Residual scale was observed less than pickling time of 20s and smudge was examined beyond the pickling time of 180s. The change of surface roughness with the increase of pickling time was examined as shown in Figure 2. Surface roughness was relatively small in the pickling range of 30 to 120s which showed good Zn coating appearance and adhesion. Figure 3 shows the cross section of Zn coating in the pickling time of 60s and 150s measured by optical microscope and EPMA. More homogeneous Zn coating layer was obtained in the pickling time of 60s, while partially thin Zn layer on the rough substrate was examined in the pickling time of 150s. The distribution of Al at the interface between Zn coating and substrate was also inhomogeneous in the pickling time of 150s. It seemed that the Fe-Al-Zn barrier layer formed on the interface between Zn coating and substrate at initial galvanizing reaction was affected by surface roughness of substrate.

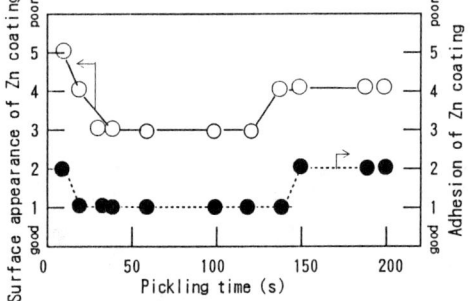

Fig. 1 Effect of pickling time on the surface appearance and adhesion of Zn coating.
(Al-k, 15%H_2+N_2 550℃30s, 0.2%Al, 450℃, 3s, 135+10g/m^2)

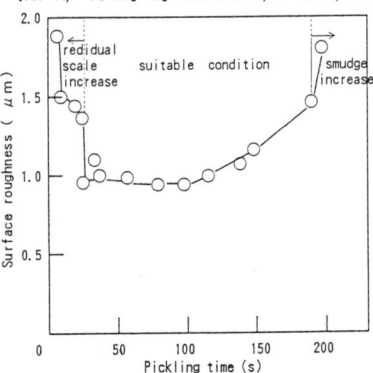

Fig. 2 Effect of pickling time for hot rolled steel sheet on the surface roughness.
(Al-killed, 90℃ 8%HCl)

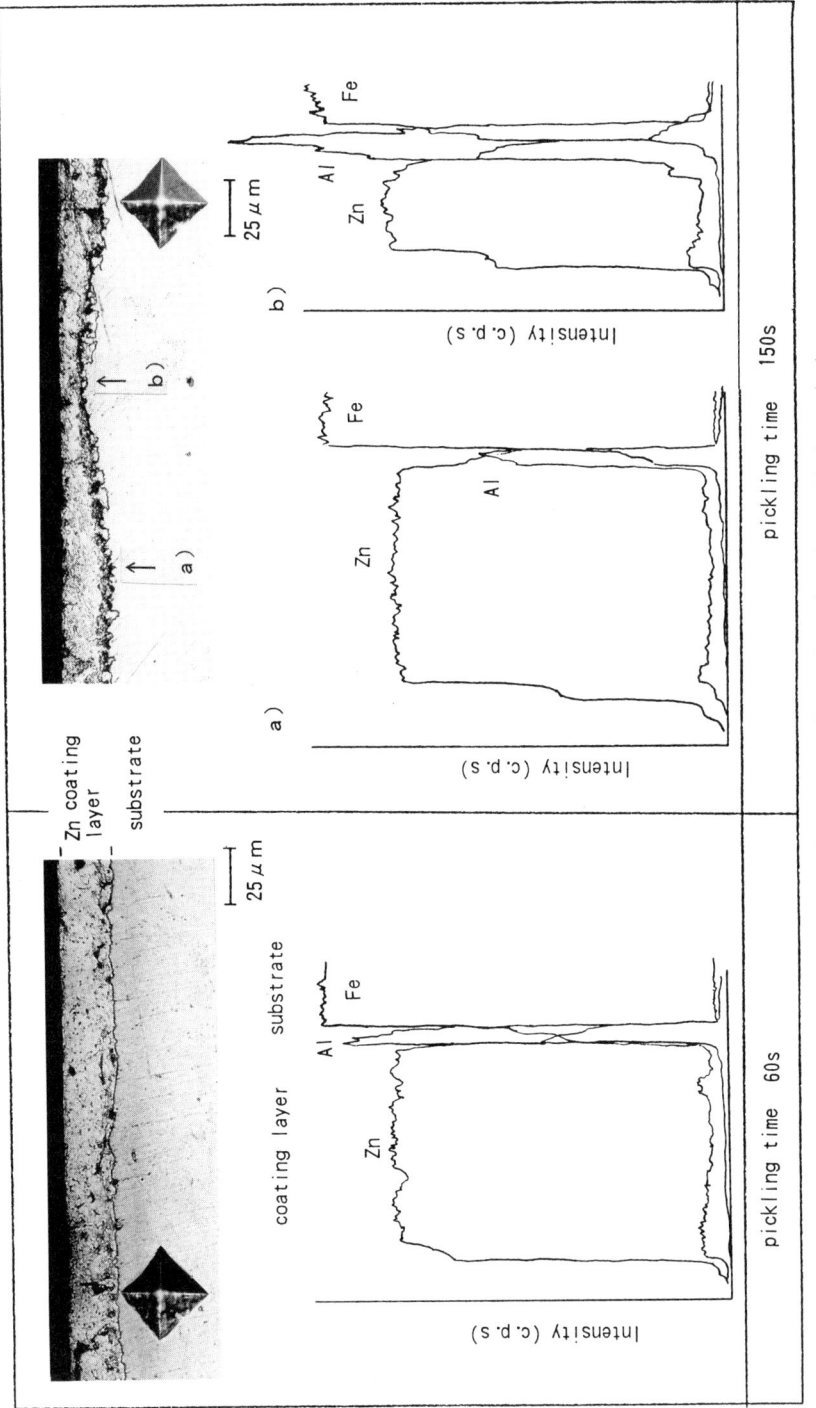

Fig.3 Cross section of Zn coated layer formed on hot rolled steel sheets with the change of pickling time. (Al-k, Al 0.2%,)

Effect of surface roughness. The effect of surface roughness on the coating appearance using various surface pretreatment methods after pickling of steels was also examined as illustrated in Figure 4. Pickling time before these pretreatments is 60 s. This time is best condition for excluding the effect of scale or smudge. Coating appearance became better in a smaller range of surface roughness. It was because the wettability of molten Zn was improved on the smooth, active surface obtained in this range of surface roughness.

Fig.5 shows the effect of the surface roughness of substrates on the Fe content of Zn coating layer which shows the reactivity of substrate. Amount of Fe in Zn coating layer increased gradually with the decrease of surface roughness. More remarkable results was that grinding after pickling and skinpass rolling before pickling of substrate improved the reactivity of molten Zn more than just pickling and electropolishing in the same range of surface roughness. This suggested the other factor of mechanical pretreatment on galvanizing reaction besides surface roughness.

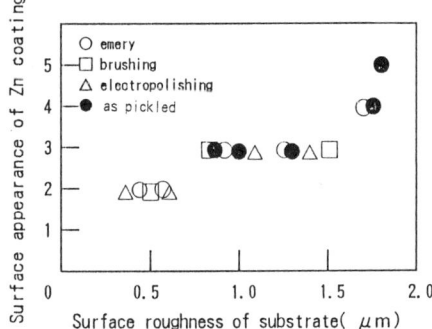

Fig. 4 Relationship between appearance of Zn coating and surface roughness of substrate pre-treated by various methods.
(Al-k, 15%H_2+N_2 550°C 30s, 0.2%Al, 450°C, 3s, 135+10g/m^2)

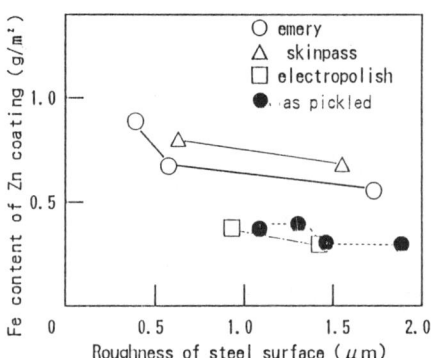

Fig. 5 Effect of roughness of steel surface on galvanizing reaction with various grinding methods. (Al-k, pickling 60s, 550 °C, 30s, 0.2%Al, 450 °C, 3s, 135+10g/m^2)

Activation of substrate by mechanical pretreatment. The relationship between molten Zn-substrate reaction and the grinding depth of various grinding methods were given in Figure 6. Reactivity of substrate in molten Zn depended on the pretreated methods. Mechanical grinding was more effective in increasing Zn-Fe reaction than electropolishing. The effect of grinding was also remarkable in galvannealing process. Zn-Fe reaction was also increased by skinpass rolling of hot rolled steel sheets as shown in Figure 7. These results suggested that the mechanical grinding has another effect other than removing the residual scale and smoothing the surface roughness. It has been known that the surface strains, lattice defects increased by mechanical treatment and these defects become the active site of reactions. [6] It seemed that Zn-Fe reaction increased with the increase of active sites on the substrate surface by mechanical pretreatment.

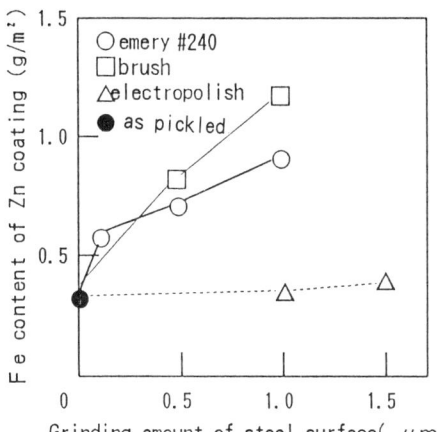

Fig. 6 Effect of grinding of steel surface on galvanizing reaction with various grinding methods.
(Al-k, pickling 60s, 15%H₂+N₂ 550 °C30 s 0.2%Al, 450 °C, 3s, 135+10g/m²)

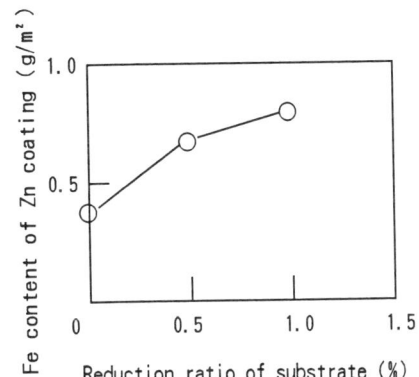

Fig. 7 Effect of reduction ratio of substrate on galvanizing reaction with skinpass rolling.
(Al-k, 15%H₂+N₂ 550 °C30s, 0.2%Al, 450°C, 3s, 135+10g/m²)

Furthermore, to examine the change of surface strains after grinding, the residual stress of steel surface with various pretreated methods was examined by an X-ray diffraction method(Target Cr K α, Fe(211))) as given in Figure 8. The residual stress (σ) is caused by the plastic strains of surface. As the distance of each crystal planes (d) changed with the surface strains, diffraction angle (2θ) also changed according to Bragg's diffraction law. Therefore, the value of σ was obtained from following formula. [7]

$$\sigma = -\frac{E}{2(1+\nu)} \cdot \cot\theta_0 \cdot \frac{\pi}{180} \cdot \frac{\partial(2\theta)}{\partial(\sin^2\psi)} = K \cdot \frac{\partial(2\theta)}{\partial(\sin^2\psi)} \quad (1)$$

(E:Young modulus, ν : Poisson ratio 2θ , ϕ , :diffraction angle changed with ϕ , ϕ ; angle between surface direction and each crystal direction ($=\phi_0+\eta$), ϕ_0: angle of incidence)

The residual stress of substrate after mechanical grinding and skinpass rolling were larger than that of just pickling. This means that mechanical grinded surface had large strains. It has been known that residual stress also depends on pre-heating (reduction) condition after grinding.[8] In this experiment, the residual stress decreased a little after heating at 550℃ 30s as shown in Figure 9.

From these results, measuring the residual stress of substrate surface might become one index which shows the change of activity of surface. The increase of residual stress means the increase of density of lattice defects. This activation of substrate surface seemed to affect the molten Zn-substrate reaction and Zn-Fe alloying reaction.

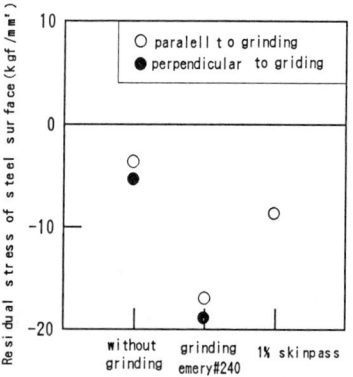

Fig. 8 Effect of grinding on the residual stress of surface of hot rolled steel after pickling. (Al-k, pickling 90 ℃8%HCl, without heating)

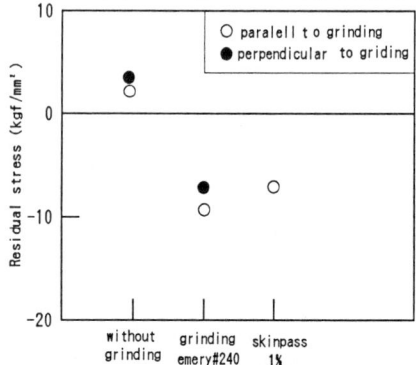

Fig. 9 Redidual stress of substrate surface pre-treated with various methods. (Al-k, pickling 90 ℃8%HCl, heating 550℃30s)

Diffusion behavior of Zn on substrate. According to the diffusion coupling test, using the Zn electroplating layer-steel substrate in previous work [9], the following diffusion theory was drawn. Zn diffuses easily through the Zn-Fe alloyed layer formed between Zn and substrate to substrate interface. At the interface between alloyed layer and substrate, diffused Zn atoms are exchanged by Fe atoms of substrate and the Zn-Fe alloyed layer gradually grows. A well-known Mo marker [10] was used to examine the behavior of Zn diffusion during the hot dip of hot rolled Al-killed steel sheet in this experiment. It is shown in Photo.1. The Mo marker was always positioned on the interface between pure Zn and Zn-Fe alloyed layer in spite of the increase of immersion time. The Zn-Fe alloyed layer grew gradually under the marker. It suggested that Zn diffuses mainly toward substrate through the Zn-Fe alloyed layer by Kirkendall effect in the galvanizing process of hot rolled steel sheet. The same behavior of Mo was also investigated in Zn-Fe alloying process in this experiment.

These results correspond to the diffusion coupling test of previous work[9]. It suggests that reactivity of substrate plays an important role in the diffused Zn-substrate reaction in alloying process. It seemed that activation of substrate by grinding improved not only molten Zn-substrate reaction in hot dip but also diffused Zn-substrate reaction on substrate interface in the alloying process.

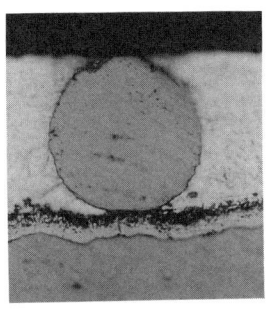 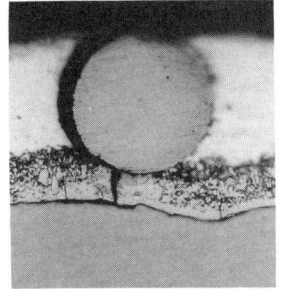 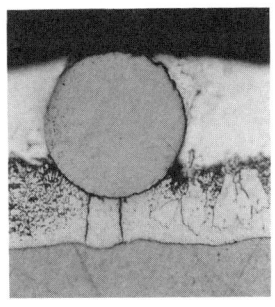

3 sec   30 sec   60 sec   20 μm

Photo. 1   Behavior of Mo marker in the Zn coating of hot rolled steel sheet with immersion time. (Hot rolled Al-killed, Al 0.05%, 450 °C)

Effect of steel type on galvanizing and galvannealing reaction

Behavior of Si-added hot rolled steel sheets. The results of Zn coating adhesion on various steel substrates with the change of grinding amounts were given in Figure 10. It has been known that Si, which partially exchange Fe atoms of lattice, is effective to get the high strength steel sheet. These Si-added steel sheets are well known to be difficult to be galvanized, because partial non-coating section are often observed. [1] The effect of mechanical grinding on Zn coating appearance depended on the type of substrates. The Zn coating adhesion was improved by grinding after pickling remarkably in Si-added hot steel sheets. Wettability of molten Zn on Si-added hot rolled steel sheets seemed to be improved by grinding.

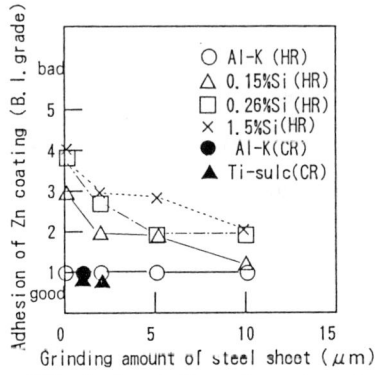

Fig. 10 Effect of grinding of steel surface on Zn coating adhesion with various steel substrates.
(pickling 60s, 15%H_2+N_2 600°C, 30s
0.2%Al, 450 °C, 3s, 135+10 g/m²)

The effect of Si content on the galvannealing reaction of hot rolled steel sheets at alloying condition of Al 0.15%, 500°C, 15s was given in Figure 11. The Zn-Fe reaction decreased with the increase of Si. The non-coated area also increased remarkably with increasing of Si content in the substrates from the investigation of coating appearance. It seemed to be due to the residual Si contained subscale after pickling or surface structure of Si-added steel itself. Figure 12 shows the effect of grinding after pickling on the galvannealing reaction of hot rolled steel sheets with various Si content. The effect of grinding on the Zn-Fe alloying reaction in coated area was remarkable. Fe content of Zn coating on Si-added hot rolled steels increased remarkably with grinding.

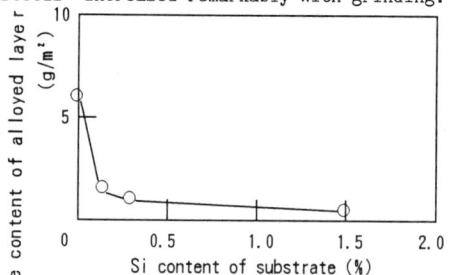

Fig. 11 Effect of Si content of hot rolled steel sheets on galvannealing reaction.
(As pickling, Al 0.15%, 450 °C 3s, alloying at 20 °C/s, 500°C, 15s)

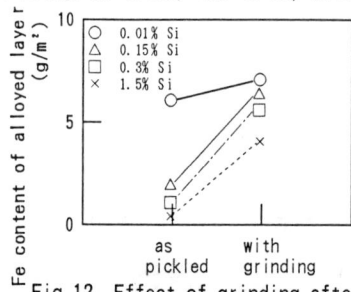

Fig. 12 Effect of grinding after pickling on the galvannealing reaction of substrate with various Si content.
(Al 0.15%, 450°C 3s, alloying at 20°C/s, 500°C, 15s)

Cross sections of alloyed layer on coated area of 1.5%Si-added hot rolled steels with grinding and without grinding were illustrated in Photo.2. EPMA profile is also shown in Figure 13. It can be seen that Zn-Fe reaction in grinded substrate increased and Zn-Fe nucleations occured more homogeneously than that of substrate without grinding(as pickled). EPMA profile showed that Fe content of Zn-Fe alloyed layer increased with grinding. Fe content of alloyed layer was about 10% in ground substrate. Furthermore, structure of alloyed layer was investigated by X-ray analysis. This result showed that Zn-Fe alloyed layer of ground substrate mainly contained $\delta_1(FeZn_7)$. On the other hand, without grinding, only ζ (FeZn$_{13}$) phase was observed.

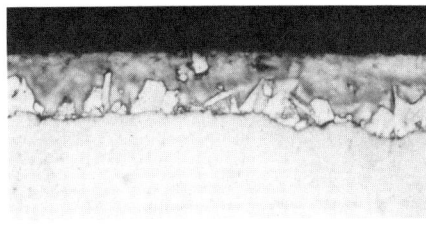
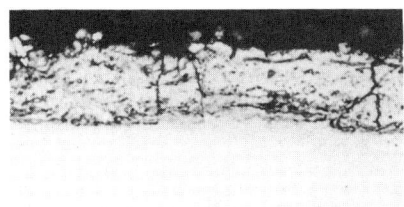

without grinding
(as pickled)

with grinding

Photo. 2 Cross section Zn alloyed layer of Si-added hot rolled steel sheets with grinding and without grinding.
(Al 0.15%, alloying 500 °C, 15s)

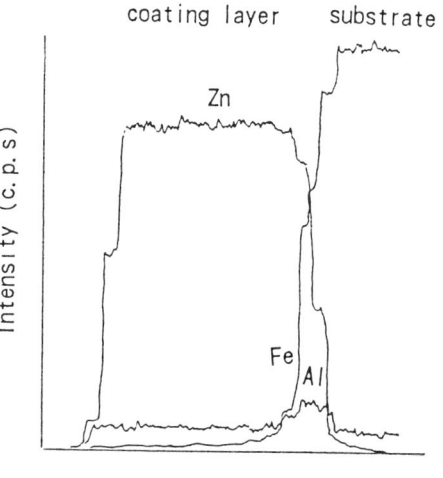
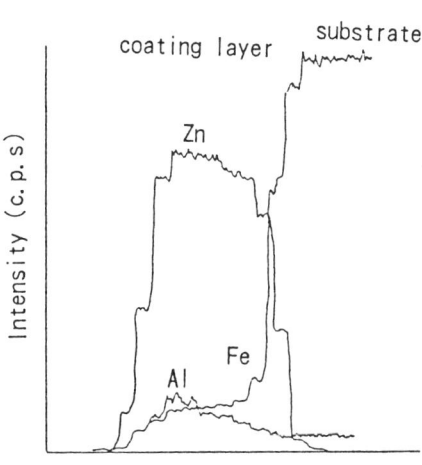

without grinding
(as pickled)

with grinding

Fig.13 Depth profile of alloyed layer formed on the 1.5% hot rolled steel sheet with and without grinding.
(Al 0.15%, alloying 500 °C, 15s)

To clarify the effect of grinding on Zn coating and Zn-Fe alloying process in high Si added hot rolled steel, surface conditions as hot rolled, after pickling and after grinding were examined by ESCA and AES. Figure 14 and Figure 15 showed the depth profile of scale layer as hot rolled, and after pickling. In hot rolled 1.5% Si added steel sheet, scale layer consisted of Fe_2O_3, Fe_3O_4 (Top layer) and $FeO+Fe_2SiO_4$ (Inner layer). Fe_2SiO_4 appeared under about 2 μ m of top layer in this case. After pickling Fe_2SiO_4 also almost disappeared from the result of ESCA. This scale layer was almost removed and thin Fe-O oxide layer containing Si appeared. This layer seemed to be so-called subscale layer (Oxygen rich inner oxidative layer) of high Si-added hot rolled steel. It has been known that this inner oxidation layer of high Si added hot rolled steel is formed by Oxygen diffusion from scale layer to substrate. Differences of surface condition as pickled and after grinding in 1.5%Si added steel were examined by AES as illustrated in Figure 16. Oxygen of surface after grinding decreased noticeably compared to that of just pickling. It means that residual subscale layer decreased with grinding. Complete removing of the residual subscale layer by grinding seemed to be one factor of improving the Zn coating of high Si contained steel sheets.

Furthermore, the change of the grain size and phases in 1.5%Si added steel by grinding were investigated as another factors of increasing of Zn-Fe reaction by grinding. Grain size near the surface measured by optical microscope was about 20 μ m (inner about 10-20μ m). Grinding thickness was rather smaller than each grain size near surface, therefore grain size cannot be changed by grinding. As the volume fraction of γ - phase in the bulk (α —p h ase) was very small near surface and was not changed in the direction of depth, the volume fraction was not changed by grinding. From these experimental results, it seemed that the change of grain size and phases by grinding on Zn-Fe reaction were relatively small.

From these results, it can be hypothesized that surface activation of Si added steels was increased by grinding. That was caused by removal of residual subscale layer (inner oxidative layer) and by increase of surface strains and lattice defects. This activation of surface by grinding affected to improve Zn-Fe reactions during hot-dip and diffused Zn -substrate reactions in alloying process.

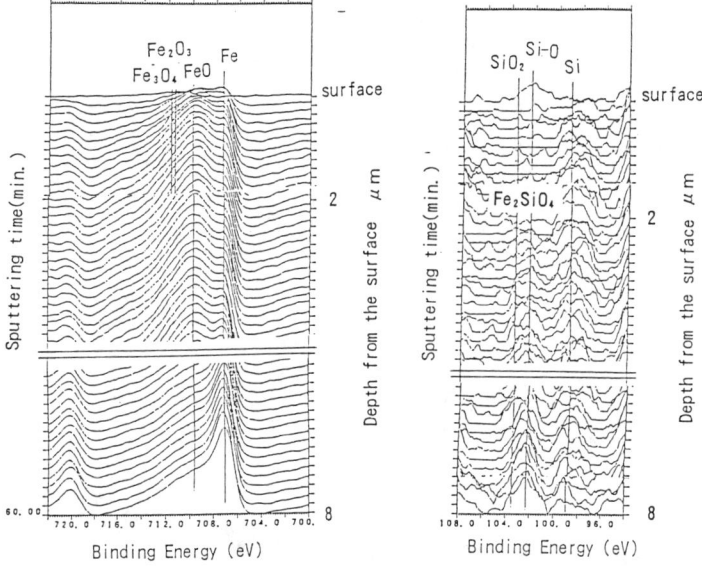

Fig. 14 Depth profile of the scale layer on 1.5%Si-added steel formed after hot rolling measured by ESCA.

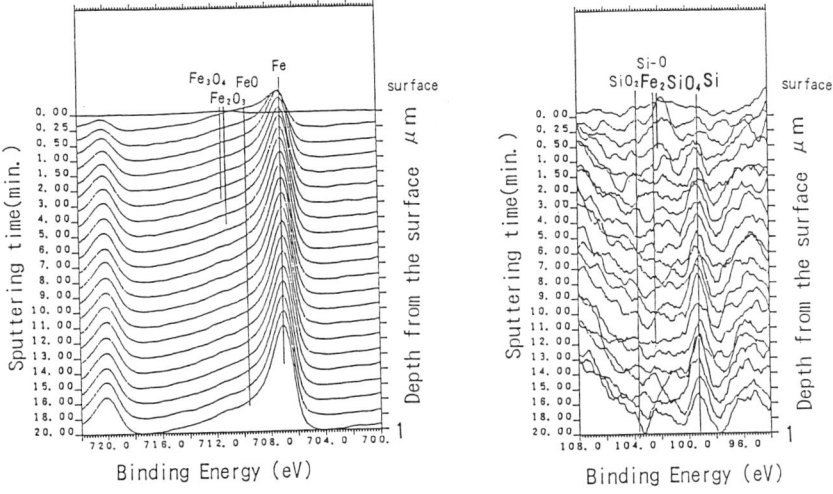

Fig. 15 Depth profile of the surface on 1.5%Si-added steel after removing the hot rolled scale layer by pickling measured by ESCA.

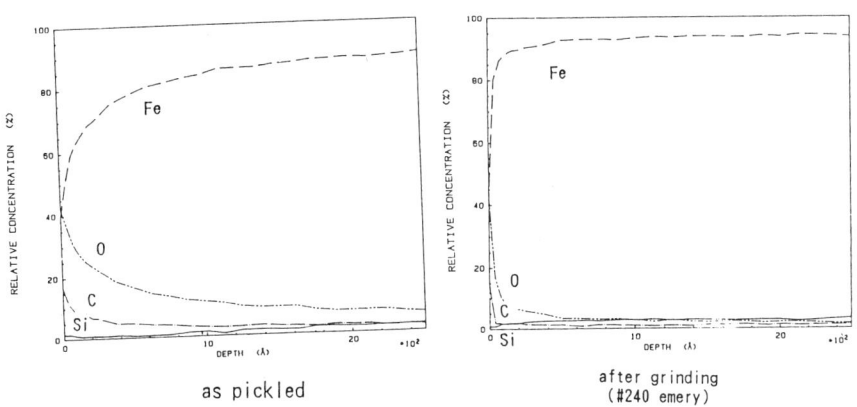

as pickled

after grinding (#240 emery)

Fig. 16 Depth profile of Si-added hot rolled steel sheet with grinding and without grinding. (AES) (1.5%Si, 90 °C 8% HCl)

Zn-Fe reaction on various cold rolled steel sheets . As cold rolled steel sheets had relatively smoother and more active surfaces than hot rolled steel sheets, the Zn coating appearance was better and the reaction of interface between molten zinc and substrate occured more easily than on hot rolled steel sheets.
Figure 17 shows the Fe content of Zn coatings on various cold rolled steel sheets at various length of immersion time. The amount of Fe reaction on Ti-added low carbon steel sheet was larger than that of Al-killed steel. It suggested that low carbon steel sheet had a more active surface than high carbon steel. In Si or P added steel, the Fe dissolution during Zn coating was rather small. These results during Zn coating correspond to the results of alloying experiment in previous papers.[4,5]
Furthermore, to examine the behavior of Zn diffusion on the surface of these steel sheets, cross section of the interface between Zn coating and substrate at the initial stage of alloying was investigated in detail in this paper. The results of cross section examined by optical microscope were given in Photo 3 and SEM-EPMA analysis of cross section were illustrated in Figure 18. In the case of Ti-added low carbon steel sheet, Zn-Fe reaction occurred everywhere on the active surface and a partial Zn enriched zone existed through grain boundaries under the Zn-Fe alloyed layer. This alloyed layer formed on the grain boundary is so called outburst layer. Fe-Al-Zn peak near interface partly disappeared. It seemed that Zn diffused through the grain boundaries, whose Fe-Al-Zn barrier layer was partially destroyed, toward the interface of substrate.
On the contrary, in Al-killed steel, Zn diffusion through the grain boundaries was not obviously observed and the Fe-Al-Zn barrier layer was retained clearly in such an initial alloying time.
This result suggested that as grain boundaries were the most active site in such a active surface of low carbon steel sheets; they played an important role on the reaction of molten Zn-substrate and diffused Zn -substrate .

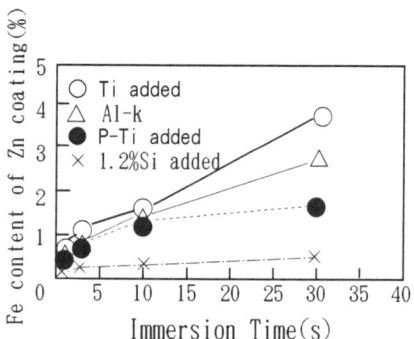

Fig. 17 Effect of Immertion time in the Zn bath on the Fe content of Zn coating with various cold rolled steel sheets.
(H_2 15%+N_2, 550 °C, 30s, Al 0.1%, 450 ° C)

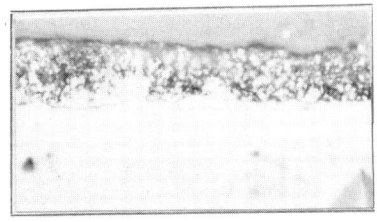
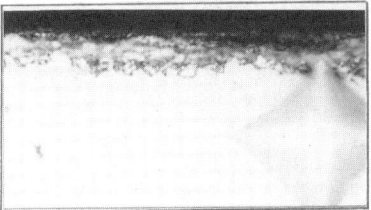

Ti-added low carbon steel   Al-killed steel $10\mu m$

Photo. 3   Cross section of initial Zn alloyed layer near grain boundaries of substrate. (Al 0.15%, alloying 500°C, 1s)

SEM ×2000

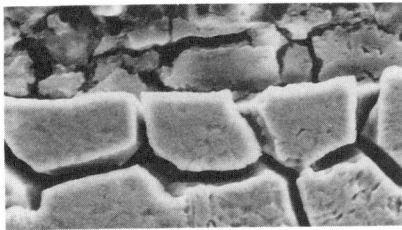
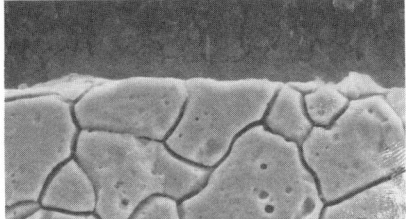

Zn

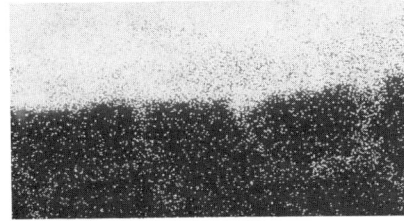

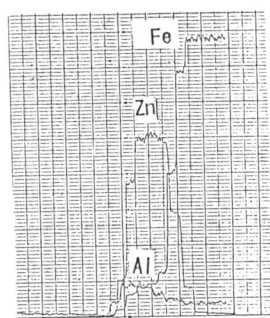

Ti-added low carbon steel   Al-killed steel

Fig. 18   EPMA profile of initial Zn alloyed layer near grain boundaries of substrate. (Al 0.15%, alloying 500°C, 1s)

The surface activity of 1.2%Si-added cold rolled steels seemed to be especially small in spite of the fact that this steel sheet had a smaller surface roughness than that of Si-added hot rolled sheet. To clarify the reason why Zn-Fe reaction was restricted in Si-added cold rolled steel sheet, surface condition of the substrate was examined by GDS. A depth profile of 1.2%Si-added cold rolled steel with and without grinding was given in Figure 19. A concentrated zone of Si,Mn and C which seemed to occur during the annealing process after cold rolling was observed near surface in this steel sheet. This zone did not appear in hot rolled Si-added steel sheets. It seemed that this concentration zone of Si-added cold rolled steel weakened the reactivity of surface in hot dip and in the alloying reaction. This zone was removed by grinding completely.

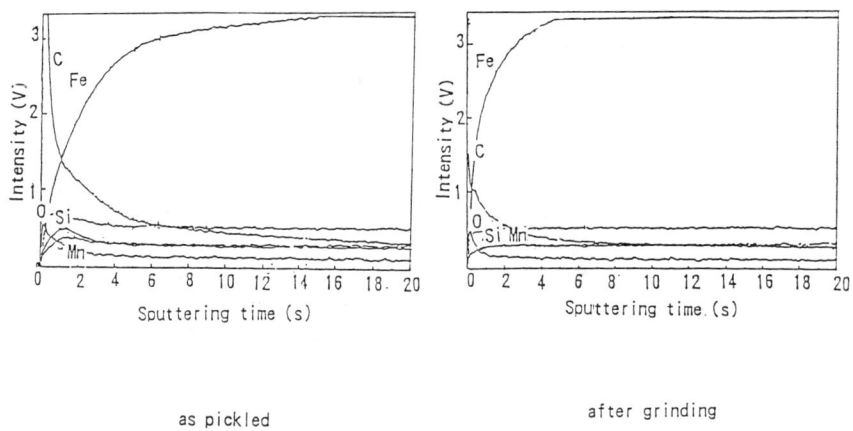

as pickled   after grinding

Fig.19 Depth profile of Si-added annealed cold rolled steel sheet with grinding and without grinding.

These results suggested the galvanizing and galvannealing mechanism of cold rolled steel sheets which had such a smoothed and relative active surface. Especially, in low carbon steel sheets, it seemed that Zn diffused toward the coating-substrate interface partially through very active grain boundaries which had many dislocations and vacancies. Fe dissolved and saturated in the Zn and the so called outburst layer grew from grain boundaries at the initial alloying stage. In Al killed cold rolled steels, they had relatively inactive grain boundaries which contained higher amounts of C than Ti-added low carbon steel, Zn-Fe reaction can not occur easily in such inactive grain boundaries. In Si-added cold rolled steels, as they had an inactive surface due to the concentration of Si,Mn,C during annealing process, Zn-Fe reaction was restricted. Activation of such a steel surface by mechanical treatment was very effective in improving the reaction of the molten Zn-substrate and diffused Zn-substrate.

Conclusions

From these experimental results, the following conclusions were drawn.

I. Galvanizing and galvannealing were strongly affected by surface conditions of hot rolled steel sheets. Main factors of surface conditions were the amount of residual scale and subscale, surface roughness, and residual strain(stress).

1) The appearance and adhesion of Zn coatings of hot rolled Al-killed steel sheets were improved in suitable pickling time range of 30-120s (8%HCl) because the residual scale layer and smudge were less and the surface roughness was relatively small in this range.

2) In the range of small surface roughness with various surface treatments, the wettability of molten Zn on substrate was improved.

3) Reactivity of substrate in hot dip galvanizing and galvannealing depended on the pretreatment methods. Mechanical grinding and skinpass rolling improved Zn-Fe reactions more than electropolishing and pickling. It seemed that surface strains(stresses) and lattice defects affected the reactivity of substrate.

4) Increase of residual stress of steel surface by mechanical treatment was observed from the result of X-ray diffraction methods. It seemed to reflect the increase of surface activity by mechanical treatment.

5) The Mo marker was always positioned on the interface between Zn and Zn-Fe alloyed layer on hot rolled steel sheets with the increase of immersion time. It suggested Zn diffusion through the Zn-Fe alloyed layer to substrate. Reactivity of substrate also seemed to play an important role in the diffused Zn-substrate reaction .

6) In high Si-added hot rolled steel sheets, the coating appearance and Zn-Fe reactions were improved remarkably by grinding after pickling. The nucleation of Zn-Fe alloyed layer also increased on such a ground surface. It seems likely that residual subscale layer (inner oxidative layer) was removed completely by grinding.

II. In cold rolled steel sheets, molten Zn-substrate reactions occured relative easily, because they had relatively smoother and more active surfaces than hot rolled steel sheets.

1) The amount of Zn-Fe reaction of Ti-added low carbon sheets was large in cold rolled steel sheets. This reaction was restricted in Si or P added steels.

2) Condition of grain boundaries played an important role in the galvanizing and galvannealing reaction in Ti-added low carbon cold rolled steel sheet. In that steel sheet, Zn diffusion through the grain boundaries where the Fe-Al-Zn barrier layer was easily destroyed occurred more easily than that of Al killed steel sheet during galvanizing and galvannealing process.

3) Zn-Fe reaction on high Si-added cold rolled steel sheets was restricted by the concentration of specific steel elements (Si,Mn,C) on the surface during the annealing process. Mechanical grinding was also effective in increasing the Zn-Fe reaction of these types of steel.

References

1. T.Gradman, B. Holmes,and F.B. Pickling," Some Effects of Steel Composition on the Formation and Adherence of Galvanized Coatings," Journal of The Iron and Steel Institute, 11(1973),765-777

2. Y.Hisamatsu," Science and Technology of Zinc and Zinc Alloy Coated Steel Sheet," Proceedings of The International Conference on Zinc and Zinc Alloy Coated Steel Sheet,(1989),3-12

3. Akihiko Nishimoto, Jun-ichi Inagaki and Kazuhide Nakaoka,"The effects of Microstructure and Chemical Compositions of Steels on the Reaction between Iron and Zinc," Tetsu to Hagane, 72(1986),989-996

4. Yoshikuni Tokunaga, Masato Yamada and Takashi Hada, "The Relation between the Powdering and the Microstructure of Coating of Galvannealed Steel Sheets" Tetsu to Hagane, 72(1986), 997-1004

5. Y.Numakura, Tadashi Honda,and Takashi Hada,"The effect of coating structure on the formability of galvannealed sheets,Proceeding of 15 th International conference on Galvanizing, (1988)

6. Namio Ootani, Theory of Surface Finishing of Metals(Japan, Maki Publishing Company, 1967), 49-63

7. Residual Stress, (Japan, Heat treatment Society, 1963),72-85

8. Tamotsu Toki, Ken Abe and Toshio Nakamori,"Effect of residual stress of steel surface on alloying behavior of high Al-content galvanized steel," CAMP-ISIJ , 4(1991),666

9. Masami Onishi, Yoshinori Wakamatsu, Koichi Fukumoto and Manabu Sagara,"Reaction-Diffusion of Solid Iron with Solid Zinc,"J.Metal Institute of Japan, 36(1972),150-156

10. C.Allen and J.Mackowiak,"The application of the Inert-Marker Technique to solid/solid and Solid/Liquid Iron/Zinc Couples," Journal of the Institute of Metals, 91(1963),369-372

DESIGN AND APPLICATION OF A CONTINUOUS ANNEALING SIMULATOR

D.M. Haezebrouck, J.W. Sinclair and D.A. White

Inland Steel Company
East Chicago, Indiana

Abstract

In order to support product development and product optimization, a continuous annealing simulator has been constructed at Inland Steel Research Laboratories. The simulator can reproduce the thermal cycles of the CAPL, CAL, galvanizing or any other type of continuous anneal line. The unique equipment and computer control system of this unit provides a degree of experimental control and flexibility superior to other laboratory heat treating facilities. The purpose of this paper is to provide a general description of the design, capabilities and operation of the simulator, as well as, several examples of simulator applications.

Introduction

A variety of continuous annealing (CA) processes are used in modern steel mills to produce cold rolled and coated sheet steel. Rapid heating and short-time soak are features common to all CA processes as compared to batch annealing. Differences between specific CA processes are found mainly in cooling path, i.e., the thermal treatment following high-temperature annealing. Widely different cooling methods are used ranging from forced convection or gas-jet cooling to water quenching. Each cooling method exhibits inherent limitations on the range of achievable cooling rates. In addition, the end point of cooling and subsequent thermal processing depends on the particular CA process. For example certain processes include in-line overaging after primary cooling. These features can have a significant affect on properties and stability of properties; sensitivity to thermal treatment following soak depends on steel grade and microstructure.

Simulation of CA processes in the laboratory is often desirable to avoid high costs associated with in-plant development of new products. Ideally the simulation method should be capable of accurately reproducing any type of CA process available in the plant. Inland Steel Company used this approach in the design and construction of a dedicated CA simulator. This paper describes the design, operation and applications of this equipment.

Need for Continuous Annealing Simulator

In early 1990, Inland Steel Company (ISC) and Nippon Steel Corp. (NSC) initiated start-up of the joint venture, I/N Tek, the world's most advanced continuous cold mill. At the heart of this facility is the continuous annealing process line (CAPL) which was engineered, designed and supplied by NSC. Details of the equipment and layout of I/N Tek can be found in ref. (1). Well before start-up of I/N Tek, ISC undertook to design and construct a CA simulator in order to support product development for I/N Tek.

Although the ISC CA simulator is capable of simulating virtually any CA process, the following will focus on simulating the annealing part of the CAPL process. Figure 1a illustrates the various furnaces and cooling sections comprising the CAPL process. The corresponding thermal history of a strip element is depicted in Figure 1b. The heating section includes a preheat furnace (PRE), two heating furnaces (NONOX and RT) and two soak furnaces (1S and 2S). Nitrogen gas-jet cooling capability is included in the second soak furnace. Accelerated cooling (ACC) is accomplished by nitrogen/water mist. End point of accelerated cooling is controlled as the strip enters the overaging section (O) which is followed by a final cooling (FC) section.

Generally CA simulation involves approximating a continuous process by a series of batch processes using a convenient specimen size. A common technique utilizes molten salt bathes (pots) for heating and soaking. Cooling is accomplished by transferring the specimen into various cooling media, including molten salt, air or water, depending on desired cooling rate. This technique is adequate for many CA processes, but it is not always applicable to simulation of the CAPL annealing process.

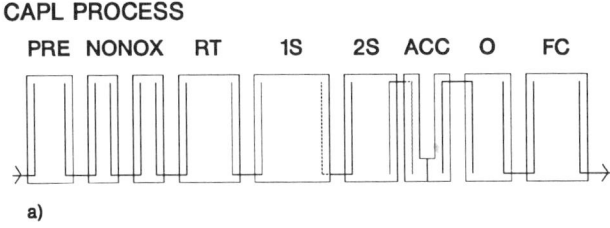

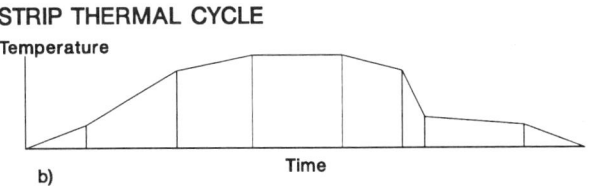

Figure 1: Schematic of CAPL annealing process (a) and corresponding temperature history of a strip (b). PRE=preheat furnace, NONOX=non-oxidizing furnace, RT=radiant-tube furnace, 1S=first soak, 2S=second soak, ACC=accelerated cooling, O=overage furnace, FC=final cooling.

To illustrate this point, Figure 2 contrasts a typical CAPL annealing cycle with that obtained through salt-pot simulation of the same cycle. Using two salt pots at two different temperatures and utilizing air cooling between immersion steps, a fairly good

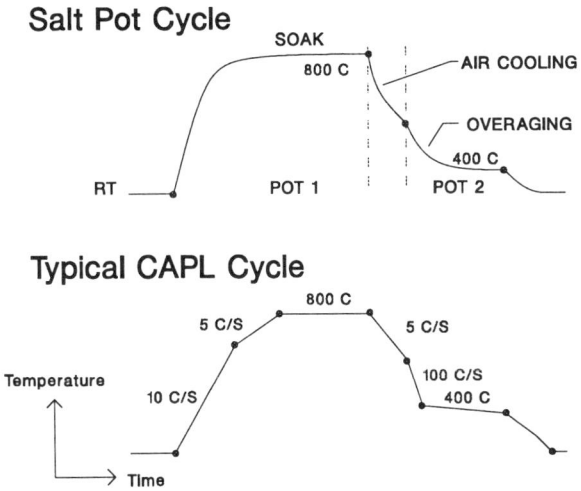

Figure 2: Salt-pot simulation of a typical CAPL annealing cycle.

simulation of the CAPL cycle is possible. However, heating and cooling rates are not easily controlled. For those cases where accurate heating and/or cooling rate control is necessary, a specialized CA simulator represents a valuable laboratory annealing tool.

Simulator Design

Design Objectives

The primary objective of the Simulator is to duplicate the CAPL process by cycling a steel specimen through a series of "batch" processing steps. Other design considerations include:

- The specimen size must be suitable for tensile testing.

- Control of the specimen thermal path, including heating and cooling rates, must be excellent.

- The temperature uniformity over the specimen must be excellent.

- The Simulator must duplicate many types of continuous annealing processes, both standard and non-standard.

- A rapid turn-around time between simulations.

Table I lists the general thermal design specifications of the Simulator and compares them with the I/N Tek CAPL process. The Simulator, shown in the photograph of Figure 3, was designed to provide better temperature control and far greater processing flexibility than the actual CAPL process. In order to achieve these goals, the Simulator was designed using unique equipment components and a sophisticated computer control system.

Table 1:

Comparison of CA Simulator Design Specifications with I/N Tek CAPL Process Capability

	I/N Tek CAPL	CA Simulator
Gauge (mm):	0.4 to 2.0	0.38 to 2.5
Soak Temperature (°C):	890 max.	1000 max.
Heating Rate (°C/s):	10 avg.	0 to 50
Cooling Rate (°C/s):	100 avg.	0 to 1000
Temp. Control Tolerance (°C):	±15	±5
Temperature Uniformity (°C):		±5
Type of Thermal Cycle:	CAPL	CAPL, CAL, CGL

Figure 3: Photograph of CA Simulator.

Simulator Equipment

A schematic of the Simulator and a typical CAPL thermal path is shown in Figure 4. The Simulator consists of three basic equipment components: 1) hydraulic cylinder, 2) furnace and 3) cooling zone. A thermal path is simulated by cycling a 100 mm by 300 mm size sheet specimen back and forth between the furnace and the cooling zone.

Rapid specimen transfer between the furnace and the cooling zone is required for accurate CAPL simulation. The sample transfer function is achieved using a custom 4 m long servo-controlled hydraulic cylinder (Moog Controls). The specimen rests in a holder made of thin gauge Inconel tubing which is attached to the end of the cylinder rod. The rod (and specimen) can be accelerated from a stopped position up to the transfer speed in a fraction of a second. Transfer speed may be controlled up to 90 m/min.

The furnace used to heat the specimen is highly responsive in order to simulate the various temperatures and heating rates experienced by the strip in a continuous annealing line. A specially designed infrared furnace (Sinku-Riko Inc.) performs this function. This furnace consists of an array of tungsten filament lamps backed by gold-plated, parabolic mirrors to uniformly direct short-wavelength infrared radiation to the specimen. The furnace body is water cooled so there is little thermal inertia which must be overcome when the heating rate is changed. The radiant energy to the specimen may be cycled from zero to full power in a few seconds.

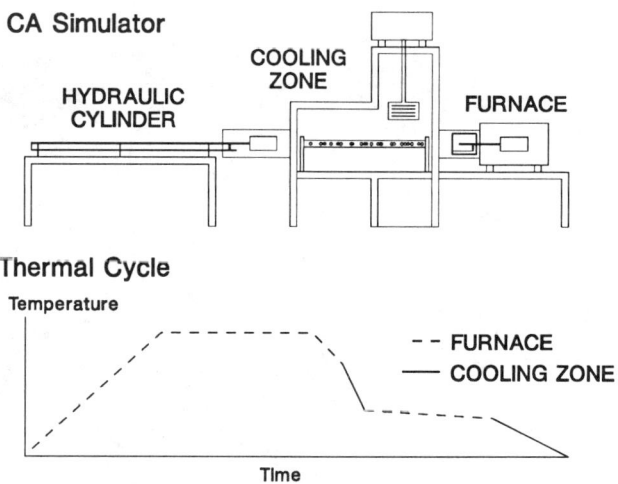

Figure 4: Schematic illustration of the basic components of the CA Simulator and corresponding specimen temperature history.

In order to compensate for temperature gradients on the specimen, the furnace lamps are arranged in a "cross-hatch" design, i.e., vertically oriented lamps on one side of the furnace and horizontally oriented lamps on the opposite side of the furnace. By selective "zoning" of the furnace lamps, both vertical and horizontal temperature gradients can be minimized. The specimen is enclosed in a quartz jacket which permits atmosphere control while the specimen is in the furnace. Gas-jet nozzles made of perforated quartz tubing run the length of the furnace and provide low level (<30°C/s) cooling of the specimen.

The cooling zone employs multiple cooling methods to achieve a wide range of specimen cooling rates. An array of high pressure nozzles runs along both sides of the 1.5 m long cooling zone. The cooling array contains gas-water mist nozzles plus water-spray nozzles. The mist nozzles provide a cooling rate capability of approximately 30°C/s to 300°C/s (0.8 mm thick sample) while the water-spray nozzles provide cooling rates up to 1000°C/s. In addition, the cooling zone contains a pair of gravity fed water nozzles which can "flood" the specimen with water for simulating an uncontrolled water quench.

The gas and water flow rates to the cooling nozzles are controlled by servo-hydraulic actuators (Moog Controls) coupled to process valves. This arrangement allows the flow rates to be rapidly changed in order to control the specimen cooling rate over a wide temperature range. An in-line water heater controls the spray water temperature from 25°C to 50°C. Because of the nonuniform spray densities of the cooling nozzles, the specimen is "oscillated" while inside the cooling zone to achieve good temperature uniformity.

Two interlock chambers isolate the cooling zone from the other sections of the Simulator. The furnace interlock, located between the furnace and the cooling zone, isolates the furnace from the high dewpoint atmosphere in the cooling zone. A nitrogen gas curtain forms a barrier between the cooling zone and the furnace when the specimen is in the furnace. When the specimen is in the cooling zone, a solid door closes off the furnace from the water spray. Hot furnace gases are exhausted from the top of the interlock to help control specimen temperature.

The rear interlock chamber is used as a loading and unloading point for the specimen. When the sample is fully retracted into the loading/unloading position, a solid door seals off the interlock from the cooling zone. The interlock may then be opened to handle the specimen. An exhaust fan purges the interlock to prevent the inert gas atmosphere from escaping into the Operator's pulpit. When the specimen moves into the cooling zone, a split door closes around the cylinder rod to seal off the interlock from the cooling zone.

Control and Data Analysis System

The most important component of the Simulator is the Control and Data Analysis System. This system (Figure 5) automatically controls and integrates all the major functions of the Simulator. The Control and Data Analysis System consists of four major subsystems: 1) Simulator Control Computer, 2) Process Input/Output, 3) Operator's Console and 4) Data Analysis Computer.

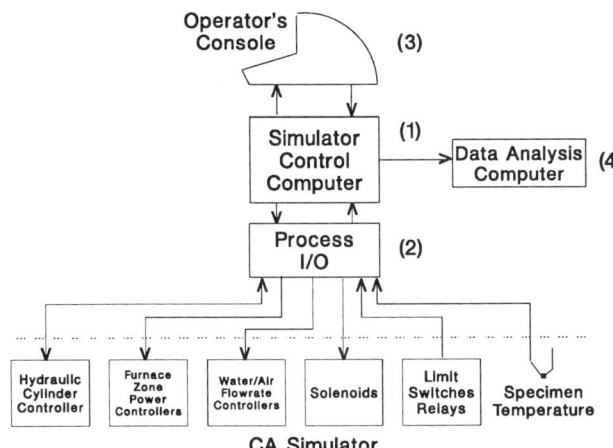

Figure 5: Block diagram depicting the control and data analysis system.

The Simulator Control Computer (RTVAX 1000) is responsible for the real-time sequencing and control of all major functions. The computer controls the following:

- Speed and position of the hydraulic cylinder.
- Power to all zones of the furnace.
- Gas and water flow rates to the cooling nozzles.
- Furnace purge gas, interlock gas curtains and furnace cooling gas.
- Interlock doors.

The process Input/Output (CAMAC) provides the interface between the Control Computer and the lower level controllers. A thermocouple spot welded to the surface of the specimen provides the temperature feedback necessary for control.

The Operator's Console (Figure 6) provides the interface between the Operator, the Control Computer and the Data Analysis Computer. Here, simulation parameters are entered and simulation progress is monitored. Software tools allow the Operator to create and edit the desired temperature cycle in a "user friendly" environment. The Console displays real-time data such as specimen temperature, temperature deviation from aim and furnace power, as well as, diagnostic information.

Figure 6: Photograph of computer console and sample loading chamber.

A data link allows pertinent test information to be transferred to the Data Analysis Computer (VAX 8800). Here, software tools have been developed to plot and analyze the test information.

Simulator Operation

Specimen Preparation

Each specimen must have a thermocouple (TC) attached to its surface in order for the Simulator to function. The location of the TC on the specimen is important for repeatable performance. A small spot welder (Duffers Scientific) is used for quick and sturdy attachment of the TC wires to the specimen with minimal disturbance of the steel surface. The specimen is then placed in the holder and the TC wires connected to a terminal block mounted on the holder.

Temperature uniformity of the specimen is strongly dependent on its surface uniformity. Oxidation, dirt or oil on the specimen will likely cause local hot spots. Oxide films must be removed by pickling and the specimen must be thoroughly cleaned with a solvent before processing.

Running a Simulation

A thermal cycle is defined by a sequence of heating, soaking and cooling segments. The segments may be arranged in any order and any value of rate, temperature or time may be used in each segment within the equipment operating constraints. The mode of cooling (e.g., gas, mist, water spray or water quench) must also be selected. Cycles which are used on a frequent basis (e.g., CAPL and CAL cycles) are stored in the system where they can be easily recalled for use and/or modification.

Once a thermal cycle is defined, the simulation is performed on command. During the simulation, the Operator may monitor the temperature history of the specimen on the real-time display. The Operator may terminate the simulation at any time. At the Operator's command, the process data collected during a simulation is transferred to a file storage area. At the end of a test session this data may be retrieved and analysed as desired.

Simulator Performance

Figure 7 shows two consecutive test runs of a CAPL cycle simulation. The CAPL cycle is one of the most difficult to simulate because accelerated cooling must be halted at the temperature "stop-point" before overaging. The Simulator is able to control the desired cooling rate of 100°C/s and stop the cooling at the desired overaging temperature of 400°C.

In addition to its excellent temperature control and repeatability, the Simulator is capable of a rapid turn around time between runs. Depending on the length of the cycle, six to ten samples may be processed in one hour.

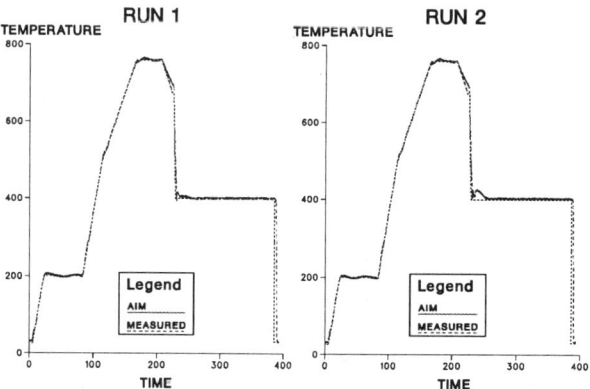

Figure 7: Temperature history of a specimen during two consecutive identical simulations.

Applications

The wide range of control capabilities enables the Simulator to function in a variety of applications. The following is a non-exhaustive list of potential applications:

- Product development

- Process development

- Verification of CA line variables.

- Process/Properties sensitivity analysis.

- Basic Research

Figure 8 schematically illustrates the sensitivity analysis application. All processes exhibit some degree of inherent variability, for example, temperature variations during annealing as indicated in Figure 8a. The Simulator is useful to define the corresponding variability in a product property (Figure 8b). Then the acceptable range of product property defines the tolerance limits of process variable, for example, annealing temperature. The accuracy of temperature control possible with the Simulator permits separate analysis of each segment of a CA cycle including heating and cooling segments.

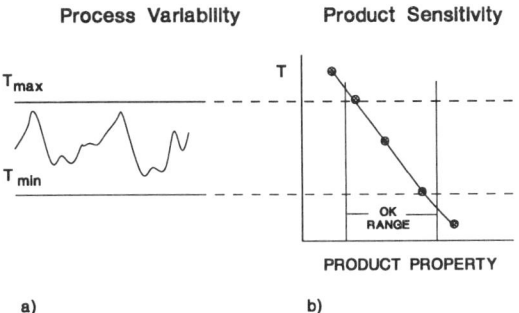

Figure 8: Example of how the CA simulator can be used in process/property sensitivity analysis.

In the application of product development the Simulator is especially suitable to development of high strength dual-phase or triple-phase steels. These steels contain controlled amounts of some or all of the following phases: ferrite, martensite, bainite, carbide and retained austenite. In particular, triple phase steels, which contain ferrite, bainite and retained austenite can be produced by controlled cooling of intercritically annealed C-Mn-Si steels to a specific temperature range for the bainitic transformation (typically 400°C). The resultant microstructure exhibits a combination of strength and ductility superior to that of conventional dual-phase steels. It has been shown that properties of this type of steel are highly sensitive to cooling path (rate) from the intercritical annealing temperature (2). Thus, accurate temperature control attainable using a sophisticated CA simulator is invaluable in developing products for production on a full scale CA line.

Summary

In response to a need for improved continuous annealing simulation in the laboratory to support product development for its joint venture with Nippon Steel Corporation, Inland Steel Company designed and constructed a computer controlled continuous annealing simulator. Temperature control, including heating and cooling rate control, exceeds the capabilities of most, if not all, production annealing processes. Thus, the simulator represents a valuable product- and process-development tool. Although much less costly alternatives are available for simulation of continuous annealing processes, a highly flexible annealing simulator is necessary for development of products that require controlled cooling from the annealing temperature to develop desired microstructure and properties.

References

1. R.A Joyce, "Design, Installation and Start-up of I/N Tek", *Iron and Steel Engineer*, May 1991, 15-21.

2. O. Matsumura, Y. Sakuma and H. Takechi, "Enhancement of Elongation by Retained Austenite in Intercritical Annealed 0.4C-1.5Si-0.8Mn Steel", *Trans ISIJ*, 27 (12) (1987), 570-579.

III. Continuous Annealing: Product Metallurgy

B. Tin Plate

METALLURGY FOR HOOGOVENS' NEW CA LINE FOR TIN PLATE

Th.M. Hoogendoorn and A.J. van den Hoogen

HOOGOVENS IJMUIDEN
Product Metallurgy Dept

Abstract

At Hoogovens IJmuiden a new continuous annealing line for tin plate has recently started production with technological know how of NSC. Additional know how was developed for assuring the economical production of specially the soft qualities in the Hoogovens situation.
In a cooperation between Hoogovens and NSC research was done to solve the problems. This research consisted of Laboratory experiments and Field tests, and was supported by metallurgical modelling at Hoogovens. As a result several adaptations were made, in the CA line itself, but also outside the line, in the steel mill and in the hot strip mill. Examples of these adaptations will be discussed.
The whole project shows, that a good coordination of all disciplines in the total production chain is of utmost importance when introducing a new technology in this chain.

1. Introduction

The introduction of continuous casting was of great importance for the production of tin plate. The amount of segregation and the number of inclusions improved dramatically. The customer complaints for these reasons virtually disappeared.
This switch in casting technique also meant a switch from rimming steel to aluminium killed qualities for tin plate. This introduced a new problem. The T4 quality which we produced by batch annealing became too soft because the ageing by solute nitrogen was excluded by using Al-killed steel.
The only way for producing a good T4 was by continuous annealing. This meant a capacity problem, as T4 is the bulk of our deliveries.
To solve this, Hoogovens decided to build a new continuous annealing line. It was also decided to make use of the expertise of NSC in the field of their CA technology for tin plate.

Of course the new line had to be fit for meeting the newest demands of the market. It also had to fit in the Hoogovens production route with our philosophy of producing and had to deliver products according the Hoogovens product mix.

One of the new developments in tin plate is, that more and more of our customers want their incoming materials classified in terms of yield points instead of hardness. For that reason we decided to introduce this way of qualification for the materials from the new line.
Hoogovens produces all its temper classes, T2 - T6, from two base compositions by two annealing methods, batch annealing and continuous annealing. We did not want to expand the number of base steel compositions. These compositions are Al-killed low carbon steels.
Moreover we wanted to be able to produce all our tin plate qualities with the new line. As there is a development in the market demand to wider strips this meant, that the production of 1230 mm wide DI qualities had to be possible.

It proved to be very difficult to meet all these demands. The situation for NSC in Yawata and Hoogovens in IJmuiden was different in many aspects. The incoming material for the line was produced in steel mills and hot strip mills with different possibilities in the two companies. The product mix and the customers' demands differed. Additional know how for producing soft tempers from Al-killed steels by continuous annealing had to be developed. At that time there was hardly any experience in qualifying tin plate products in terms of yield points.
The first concept for the new line was an extrapolation from NSC's Yawata Nº 2 CAPL with our product mix with its many change overs. After long discussions both companies decided that this line would be too long to be practical. Moreover the incoming material could not be produced by the Hoogovens' HSM for those cases which require a low solute nitrogen content. New metallurgical solutions had to be found.
So it was decided to do a Joint Metallurgical Study NSC-HO to solve the problems. In this study the know how and facilities in the field of CA of tin plate were combined. The study consisted of laboratory simulations and field tests.
Although not all of the questions are answered up till now we are now on the eve of regular production with our CA12, as our installation is named.

2. Metallurgical aspects of CA for tin plate

The main purpose of the CA heat cycle is the full recrystallization of the heavily deformed, fullhard thin material. The most elementary form of the continuous annealing process is the Mohri cycle, which consists of a heating, a soaking and a cooling section. On low carbon compositions this cycle results in a T4 quality. Carbon and nitrogen in solution after soaking, have no opportunity to precipitate during cooling. Ageing phenomena prevent the production of softer qualities. For harder products heavy temper rolling is used.

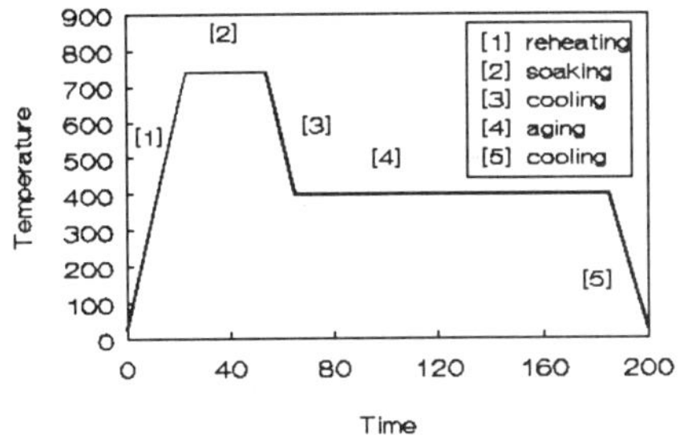

For softer tempers an overageing section is introduced, like for the production of soft sheets. This is also the case in the Yawata N° 2 CAPL of NSC. This cycle, at the start of our Joint Metallurgical Study consists of heating, soaking, cooling, over-ageing and cooling (see fig. 1).

Fig. 1: CA temperature-time cycle. Basic concept.

During soaking full recrystallization is attained, if the material stays long enough above the recrystallization temperature, normally around 640°C. With higher soaking temperatures a coarser grain can be obtained. Tin plate is characterized by the low thickness of the strip. This limits the soaking temperature to 750°C because of the risk of strip breakage. This is an important limitation compared to the annealing of sheet. Due to the high cold rolling percentage, usually > 80 % and the limitation in the soaking temperature, CA tin plate always shows a relatively fine grain size. This makes it difficult to produce soft qualities. Up to temperatures of around 720°C the amount of carbon which goes into solution rises. If this stays in solution it will give more ageing.

During overageing at around 400°C the carbon has the opportunity to precipitate, given enough time. The amount of carbon staying in solution is thus limited by the equilibrium at 400°C. At lower OA temperatures the carbon precipitation is very slow.

After the annealing cycle, the strip is temper rolled. This is meant for:
1) suppression of the Lüders phenomenon
2) getting the right roughness
3) correction of the shape
4) getting higher strength for hard tempers.

Table I: Classification of LYP.

AISI	EU	LYP (Baked)
T2	T52	260 MPa
T3	T57	350 MPa
T4	T61	400 MPa
T5	T65	450 MPa
T6	T70	500 MPa

The yield points of tin plate qualities, according to the ISO working group, which is concerned with this subject, have to be measured after baking (20' 200°C). This means that the effect of the amount of solute C and N and the dislocations introduced by TR on the yield point has to be known. The values we used as objectives for our qualities are given in Table I.

Solute C content can be controlled by OA up to a certain limit.

More or less, the same phenomena as described for the precipitation of carbon, occur for Nitrogen. However, the Nitrogen concentration in steel is much lower than the Carbon concentration. It can precipitate with Aluminium which is present as a very dilute solution in iron. All this

Fig. 2: Overview of Continuous Annealing SIMulator (CASIM).

Fig. 3: Specimen positioned in CASIM.

makes that the overageing section is not effective enough to remove sufficient Nitrogen from solution to prevent ageing. So, C can be removed from solution in the OA section for a great deal. For Nitrogen this is not the case. To obtain a material with low solute N the incoming material of the CA line already has to have the low solute N content.
In other words, for soft tin plate products with minimum ageing, after CA, the incoming material has to be produced in such a way, that it contains no free Nitrogen.
Both solute contents have to be controlled very accurately.

3. Metallurgical investigations

3.1 Laboratory investigations on the CA heat cycle

Investigating the effect of variations in the CA heat cycle thus means quantifying the effects of recrystallized grain sizes and of solute carbon on the yield point.
At Hoogovens R&D a Continuous Annealing Simulator -CASIM- was developed (fig. 2). A specimen of 450 x 100 mm can be subjected to an arbitrary temperature-time cycle.
For this purpose the fullhard specimen is positioned in two copper clamps (fig. 3). The heat cycle is given into a micro processor.
The heating is done by direct current through the clamps and the specimen. Cooling is done by feeding compressed air into the shower type gasjets. By adjusting the amount of air, the distance of the gasjets to the specimen and the thickness of the sheet, cooling speeds of over 300°C/s can be achieved on tin plate gauges. The whole cycle is temperature controlled by the micro processor, through a thermocouple which is spotwelded on the specimen.

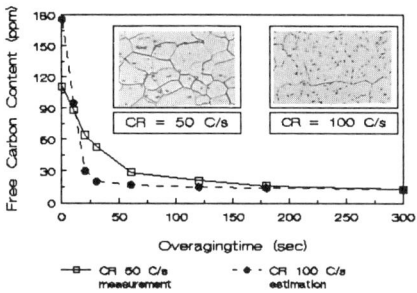

Fig. 4: Preliminary Laboratory Tests. Influence of overageing time on free Carbon for different cooling rates.

The homogeneity of the temperature is better than ± 5°C over the gage length of two tensile test specimens which can be cut from the testpiece.
A first orientation was done on the effect of the primary cooling speed from the soaking temperature to the over-ageing temperature. For a low carbon steel with 0.02 % C in fullhard condition the effect of cooling speeds of 50°C/s -the normal cooling speed-, and 100°C/s were compared for different over-ageing times. The completely different micro structures are given in fig. 4. with an estimation of the effect on C precipitation.

The conclusion from this orientation is, that for normal cooling speeds a very long over-ageing section is required. With a cooling speed of 100°C/s the situation has stabilized after a much shorter time.
The fast cooling causes a high sursaturation of solute carbon, which is the driving force for cementite precipitation. This gives a high nucleation rate and a rapid formation of fine carbides during over-ageing at 400°C.
Slow cooling gives rise to a smaller amount of coarser carbides. Not only does it take more time for a full precipitation due to longer diffusion ways. The coarser carbides also might have a negative influence on the corrosion properties of tin plate. These results were confirmed in terms of yield points by tests on the Howaq pilot line of CRM in Belgium.

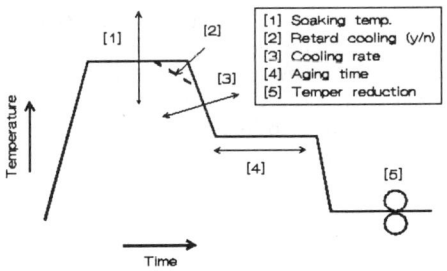

Fig. 5: Investigated aspects of CA heat cycle.

After this first orientation the decision for the Joint Metallurgical Study was taken. A more complete study on the effect of possible variations in the CA cycle was done. Aim was to see how it would be possible to produce a T2 quality with a lower yield point of 260 MPa after baking (10' at 200°C).

Two steel compositions with carbon contents of 0.02 % and 0.04 % were produced with two coiling temperatures in the HSM. After cold rolling the steels were subjected to CA simulation in Yawata and in IJmuiden.

The following variations were investigated, see figure 5:
1) soaking temp.   710°C, 740°C   4) OA time   up to 120 s
2) retard cool   yes/no   5) temper reduction   up to 3 %
3) fast cool   100-300°C/s   6) baking (10' 200°C) yes/no.

The main conclusions concerning the CA cycle from this parameter study were, (fig. 6a) that cooling at higher speeds than 100°C/s was not necessary. At these speeds, the largest effect had taken place after 30 s.
The softest quality was obtained with a temper reduction of 0.8 % (lab. conditions, fig. 6b). A yield point level of 260 MPa after baking however was not achieved.

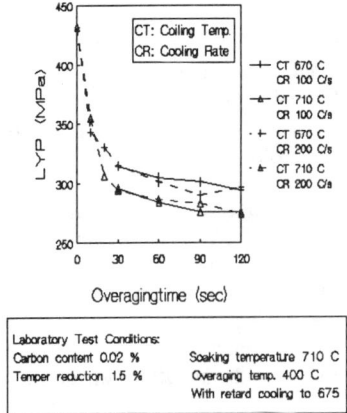

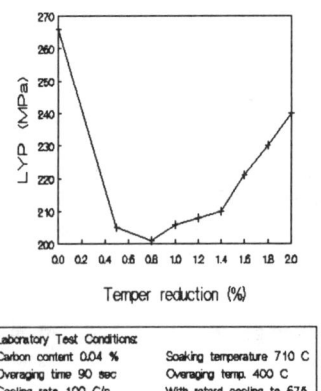

Fig. 6a: Laboratory tests on CASIM. Influence of overageing time on LYP for different cooling rates and coiling temperatures (baked).

Fig. 6b: Laboratory tests on CASIM. Influence of temper reduction on LYP (not baked).

The influence of the coiling temperature can be explained by the coarser grain and the lower content of solute nitrogen at higher coiling temperatures.

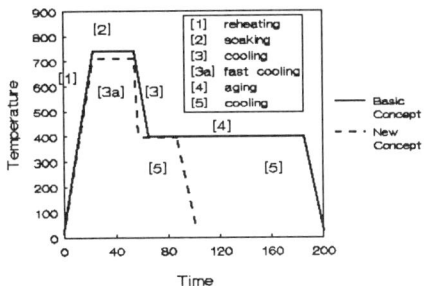

Fig. 7: Comparison of basic and new concept of the CA line.

Based on these conclusions from the laboratory experiments, discussions were started, whether a fast cooling in our new line could be possible.

In fig. 7 you see the original Yawata heat cycle and the new concept in one figure. The advantages of the fast primary cooling in terms of cost and operatability can be easily thought up.

As a result of these discussions, NSC decided to develop further a system for a fast primary cooling for our new CA line for tin plate, the so called HGJC system (High cooling rate Gas Jet Cooling system).

3.2 Nitrogen content of incoming material

We mentioned earlier, that also the incoming material has to meet certain conditions. To produce a blackplate with low ageing characteristics the content of solute C and N must be very low. The solute C content can be influenced by the CA heat cycle, although the solute level has a lower boundary which is determined by the thermo dynamic equilibrium content at the overageing temperature.

For the solute nitrogen level the CA cycle gives no effective means. The incoming material has to show a solute N level, which is homogeneously low.

When we look backwards from the CA into the production line, the HSM is the first installation, which can influence the solute N content.

This content is determined by the equilibrium

$$Al + N \leftrightarrows AlN.$$

AlN will be formed during heating of the slab, dissolve partly or completely in the slab furnace, depending on the slab temperature, and the Al and N contents. According to Hoogovens' experience, for our compositions, the kinetics of AlN precipitation is too low to give AlN formation during the rolling process and the cooling on the run-out table. Precipitation can take place however during coil cooling, depending on coiling temperature and cooling cycle of each place of the coil. This is shown schematically in fig. 8.

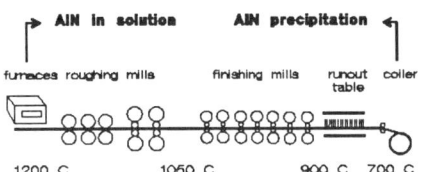

Fig. 8: Lay out of Hot Strip Mill No. 2 of Hoogovens.

In our normal practice all AlN goes into solution in the slab furnace. Using lower furnace temperatures could be benificial. The aluminium nitrides which do not dissolve do not have to precipitate later. The possibility of lowering the slab temperature however is limited by the temperature losses during the rolling process. The finishing temperature has to stay in the austenitic region. This has a strong relation with the lay out of the mill.

The other way for letting precipitate AlN is using a high coiling temperature. With this one has to consider, that coil cooling is an inhomogeneous process. Every part of a coil cools in a different way. Coiling temperatures are bound to a maximum through scale formation, important for pickling capacity and the formation of coarse carbides, important for the corrosion behaviour and formability.

All this makes the AlN formation in the HSM very complicated. To quantify the precipitation one must be capable of determining the solute N content at low levels. Happily we are capable of this by a hot hydrogen extraction method, which was developed in house. Solute nitrogen is extracted from finely milled steel by means of hydrogen and forms ammonia. The amount of ammonia is determined. This method is able to give results in the ppm range.

3.2.1 Quantification of the nitrogen precipitation

To develop a good processing routine for the production of low solute nitrogen hot strip, we had to quantify the effects in the various production stages. We had the great advantage that we had already spent a lot of effort in making physical-mathematical models for the AlN precipitation.

AlN dissolution could be measured by laboratory tests. This was modelized using solubility equations as a base. Precipitation during cooling was investigated by simulation of the cooling pattern of several places in a coil on sheets in a furnace in the laboratory. This was verified in mill tests. A good prediction could be given, if the cooling pattern of the testpiece taken from a coil was known (fig. 9a + b).

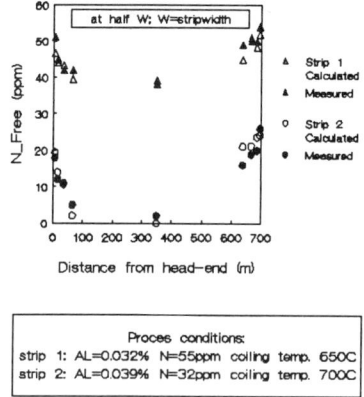
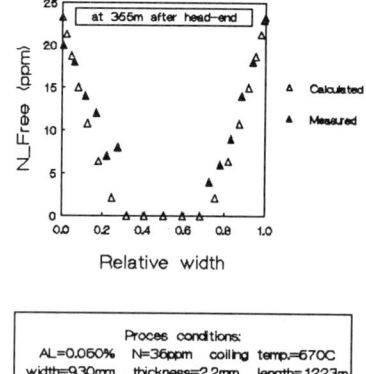

Fig. 9a: Measurements and calculations of free Nitrogen over the length of a strip after coiling for different strips.

Fig. 9b: Measurements and calculations of free Nitrogen over the width of a strip after coiling for different strips.

The cooling cycle of any place in a coil is given by the coil cool model. This model predicts the temperature of any place in the coil at any moment. Geometry like width, length, thickness and crown of the strip, are taken into account at the start but also during the cooling process.

Comparison of the calculations with earlier published measurements on a cooling coil with thermocouples [1] showed after 8 hours no greater deviation than 20°C.

To compensate for inhomogeneities in the coiling temperature caused by the actual run-out table cooling we installed a scanning measurement device for the coiling temperature. The results of this measurement can be fed in the coil cooling model.

In this way we have developed a complete AlN precipitation model for the HSM, by integrating
1) the dissolution model
2) the precipitation model [2]
3) the coil cool model.

3.2.2 Parameter study

The integrated model gives us the opportunity to calculate the amount of solute N as a function of chemical composition and processing conditions over the length and width of the strip. In this way the model can be used for developing the optimal processing conditions.

For our studies we assumed a constant crown of 0.05 mm, a fixed distance between the coil rings of 0.04 mm. So deviations in crown and other coiling practices could influence the results. We also assumed a constant coiling temperature over the width of the strip. Eventual cold sides are not taken into account. We started with a constant coiling temperature over the length also.

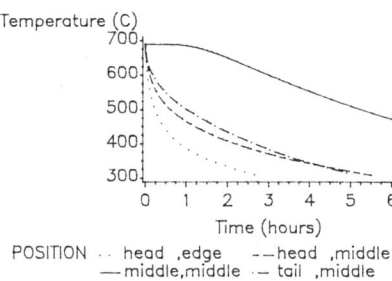

Fig. 10: Inhomogeneous cooling of a coil

Figure 10 is a typical example of a coil cooling. When a strip is coiled, head and tail of the strip, and also the edges cool faster than the middle of the strip. In fig. 11 you see a typical calculated solute N distribution for a normal steel used for tin plate application. The slab temperature is chosen sufficiently high to have all AlN dissolved in the furnace.

Due to the inhomogeneous cooling of the coil a non-homogeneous distribution of solute N is obtained over the length and width of the coil.

The level of solute N is higher at head, tail and edge positions than in the middle. If we set the condition that a level of 5 ppm of solute N may not be exceeded, 30 % of the strip is not meeting that condition.

Using the model we now can indicate how changes in chemical composition or processing conditions influence the amount of solute N (Table II).

Table II: Results of parameter study.

Chemical composition and Process hot strip mill	Variation	Percentage of strip with N-free > 5 ppm
Base	-	30 %
N	- 10 ppm	7 %
Al	+ 0.02 %	12 %
Furnace temperature in combination with Al	- 40°C + 0.02 %	5 %
Coiling temperature without 'U-shape' cooling pattern	+ 20°C	20 %

Chemical composition. Lowering the nitrogen content in the steel base with 10 ppm gives a considerable reduction of the area with solute N $\geq$ 5 ppm. However, this reduction is at the moment unacceptable for the steel mill due to the nitrogen pick-up during casting.
Enlarging the Al percentage with 0.02 % improves the homogeneity clearly. It seems obvious to make use of a still higher Al content. Experiences at Hoogovens are however that over 0.1 % Al gives rise to excessive roll wear during cold rolling. Moreover it is expensive, due to the price of Al, but also due to the lack of switch possibilities.

Hot Strip Processing. As we mentioned earlier it is possible to obtain incomplete AlN dissolution by lowering the slab temperature. This is favourable because these nitrides are not to be precipitated in the coil. This part will not attribute to the inhomogeneity.
We can make maximal use of this solubility effect in the furnace at higher Al contents. Lowering the slab temperature from a normal temperature with 40°C shows a large improvement in homogeneity when the Al content is enlarged with 0.02 % at the same time. The area with solute N > 5 ppm decreases from 30 % to 5 %.
Realisation of low slab temperatures is restricted by the lay out of the Hot Strip Mill and the size of the package of the product mix which needs to be processed with low slab temperatures.
Boundary condition is the finishing temperature, which has to stay in the austenitic region. In the production of thin gauges the strip loses more temperature up to the last finishing stand, than in the case of thick gauges. The same is the case when large width reductions have to be applied in the roughing area. They require reversing, which costs time, which is temperature.
Finally, if the amount of slabs is limited, the programming of the furnaces can give problems.

Since the precipitation of Al and N to AlN is a function of temperature and time, the amount of solute N can also be reduced by coiling at higher temperatures. Increasing the normal coiling temperature with 20°C gives an increase of 10 % of the approved area with solute N under 5 ppm. This improvement is reached both at the edges and the ends of the strip.

At this point we left the philosophy of a constant coiling temperature over the length of the strip. Instead we studied the effect of a U shaped cooling pattern for the coiling temperature. Head and tail are coiled at a higher temperature than the middle section of the coil.

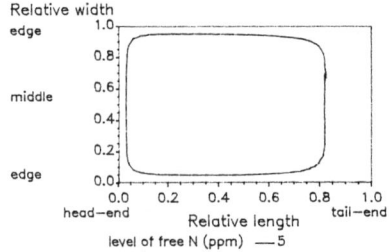
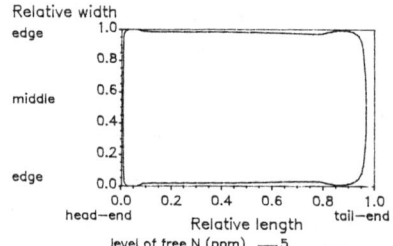

Fig. 11: Level of free Nitrogen after coiling at a normal coiling temperature without a 'U-shape' cooling pattern.

Fig. 12: Level of free Nitrogen after coiling at a high coiling temperature with a 'U-shape' cooling pattern.

Comparing fig. 11 and 12 shows the effect of the higher coiling temperature in combination with a U shape cooling pattern (see fig. 13).

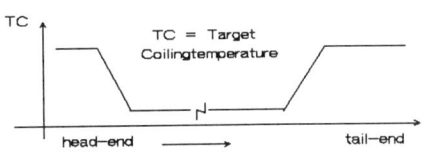

Fig. 13: A typical 'U shape' cooling pattern.

The increased coiling temperature for the middle section gives rise to an acceptable solute N level over the width, since there will always be some sidetrimming in further processing. The shorter cooling times of head and tail of the coil can be compensated by coiling these parts at higher temperatures.

The preference for a U shaped coiling temperature to a higher coiling temperature for the whole strip is given by the fact that higher coiling temperatures give rise to critical grain growth caused by deformation by the coiler tension. Also pickling gives problems after high coiling.

In summary, this parameter study shows that a low and homogeneous level of solute N can be achieved by adapting the chemical composition in the steel mill and by adapting the processing in the hot strip mill.

The chemical composition is characterized by a low level of total nitrogen and a relatively high level of aluminium. In the Hot Strip Mill there has to be aimed at low slab temperatures and high U shaped coiling temperatures.

The separate effects have been quantitatively calculated using physical-mathematical models.

In real practice the possibilities for realization are largely restricted by the specific lay out of the mill and the product mix.

The possibilities for variations in chemical compositions is limited by possibilities of the steel mill and cold roll wear in the cold mill.

Our integrated model for predicting the solute N has been of great importance and is a powerful mean in developing an optimal processing practice for soft temper CA products.

4. Field Tests Hoogovens-NSC

"The proof of the pudding is in the eating" as an old phrase says. The starting point of the discussions around Hoogovens' new continuous annealing line for tin plate were our demands:
- All tempers classified in yield points in baked condition.
- Al killed low carbon steel.
- Hoogovens' incoming material.
- Hoogovens' product mix, including wide DI.

To demonstrate that this was possible a large Field Test was organized. For this test hot strips from the normal Hoogovens production for tin plate were cold rolled and annealed in Yawata Works of NSC. Several heat cycles and temper reductions were used.

The general conclusion was that it was possible to produce a range of tempers. For hard tempers no special problems were foreseen using the HRT of NSC, but the softer qualities required adaptations of the incoming material. Moreover long overageing times were necessary. This is in agreement with earlier laboratory experiments and our own CA experience on our Mohri line.

This led to the decision to conduct a combined study Hoogovens-NSC, the Joint Metallurgical Study, for studying the possibilities for the production of soft tempers CA.

As mentioned earlier, one of the results of the laboratory investigations was the conclusion that a fast primary cooling was very beneficial for the effectiveness of the over-ageing section and the operatability of the line. This last point became more or less imperative by the product mix of Hoogovens with many change-overs. Nippon Steel decided to develop their HGJC system for tin plate and equipped their Yawata Nº 2 CAPL for tin plate with this system.

Then a second Field Test Hoogovens-NSC was performed, specially aimed at testing the HGJC system and confirming the possibilities of this system for softer tempers as they appeared from the laboratory experiments. All based on hot strip material produced at Hoogovens.

The conclusions from this second Field Test were as follows:
- The HGJC system performed very well, also at high annealing temperatures, high speeds and thin strips.
- The softest tempers were produced from hot strip coiled at high temperatures (710°C) and low temper reductions of 1 %.
- The yield point level of 260 ± 20 MPa in temper rolled and baked condition was not achieved. The lowest values were somewhat above 280 MPa, which is significantly lower than the 350 MPa of a T3 quality.
- The variation in yield points over length and width of the strip were unacceptably high. Variations of far more than 20 MPa could originate from variations in solute N content.

All this meant that Hoogovens had to adapt its original aims for the line.
- T2 with a lower yield point of 260 ± 20 MPa could not be produced, so we set our target at as soft as possible, suitable for DI can production. This was suitable because DI can production makes use of not baked material.
- The incoming material had to be produced in such a way, that a more uniform distribution of low solute N was obtained. This means that the normal way of processing had to be changed.

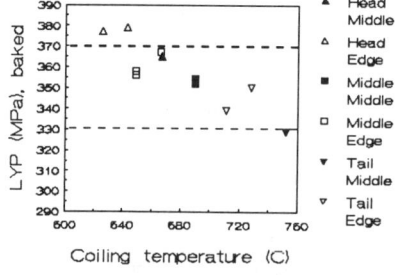 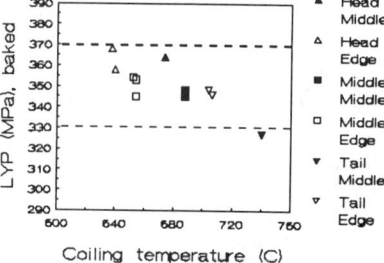

Fig. 14a: Results of the third Field Test. LYP as a function of local coiling temperature for the strips with an Al content of 0.085 %.

Fig. 14b: Results of the third Field Test. LYP as a function of local coiling temperature for the strips with an Al content of 0.13 %.

As a logical consequence of this, a third Field Test was set up. Aim of this test was to find ways to reduce the variations in yield points of the annealed, temper rolled and baked material. For this test again hot strip had to be produced at Hoogovens with adapted

processing. Test quality this time was T57 with yield points of 350 ± 20 MPa.
Based on the parameter study we discussed earlier steel composition and hot strip processing were changed.
A total of 13 coils were produced. We chose an aluminium content of 0.085 %. Because NSC had good results with higher Al contents also 0.13 % Al was added to the programme. Nitrogen was limited to 30 ppm maximum.
In the Hot Strip Mill we tried to achieve slab temperatures of 1160°C. This was not our normal practice!
We tried to realise a U shaped coiling temperature. This also had to be done on an experimental base. At that time we could only realise a hotter coiling of the tail of the strip.
You will find the results for both aluminium contents in fig. 14a and 14b.
In these figures the yield point in baked condition is plotted against the "local" coiling temperature. This is the temperature at which the testpiece was coiled as measured by the scanning coiling temperature device. This is done for edge and middle positions for head, middle and tail of the hot strip. Also the tolerance limits are given.
Head and tail fall just outside these limits.
Yield points from head positions are too high because the U shaped coiling temperature was not realized. Yield points at the tail position are too low because the U shaped coiling temperature was not yet fully operational and too high coiling temperatures were obtained in that position. These types of problems have been solved now. The U shaped cooling is now integrated in the control of the run-out table cooling.
In the strips with the very high aluminium contents of 0.13 % Al the variation looks somewhat lower, especially between middle and sides. This could be caused by the limited dissolution of AlN in the slab furnace.
One must keep in mind, however, that the number of test slabs was very limited in this case.
For a number of coils with the lower Al content, solute N was determined in mid width positions over the length of the strip and compared with calculations making use of the measured local coiling temperatures (fig. 15).

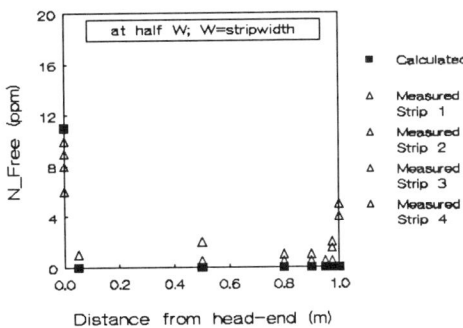

Fig. 15: Results of the third Field Test. Measurements and calculations of free Nitrogen over the length of a strip after coiling.

In the head positions too high solute N contents were found because the U coiling was not realized there. Otherwise the agreement is very good.

5. Conclusions

Transferring a copy of an existing CA installation in the also existing processing line of Hoogovens proved to be an illusion. Special requirements based on market demand (product mix, quality classification) and practical possibilities of the earlier installations (BOF and HSM) made major changes necessary, in the CA line itself but also in the HSM.
T2 CA production appeared not to be possible. We had to adapt our preconditions. This quality we will produce by batch annealing.
The steel mill had to produce tin plate compositions with higher Al contents and nitrogen contents with a low maximum.

The HSM had to adopt a processing routine with slab temperatures with a maximum at a low level and to realize U shaped coiling temperatures at a high level. The CA line itself had to be given a new fast cooling system.

All this has been realized in a relatively short time. This would not have been possible without a very close cooperation between the two companies and the synergistic effect of metallurgy translated into physical models and CA technology.

We have the feeling that we are very near the optimal solution in terms of quality and cost for the whole processing route for tin plate, given today's possibilities.

We also have the feeling, that with our physical models, we are in a good position to evaluate future developments. By pursuing the most promising ones we can stay in the front line of materials quality and economy.

6. References

1. P. Funke, K. Bösenberg, "Untersuchungen über die Abkühlung des Walzgutes auf dem Auslaufrollengang," Stahl u. Eisen, 89 (1969) 26, 1446-1452.

2. G.A. Duit et.al., "A model for the kinetics of aluminum-nitride precipitation," (Paper presented at Thermec-88, Tokyo, 6-10 June 1988), 114.

ADVANCED MANUFACTURING PROCESS FOR TIN MILL BLACKPLATE WITH

ALL TEMPER DESIGNATIONS BY CONTINUOUS ANNEALING

H. Kuguminato, T. Kato, T. Sekine, A. Tosaka*, C. Fujinaga*
Y. Shimoyama, H. Ohno, R. Asaho,

Chiba Works
* Technical Research Division
Kawasaki Steel Corporation,
1 Kawasaki-cho, Chiba 260, Japan

Abstract

The softest T1 grade tin mill blackplate (HR30T:49) has not been successfully produced by the conventional continuous annealing (CAL) process. Kawasaki Steel Corporation (KSC) has developed a new process which makes it possible to produce the T1 grade from extra-low-carbon steel material with a small addition of niobium using CAL. The new process features control of the hot-rolling finishing delivery temperature in the low range (aimed value, 830°C), control of the hot coiling temperature in the medium range (aimed value, 600°C), and the application of a new heat cycle at the CAL itself in order to obtain appropriate recrystallization of the cold-rolled tin mill blackplate. A medium annealing temperature (750°C) and short time soaking (under 10 sec) are used.
To obtain both high productivity and excellent quality with this steel grade, KSC constructed No. 4 CAL at Chiba Works. The line, which was put into operation in March 1990, has a top in-furnace speed of 1000 m/min and a multipurpose cold-rolling mill at the exit side of the furnace, and can handle extra-low-carbon steel modified with niobium. A noteworthy feature of No.4 CAL is that KSC has successfully established the technical conditions necessary for the manufacture of all grades of tin mill blackplate (T1 to T6, DR8 to DR10) using only this one line.

Introduction

The tinplates used for the manufacture of beverage cans, food cans, and various types of caps are specified by temper. The temper is represented by the value of Rockwell T hardness (HR30T) in JIS G3303. The temper designations are classified into the softness grades T1 - T6 for single cold-reduced products and DR8 - DR10 for double cold-reduced products.

Tin mill blackplate (TMBP) having the softness of T3 or less is generally manufactured by batch annealing using low-C Al-killed steel. TMBP having a hardness of T4 or more is manufactured by CAL, while DR TMBP is manufactured by the double cold-reduced method (DR) after being annealed by either batch annealing or CAL. However, the application of the CAL method to soft TMBP had been desired as it is capable of providing homogeneous distribution of the mechanical properties and is extremely fast in comparison with batch annealing.

For the first time in the world, KSC established a technology to produce softer temper designations in a Kawasaki Steel Multipurpose Continuous Annealing (KM-CAL) method that has a rapid cooling-overaging treatment, using low-C Al-killed steel as material; products up to T2 have been manufactured commercially(1-5). This was widely expected to rationalize the manufacturing process because temper designations of T2 or more can be manufactured by modifying the hardening method, provided the softest T1 can be manufactured.

However, the manufacture of T1 was difficult using the CAL method(1). The optimum conditions for producing soft-temper blackplates by the KM-CAL are summarized in **Table I**. For obtaining soft-temper blackplates, it is important to make ferrite grains coarser and to reduce the contents of solute C and N. The conditions of CAL should be such that the soaking temperature is set higher so as to promote the grain growth, with a heat cycle adopted so as to facility the precipitation of carbide in the rapid cooling-overaging zone. The conditions of steel should be such that it has dense distribution of fine carbide particles serving the precipitation sites of solute C, with lesser amount of solute N and with crystal grains made coarser. However, hardness increases when the C content is either lower on higher than a low range of 0.02 - 0.07%.

Table I Conditions suitable for producing soft temper tinplate in KM-CAL

Material steel	Steel type : C.C. low-carbon aluminum killed steel C : 0.02 - 0.07 (wt%) $N_{Total} \leq 0.003$ (wt%) $N_{Total} - N$ as $AlN \leq 0.001$ (wt%)
Hot rolling	Coiling temp. : 620°C
Heating cycle	Annealing (700°C x 40s) Cooling rate (40 - 70°C/s) Overaging (400 - 450°C x 60s)
Temper rolling	Rolling reduction : 0.8 (%)

In particular, in areas where the C content, which is expected to soften the TMBP, was small, the hardness became higher although the grain size increased.

When steels having lower C content are overaged for a short time in the CAL process, their hardness decreases as the ferrite grains turn coarser. In addition to this, since the C content is too low, fine carbide particles which serve as precipitation sites of solute C turn sparse, making grain boundaries more widely spaced. This means that the diffusion distance required for the carbide precipitation is extended. Consequently, the precipitation of C after overaging is delayed to augment the aged hardening and increase the hardness. It is important, therefore, for manufacturing soft-temper blackplates to increase the C content to such an extent that fine carbide particles exist adequately dense even if the grain size becomes smaller to some extent. The presence of solute N promotes solute hardening. In the batch annealing, N precipitates as AlN without leaving solute N. In the CAL process, however, N does not precipitate fully as AlN, and a certain amount of solute N persists because of short annealing time. It is essential, therefore, to reduce the solute N before annealing. For this purpose, it seems effective to keep the N content in slab to 0.003% or less and to promote the precipiration of AlN during hot rolling by coiling at medium temperature so as to prevent carbide particles from coarsening.

The ferrite grains grow and the carbide particles become coarser when coiling is done at high, and after CAL, a large amount of solute C remains. In addition, as this also deteriorates the corrosion resistance of tinplate, high temperature coiling was not deemed appropriate(2). Though the ferrite grains grow and the carbide particles become coarser to some extent at the medium coiling temperature (CT), the hardness is lower than that in steels hot rolled at lower CT. This is attributed to the fact that hardening by sparse distribution of carbide particles is overcome by the greater effect of increased grain size and AlN precipitation.

If the CAL temperature is quickly cooled from a temperature higher than A_1 transformation, the grain size increases. At the same time, however, its composition becomes such that pearlite distribution in ferrite, resulting in ineffective softening. Therefore, low temperature annealing immediately below transformation A_1 point was used. Thus, high temperature annealing did not soften TMBP by increasing the grain size.

As outlined above, in the TMBP softening method that uses low-C Al-killed steel as material and provides rapid-cooling and overaging in CAL, there is a limit to the obtainable softness due to metallurgical conditions inherent in TMBP. This indicates that manufacturing T1 would be difficult. In the meantime, recent progress in steelmaking technology, particularly degassing-refining technology, has been remarkable, enabling the economical and stable manufacture of extra-low-C steel having less than 0.003% of C content(6,7). Therefore, a study of the application of extra-low-C steel with a C content reduced to the minimum technical level was judged necessary in order to allow the manufacture of T1 using the CAL method.

KSC has developed conditions appropriate for manufacturing TMBP using the CAL method by utilizing the features of extra-low-C steel(8-13). In addition, conditions for producing different temper designations with a modified hardening method were studied, and No. 4 Continuous Annealing Line (No. 4 CAL), which is provided with a multipurpose cold-rolling mill directly connected to the rear portion of the annealing furnace(14-17), began operating in March 1990. **Figure 1** shows the appearance of the multi-purpose cold-rolling mill. KSC has succeeded in commercially manufacturing TMBP with all temper designations using No. 4 CAL alone, as outlined below.

Fig. 1 The multipurpose-cold rolling mill and exit looper of No.4 CAL

Study of Production of TMBP with Various Temper Designations by CAL

Study of Steel Compositions Adequate for Low-temperature Annealing

Manufacturing T1 TMBP with the CAL method using low-C Al-killed steel is difficult since the reduction of solute C during annealing is limited and the recrystallization grain size is small. In addition, material having a low recrystallized temperature is necessary in order to effectively pass thin TMBP through in the CAL furnace. Thus, a study was conducted on the compositions of extra-low-C steel with a small amount of solute C, which would be suitable for fast growth of recrystallized grains and also for achieving T1 at the lowest annealing temperature possible. **Figure 2** shows the effect of the C content on tinplate hardness: the effect is noticed after a steel sheet made of the components shown in **Table II** is hot rolled at the temperature shown in the same table and then cold rolled to 0.32 mm thick to produce a TMBP. The required hardness for T1 was obtained by reducing the amount of C to its lowest limit. However, the desired hardness can be obtained and the Lankford value (r-value) increased with extra-low-C steel sheet alone, but at the same time, the planar anisotropy of r-value (Δr) worsens as the recrystallization texture of {uvo} <001> increases, which lowers the r-value in an angle of about 45° to the rolling direction. Improvement occurs when the grains of hot bands become fine and solute C is reduced by the addition of an extremely small amount of Nb so that the development of the {uvo}<001> recrystallization texture that worsen r-value in the 45° directions during recrystallization can be restrained. The effect of this method of improvement is known to be significant(18). **Figure 3** shows the effect of the Nb content on Δr: the effect is noticed after the steel sheet made of components shown in **Table III** is hot rolled and then cold-rolled under the conditions shown in the same table to produce a cold-rolled steel sheet for deep-drawing quality. Note that Δr was obtained in the following equation by measuring the r-value in the 0° (L), 90° (T), and 45° (D) directions to the rolling direction: $\Delta r = (r_L + r_T - 2r_D)/2$

Table II Chemical compositions and hot rolling temperatures of steels used to study the effect of carbon content

Steel type	Chemical compositions (wt.%)				Hot rolling temp. (°C)		
	C	Mn	Al	N	SRT	FDT	CT
Low-C Al-killed steel	0.011~0.090	0.26	0.030	0.0025	1255	860	610
Extra low-C Al-killed steel	0.002	0.15	0.052	0.0015	1160	840	580

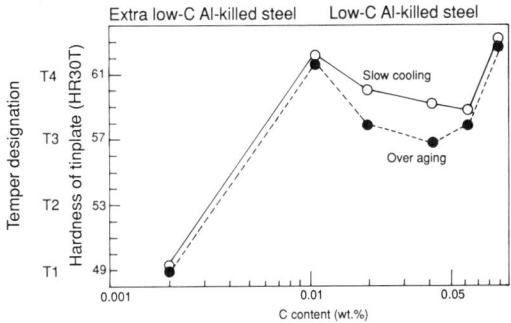

Fig. 2   Effect of carbon content and heat cycles on hardness of tinplate

Table III Chemical compositions, hot rolling temperatures, and cold rolling reduction of steels used to study the effect of Nb content and steel sheet process

Sample steel sheets	Chemical compositions (wt.%)				Hot rolling temp. (°C)			Cold rolling reduction (%)
	C	Mn	Al	N	SRT	FDT	CT	
Tin mill blackplate	0.003	0.18	0.051	0.002	1170	850	570	88
Deep-drawing quality	0.002	0.15	0.040	0.003	1250	880	680	75

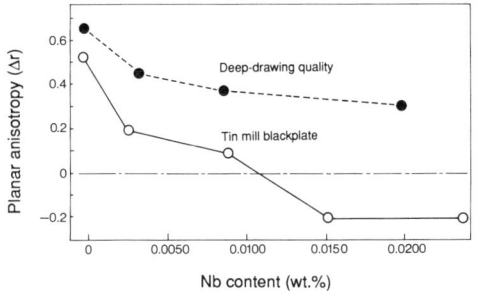

Fig. 3   Effect of Nb on planar anisotropy of r-value (Δr) of cold rolled steels in the mill blackplate, and deep-drawing quality

Adding an extremely small amount of Nb improves Δr. Its effect is particularly large on TMBP. The reason for this is considered to be that the cold-rolled reduction rate is high because the TMBP thickness becomes smaller. As a result, since the non-earing properties is required for TMBP for drawing in the two-piece can manufacturing method (in which a can is finished by filling the can with food and fitting double seaming for top end after a cup has been formed by drawing), improvement of Δr is advantageous.

In the Nb-added steel, the combination of the pinning effect on the grain boundary migration due to precipitation of NbC and the drag effect of solution Nb delays the growth of recrystallized grains(19), thus requiring high temperature for recrystallization. Hence, poor passing results in the furnace. Therefore, it was necessary to study the optimum amount of Nb to add giving consideration to smooth passing. **Figure 4** shows the effects of Nb addition on the relationship between the annealing temperature and hardness (recrystallization curve) in order to obtain the necessary temperature for recrystallization. The test material was a small vacuum-melted steel ingots having the basic composition of 0.003% C - 0.18% Mn - 0.05% Al with Nb added. After hot forging to a thickness of 50 mm in the lab, the material was rolled to a thickness of 5 mm by hot-rolling (slab reheating temperature (SRT) 1100°C, finishing delivery temperature (FDT) 840°C, and CT 600°C), and this hot-rolled strip was ground on both sides to a thickness of 2.6 mm.

After cold-rolling (reduction rate: 88%), the cold-rolled sheets were annealed by direct resistance-heating CAL simulator. In this testing device, a precise heat history can be reproduced since the steel is heated by direct charging. In addition, it uses the controlled cooling method, in which nitrogen gas is blown at both sides of the plate to cool it. The recrystallization curve shifts to the higher temperature rises further when more Nb is added.

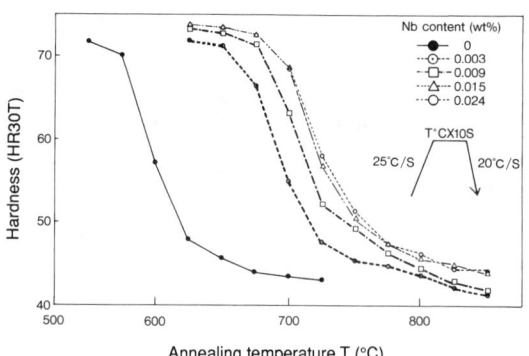

Fig. 4 Effect of Nb content on relation between hardness and annealing temperature

In studying the No.4 CAL equipment, technology was developed for passing ultra-thin gauge steel sheet through in the furnace economically and at stable high-speed operation. These ultra-thin gauge steel sheets are made from extra-low-C steel with a minimal addition of Nb to increase the recrystallization temperature. **Figure 5** shows the sheet thickness, the operable annealing temperature range, and the actual operation temperature range obtained with the techniques(16,17). The temperature for passing steel strip can be reduced as the material becomes thinner. Material with a thickness of 0.15 mm is the most difficult to pass.

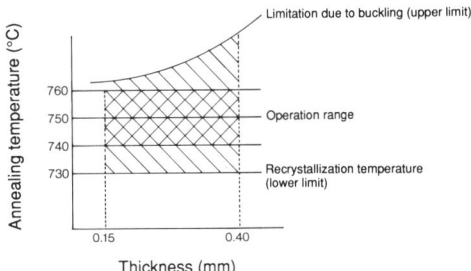

Fig. 5　Optimum conditions of annealing temperature range at No.4 CAL for manufacturing temper designation T1 products using extra-low-carbon steel

However, improved technology has made it possible to effectively pass this strip if the annealing temperature is 760°C or lower. Therefore, 0.009% was set as the upper limit of Nb addition, which maintains the recrystallization temperature at 760°C or lower, and 0.003% was set as the lower limit, stabilizing Δr within ± 0.2.

Study of Hot Rolling Temperature to Obtain Homogeneous Distribution of the Mechanical Properties

TMBP strip having excellent deep-drawability (high r-values) for use in cans was formerly manufactured at the Ar_3 transformation temperature or higher. This is because if steel is processed in the ferrite (α) phase range below Ar_3 point, a marked {100} <011> texture is formed at the center of the thickness of the hot-rolled strip and the deep-drawability after cold-rolling and recrystallization deteriorate. However, since Ar_3 point of extra-low-C steel is a relatively high temperature, it is difficult to maintain homogeneous distribution of the mechanical properties over the entire width and length of hot-rolled coil for thin TMBP. It is also difficult to maintain high temperatures for CT. Therefore, low temperatures are preferred for both FDT and CT, and hot-rolling temperatures adequate for TMBP were studied. **Figure 6** shows the effect of FDT and CT on the r-value of TMBP made from extra-low-C steel. The $\bar{r}$-value was obtained from the following equation: $\bar{r} = (r_L + r_T + 2r_D)/4$

Samples having the basic composition of 0.0028% C - 0.18% Mn - 0.04% Al - 0.005% Nb were produced at the shop. Hot rolling at the shop was carried out under the following three conditions: SRT, at low temperatures of 1180 to 1070°C, FDT of 830, 800, and 770°C, and CT of 650, 600, and 550°C. After cold rolling (reduction rate: 88%), CAL was carried out at a soaking temperature of 750°C. The $\bar{r}$-value improves significantly as both FDT and CT increase. The value of Δr approaches zero (non-earing properties) as FDT becomes higher, and CT also improves as the temperature increases. Since both 600°C and 650°C are equally favorable, it was judged that a stable temperature was easier to maintain with a medium temperature of 600°C. Although its r-value is lower than cold-rolled steel for deep drawing, for which FDT is carried out at Ar_3 point or higher, it is better than batch annealed material ($\bar{r}$-value: 1.1 - 1.6) manufactured using the conventional method. In addition, as the non-earing properties was obtained for Δr, TMBP satisfactory for the intended application could be manufactured. Generally this is because facilitating rapid grain growth during the recrystallization annealing process due to the presence of coarse precipitates and the SRT temperature can be kept low, thus creating advantages as material. Therefore, low temperatures 830°C or thereabout for FDT and medium temperatures 600°C or thereabout for CT are most

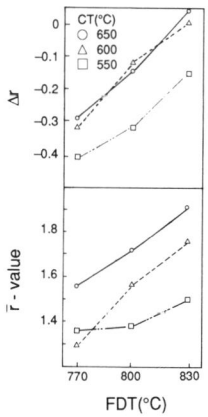

Fig. 6   Effect of hot rolling conditions, CT and FDT, on r̄-values and their planar anisotropy Δr on commercially produced Nb-added extra-low-C tinplate

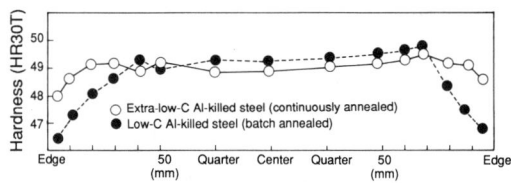

Fig. 7   Distributions of hardness in transverse direction of two kinds of tinplates

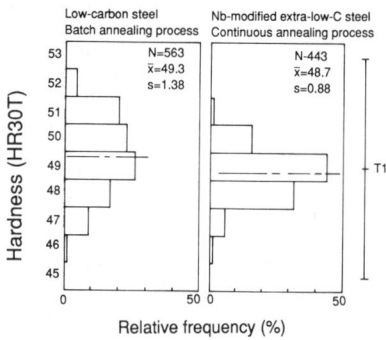

Fig. 8   Effect of steel type and annealing process on hardness variation of T1 temper tinplate

appropriate as the hot-rolling temperatures for TMBP for two-piece cans, which require both deep-drawability and non-earing properties.

Figure 7 shows the distribution of tinplate hardness in the transverse direction when the material is manufactured under the optimum conditions mentioned above, in comparison with products of the batch annealing method. **Figure 8** shows the variation of hardness of tinplate coil throughout the entire width and length, in comparison with the batch annealing method. Continuously annealed material has a narrower distribution and few variations, which shows the effects of the low FDT and medium CT.

Study of Heat Cycle of CAL for Passing Ultra-thin Gauge Steel Strip at Stable High-Speeds

In manufacturing TMBP for cans while meeting requirements for temper and ever thinner sheet thicknesses, it was necessary to increase the speed at which the cold-rolled steel strip pass through in the CAL furnace in order to maintain profitability. Since extra-low-C steel with a small Nb addition, which can be annealed at medium temperatures, was established as the material, a study was made of the optimum heat cycle. Material of 0.0016% C - 0.015% Si - 0.16% Mn - 0.008% P - 0.011% S - 0.052% Al - 0.006% Nb - 0.0022% N was hot rolled at 840°C FDT and 580°C CT before cold-rolling to achieve a reduction rate of 88% and a thickness of 0.32 mm. After heat treatment was applied to the cold-rolled sheets were annealed by direct resistance-heating CAL simulator and samples having different annealing temperatures and times were prepared, hardness was measured. **Figure 9** shows the results in equal hardness curves. When the annealing time is shorter than ten seconds, the effect of annealing time on hardness is large, but its effect is small when the annealing time is longer than ten seconds. The temperature at which material thickness of 0.15 mm can be stably passed through in the furnace at stable high-speeds is a medium temperature of lower than 760°C. If an annealing time of ten seconds is maintained at this temperature, recrystallization is thoroughly completed and T1 TMBP can be effectively manufactured using the CAL method.

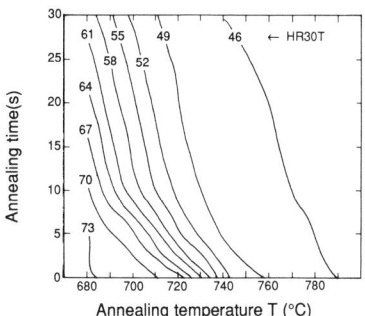

Fig. 9   Effect of annealing temperature and annealing time on hardness of an extra-low-carbon Nb-added sheet steel

Method of Manufacturing TMBP of All Temper Designations by No. 4 CAL

A T1 manufacturing technology was developed using extra-low-C Al-killed steel with a small addition of Nb (Nb: 0.003 - 0.009%, C :0.003% or less). FDT and CT are set at low temperatures (aimed value 830°C) and medium temperatures (aimed value 600°C), respectively. CAL is controlled in a heat cycle with a medium temperature immediately above the recrystallization temperature (750°C) and a short soaking time (ten seconds).

Chiba Works No. 4 CAL, shown in **Figure 10,** was put into operation in

March 1990 and has performed smoothly since startup. Because a multi-purpose cold-rolling mill is installed immediately following the annealing furnace, for the first time in the world it is now possible to produce both single cold-reduced products T1-T6 and double cold-reduced products DR8-DR10 on a single line as shown in **Figure 11**. Using extra-low-C Al-killed steel with a small Nb addition, T1-T3 are produced using the modified hardening method, while T5-DR8 are obtained by increasing the level of modification. In addition, T3-DR10 can also by produced using N-adjusted low-C steel and an appropriate combination of heat cycles and hardening practice. As shown in **Figure 12**, a technology for manufacturing TMBP of all temper designations using only No. 4 CAL has thus been established.

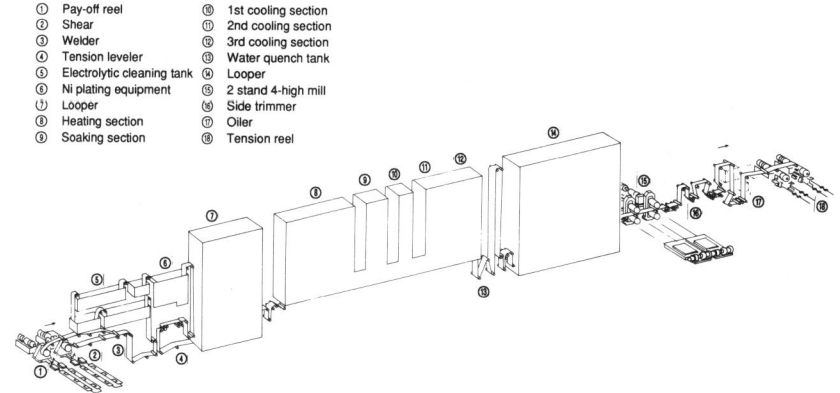

Fig. 10 Layout of No.4 CAL at Chiba Works

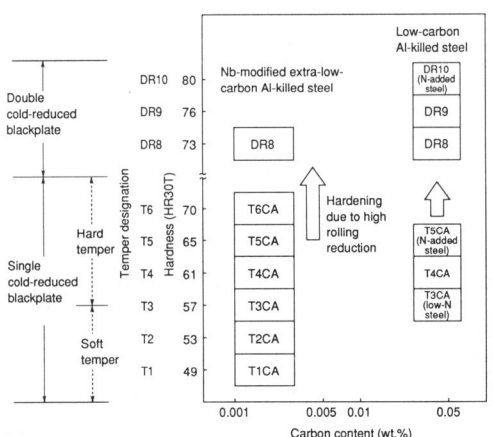

Fig. 11 Manufacturing process for tin mill blackplate of all temper designations by No.4 CAL

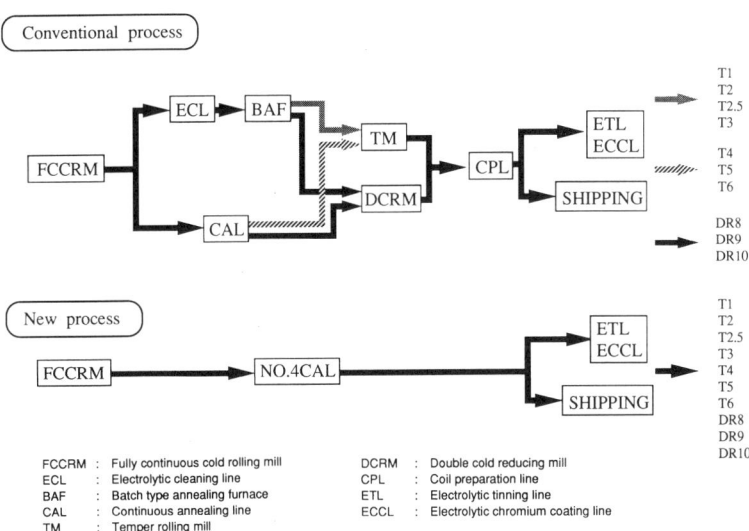

Fig. 12 Comparison of conventional process and new No.4 CAL process for producing tinplate

Features of Equipment of No. 4 CAL

Table IV shows the major specifications of No. 4 CAL. Although there was some likelihood heat buckling would occur because the recrystallization temperature of extra-low-C Al-killed steel containing Nb is about 100°C higher than that of low-C steel material and its yield strength is low, as a result of a study of techniques to prevent buckling, ultra-thin-gauge blackplate 0.15 mm thick can be passed through in the furnace effectively. Techniques for passing the material at stable high-speeds were also developed, and a speed of 1000 m/min has been achieved. This speed is about 30% higher than that of the world's fastest conventional CAL. A multipurpose cold-rolling mill was installed directly adjacent to the exit side of the furnace so that the reduction rate can be controlled over a wide range. Operations are thoroughly automated, and an automatic transfer system controls all aspects of transportation from the receiving of cold-rolled coils to delivery of the annealed coils to next process.
This automatic system is outlined below.

Table IV Specifications of No.4 CAL at Chiba Works

Annual production		(t/year)	560000
Operating speed	Entry section	(m/min)	1200
	Furnace section	(m/min)	1000
	Exit section	(m/min)	1400
Strip	Thickness	(mm)	0.15 ~ 0.40
	Width	(mm)	600 ~ 1067
Coil	Weight	(t)	Max. 22.0
	Diameter	(mm)	Max. 2134
Steel type			Tin mill blackplate
Inline mill type			2-stand 4-high multipurpose cold-rolling mill

(1) Entry side equipment
   (a) The coil end is automatically passed through in the furnace.
   (b) Cleaning equipment adequate for high-speed passing, provided with a support roll to prevent catenary, was adopted.
   (c) A tension leveler is used to correct the flatness of cold-rolled strip in order to prevent snaking in the furnace.
   (d) Ni plating equipment is available for reforming of base steel.

(2) Furnace
   (a) Snaking was feared as the crown of the helper roll in the heating zone become concave due to the decrease in temperature when cold-rolled steel sheets are passed. To prevent this, a thermal crown control device uses a gas jet to cool the roll barrel edges, which are not in contact with the strip.
   (b) Buckling was also possible because the crown of the helper roll in the cooling zone become convex due to the increases in temperature when hot steel strip is passed. An electric thermal crown control device was adopted to heat the roll barrel edges.
   (c) A helper roll with a roll shape adequate to prevent buckling was developed.

(3) Multipurpose cold-rolling mill
   (a) Practical use is made of a multipurpose cold-rolling mill capable of controlling reduction rates over a wide range by three methods: dry type temper rolling, wet type temper rolling, and DR rolling.
   (b) To finish TMBP with excellent flatness, the chock portion of the work roll is cooled to reduce the heat crowns in the dry type temper rolling method.
   (c) A zone coolant device was installed at No. 2 stand on the delivery side in order to reduce the heat crowns in the DR rolling method.
   (d) A variable crown back up rolls was adopted. (Taper piston rolls)
   (e) Automatic work roll changing equipment was adopted

(4) Delivery side finishing equipment
   (a) Various types of automatic equipment to finish TMBP and a measuring device for quality assurance were adopted.
   (b) A device to automatically insert paper to indicate weld point.
   (c) Sleeves for the inner diameter of product coils are automatically stored, positioned and device to automatically insert sleeves.
   (d) The coil tail end is automatically stopped with tape. (Coil tail end stopper)

(5) Stable high-speed operation techniques
   (a) A low inertia, high response tension device was installed in the entry side of the furnace to prevent tensile fluctuation generated during the speed changes at the entry side.
   (b) A newly developed vector inverter device with an FF (Feed Forward) function precisely controls furnace tension fluctuations generated by changes in line speed.
   (c) Helper rolls with a set roll surface roughness have been adopted to prevent slippage caused by the ambient gas fluid film between the roll and steel strip in the furnace.

(6) Automatic coil transportation system
   Cold-rolled steel coils are conveyed to No. 4 CAL by an automatic transportation system consisting of shuttle cars and an automatic storage yard. Coils that have been treated are sorted into those to be sent to the plating line by the same transportation system and those to be packed for export as TMBP.

Conclusions

As a result of a study of the Nb addition, hot-rolled temperature, and the heat cycle of CAL using the features of extra-low-C steel, a technology was developed to economically manufacture T1, which had been difficult to using the CAL method. No. 4 CAL, which can pass cold rolled steel strip at the fastest speed in the world, has been put in operation at Chiba Works. This new CAL produces T1, and because a multipurpose-cold rolling mill is directly connected to the exit side of the furnace, commercial production of TMBP of all temper designations is possible by the modified hardening method.

As a particularly important feature of Chiba Works No.4 CAL, the manufacturing conditions for the softest T1 TMBP are outlined below.

(1) Annealing at medium temperatures is possible using extra-low-C steel containing 0.003 - 0.009% of added Nb, greatly contributing to improved passing.

(2) Because the hot-rolling FDT is set at a low temperature of 830°C or thereabout and the hot-rolled CT is set at a medium temperature of 600°C or thereabout, a uniform temperature can be stably maintained throughout the entire width and length of the hot-rolled steel strip. Therefore, TMBP having homogeneous distribution of the mechanical properties can be obtained.

(3) In the heat cycle of CAL, annealing is conducted at a medium temperature of 750°C; the soaking time necessary for sufficient grain growth is short at 10 seconds or below. Therefore, TMBP 0.15 mm thick can be effectively passed through in the furnace at a stable high-speed of 1000 m/min.

References

1) H. Kuguminato, Y. Izumiyama, H. Sunami, F. Yanagishima, Y. Nakazato, and T. Obara: Kawasaki Steel Technical Report, 14 (1982) 4, 466.

2) K. Mochizuki, A. Yasuda, T. Ichida, H. Kuguminato, Y. Izumiyama, and T. Ukena: Tetsu-to-Hagané, 68 (1982) 11, S1176.

3) K. Sakata, M. Nagano, M. Nishida, and H. Kuguminato: Tetsu-to-Hagané, 68 (1982) 11, S1177.

4) H. Kuguminato, Y. Izumiyama, T. Ono, F. Yanagishima, N. Ohta, and T. Obara: Tetsu-to-Hagané, 68 (1982) 11, S1178.

5) F. Yanagishima, Y. Nakazato, Y. Shimoyama, H. Sunami, Y. Ida, T. Haga, and T. Irie: "Development of a Multipurpose Continous Annealing Line for Cold Rolled Sheet Steels," 1982 AISE Spring Rolling Mill Conference, Baltimore (USA) 36.

6) J. Takasaki, T. Irie, T. Haga, F. Yanagishima, and K. Komamura: Tetsu-to-Hagané, 68 (1982) 9, 150.

7) N. Sumida, T. Fujii, Y. Oguchi, H. Morishita, K. Yoshimura, and F. Sudo: Kawasaki Steel Technical Report, 8 (1983), 69.

8) Kawasaki Steel Corporation: Toku-Ko-Hei 1-52450.

9) Kawasaki Steel Corporation: Toku-Ko-Hei 1-52451 (Reg. No. 1569910).

10) Kawasaki Steel Corporation: Toku-Ko-Hei 1-52452 (Reg. No. 1569990).

11) T. Obara, K. Sakata, M. Nishida, H. Kuguminato, T. Akiyama, and N. Ohta: Tetsu-to-Hagané, 69 (1983) 5, S409.

12) H. Kuguminato, T. Akiyama, T. Ukena, and N. Ohta: Tetsu-to-Hagané, 69 (1983) 5, S410.

13) T. Obara, K. Sakata, K. Osawa, M. Nishida, and T. Irie: Proceedings of the Symposium on "Technology of Continuously Annealed Cold-Rolled Sheet Steel", ed. R. Pradhan, TMS-AIME, (1984), 363.

14) Y. Shimoyama, T. Yasumi, M. Yoshida, H. Ohno, T. Nakamura, S. Ikeda, Y. Mashino, and M. Sakuta: Material and Process, 4 (1991) 2, 597.

15) S. Ideda, Y. Ichi, T. Chino, T. Nakamura, Y. Shimoyama, and H. Ohno: Material and Process, 4 (1991) 2, S598.

16) H. Kawahara, H. Ohno, H. Ogawa, M. Ehara, Y. Nakajima, and T. Hira: Material and Process, 4 (1991) 2, S599.

17) Y. Shimoyama, T. Yasumi, H. Ohno, T. Ohnishi, T. Nakamura, and T. Chino: Kawasaki Steel Technical Report, 23 (1992) 4,

18) S. Satoh, T. Obara, J. Takasaki, A. Yasuda, and M. Nishida: Kawasaki Steel Technical, Report, 16 (1984) 4, 273.

19) T. Nishizawa: Iron and Steel, 70 (1984) 15, 1984.

HARDNESS AND RECRYSTALLIZATION BEHAVIOR OF CONTINUOUSLY ANNEALED SOFT-TEMPER BLACKPLATES CONTAINING ZIRCONIUM OR NIOBIUM

J.H. Kwak and S.K. Chang

Research Institute of Industrial Science and Technology,
699 Kumho-Dong, Dongkwangyang, 544-090, Korea

Abstract

This study outlines how precipitates, grain size and solute atoms affect the hardness of soft-temper blackplates made of an extra-low carbon steel containing Zr or Nb. As a result, it is difficult to produce T1(HR30T; 49±3) by continuous annealing of the extra-low carbon steel without strong carbonitride formers such as Zr or Nb. The decrease of hardness in the extra-low carbon steel containing 0.017% Zr is considered mainly due to a less amount of interstitial solute atoms as well as a higher grain growth rate than those of the extra-low carbon steel and that containing 0.028% Nb. Although Zr content was not enough to scavenge solute atoms in the extra-low carbon steel containing 0.017% Zr(atomic Zr/C; 0.86), aging index could be lowered with increasing coiling temperature.

The recrystallization temperature of the extra-low carbon steel containing Zr is lower than that of the Nb containing steel, since Zr-precipitates are not only coarser but also less densely dispersed than Nb-ones. Therefore, the extra-low carbon steel containing Zr is preferable in manufacturing the soft-temper blackplates of T1 and T2 by continuous annealing.

Introduction

In order to manufacture blackplates of different tempers, it used to customary adopted the batch annealing process for soft-temper blackplates equal to or milder than T3 and the continuous annealing process for hard-temper ones equal to or harder than T3. Since 'continuing' a manufacturing process is a general trend in the modern steel making industry, it is desirable to produce soft-temper blackplates by the continuous annealing process. However, continuous annealing facilities have not been competitive with batch annealing in producing the soft-temper blackplates, although highly ductile sheet steels made by continuous annealing have been extensively used for automotive panels.

In the metallurgical point of view, hardness of blackplates depends on grain size, solute atoms, dispersed precipitates or clusters, and dislocation density (1). Since the heat cycle of continuous annealing, however, is composed of short time annealing and rapid cooling, the annealing temperature should be high enough to produce fully grown grains and the solute atoms should be reduced by overaging or addition of scavenging elements.

Recently, from many investigations on overaging, Nb-interstitial(IF) steel(2,3) and extra-low carbon steel(4), hardness has been reported to be decreased by reducing solute atoms. But niobium has two different effects on the hardness of Nb-IF steel. The one is hardness decrement by reducing the number of interstitial solute atoms(5). The other, on the contrary, is hardness increment by grain refining because dispersed fine Nb- precipitates inhibit grain growth during continuous annealing(6,7). In other words, coarsening the precipitates (carbonitrides) is remarkably effective in accelerating the grain growth.

This paper describes the effects of processing parameters such as hot-mill coiling temperature, annealing temperature, and reflowing on the hardness of extra-low carbon steels containing Nb or Zr to produce the soft-temper (T1,T2) blackplates by continuous annealing. Not only the effects of distribution and size of Nb, Zr precipitates on the recrytallization-grain growth kinetics, but also the effects of solute atoms and precipitates on the hardness of these steels are discussed.

Experimental procedure

The steels with a basic composition of 0.003 wt.%C were vacuum melted. The chemical compositions of the steels investigated are given in Table 1. Then, the specimens were hot-rolled to the final thickness of 2.0mm in a laboratory simulator. Soaking and finishing temperatures were kept constant at 1250℃ for an hour and 900℃, respectively. Coiling temperatures were varied between 550 and 700℃ at intervals of 50℃. The hot-rolled specimens were cold-reduced to 0.3mm in thickness.

The continuous annealing was simulated in an infrared furnace followed by temper-rolling of annealed sheets by the elongation of one percent

Table 1. Chemical compositions of steels used

Steel	Chemical Composition (wt.%)							Atomic Ratio	
	C	Si	Mn	P	S	Sol.Al	N	Others	X/C	X / C,N
C	0.0023	Tr	0.19	0.01	0.0014	0.047	0.0029	-	-	-
Nb	0.0029	"	0.19	0.012	0.0017	0.040	0.0014	Nb : 0.028	1.24	0.88
Zr	0.0026	"	0.19	0.012	0.0014	0.062	0.0036	Zr : 0.017	0.86	0.39

for reflowing treatment. The sheets were heat-treated corresponding to reflowing (250℃ X 5sec) in tin-plating process. The hardness was measured by the Rockwell superficial harness tester (HR30T). Aging indices(Specimen; ASTM Subsize) were measured by the difference between the flow stress at 7.5% prestrain and yield stress after subsequent heating at 100℃ for 30min. Electrical resistivity of the steels was measured by the '4-probe' method at 77K with the applied current of 250mV, and the isothermal aging temperature was 150℃. The cold-rolled specimens for the electrical resistivity test were prepared in the dimension of 5mm X 100mm and then annealed at 720℃ (steel C, steel Zr) and 760℃(steel Nb) followed by water quenching to room temperature.

Results

Effect of precipitates on recrystallization

Figure 1 shows transmission electron micrographs of the hot bands of Zr and Nb bearing steels. In the hot band of steel Nb coiled at 700℃, Nb precipitates were observed densely with a mean size of 0.01 to 0.02μm diameter. Most of precipitates were identified as Nb(C,N) particles by EDS analysis. In the hot band of steel Zr coiled at 700℃, the size of precipitates was 0.05 - 0.1μm with an average of about 0.08μm. Although zirconium sulfide was found in rare cases, Zr(C,N) precipitates were mostly observable. The density of Zr precipitates was considerably lower than that of Nb ones. Transmission electron micrographs in Figure 2 show the size and distribution of two different precipitates at same magnification. The size of Nb precipitates was considerably small compared those of Zr ones.

It is considered that the recrystallization behavior of each steel is quite different due to the difference in size and distribution(5) between Zr and Nb

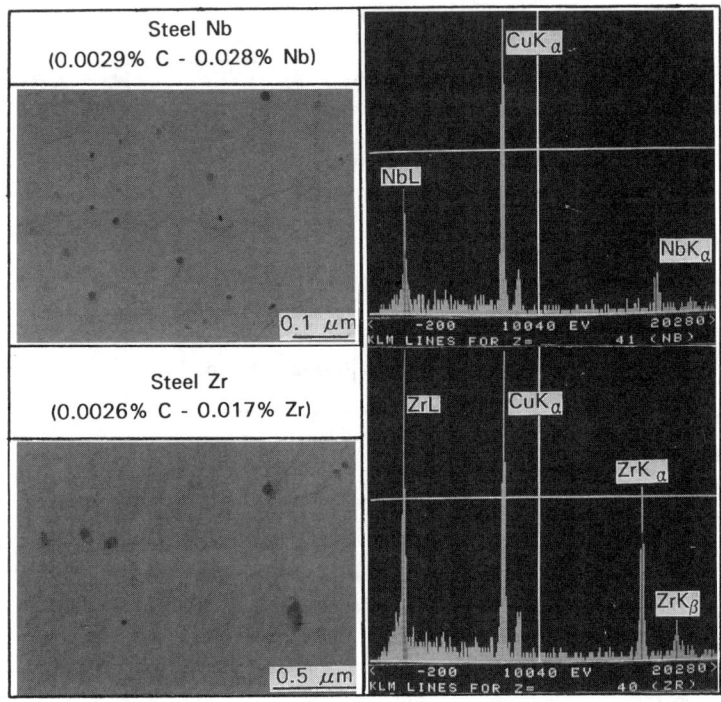

Figure 1. Transmission electron micrographs showing precipitates in hot bands of steel Nb,Zr and EDS analysis of each precipitates.

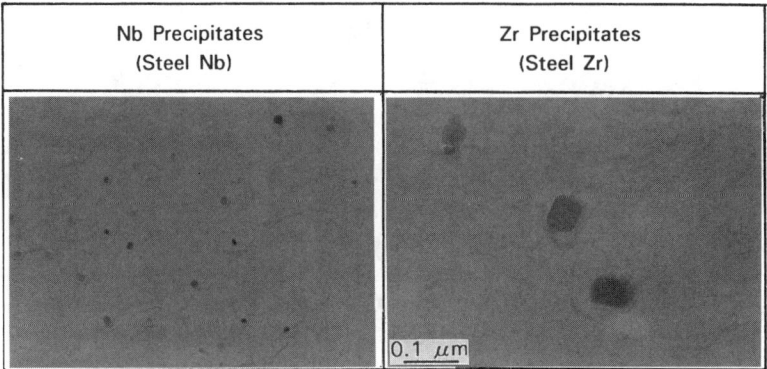

Figure 2. Comparison the size and distribution of two different precipitates at same magnification.

precipitates as mentioned above. Figure 3 shows the recrystallization behavior with annealing temperature. In the steel C, recrystallization finished at lower temperature than in the steel Zr or the steel Nb. Since fine and dense Nb precipitates inhibit recrystallization and the grain growth, the recrystallization temperature in the steel Nb appeared to be highest. But the recrystallization temperature of the steel Zr with coarse and less dense precipitates was a little higher than that of steel C.

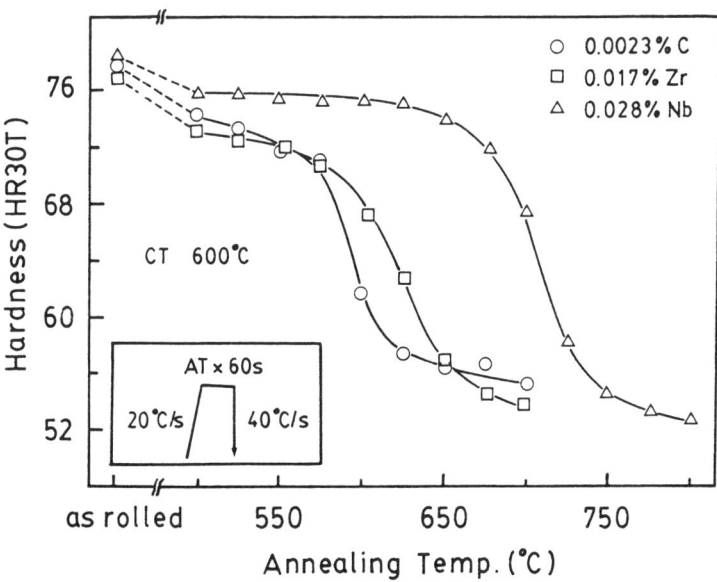

Figure 3. Effect of annealing temperature on the hardness as a measure of the recrystallization behavior.

Effect of coiling temperature

Figure 4 shows the hardness changes of the annealed specimens as a function of the coiling temperatures. The coiling of hot strips was simulated in the furnace in which hot bands were kept constantly at given temperatures for an hour and then cooled in the furnace. The higher coiling temperature lowered hardness of both the annealed steels Nb and Zr, but the hardness dependence on the coiling temperature was negligible in the steel C. This phenomenon can be explained as follows; the grain size of hot bands increases with raising the coiling temperature. Especially, a higher temperature makes Ostwald ripening of fine precipitates easier(8,9). In other words, precipitates in hot bands become coarser with increasing coiling tmeperature and, as a result, the inhibition effect of the grain growth decreases during annealing. In this point of view, it seems to be

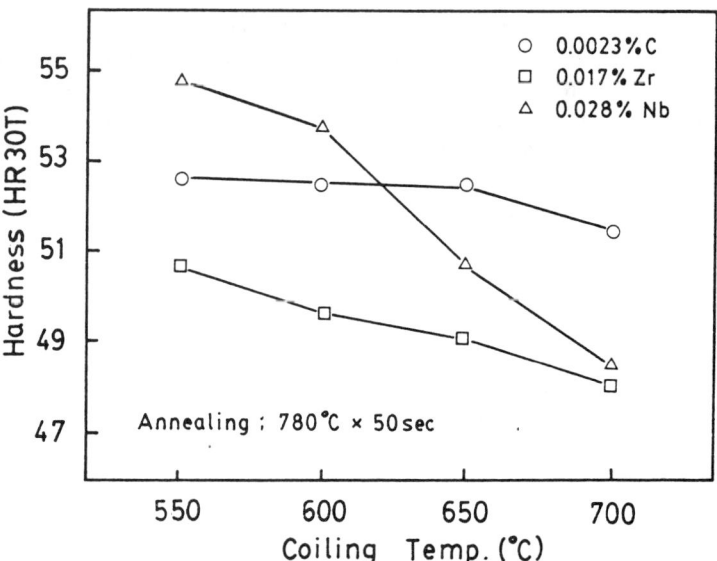

Figure 4. Effect of coiling temperature on the hardness of annealed steel sheets.

natural that the hardness dependence of the steel Nb on the coiling temperature was more remarkable than that of the steel Zr.

Aging indices of the steel Zr and steel C decreased with increasing coiling temperature as shown in Figure 5, and this tendency is remarkable in the steel Zr. It seems that the solute carbon and nitrogen are mostly scavenged by

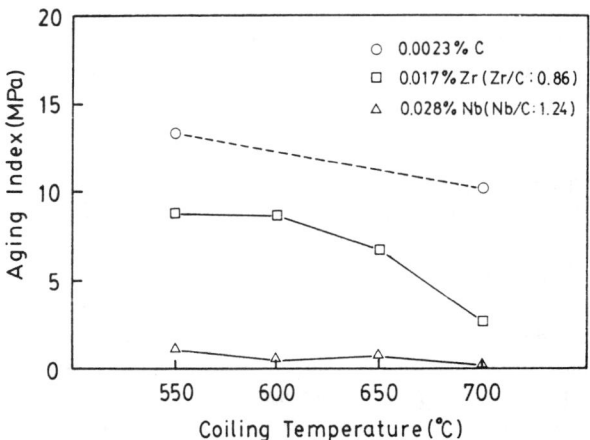

Figure 5. Effect of coiling temperature on the aging index of annealed steel sheets.

the sufficient amount of Nb in the steel Nb, but not by Zr in the Steel Zr. Although the amount of Zr was not enough to scavenge solute atoms in the steel Zr, aging index could be lowered to a great extent by increasing coiling temperature.

Interstitial solute atoms

The hardness increase in the steel Zr by insufficient scavenging is possible after reflowing in tinplating process. Figure 6 represents the hardness changes of specimens of which one was overaged at 400°C for 2minutes and the other was

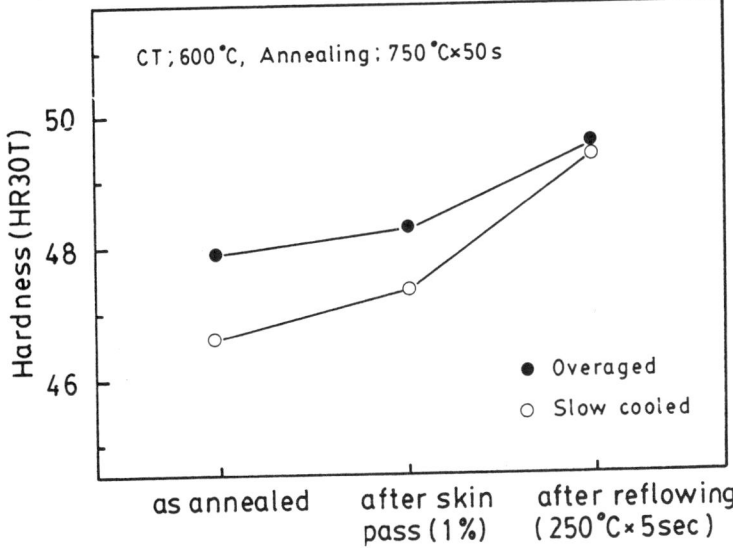

Figure 6. Hardness changes after reflowing between two different heat-treated specimens of steel Zr.

slowly cooled by 3°C/s. Comparing to the hardness of the skin passed specimen, the increment of hardness after reflowing was less than 2 in HR30T. Since the hardness allowance of a temper-grade in tinplates is ±3 in HR30T, the slight increment of hardness by strain aging in the steel Zr may not be problem after tinplating.

Figure 7 shows the resistivity changes during aging at 150°C. Resistivity change($\Delta \rho$) at 77K was obtained as follows;
$$\Delta \rho = \rho_t - \rho_0 \quad - [1]$$

where ρ_0 is the resistivity of annealed specimen followed by water quenching and ρ_t is the resistivity of aged one. Since the resistivity decrease of each steel is in inverse proportion to the number of solute atoms, the results of Figure 7 show a similar tendency to those of Figure 5. The decrease of resistivity during aging is considered mainly due to the precipitation of ε-carbides or Fe-C-N precipitates(10). In this study, thin and rod-like precipitates were observed in the aged steel C in a transmission electron microscopy.

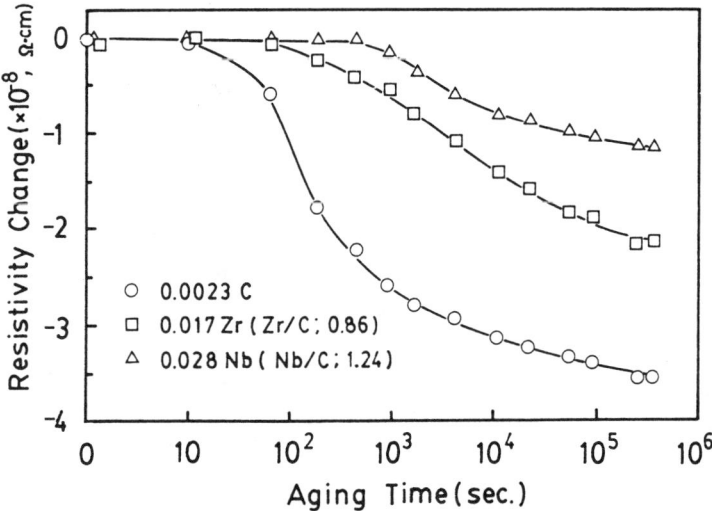

Figure 7. Electrical resistivity changes at 77K during aging at 150℃(annealing: steel C and steel Zr;720℃ for 1min, steel Nb;760℃ for 1min).

Effect of annealing temperature

Figure 8 shows hardness changes with annealing temperatures. The hardness of specimens was lessened by coarsening of grain as the annealing temperature increases. But the decrease of hardness in the steel C was not remarkable compared with those in the steel Zr and Nb. The grain growth was sluggish in the steel Nb compared to the steel Zr, as shown in Figure 9, so that the deformed structure still remained in the steel Nb at the annealing temperature of 700℃.

Effect of precipitates on grain growth

Supposing that driving force for grain growth is equivalent to dragging force by particles for grain boundary migration in normal grain growth, the average grain radius ($\bar{R}$) can be expressed by Zener's equation (8,11) as follows ;
$$\bar{R} = \beta \cdot \bar{r} / f \quad - [2]$$
where $\bar{r}$ is the average radius of precipitates, f is the volume fraction of precipitates and β is the constant. Assuming that the normal grain growth is

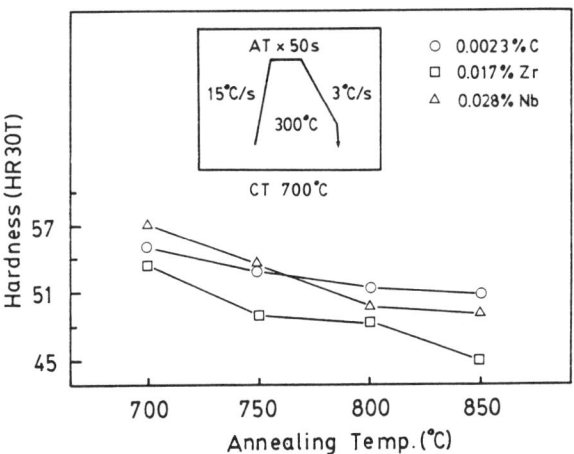

Figure 8. Effect of annealing temperature on the hardness.

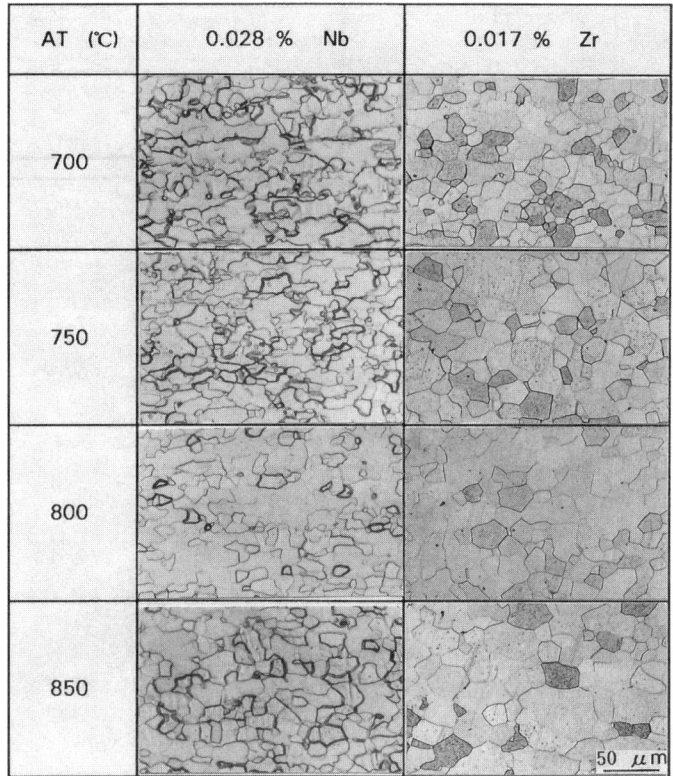

Figure 9. Optical microstructures showing the effect of annealing temperature in steel Nb,Zr (coiled at 700°C).

dependent on Ostwald ripening of precipitates, the relationship between average radius of precipitates ($\bar{r}$) and isothermal soaking time (t) can be expressed as follows(8,11) ;

$$\bar{r}^3 - \bar{r}_0^3 = K \cdot t \quad - [3]$$

where K is constant and $\bar{r}_0$ is the initial particle radius. Replacing $\bar{r}$ in equation[3] with equation[2], equation[3] becomes

$$\bar{R}^3 - \bar{R}_0^3 = K' \cdot t - [4]$$

where K', grain growth rate constant, is $(\beta/f)^3$.

Figure 10 shows a relation between $\bar{R}^3$ and soaking time. The slope (K') of the steel Nb with fine precipitates was considerably smaller than that of steels Zr and C.

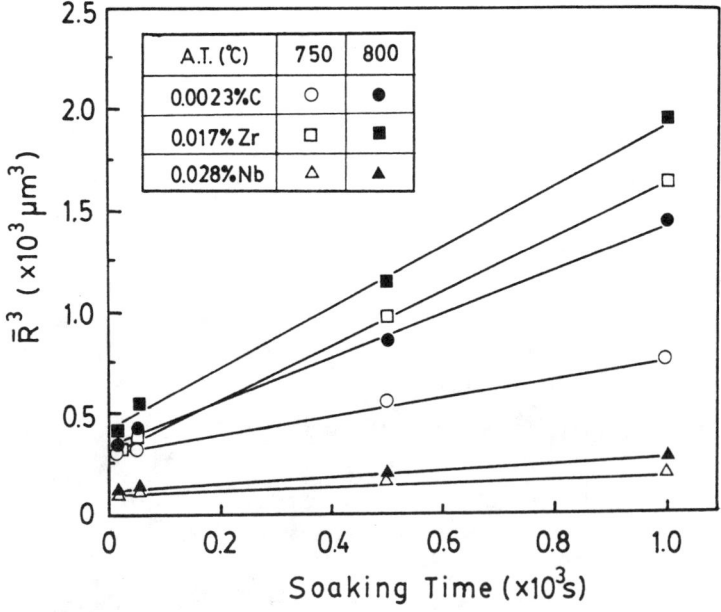

Figure 10. Relation between the cube of average grain radius(R^3) and soaking time.

K' was calculated from equation[4] and plotted in Figure 11. The grain growth rate constant (K') was considerably lower in the steel Nb than in steels Zr and C. K' of steel Zr is higher than that of the steel C up to the soaking temperature of 800°C. From the above results, the grain growth rate of the steel Zr tends to be faster than that of the steel C.

Consequently, the grain growth of steel sheets greatly depends not only on soaking temperature and time but also on size and distribution of precipitates.

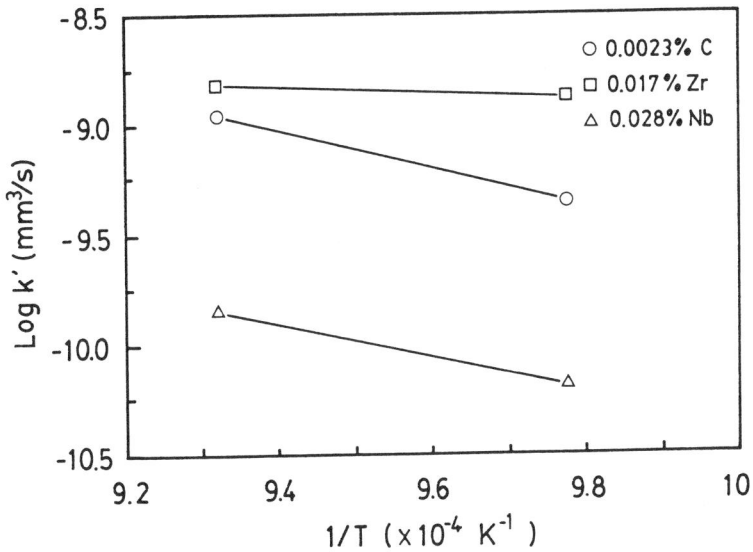

Figure 11. Relation between the grain growth rate constant(K') and the inverse of absolute temperature.

Discussion

The relationship between hardness and ASTM grain size number in the steels presently used as well as some commercial steels was summarized in Figure 12. The hardness of blackplates produced by batch annealing becomes lower as the grain size is steadily increased(3). But the hardness of continuously annealed ones greatly depends on the grain size as well as the amount of solute atoms(2).

Although the grain size of the 0.02%C steel was the same as that of the 0.03-0.075%C steels, the hardness of the 0.02%C steel was higher than that of the 0.03-0.075% steels. The difference in the hardness comes from the amount of solute carbon remained after continuous annealing(2,3).

The hardness of the steel Zr was lower compared with the steel C, and the ASTM grain size number of this steel was shifted to the left side compared with the steel C. In other words, the grain size of steel Zr tends to be bigger than that of the steel C in the same annealing conditions. Thus, it is considered that the decrease of hardness in the steel Zr is mainly due to the less amount of interstitial solute atoms(Fig. 7) as well as the higher grain growth rate(Fig. 11), when compared with the steel C.

The hardness of the steel Nb with little amount of interstitial solute atoms remarkably depends on grain size compared with the other steels. The hardness difference between the steel Nb and the steel C increased with increasing the grain size. Therefore, it can be said that the hardness increase by interstitial solute atoms in continuous-annealed blackplates becomes more serious as the grain coarsens.

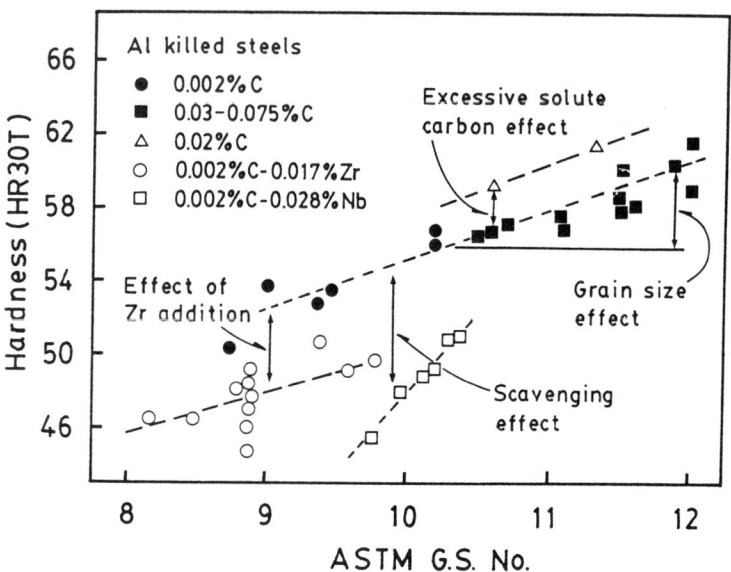

Figure 12. Effects of grain size, carbon content and addition of carbonitrides forming elements on the hardness of continuously annealed blackplates.

Consequently, reducing the number of interstitial solute atoms by adding an appropriate quantity of carbonitride formers in the extra-low carbon steel is quite effective in decreasing hardness of the continuously annealed blackplates. Since short time soaking is the characteristics of continuous annealing, fully grown grain is necessary for soft-temper blackplates manufactured by continuous annealing. Therefore, it seems to be more effective to add Zr instead of Nb in the extra-low carbon steel in order to produce the soft-temper blackplates by continuous annealing, because Zr- carbonitrides are not only coarser but also less densely dispersed than Nb ones and the binding force of Zr with solute carbon and nitrogen is stronger than that of Nb in austenite region(5,12)

Conclusions

For the purpose of producing soft-temper blackplates by continuous annealing, the role of precipitates, grain size and solute atoms on the hardness of the extra-low carbon steel containing Zr or Nb were investigated. The results obtained are as follows;

1. It is difficult to produce T1(HR30T; 49±3) by continuous annealing of the extra-low carbon steel without the strong carbonitride formers such as Zr or Nb.

2. The higher coiling temperature lowered hardness of both the annealed extra-low carbon steels containing 0.028%Nb and 0.017%Zr, but hardness dependence on coiling temperature was negligible in the extra-low plain carbon steel.

3. Although Zr content was not enough to scavenge solute atoms in the extra-low carbon steel, aging index could be lowered to a great extent with increasing coiling temperature.

4. It is considered that the decrease of hardness in the extra-low carbon steel containing 0.017%Zr is mainly due to the less amount of interstitial solute atoms as well as the higher grain growth rate, when compared with the extra-low plain carbon steel and the 0.028%Nb containing extra-low carbon steel.

5. The recrystallization temperature of the extra-low carbon steel containing Zr is lower than that of the Nb containing steel, since Zr carbonitrides is not only coaser but also less densely distributed than Nb ones. Therefore, extra-low carbon steel containing Zr is preferable in the production of the soft-temper blackplates of T1-T2 by continuous annealing.

Acknowledgements

The authors would like to thank Dr. Z.S. Lim for helpful advices on the electrical resistivity measurements. Thanks are also extended to Y.R. Cho and J.C. Kim for their useful comments and assistance.

References

1. E.Shuto, "On the Hardness of Low Carbon Steel Sheet after Continuous Annealing", Tetsu-to-Hagane, 50(1964)5 pp.766-773.
2. T.Obara, K.Sakada, K.Osawa, M.Nishida and T.Irie, "Metallurgical Basis of Producing Continuous-Annealed Low-Temper Tinplate", pp.363-383 in Technology of Continuous Annealed Cold-Rolled Sheet Steel, R.Pradhan eds., AIME, Warrendale, PA, 1984.
3. H.Kuguminato et al,"Manufacture of Soft-Temper Blackplates in Kawatetsu Multipurpose Continuous Annealing Line", Kawasaki Steel Tech. Report, No.7(1983) pp.34-43.
4. S.Aoki et al, " Production of Soft-Temper Tinplate by Continuous Annealing", CAMP-ISIJ, 1(1988) p.723.
5. I.Kokubo, M.Sudo, K.Kameno, S.Hasimoto, I.Tsukatani and T.Iwai,"On the Drawablity and Recrystallization Texture of Low Carbon Sheets Containing Carbide Forming Elements", Tetsu-to-Hagane, 59(1973)3 pp.469-492.
6. C.Brun, M.L.Gac, F.Moliexe, P.Patou and B.J.Thomas, " Metallurgy and Aging Properties of Continuous Annealed Cold-Rolled High Strength Low Carbon Steels Microalloyed with Niobium", pp.223-239 in Technology of Continuous Annealed Cold-Rolled Sheet Steel, R.Pradhan eds., AIME, Warrendale, PA, 1984.
7. H.Kubota, I.Kazasu, H.Kido and T.Shimizu, "Effect of Niobium on Cold Work-Annealing Process of Low Carbon Steels", Tetsu-to-Hagane, 54(1968)8, pp.954-966.
8. T.Nishizawa, "Grain Growth in Single- and Dual-Phase Steels", Tetsu-to-Hagane, 70(1984)15 pp.194-202.
9. H.I.Aaronson and J.K.Lee, "Diffusional Nucleation and Growth", pp.31-86 in Precipitation Process in Solids, K.C.Russell and H.I.Aaronson eds., AIME, Warrendale, PA, 1976.
10. K.Abiko, "Precipitation of Nitrogen and Effects of Alloying Elements on the Precipitation of Nitrogen in Alpha-Iron", pp.1-16 in Reports of low Carbon Sheet Steel Committee of ISIJ, ISIJ, 1987.
11. S.Satoh, Y.Yamazaki, I.Osawa and H.Abe, "Effect of Precipitates on Recrystallization Behavior of Cold-Rolled Extra-Low C Sheet Steel", CAMP-ISIJ, 3(1990) pp.1772-1775.
12. K.Narita,"Carbides in Steel(II)", Bulletin of the Japan Institute of Metals, 8(1969)1 pp.49-57.

IV. Batch Annealing: Process Technology

SURFACE IMPROVEMENTS OF STEEL STRIP ANNEALED IN

EBNER HICON/H_2 ® BATCH FURNACES

Heribert Lochner

EBNER-Industrieofenbau
Gesellschaft m.b.H.
Ruflinger Strasse 111
A-4060 Leonding/Austria

Abstract

The new EBNER HICON/H_2 ® annealer was developed in the early 1980's to meet the following goals:

- safety, high throughput, low capital and operating costs
- superior physical properties of the annealed coils
- surface cleanliness at least equal to that of CAL-annealed coils (which are degreased beforehand, though, in contrast to the HICON/H_2 ® process)

The most difficult goal was surface cleanliness. This paper therefore concentrates on this subject and surveys the reactions that affect the surface cleanliness of hydrogen-annealed strip both for carbon deposits and for partial oxidations. Their negative influence on corrosion of coated sheet is also discussed. A link is then established between effects on strip production and the annealer design required to achieve the best results.

1) **INTRODUCTION**

By the end of 1992 roughly 9 million US tons of wide cold rolled strip and roughly 1 million tons of medium width strip including high-carbon and alloy steels will be processed in EBNER HICON/H_2 annealers every year. Over 500 workbases in 60 companies will be in use worldwide. Fig 1 shows a 40,000 ton per month installation with 24 bases at NUCOR Steel in Crawfordsville/IN.

Fig. 1 HICON/H_2® bell annealer battery (40,000 tons/month), Nucor Steel, Crawfordsville, IN/USA

The HICON/H_2 process is now well known and several papers [1] [2] have described its benefits; the main one is the doubling of throughput over conventional annealers. Production experience also shows great improvements in physical properties and surface cleanliness of the annealed coils.

With these quality improvements the EBNER HICON/H_2 process provides a low-cost alternative to continuous annealing lines. The mechanical properties of HICON-annealed deep-drawing grades made of aluminium-killed continuous-cast steels are superior to those obtained in CA lines. This also has been reported in the technical literature [3] and at the TMS meeting in 1989 [2]. In the meantime the largest HICON/H_2 annealer so far has gone into production at Hoogovens in IJmuiden/Netherlands; it has a maximum charge weight of about 200 tons (Fig.2).

Fig. 2　largest HICON/H_2 ® bell annealer for 200 ton charge, Hoogovens, Ijmuiden, Netherlands

The goals of a modern hydrogen annealer must include:

- Safety
- High throughput
- Low capital and operating costs
- Superior physical properties
- Surface cleanliness at least equal to that of CAL-annealed coils (which are degreased beforehand, though, in contrast to the HICON/H_2 process).

The radical new EBNER HICON/H_2 annealer design was developed in the early 1980's with these goals in mind. Its success to date demonstrates that we have been able to meet them. The most difficult goal was surface cleanliness. This paper will therefore concentrate on this subject and survey the reactions that affect the surface cleanliness of hydrogen annealed strip. A link is then established between effects on strip production and the annealer design required to achieve the best results.

2) CHEMICAL REACTIONS DURING THE ANNEAL PROCESS

Fig 3 shows the mean temperature profile of the charge during the anneal cycle and the concentrations of various gases inside the coils. CRM Belgium [4] have investigated the various reactions involved by drilling coils, inserting gas probes and then analyzing the changing compositions of the gases during the course of the anneal. We propose quations (1) to (4) to explain their findings.

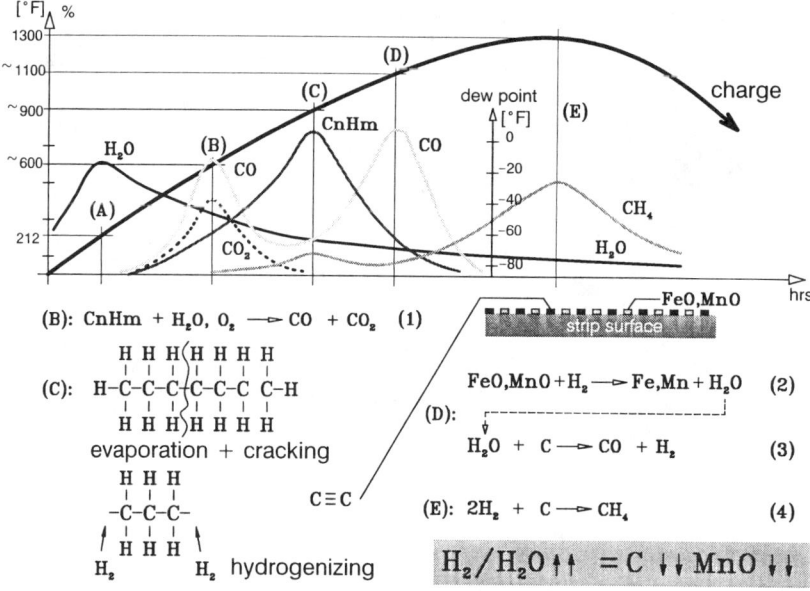

Fig 3 Chemical reactions during the anneal process

At the beginning of the anneal, the dewpoint rises as the water in the emulsion residues boils off (A) (in the metal-encased EBNER workload space it goes down to -80°F by the time cooling starts). Then CO_2 and CO evolve (B). The evaporating hydrocarbons first react with water vapor or air oxygen left between the wraps to form CO and CO_2 (Equation 1).

Then they heat up further, the long chains start to crack and radicals form. The small hydrogen molecules diffuse very quickly between the coil wraps, and the open ends of these radicals are saturated with hydrogen, yielding light hydrocarbons which boil off readily (C). Here the hydrogen helps for the first time to reduce carbon deposits. Some carbon is still left on the surface, though.

There are also some traces of metal oxides such as FeO and MnO left on the surface from prior processing. As the strip continues to heat up, the highly reducing hydrogen converts the oxides to water vapor at around 1 100°F (Equation 2). This water vapor then reacts with carbon left on the strip to form carbon monoxide (Equation 3) (D).

Towards the end of the soaking period, at around 1 300°F, hydrogen reacts directly with the residual carbon to form methane (Equation 4) (E). This reaction carries on during the first part of the cooling cycle.

3) **CARBON DEPOSITS, IRON FINES**

Both rolling and annealing practices have to be designed to achieve the lowest carbon residues on the surface of the annealed strip. Attention has to be given to the selection and maintenance of the rolling lubricant, along with the wipes on the last stand, to ensure a minimum of residual lubricants and iron fines. Realistic target values before annealing are:

carbon from emulsion residues	abt 4 to 6 mg/sq.ft.
iron fines	abt 3 to 4 mg/sq.ft.

From Equations 1 to 4 one can see that it is important to optimize the time/temperature cycle and atmosphere purge sequence to achieve the best conditions for the various reactions.

In the early stages of heating water vapor, carbon monoxide/dioxide and hydrocarbons are removed with a fairly high hydrogen flowrate, say 850 cft/h (A, B, C). Later a reduced hydrogen flowrate (say 250 cft/h) or intermittent purging with hydrogen is still necessary to avoid a build-up of carbon monoxide and methane (Equations 3, 4) (D, E), so that surface carbon can be minimized.

On top of this, an extremely low dewpoint (high ratio of hydrogen to water vapor) is important to provide enough potential for reducing FeO/MnO (Equation 2) and thus eliminating carbon deposits (Equation 3).

The direct reaction of carbon with hydrogen to methane is normally quite slow. We have noted, though, that in the case of a very low dewpoint the reaction proceeds quite rapidly. We suspect the reason is that in the highly reducing atmosphere of purge hydrogen and the absence of water vapor (high H_2/H_2O ratio) a very active metal surface that catalyzes the reaction is formed.

To sum up, with a low dewpoint (dropping to -85°F), i.e. a high H_2/H_2O ratio, both carbon deposits and partial manganese oxidation will be minimized. The amount of iron fines is decreased by sintering to the strip surface.

The values achieved with coils of extra deep-drawing quality strip (EDDQ) that have been tandem rolled but not degreased are as follows:

C	.1 to .2 mg/sq.ft. (Ford method with power wash)
	.3 to .5 mg/sq.ft. (Leco method)
MnO	increase from .2 % base level to only .3 % (measured by washing with oxalic acid)
Iron fines	.5 to .7 mg/sq.ft.

4) GRAPHITE AND CEMENTITE PRECIPITATION

In very dry hydrogen there are also no deposits of graphite or iron carbide (oriented on suitable crystal layers or grain boundaries) on the strip surface. Inokuti found this in small-scale tests in straight hydrogen with a dewpoint of -90°F [5].

With a relatively high dewpoint or relatively low hydrogen concentration (HNX gas) carbon can even be removed from the strip - quation (5):

$$C\alpha\text{-}Fe + H_2O \longrightarrow CO + H_2 \qquad (5)$$

$$2CO \longrightarrow C\ \text{graphite} + CO_2 \qquad (6)$$

$$3Fe + 2CO \longrightarrow Fe_3C + CO_2 \qquad (7)$$

In this case carbon monoxide builds up between tight wraps to a point where either graphite (quation 6) or iron carbide (quation 7) forms.

Carbon in solution in alpha iron has a different level of activity (free energy) from graphite or iron carbide on the surface; this is why carbon is first removed from the strip and then deposited on the surface again.

If the ratio of hydrogen to water vapor is high enough, the reaction in Equation (5), which starts only above 1 200°F, never gets going, so graphite and cementite do not form (Equations 6 and 7) - they would of course also have a negative effect on coating and corrosion behaviour.

5) PARTIAL OXIDATION

Along with carbon residues, the partial oxidation of the impurities and alloying elements in the steel plays a very important role when the steel is coated and in subsequent corrosion behavior.

Manganese in a concentration from .2 to 1.0 % (depending on the steel grade) has a high affinity to oxygen, and its partial oxidation is fairly well known, which is why we used it as an example earlier.

Recent research [6] suggests that the other elements with high affinity to oxygen must also be investigated if we are to understand the reactions during phosphating/metal coating and corrosion mechanisms. The general equations (8) and (9) show that the formation of metal oxides is a function of the metal's reduction potential and the of H_2/H_2O ratio.

$$Me + H_2O \longrightarrow MeO + H_2 \qquad (8)$$

$$k_{MeO} = f\ (\text{reduction potential of pure Me},\ H_2/H_2O) \qquad (9a)$$

$$k_{MeO} = f\ (\text{activity of reduction potential of Me in low concentration},\ H_2/H_2O) \qquad (9b)$$

Figure 4 shows this correlation. All the metal oxidation reactions are exothermic; therefore reduction to metal is easier at higher temperatures. The figure also reveals that a low dewpoint is required to reduce oxides or avoid oxidation. The figure is valid for the pure metals. It shows that pure Al, Ti and Si have the highest affinities to oxygen. Next come Mn, Cr and P. When the elements are present in low concentrations their activity has to be used in Equation (9b), they have less affinity to oxygen and the lines in Figure 4 shift to the left.

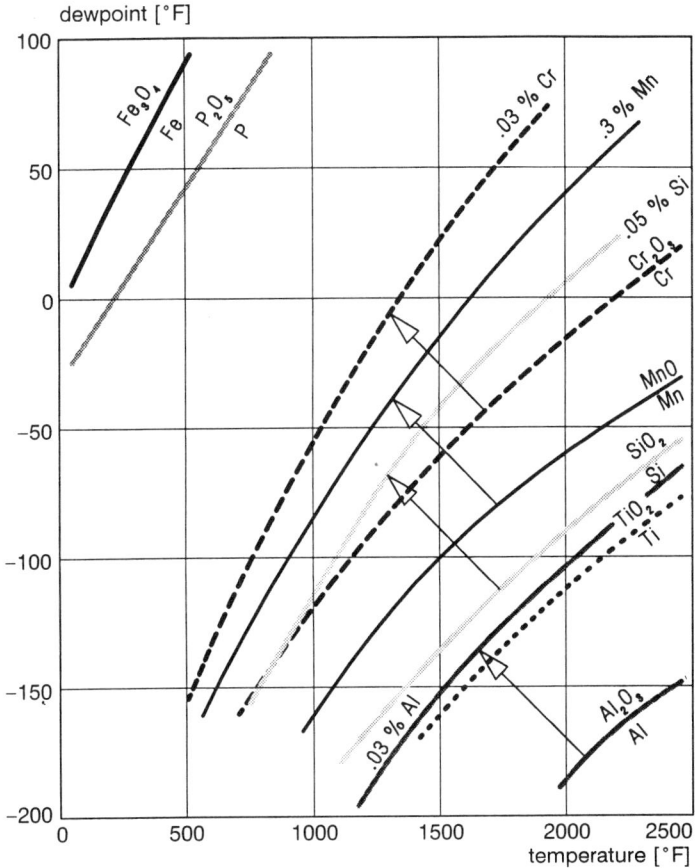

Fig. 4   Influence of dewpoint and temperature on metal oxidation in hydrogen

To sum up, it is possible to inhibit the partial oxidation of steel containing elements such as Mn, Si, Ti, and Al to a certain extent, by maintaining very low dewpoints matched to the time/temperature cycle. The distribution of the various oxides formed can be predicted. With the lowest possible dewpoint, surface detection will show some enrichment on aluminum, less on silicon and very little on manganese, chromium and phosphorus. The duration of gas action at temperature is also important. The differences in partial oxidation between the CA anneal and the HICON/H$_2$ anneal, and thus coating and corrosion behavior, can be understood better on this basis.

6) SURFACE REACTIONS ON THE ANNEALED STRIP

Long-term studies [7] [8] confirm that carbon deposits and partial oxidations have a negative effect on the corrosion resistance of phosphatized, painted sheet.

Car body panel sheets
Material HICON/H_2 annealed at Klöckner/Germany was coated with phosphate from Chemetall and the resulting automobile components were tested for corrosion resistance at Mercedes. This included 12 months' weathering exposure.

First the annealed sheets were checked for cleanliness: carbon contamination and manganese enrichment were very low. The surfaces were then examined by glow-discharge optical spectroscopy (GDOS). Figure 5 shows that the surface was much less enriched with base elements such as Mn, Si, P and C after hydrogen annealing on a metal-clad EBNER workbase than after annealing in DX gas on a conventional open workbase.

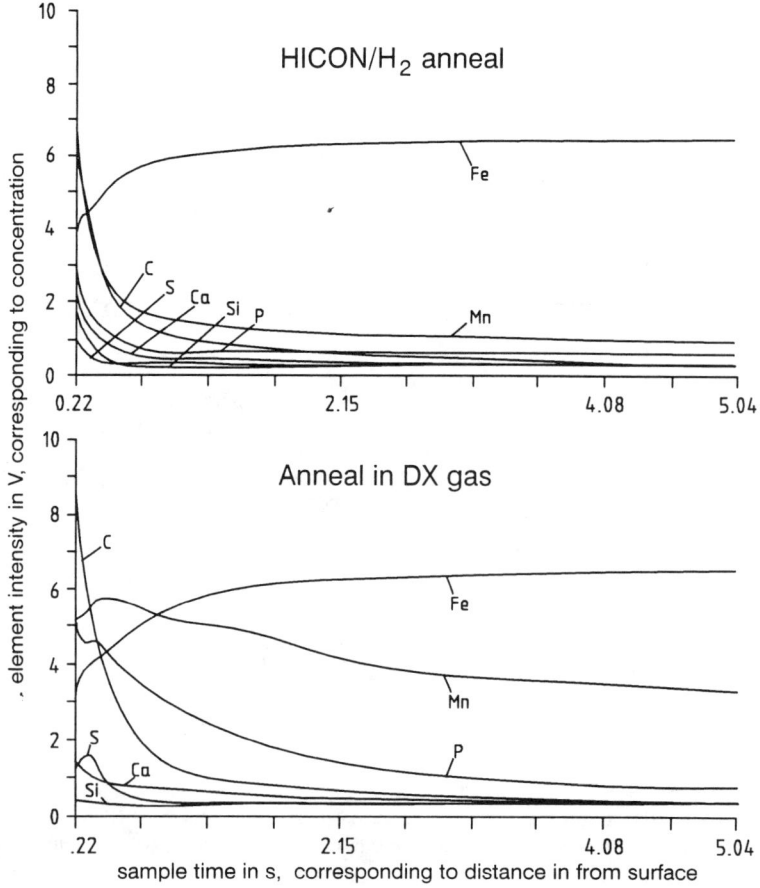

Fig. 5   GDOS plots of metal surface condition

In addition to the evaluation of the bright sheet surfaces after annealing and temper rolling, the phosphated surface was tested for pore area. The cathode current density (in A/dm²) with phosphatized specimens was examined in an electrolyte (Na_2SO_4) at -1300 mV; a comparison with the bright specimen is shown in a diagram (figure 6).

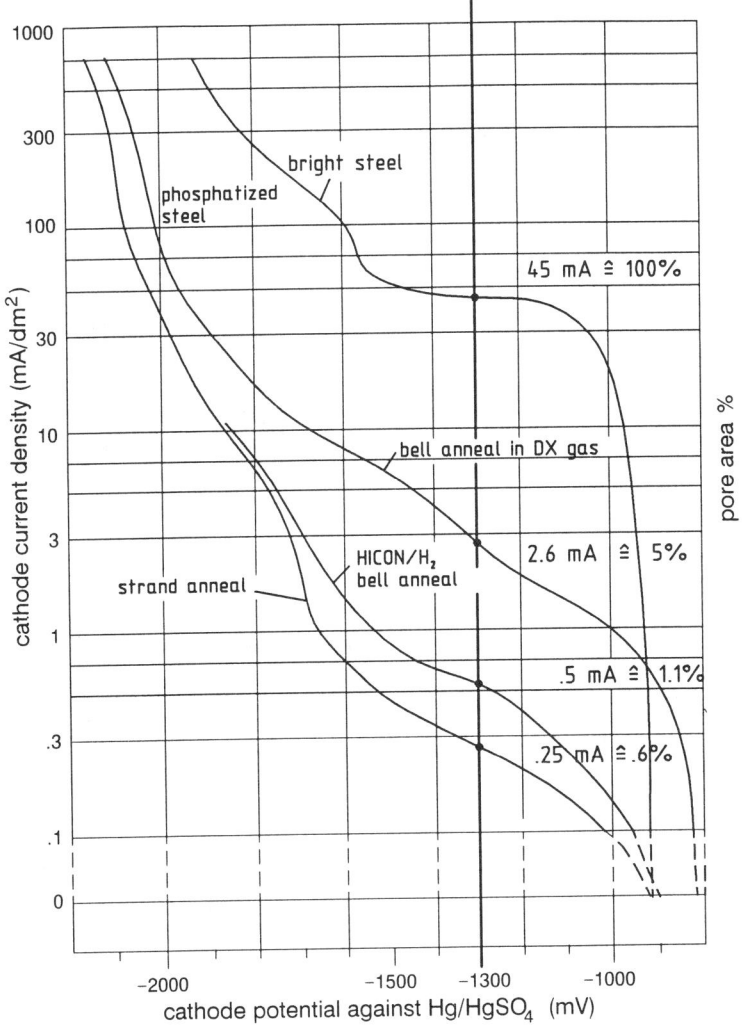

Fig. 6   Cathode current/potential plots for bright and phosphatized steel specimens in 0.1M Na_2SO_4

One can see that the pore area of the material annealed in DX gas is around 5 %, as against a fifth of this for annealing in hydrogen, where the results are very close to those for CAL material.

After these basic tests, various corrosion tests were carried out on phosphated and painted specimens. This included weathering of car parts, which provided clear evidence that a scratch on a sheet annealed in hydrogen under-rusted to around .08 in, as against more than three times as much - i.e. .3 in - on a conventionally annealed sheet (figure 7). If rusting begins at one point, as often happens when a stone hits a car, underrusting on a conventionally annealed sheet affects ten times the area.

Fig. 7   Underrusting in phosphatized, painted sheet

Domestic appliances industry [6)]
Cold-rolled strip for domestic appliances is normally given 2 minutes of low-zinc phosphatizing, and then coated with a single layer of acrylic paint. Here too, it turns out that the less surface enrichment takes place - i.e. the lower the total of all alloying element concentrations - the less the material corrodes in an ASTM salt-spray test, even though, with the same phosphatizing time, the phosphate layer was no longer quite complete.

Partial oxidations influence corrosion behavior greatly here, while surface carbon has relatively little effect.

Continuous coating with Epoxy system[6)]
In contrast to the methods of coating described above, resistance to corrosion goes down only slightly if there is considerable surface enrichment. Instead, the strip surface becomes so much more reactive during phosphatizing, which lasts only a few seconds here, that the layer of phosphates and various iron oxides formed is very thick in proportion to the layer of paint. As a result, the paint does not adhere properly, and can lift off the metal surface when the coated material is shaped. It turns out that the loss of adhesion in a bending test is directly proportional to the total of all surface enrichment (partial oxidations).

Metallic coating
The surface should also be as clean as possible for applying metallic coatings to sheet or tin plate in order to avoid blisters during (say) electrolytic zinc coating. When tin plate is plated electrolytically, the necessary step of washing with weak acid is made easier if the strip surface is really clean; in that case the surface will stay relatively smooth and the tin layer need not be as thick.

Alloyed qualities

None of the grades described so far exhibit visible oxidation; a GDOS or SIMS system is needed to reveal oxides.
In two cases the oxides are visible to the naked eye:

HSLA XK 50 with approx. 1 % Mn, .3 % Si, .05 % Nb
SAE 400 grades with 12 to 17 % Cr and some Ti or Mo

The HSLA grades are used for car bodywork reinforcements, the Cr steels for exhaust manifolds and catalyst chambers. With straight hydrogen and very low dewpoints any snaky edges or discoloration of HSLA grades can be avoided; with special anneal cycles the Cr steels can be kept bright, too (previously regarded as impossible).

7) **FURNACE DESIGN REQUIREMENTS**

All the major improvements in surface finish described can be realized only if the hydrogen atmosphere is kept extremely dry.
To achieve this we had to develop a design completely different from conventional batch annealers. Figure 8 compares the new EBNER HICON/H₂ metal-clad base with an old-style flat open base, with external cooler, also using hydrogen and equipped with a new impeller.

The workbase features a gas-tight base plate with a heat-resistant alloy steel structure above; the thermal insulation is sealed in under a bubble-tight casing. This means that no air with its oxygen and moisture can seep into the porous base insulation during charging, and then diffuse back into the anneal chamber during the anneal. This is the only way to ensure the low dewpoints required by the process. This design was created about 15 years ago and has been developed intensively since then. The structure must accommodate thermal expansion and contraction during heating and cooling without leaks appearing, and support loads of up to 200 US tons. The solution we found was to use a concave casing supported by internal columns on which the casing can slide.

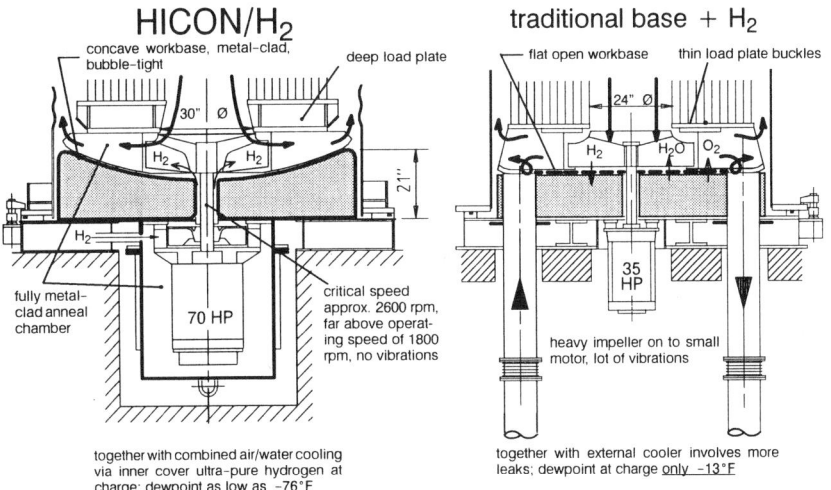

Fig. 8   Differences in design for clean atmosphere and convection

Fig. 9   Fabricating workbases at EBNER in Linz/Austria

Figure 9 shows bases in different stages of manufacturing. The large base here is 180" in diameter. In the rear we see another one being welded with a robot.

The rubber O-ring seal between the base and the inner cover is water-cooled on both sides. The fan motor is encased with a steel cover which has an O-ring seal. This makes a sliding seal on the shaft unnecessary.

All our investigations have shown that the external cooler forms a dead space which provides many sources of leakage. The connections to and from the base cannot be kept bubble-tight. Therefore we have not adopted this technique. The sequenced use of air followed by water on the inner cover gives faster cooling without compromising the integrity of the anneal chamber.

The metal-clad base and inner cover form an anneal chamber where the atmosphere will not be contaminated by diffusion of wet gases from the base insulation or by leaks. The system is checked for leaks twice during each anneal: once in cold condition before the anneal starts, and once at the end of the soak, before cooling starts. This second test is very revealing, since the hot hydrogen molecules penetrate very easily through the smallest of cracks or pinholes.

Figure 10 shows the dewpoint profile over a full cycle obtained in tests of both systems. In a HICON/H_2 bell annealer the dewpoint is already pretty low after the safety purge, at about -50°F; it then rises to roughly -20°F, and goes down again to -76°F (10 ppm H_2O) by the end of soaking - the H_2/H_2O ratio is then 100,000 to 1.

With the same hydrogen consumption of around 70 cft/ton the dewpoint never dropped below +6°F with the open workbase. The H_2/H_2O ratio is only 500 to 1 - even worse than for an anneal in HNX gas with 6 % hydrogen and a high gas flow of 580 cft/ton. This is why hydrogen annealing on an open base using the low hydrogen flows of a metal-clad base will not yield much better results than an HNX annealer.

Even with 4 times this hydrogen consumption the dewpoint drops only to -30°F on an open workbase. The H_2/H_2O ratio of 3,500 to 1 is still 1.5 orders of magnitude lower than the 100,000 to 1 ratio of the metal-clad base.

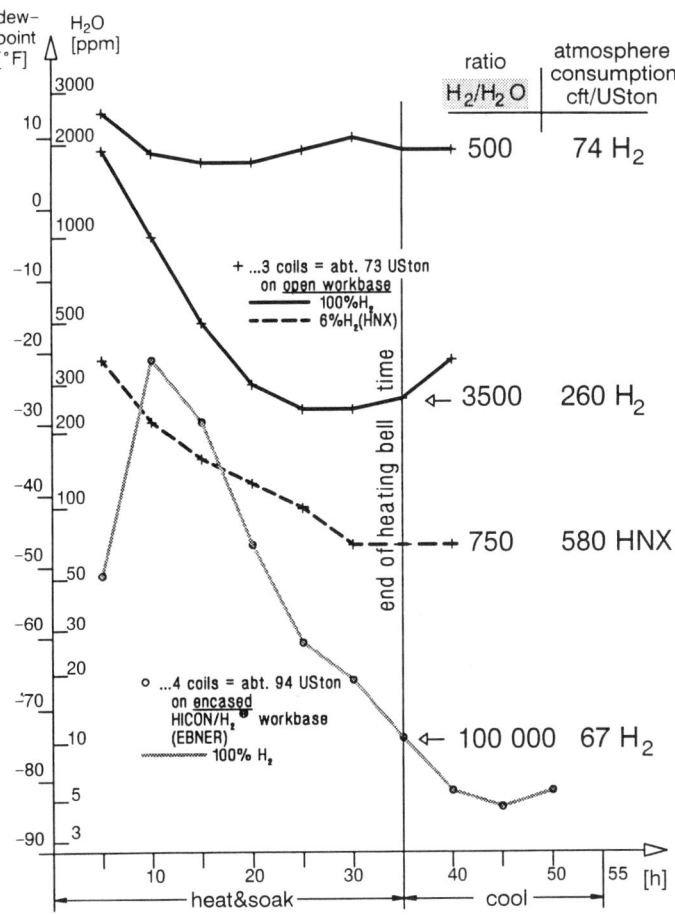

Fig. 10 Dewpoint profile during anneal

Figure 11 gives a table comparing the information available to date on the surface quality results obtainable with an EBNER HICON/H₂ closed base, and with a conventional open base, also with a hydrogen atmosphere. In the case of the HICON/H₂ base hydrogen consumption was 70 cft/ton, and for the open base it was 140 cft/ton.

The dewpoints were -76°F versus -13°F and the H₂/H₂O ratios were 100,000 to 1 versus 1,600 to 1. The carbon contamination found was twice as high, and there was much more manganese enrichment..

Corrosion was more than twice as wide for a scratch, and 5 times the area for a spot.

	HICON/H₂	conventional + H₂
	metal clad base	flat open base
H₂ consumption	70 cft/US ton	140 cft/US ton
dewpoint during soaking H₂/H₂O ratio	-76°F abt 10 ppm H₂O 100 000	abt -13°F abt 630 ppm H₂ 1600
Carbon contamination Ford power wash: Leco method:	.2 mg/sft .5 to .6 mg/sft	.3 to .4 mg/sft 1.0 to 1.2 mg/sft
Partial oxidation of Mn, rising at the surface from .2% inside to	abt .3 % Mn	abt .6 % Mn
Corrosion of phosphatized, painted sheet	.08 in .005 sq.in	.2 in .03 sq.in.
paint adhesion in bending test	excellent	
metal coating	best conditions no blistering thinnest layer	

Fig. 11   Differences in surface quality of coils annealed

8) FURTHER ADVANTAGES OF METAL-CLAD BASE

The explanation so far make it clear that only the metal-clad base can provide the conditions needed for top-class surface finish.

In addition, the metal-clad base has further advantages (fig. 8):

- Best possible convection with aerodynamic inlet and outlet diffusers and large impeller with wide intake

- Lightweight impeller on oversized shaft and bearings with powerful motor, thus high critical speed and no vibration - rebalancing every 2 years

- High rim because of concave shape; low turbulence at outlet, so little heat migration to workbase and inner cover flanges - no overheating and no warpage

- Low heat losses because highly heat-conducting hydrogen does not seep into the base insulation or into an external cooler opening

- Maximum safety because hydrogen is contained in a metal-clad sealed anneal chamber which is leak tested twice per anneal.

9) **SUMMARY**

We have shown in this paper that extremely low dewpoints must be maintained in the anneal chamber in order to obtain optimum surface cleanliness. Precision control of ultra-pure process atmospheres is becoming increasingly important.
The EBNER HICON/H_2 annealer with its metal-clad closed base is a major step forward in ensuring that we are prepared for whatever technological requirements the future will bring.

References

1) H. Lochner, G. Schweiger, Iron and Steel Engineer, 64(1988) 4, p. 45 to 51

2) H. Lochner, Iron and Steel Engineer, 67(1990) 3, p. 43 to 50

3) F. Welser, H.T. Junius, Stahl und Eisen, 111(1991)4, p. 65 to 74

4) B. Chatelain, V. Leroy, La Métallurgie, 86 (1989) 2, p. 173 to 180

5) Y. Inokuti, Research Article, Transactions ISIJ, 1975, pp. 314 to 323

6) D. Paesold, K. Köster, Vöest Alpine Stahl, Linz/Austria, Paper at the Austrian Iron & Steel, Convention, Leoben 1991

7) R. Eylens, W. Rausch, B. Voigt, Stahl und Eisen, 110(1990) Nr. 3, p. 73 to 78

8) R.W. Kessler, M. Brögeler, M. Tubach, W. Degen, W. Zwick, Werkstoffe und Korrosion 40, 539-544 (1989)

Modern concepts of process control and optimization in hydrogen batch annealing shops

Michael Bock
Manager of LPE Process Control
Division of LOI INC.
Pittsburgh Pa.
USA

Dr. Karl Nolte
General Manager of LVE Verfahrenselektronik GmbH
4300 Essen 1
Germany

Peter Wittler
Director of LOI Industrieofenanlagen GmbH
4300 Essen 1
Germany

Abstract

The use of hydrogen instead of nitrogen as the atmosphere in batch annealing has led to several major developments. The achievements attributed to hydrogen annealing are improvement of shop capacity, mechanical properties and surface cleanliness.

The advantages of hydrogen annealing have required modern and precise concepts for process control and shop optimization in order to achieve the maximum quality and quantity output. LOI satisfied these demands by developing control systems for several aspects of the annealing process. This paper describes four of these as follows:

1. Hardware and software design for base control units and supervisory control computers

2. Computer aided process optimization

3. Computer aided quality control in an annealing shop

4. Computer assisted material handling using digital image control

1. Hardware and Software Design for Base Control Units and Supervisory Control Computers

The Process and Safety Strategy

Industrial use of hydrogen as the protective gas in bell type furnace plants with 30 or more stationary bases (see **Figure 1**) requires greater expense for detecting and avoiding non-permissible working states, when compared with more common nitrogen based equipment. **Figure 2** shows the processing scheme of an HPH (High Performance Hydrogen) bell type annealing furnace.

Figure 1: HPH facility

An integrated control system was developed by LOI specifically for HPH bell-type annealing furnace plants. This integrated control system is composed of two major subsystems, namely the Base Control Unit (BCU) and the Supervisory Control Computer (SCC). In a typical installation (see **Figure 3** and **Figure 4**), there will be several BCU's (one for each individual base) and a single SCC (sometimes doubled for redundancy) providing centralized control of the shop.

This concept allows centralized handling of each base without the loss of individual control required for safety purposes. Due to the safety considerations and the high value of the material, the availability of the control system must be very high. Hardware and software comprising the BCU's have been designed to continue to control the annealing cycle (without manual intervention) even if the link to the SCC malfunctions.

Base Control Unit

Each annealing base is equipped with its own independent control unit, which is interfaced to the supervisory control computer installed in the plant control room. For annealing shops in which product specifications and annealing cycle configurations may change frequently, this strategy ensures maximum efficiency and high productivity. Apart from adaptability, however, reliability, safety and availability have now become vitally important criteria as hydrogen is used as a furnace atmosphere at modern annealing shops. At such plants, all annealing cycle data must be logged to

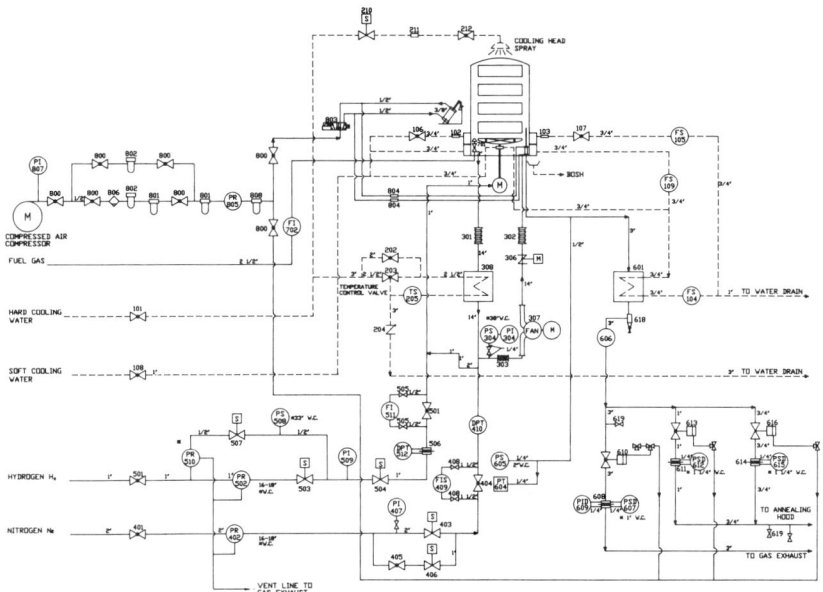

Figure 2: HPH processing scheme

Figure 3: Supervisory Control Computer

demonstrate product properties. A loss of data necessitates, in some cases, the repetition of an entire annealing cycle.

For this reason, all data for each cycle (such as temperature or flow profiles) are stored in the semi-conductor memory of each base control unit from which they may be retrieved by the supervisory control computer for further processing or for logging. If communication between the

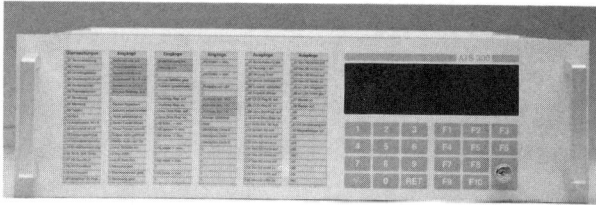

Figure 4: Front panel of the Base Control Unit

temperatures and flows for the entire cycle. When communication is re-established, the supervisory computer can collect all of the stored data from the BCU. Communication between the base control units and the supervisory computer is accomplished by use of a local area network (LAN).

Each base control unit features analog input channels and may be programmed with different algorithms for temperature, flow, and pressure control. In fact, different temperature control functions may be triggered as different process temperature thresholds or phase durations are reached, therein optimizing base control for all process conditions. For temperature control, each base is fitted with two thermocouples. Depending on the annealing phase, output from one of the thermocouples is used in the temperature control loop. Control can be switched over to the thermocouple used for charge temperature measurement, as soon as the temperature, as indicated by the thermocouple provided for control atmosphere measurement, has reached the setpoint temperature. The temperature detected by this thermocouple then becomes the controlling variable. The temperature control loop is started with burner ignition. The temperature is controlled in response to the rate of temperature change. Ten or more setpoint rates of change may be programmed for each thermocouple.

A number of different approaches have been developed for controlling temperature. ON/OFF sequential burner control is the preferred technique as it adapts heat output easily to operational needs and enhances even temperature distribution. With this technique, it is possible to employ low temperature ($<500\,°F$) annealing. If the furnace is equipped with ON/OFF burners, cooling may be controlled (to prevent sticking) using these same burners since air supply may be independently controlled by the ON/OFF switching.

The essential features of each BCU are in short:

Hardware:
- Independent base control even without interfacing to a supervisory control computer
- Display of all input and output data as well as status information on the base control unit front panel
- Standard microcomputer system built into a plug-in rack for quick replacement and for ease of maintenance
- Fail-safe system in that control resets to a completely safe position in the event of an internal failure
- Longtime battery-backed memory buffering for all control unit parameters as well as recorded data

Software:
- Storage of all temperature and flow profiles (three temperatures, three flows, one pressure and one setpoint) for as many as 100 hours of operation
- Storage of all setpoints for a complete annealing and cooling cycle (e.g. 10 rates of change for each thermocouple)
- Manual operation for testing and maintenance purposes

Supervisory Control Computer

Modern plant automation requires information exchange in all directions. To accomplish the supervision of an annealing shop, many different tasks must run simultaneously. The annealing process is only one step in a much larger and more complex process. A great deal of information accompanies the product when it enters the annealing shop and even more information is added thereto.
For these reasons, the supervisory control computer must have links not only to the base control units in the annealing shop, but must provide an interface for connection to a host computer as well. To assist shop operators, the SCC also requires input from other devices to give and accept further information. **Figure 5** shows a LOI HPH bell-type annealing plant control system.

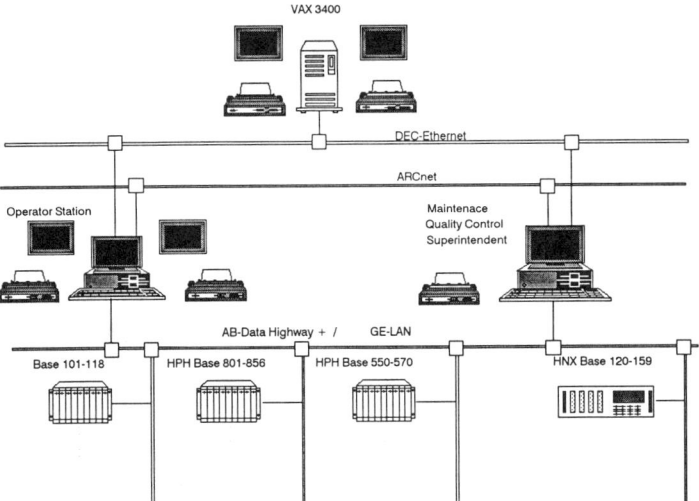

Figure 5: HPH Plant control system

The supervisory control computer has access to all BCU data memory. In the automatic mode, the supervisory control computer will retrieve data logged by the base control units and will down-load setpoints to them. Flags are used to determine if new data has been logged in the base control unit or if manual changes have been made at the furnace base.
All supervisory control computer functions are menu-driven. They are very easily accessed by function keys, making the system user-friendly even for those operators with little or no previous computer experience.

Main SCC functions:
- Display of base data such as material quality, type of annealing cycle, atmosphere and material temperatures, time and temperature setpoints, H$_2$ and N$_2$ flow, current phase (e.g. heating or cooling) of the annealing cycle, etc. **Figure 6** shows a detailed base information mask.

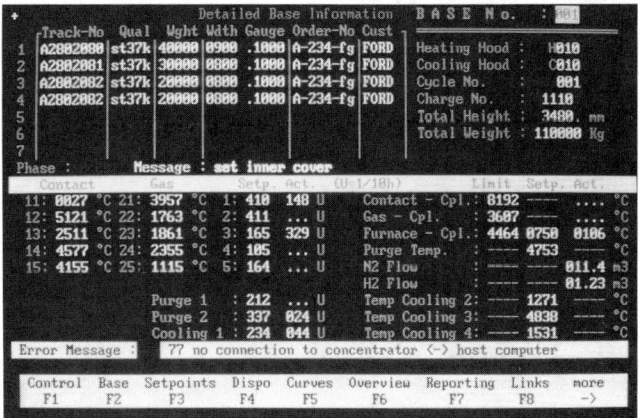

Figure 6: Single base detailed information mask

- Display of a base management table showing assignment of furnaces and cooling hoods to the various bases and the detailed time schedules which can be used for planning overall shop operations.
- Display of all bases on one or more color monitors for a quick view of the overall plant status. In larger plants (e.g. 30 bases) this mask should steadily appear on an special monitor for a fast overview or rapid trouble shooting (see **Figure 7**).
- Detailed color-graphical schematic representation of each individual base showing valve positions, motor status, temperatures, faults, etc. These masks (see **Figure 8** and **Figure 9**) show the actual operation of each base and help the operator to understand the system as a whole.

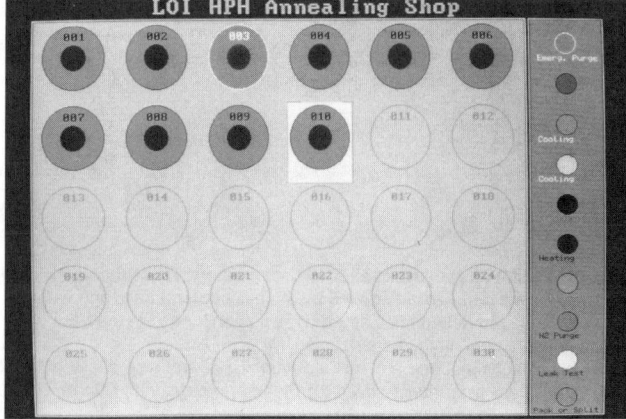

Figure 7: HPH annealing plant overview

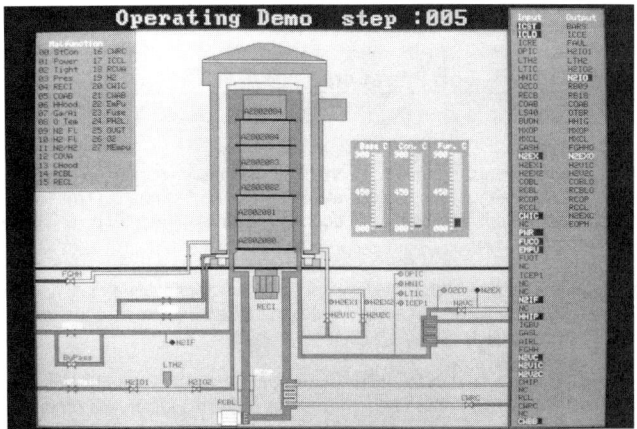

Figure 8: Schematic base representation during heating

- Intelligent data transmission to and from sub- stations and higher-level control computers
- Direct control of individual annealing bases (set points for temperatures and durations stored in system memory and retrievable as recipes of set points via reference number)
- Process display of the complete furnace plant and individual annealing base to determine plant status
- Display of curves including facility for highlighting temperatures flows, pressures etc.
- Provides annealing protocol including coil data (coil identification, weight, size etc.), listing of process events, major temperatures and specific performance data
- Provides individual annealing base log detailing every annealing cycle
- Control of utilities, inner covers, heating and cooling hoods (marked "available", "in use" or "under repair"), sorted in terms of availability
- Context sensitive help available on screen at all times
- Configuration of requirements specified by customer

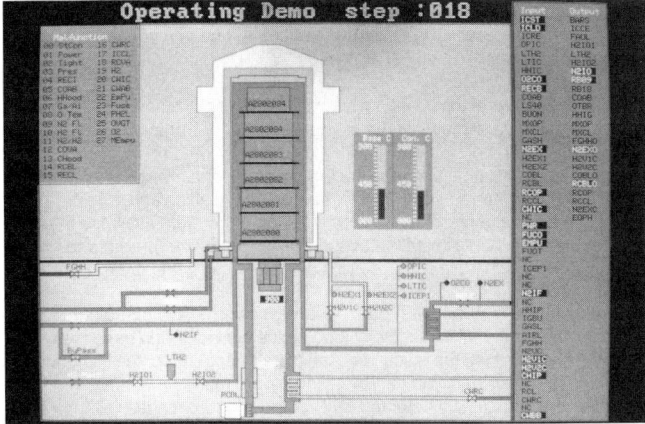

Figure 9: Schematic base representation during cooling

2. Computer aided process optimization

Introduction

In general, an annealing facility is constructed and is provided with a given number of heating hoods, so that the heating hood to base ratio is nearly equivalent to the predicted heating time to total time ratio of the material to be annealed. The number of cooling hoods required can be calculated in a similar manner.

LOI developed a program (ProOpt$_c$) which provides a planning tool for attaining maximum material throughput based on available equipment in a box annealing facility. Equipment consists of bases, furnaces, cooling hoods and cranes.

The program schedules charges on bases to attain the maximum balanced utilization of equipment while assuring that all material is delivered on or before its due date. In addition, the program allows operator intervention for special cases which may arise, such as an accelerated delivery schedule for one or more stacks.

Due to equipment failures, human error, or a change of time in the heating to total time ratio of the material (possibly due to new customer requirements), it may become more difficult to run the shop in a "balanced" manner.

ProOpt$_{(c)}$ is able to produce the best balance for any situation, and can select stacks with correct times to reduce the difference between actual and target ratios. Troubles resulting from mistaken planning (or in many cases no planning at all), failed equipment, and human error can be reduced to a minimum using ProOpt$_{(c)}$.

Description of Operation and Performance

The Annealing Shop Balance program provides a dynamically updated operating schedule which allows a shop to be run in a balanced manner thereby increasing productivity. A balanced shop is one which operates with a minimal equipment waiting time. The balancing program considers the number of bases, heating hoods, cooling hoods, and cranes and then produces a schedule based on the input pool of stacks. By following the resultant schedule, one can consistently maintain the highest possible shop production.

The schedule produced is dynamic in that real-time events, such as an extended or shortened heating time, are detected via an interface to the HPH Annealing Supervisory Computer and the schedule is then automatically updated based on the new information. Configuration parameters are provided which allow for complete customization of the balancing program for a given facility. These parameters provide for input of basic items, such as the number of heating and cooling hoods, and additionally for certain time constraints which limit the degree of freedom available to the program for adjusting the schedule based on a real-time event. For example, it is possible to constrain the program so that it would not reschedule a charge to a different base if that charge were within one hour of being packed.

Finally, the balancing algorithm may be overridden when necessary. This can occur, for example, when a given stack is so important that it must be placed on the next available base even if this would upset the balance of the shop. In fact, the stack in question can be "tried" to see how much it upsets the balance, and then an informed decision can be made by operating personnel. In any event, an upset in the balance will be immediately detected by ProOpt$_{(c)}$ and changes will then be made to move the shop back into balance as quickly as possible given the stacks available in the pool.

Operator Interface Overview

The man-machine interface generates several screens providing various levels of information about the shop. These screens are briefly outlined below.

- A horizontal bar chart which shows for all bases the current and planned charges. Each charge is depicted by a contiguous set of colored bars, one each for packing, setting the furnace, heating, lifting the furnace, setting the cooling hood, cooling, lifting the cooling hood, and splitting. In addition, any delays due to lack of equipment are also depicted. A legend describing the meaning of each color is also displayed. The scale can be changed independently in both the horizontal (time) axis and the vertical (# of bases) axis. In addition, each base can be interrogated to see the information for the stack currently in charge or for the planned stack(s). (see **Figure 10**)

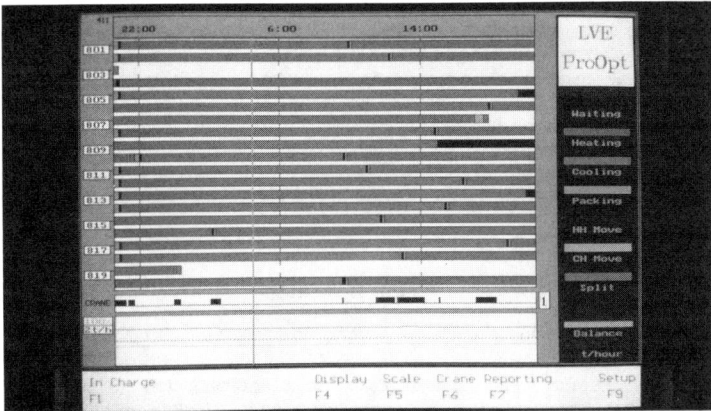

Figure 10: Main mask of optimizing program ProOpt

- A crane usage log which is used in conjunction with the schedule graph described above. The crane usage log shows for each time slice whether or not the crane is in use. A separate log can be displayed for each crane in the shop.
- Also on the main schedule display, a graph is displayed which shows, again for each time slice, the target and actual shop balance ratio, and the current average shop production in tons/hour.
- A second screen is provided which shows the distribution of the frequency of occurence of heating time vs total time for all stacks in charge and for all stacks in the availability pool. This is useful for real-time operations as it allows the operators to forsee potential changes in balance due to the arrival of new stacks from the cold mill with different properties. Moreover, it is an important tool for building annealing shops through modeling with variant numbers of heating and cooling hoods, vs. the expected distribution of heating/total times for the material.
- For planning purposes, hardcopy output will be available on demand and/or automatically at preset times such as shift changes. For instance, a time schedule plan of crane activities can be generated. If desired, the scheduling can be directly transmitted and displayed in the crane drivers' cabins.

Description of the Shop Balance Model

The following list describes the requirements which must be satisfied by the shop balancing algorithm.

- Ability to deal with the occurrence of real-time, unscheduled events such as an increase in a heating time, increase in a crane service time, and of course failure or service of equipment.
- Must produce an optimized schedule, which minimizes waiting times by utilizing equipment (heating hoods, cooling hoods, cranes, and bases) in a balanced manner. For example, the algorithm must avoid situations which require more heating hoods than are available by carefully placing charges with the appropriate heating times.
- Must completely avoid operation time "collisions" of all equipment. For example, the algorithm must never schedule one crane to perform two different operations at a given time.
- Must assure that all stacks are scheduled in sufficient time to meet delivery constraints. For example, a stack with a latest completion date of 12 p.m. on 30 August must be placed in charge in sufficient time to meet this criterion.
- Must allow operator override of the shop schedule through priority assignment to each stack. This feature must allow an operator to specify any given stack as the next stack to be placed in charge.
- Must produce data sufficiently fast to allow for dynamic shop operation. In effect, this means that all input data must be considered and a schedule produced in less than six minutes, in order to keep pace with operations in the HPH shop.

Balance Algorithm Implementation Specifics

The algorithm meets the above requirements using the decision criteria outlined below.

- Equipment availability always takes precedence over other factors so that no single piece of equipment (furnace, crane, - etc.) will be scheduled for use in two places at the same time. Necessary delay times are introduced to avoid equipment operating time collisions in all cases.
- All real-time unscheduled events are detected via an interface to the HPH Supervisory Computer database. Any change is immediately detected and the schedule adjusted accordingly within the constraints of the decision parameters. Automatically detected events are for instance increased or decreased duration times, failures of bases, furnaces and cooling hoods, crane delays because on usage at another place (cold mill service).
- To avoid unrealistic schedule changes, such as rescheduling a charge from base 801 to 835 15 minutes before it is scheduled to be packed on base 801, the algorithm contains a configurable parameter labeled "In Charge Decision Time." For example, setting this parameter to 2 hours means that any planned charge within 2 hours of being packed would be locked-in to that particular base. In this way, a crane operating schedule can be produced hourly, every four hours, or every shift simply by changing this parameter. Of course, making this decision time excessively long restricts the planning freedom of the scheduling algorithm.
- The target balancing function is one of the fundamentals of the program ProOpt$_{(c)}$. It consists of three terms, each of which is itself a varying function. Each function has an associated importance weight factor which is tuned to the specific shop environment in question. Each of the functions making up the three terms of the balancing function are described below.

    1. The first term considers the current actual shop balance ratio vs the target balance ratio which is a function of available hoods and bases. Each schedule possibility is considered to attempt to minimize the difference between actual and target balance ratios.

2. The second term considers the remaining time before a stack's latest delivery date. This assures that no stack will be left in the annealing shop beyond its latest delivery date.

3. The third term considers the priority of a given stack vs the priority of the other stacks. This term is used to shorten the delivery time or even to force the algorithm to choose a given stack next no matter what the previous terms would suggest. This term is installed for tuning the program to presently unknown conditions or special customer requests.

Operator Interface in Detail

Figure 10 shows the main screen which is presented on start-up. This screen depicts the bases and the charges assigned to them. Each charge is shown as a series of colored bars, with each color depicting a given operation (e.g. green for heating, blue for cooling, and red for delays due to lack of equipment). At the bottom of the display is a list of function keys available to the operator from the main screen. A legend on the far right of the screen describes the meaning of each color in the schedule.

The schedule begins on the left side which is the current time, and extends up to 96 hours into the future. Using the zoom function, the operator may expand or contract the horizontal and vertical axes independently. This allows the operator to concentrate his view on only a certain part of the shop (e.g. one crane zone) and/or on a given time span (e.g. one shift). Vertical lines depict shift change times for each of the three shifts.

A second window below the schedule window shows the crane activity for one of as many as five cranes. The log may be scrolled to each of the cranes in use simply by pressing a function key.

The third window displays curves for both the actual and predicted shop balance ratio and production rate. The vertical scale for these curves may be magnified by a factor of 2 for more detail, if necessary.

Reporting Options

Several reporting options are available such as hardcopies, crane planning logs, etc. The exact formats should be developed jointly with each customer, since reporting requirements differ greatly from site to site.

Interfaces and Data Transmission

The shop balancing data flow diagram shows the main paths of data flow between the various processes. The Supervisory Computer process depicts the existing control program used by the annealers to run the HPH facility. Note that the data flow diagram (DFD) is a logical representation and does not attempt to show physical devices upon which processes run or data is stored. **(see Figure 11)**

The Man-Machine Interface process produces the various graphic displays and accepts input from the operator. The Shop Balance process is responsible for producing the balanced shop schedule based on stack input, equipment availability, and real-time events such as changes to heating or cooling times.

Shop Balancing Data Flow Diagram

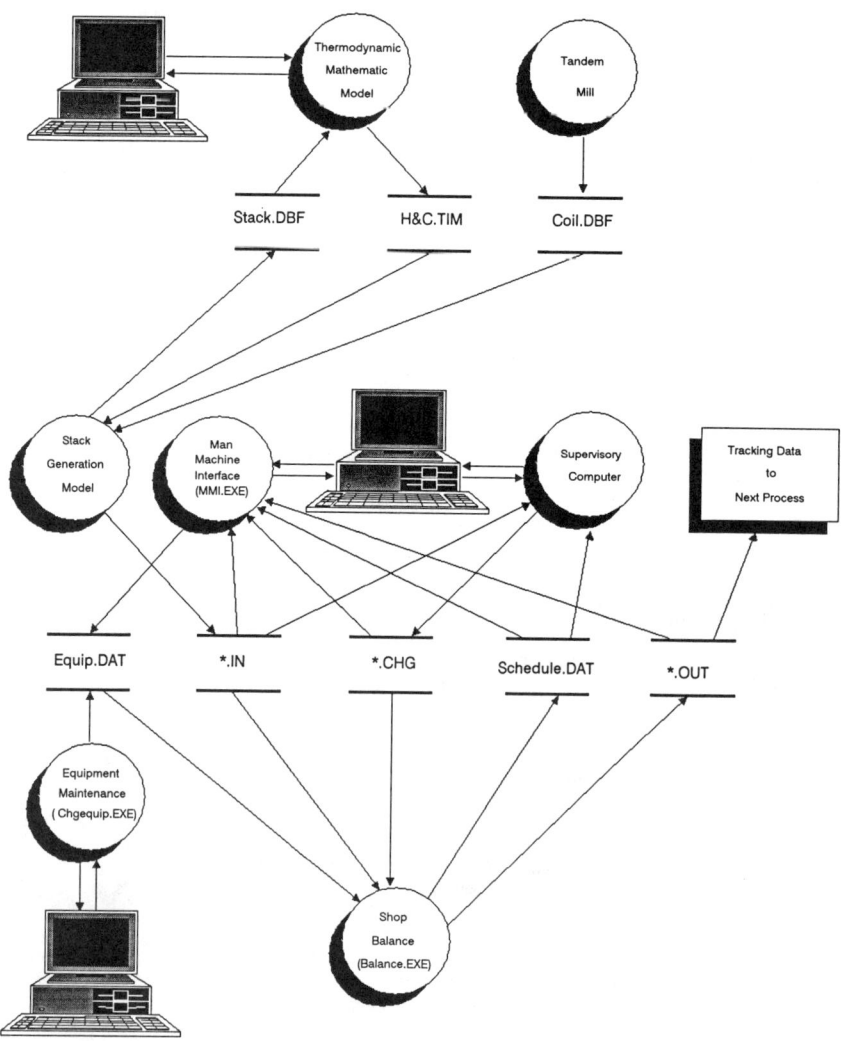

Figure 11: Data flow and Data Base of the ProOpt program

3. Computer-Aided Quality Control in an Annealing Shop

Computer-aided quality control (CAQC) is today used throughout the manufacturing industry. Many checks and testing stages including automatic checking of incoming material, automatic component checks, test equipment scheduling and final product inspection are computer-controlled, with the main objective being production optimization.

CAQC is also used today in bell-type annealing furnace plants to control process conditions including temperature, flows and time and to produce graphs and other documentation. Computers also record failures and operating times and issue instructions for immediate and proper remedial action as well as maintenance instructions, thereby contributing to quality assurance.

In bell-type furnace plants, accurate temperature control is the most important aspect of quality assurance. Although inspecting and testing thermocouples and calibrating control instruments is extremely time-consuming, this part of quality assurance is still done manually at most annealing plants.

Bell-type annealing furnace plants consist very often of many equal annealing bases, furnaces and cooling covers. Some furnace plants have more than 200 annealing bases with accompanying heating hoods and cooling covers. All are used in a multitude of combinations during the heat treatment process.

Temperature Loop Configuration

Bell type annealing bases are normally equipped with two thermocouples controlling the various stages of the annealing process. **Figure 12** shows a typical temperature loop with mV signals

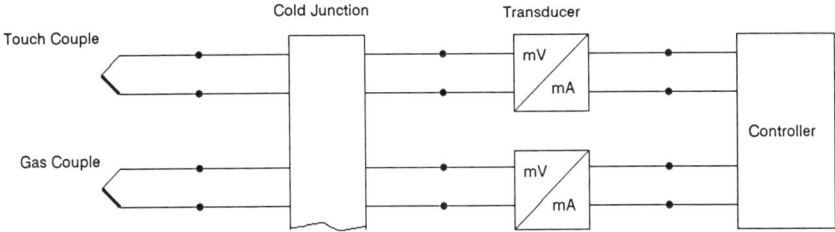

Figure 12: Principle of temperature measuring

corrected via a reference oven which then are converted to mA-signals.

In order to ensure constant accuracy for all temperature measurements, all loops have to be checked on a routine basis. All thermocouples, reference ovens, transducers and controllers as well as recorders must be checked, calibrated, and replaced individually and/or as a whole loop.

Due to the high number of thermocouples and equipment in annealing shops there is often no routine checking or instrument calibration, although temperature accuracy is critical for product quality.

operation and maintenance are a good basis for computer-aided quality control. Some of the steps involved in quality control which may be computer-controlled are described below.

Thermocouple Check

As thermocouples measure process conditions, on-line checking is indispensable. This can either be done by installing a second thermocouple for reference temperature measurement or by simulating process conditions and evaluating the signal produced due to the known conditions.

An often used method is to remove the thermocouples and check them in a test furnace. Using an automated test furnace, the thermocouples may be checked and then returned to the manufacturer together with test protocols. Test results may also be recorded on a data carrier which would allow this data to be used by the control room computer. Data can be used for instance to report the quality of thermocouples in the charge report. Another application is the compensation of thermocouple deviations. **Figure 13** shows the principle of this calibration method.

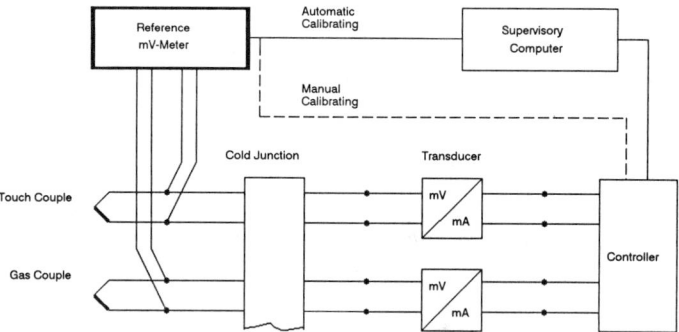

Figure 13: Calibration of thermocouples and instrument

Instrumentation Check

Although instruments are not subjected to the same aging process as thermocouples, they must be checked on a routine basis. This may be done at longer intervals if only minor drifting is observed.

A number of methods may be used for checking instruments. These are less complex then testing thermocouples because their process conditions (voltage or current) can be simulated easily and accurately with special signal sources or even with on-line signals. A practical method of testing and calibrating thermocouples and controllers is described below.

Test Signal Supply

One way of checking the accuracy of instrumentation in control systems is to disconnect individual instruments or the whole loop (with possible production losses as a consequence) and to check and calibrate the instrument as required.

This is usually done using a precision signal source providing a voltage or current signal as well as a meter to insure that the signals generated and the instrument output signals are accurate. These checks must be carried out on all the components of a temperature loop, i.e. for a bell-type annealing furnace the cold junction, transducers, panel instruments and controllers.

These checks may also be carried out by connecting a signal source with an accurate simulated reference signal. This signal will be transmitted to the controller passing-through all other loop components. Any deviation shown on the controller indicator will be the sum of all inaccuracies of that particular loop.

If, as in the case of modern controllers, calibration can be carried out via the system's software, the instrument need not be removed. It should be noted that both zero and span must be checked and adjusted as necessary to obtain full instrument accuracy.

Reference Instruments

Using reference instruments for calibrating field instruments is far easier than described for thermocouples as these checks can be carried out while the furnace is on line. For these tests a second instrument is connected in parallel to the loop and both instrument readings are compared.

If no major deviation is found, none of the equipment need be removed. If one of the loop instruments features the calibration via software described above, calibration can be carried out on the field instrument.

The instrument will record the values measured and display them as a diagram or table. Deviations between the controller installed in the loop and the connected reference instrument can be identified by comparison. If major deviations are found, the controller in question must be calibrated. The use of a reference instrument is the most suitable method of identifying instrument errors.

With a multiplexer instrument which can be connected to a number of controllers in parallel and the appropriate software on the process control computer, the instruments and the controllers can be checked automatically. The system is described below.

Computer-aided Calibration

Using the functions described above, calibration data are usually input into the controller using keys or push- buttons provided on the unit (see **Figure 13**).

The supervisory control computer described in section 1, however, also allows menu-driven computer-aided calibration. If the reference instrument is connected to this computer via an interface, any deviations could be corrected automatically. **Figure 14** shows the principle of the fully computerized calibration of annealing shop thermocouples.

In order to allow monitoring of the maintenance work, a software package designed to allow checking of the status of calibration work on a continuous basis could be installed on the control computer. The data of all thermocouples, panel instruments and controllers could be stored in a data base. This data could include

Thermocouple data:
    number of thermocouples
    manufacturing year/month

hours in service
number of annealing cycles performed
place of installation
date of last check
parameters of last calibration

Controller data:
controller number
manufacturing year/month
place of installation
date of last check
parameters of last calibration

The program is designed to record all equipment check-outs and replacements. The status of work associated with quality assurance is thus constantly updated and readily accessible.

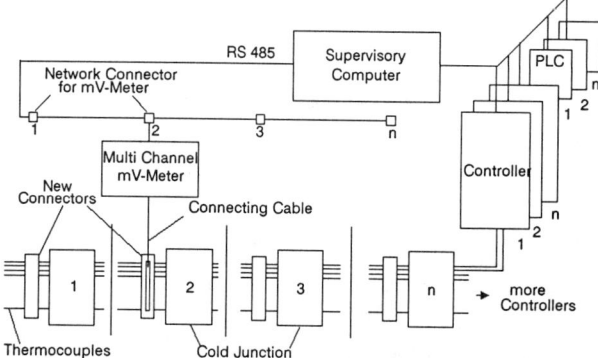

Figure 14: Full CAQC of the temperature measurement

4. Computer Assisted Material Handling using Digital Image Control

Introduction

Batch annealing facilities for wide-strip coils have found wide-spread acceptance in the steel industry. LOI has itself built approximately 5000 annealing bases worldwide.

Because of the discontinuous material flow within a batch annealing facility resulting from the packing and unpacking of the bases, including the movement of furnaces, cooling hoods and inner covers, the continuous annealing facility has been viewed as the better technology. With the introduction of new technologies in the area of box annealing (Hydrogen atmosphere together with high convection), the batch annealing process today has become important once again. The sheet quality coming from batch annealing facilities has been significantly improved. Therefore, sheets from a batch annealing facility equal in quality those from continuous annealing lines. In addition, the capacity of a modern HPH (High Performance Hydrogen) facility is double that of a traditional HNX annealing plant.

The batch method for annealing presents several advantages over its continuous counterpart as follows:

    Low investment cost
    Low operational costs
    High operational flexibility
    High shop utilization

Recently, due to these advantages, only batch furnaces have been used in new facilities or for increasing present capacity. The introduction of hydrogen and high convection technology has led to improved quality and higher capacity. Moreover, new operational practices have been necessitated by the new HPH technology, in order to accomplish higher goals in the areas of quality and safety. In short, methods used to control the annealing process in the past are no longer sufficient using the newer technology.

A further innovation resulting from the use of the new technology is the mathematical thermal modelling of the batch annealing process. With a three-part computer model, the batch annealing process becomes very efficient. In the first part, the stack composition is determined. In the second part, a heating recipe is developed for each stack taking into consideration all outside influences. Finally, in the third part, the shop throughput is optimized by balancing the utilization of equipment (bases, hoods, and cranes).

Finally, a new measurement system has been developed which is capable of measuring coil, stack and equipment configurations in three dimensions. The following sections briefly describe the batch annealing process and several ways in which such a measurement system can make the process more efficient.

Overview of the Batch Annealing Process

Basically, coils, bases, and hoods are cyllindrical objects. The cyllindrical coils are stacked on bases which are also cyllindrical. The coils must be stacked on the base so that the center of their vertical axis is concentric with that of the base. Additionally, the inner cover, furnace, and cooling hood must be similarly placed on the base. **Figure 15** shows a side view of a stack on a base.

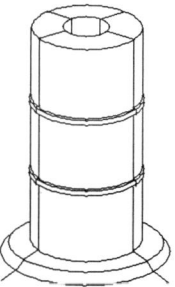

Figure 15: Side view of a coil stack

The recirculation of the atmosphere within the inner cover follows a cyllindrical pattern. In addition, the input of heat from the twelve burners (placed equi-distantly on the circumference of the heating hood) follows a similar cyllindrical path. Due to the symmetry of the heating hood, the mathematical thermal model was easier to develop since it basically needed to consider only two dimensions.

The stacking of coils and placement of hoods is done in nearly all cases with the assistance of an overhead crane. The placement of inner covers, heating hoods, and cooling hoods on the center of the base is done with the help of guideposts. In some cases, these guideposts also assist in the placement of the the convector plates which are placed between each coil. However, in most cases, the placement of the coils is done purely by manual means without assistance of any guiding devices.

Automatic Coil and Stack Measurement

Coil and stack handling in facilities today is not a state-of-the-art operation. In general, stacking of coils is done visually in a "seat of the pants" fashion by the crane driver and his floorman assistant. When the coils arrive in the annealing shop, they are normally stacked vertically and are handled by a magnet or gripper (Heppenstahl). The first positioning done by the crane driver is to align the centering device of the magnet or gripper into the "eye" of the coil. If the centering device is not aligned correctly, the edges of the wrapped coil can be damaged by the severe weight of the gripper or magnet, which can be as heavy as ten tons. In addition, the edges can also be damaged if the magnet or gripper is lowered onto the coil too quickly.

The accuracy of setting coils on the base or on top of other coils is a result in most cases of the dexterity of the crane driver. The craneman must maneuver his crane in three axes from a gondola which can be as far as thirty feet from the coils. From this perspective, he must (with or without the assistance of a floorman) accurately stack as many as seven coils. In addition, he has to place convector plates between each coil using the crane. For these reasons, the stack is often built in a non-cyllindrically symmetrical configuration. Due to the failure to build a symmetrical stack, the temperature distribution is disturbed. The flow conditions are negatively influenced by the non-symmetry of the stack. The afforementioned thermal model can not accurately predict the heating profile for a stack which has not been built in a symmetrical manner. Radial inconsistency in the heating process can have negative effects in the steps of the production process which follow. Variation in mechanical property over the length of a strip can disturb the skin-pass process. Variation in the hardness of the strip will cause oscillation in the control system, thereby placing greater stress on the equipment involved. In some cases, the symmetry of the stack is so poor that the inner cover, which is only slightly larger in diameter than the convector plates, will not fit over the stack. In this case, the coils must be re-stacked resulting in a loss of production time.

Description of the Automatic Coil and Stack Measurement System

The measurement system is based upon a camera system which is installed on the crane. The camera measures, from a top view, the position of the base, stack, and the coil which is currently held by the magnet or gripper. The camera system works using digital image processing techniques. Although the system described herein uses two cameras, a single camera system can also be used. Two cameras are installed on the trolley of the overhead crane. The system measures and tracks the axes of the base, stack, and coil to be placed, all of which is dependent on the actual configuration of equipment at the site. **Figure 16** depicts both single and double camera systems.

Coil loading with computerized image processing

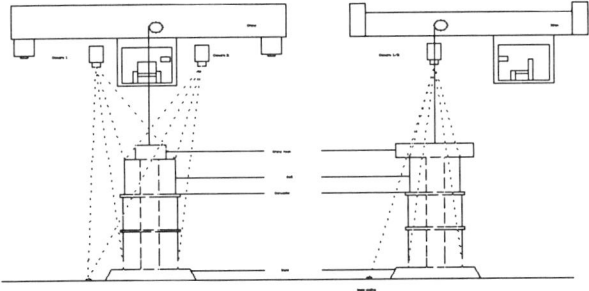

Figure 16: Principle of computerized image processing

From the video images and with the help of correctly placed markings, the geometrical data describing the facility can be gathered. The camera system can measure other data in addition to the vertical axes of the coils and the base, thus leading to other advantages which are described in the following paragraphs.

Advantages Resulting from Automatic Coil and Stack Measurement

During the stacking process, the current relative positions of the base and coils can be shown to the crane operator (e.g. using an X-Y coordinate grid). Most importantly, after placing a coil, the crane operator could check to see whether the coil was placed within allowable tolerances. One possibility resulting from having this system, would be to check, using the diameter of the coil, whether the crane has picked-up the correct coil to be placed on the stack. That is, coils could be identified by their physical dimensions.Often, time is wasted when lowering a coil onto a stack. This is because of the difficulty in determining the distance between the coil and the stack, which results in the crane operator having to slow the lowering of the coil to avoid potential damage to it or the stack. The camera system could show the crane operator precisely how far the coil is from the stack (or empty base), thus allowing him to lower the coil more quickly without sacrificing safety. This would result in less wasted time when stacking coils.

If the camera system is interfaced directly to the crane controls, then the stacking process can be automated completely. Using an interface to a supervisory computer system, the coil data (O.D., width, etc.) can be transfered and logged. Because the cameras are fixed to the crane, any swinging of the load can be detected. With the system interfaced to the crane controls directly, the swinging of the load can be automatically eliminated or prevented altogether. Optimal performance will always be guaranteed during loading and unloading of coils and hoods.

In summary, many advantages can be gained by using the coil and stack measurement system. These include less lost production time due to incorrect stacking, better quality of the annealed product due to more accurate heating, and less wear and tear on equipment resulting in lower maintenance costs.

DEVELOPMENT OF A THEORETICAL ANNEALING MODEL

FOR HYDROGEN ANNEALING OF STEEL SHEETS

S. Ramasamy[*], R. L. Simmons[**], A. P. DeVito[***] and K. G. Brickner[****]

Abstract

With the installation of 100 percent high flow hydrogen (HFH) batch annealing at the Irvin plant of Mon Valley Works, the need arose to develop a theoretical computer model which accurately predicts annealing times for this process. To achieve acceptable accuracy, the determination of radial conductivity as a function of gage, and the measurement of gas flow distributions within the stack were initiated. The results of these tests were then incorporated into the theoretical model and its performance evaluated in extensive production trials. Test results have shown that good mechanical properties with better uniformity are being achieved with a productivity increase of about 6 percent.

[*] Research Engineer, USS Technical Center,
4000 Tech Center Dr., Monroeville, PA 15146

[**] Assistant Scientist, Carnegie Mellon Research Institute,
4616 Henry Street, Pittsburgh, PA 15213

[***] Quality Engineer, Mon Valley Works,
Irvin Plant, Dravosburg, PA 15034

[****] Chief Consultant, Delafield Services, Inc.,
411 Virginia Avenue, Pittsburgh, PA 15215

Introduction

In 1988, U. S. Steel installed 52 single-stack high flow hydrogen (HFH) box-annealing bases of LOI Inc. manufacture at the Irvin Plant of Mon Valley Works (1). Additional 6 bases have been installed since that time. The six basic components of the LOI HFH annealing system are shown in Table I and a comparison of HFH and conventional box-annealing equipment is shown in Table II. The important differences between the two systems are the use of 100 percent hydrogen as the annealing gas, the high flow rate, microprocessor control, and the rapid cool heat exchanger.

Table I - LOI Annealing System

1	Inner Cover
2	Base with External Rapid Cool Heat Exchanger
3	Heating Hood (Furnace)
4	Cooling Hood
5	Convector Plates
6	Temperature Control System and Microprocessor

Table II - Comparison Between HFH and HNX Annealing

LOI HFH	HNX
Single-Stack Base	Single or Multi-Stack Base
Fiber-Lined Furnaces	Ceramic-lined Furnaces
Tangential Direct-Fired Burners with Recuperative Features	Direct-Fired or Radiant-Tube Heating
84-Inch OD Load Plate	Less Than 84-Inch OD Load Plate
Mechanical Inner Cover Seal	Sand or Fiber Seal
40-Hp Base Blower Fan	Generally 15 or 25-Hp Fans
~ 30,000 cfm Flow Rate	6,000 to 12,000 cfm Flow Rate
Microprocessor Control	Manual Control
100% Hydrogen Annealing Gas	HNX (~6% H_2 - 94% N_2) Gas
Cooling Hood	
Rapid Cool Heat Exchanger	

Hydrogen is the ideal gas in which to conduct annealing. It conducts heat about seven times more efficiently than that of HNX (~6% H_2 - 94% N_2) gas, which is usually used as the protective annealing gas in conventional box-annealing. The result is that heating and cooling rates are increased about 50 percent when 100 percent hydrogen is used as the protective gas in comparison to HNX gas, and there is less variation in temperature between the outer and inner wraps of the coils being annealed. Also, hydrogen is beneficial to steel cleanliness because it penetrates between the wraps of the coils and reacts with the rolling lubricants on the steel surface permitting them to be carried off into the system for removal.

The light weight of the small hydrogen molecule is also advantageous because hydrogen annealing gas can be moved rapidly within the inner cover of the annealing equipment by an impeller driven by a relatively small motor. By using high annealing-gas flow rates of about 30,000 cfm, markedly increased convection heat transfer rates from the inner cover to the coils can be obtained.

High flow rates and the use of 100 percent hydrogen as the protective annealing gas result in a much more efficient box-annealing system. HFH annealing achieves at least 15 percent more than the productivity of the older HNX annealing (when single-stack HFH is compared to 4-stack HNX base) and results in much more uniform temperature distribution throughout the coils being annealed. The result is more uniform mechanical properties for coils annealed by HFH box annealing (2).

Annealing Models

In operating box-annealing equipment, annealing models are used to determine the temperature and time needed to heat charges of steel coils for various customer applications. These applications require products with good flatness, surface cleanliness and uniformity of mechanical properties; and encompass a wide range of steel grades, coil sizes, widths and gages, conditions that make the use of batch annealing more attractive than that of continuous annealing.

The annealing models used are generally empirical although some theoretical models have been developed (3,4). Within U. S. Steel, an internally developed box- annealing model, called BOXAN, has been used for a number of years to determine the annealing temperatures, times, and heating rates for coils annealed in HNX annealing facilities. The BOXAN model is based on radiation, convection and conduction heat transfer theories and proved, over the years, to produce coils with very acceptable mechanical properties at reasonable production rates. Because of the previous success of the BOXAN model in HNX annealing, Irvin personnel wanted to employ the model in the new HFH annealing facility. However, initial attempts to modify the model for use in HFH annealing were unsuccessful. This lack of success was attributed to:

1. lack of precise data on the effect of sheet thickness (gage) of the coil on the heat transfer properties of the coil in hydrogen, and

2. inaccurate information on the actual hydrogen gas-flow distribution within the coil stack for a wide variety of coil and stack configuration.

Accordingly, two programs were initiated to develop the above information. The first program encompasses the determination of radial conductivity in hydrogen as a function of gage, and the second program involved the measurement of annealing gas-flow distributions within a number of different coil stacks.

In the interim, a simple empirical annealing model, developed for the new HFH facility from data obtained from the testing of numerous stacks of coils with embedded thermocouples, was used to anneal production coils. This simple model only took into consideration coil weight and width and thus, for certain stack configurations predicted longer heating times than necessary to achieve the desired mechanical properties. Also, this empirical model was strongly dependent on the annealing cycle (ΔT) and any variations from this annealing cycle would result in conservative heating-time predictions. However, the model proved to produce good quality coils with respect to mechanical properties for coils annealed using the annealing cycles for which the model was developed.

This paper summarizes the results of the two experimental programs, describes the verification of the new theoretical model (HFH BOXAN) and presents a comparison of mechanical-property data obtained with the new model with that of the previously used empirical model.

Testing

Radial Conductivity

Tests were conducted in the HFH facility at the Irvin plant, using three different test coils with 0.0722-, 0.0414- and 0.0207- inch-thick sheet. Thermocouples were inserted at mid-width of these coils at the one-quarter, center, and three-quarter positions measured from the outer diameter of these coils. Thermocouples were also positioned on the outer diameter at one-quarter distance along the width of these coils as shown in Figure 1. These thermocouples were then connected to an IBM AT computer and during annealing, temperatures were recorded at six-minute time intervals. From the recorded temperature data and using heat-transfer analyses, smooth radial conductivity curves as a function of temperature and coil sheet thickness were plotted as shown in Figure 2.

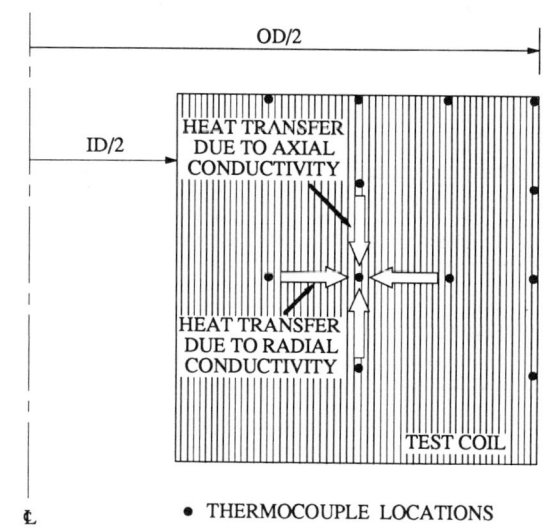

Figure 1 - Thermocouple Locations in Test Coils for Radial Conductivity Test, Sheet Thickness = 0.0722, 0.0414, 0.0207 inch.

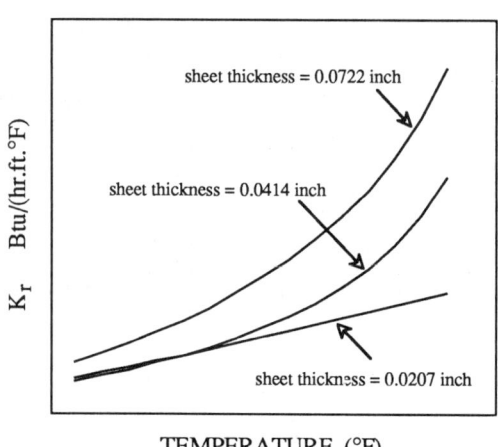

Figure 2 - Theoretical Radial Conductivity Based on Experimental Data.

From these results, it was concluded that:

1. The results confirm those in the literature (5,6).

2. Radial conductivity is markedly affected by sheet thickness and wrap-to-wrap gap distance.

3. Apparent wrap-to-wrap gap distance appears to decrease linearly with increasing temperature.

4. Coils of thicker sheets have higher radial conductivity values and therefore can be annealed for shorter times than coils of thinner sheet to attain the same level of mechanical properties.

Flow Measurements

Flow tests were conducted in the HFH facility to determine:

1. The flow distribution through the convector plates of the coil stack for various stack configurations, and

2. The total flow output by the fan in the diffuser at different fan speeds and gas contents.

In order to measure the flow distributions and total flow, three hollow cylindrical coils (test coils) 48.0 inches wide and 24.0 inches inside diameters with varying outer diameters of 60.0, 70.5 and 82.25 inches were constructed of 0.5-inch-thick steel plates. Within each test coil, stainless-steel Pitot tubes protruded into the flow region through holes in the sides (ID and OD) and bottom of the test coil as shown in Figures 3 and 4.

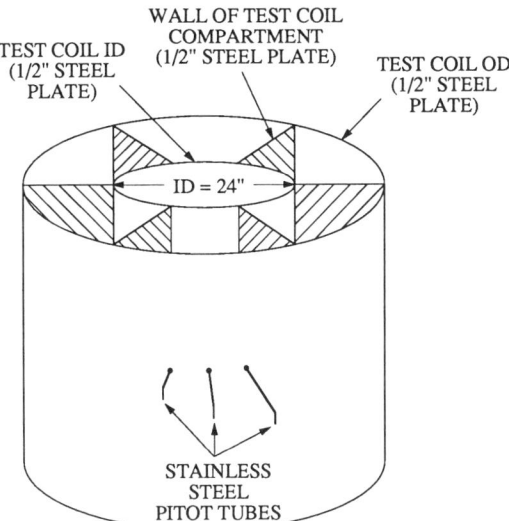

Figure 3 - Sketch of Test Coil Used in Flow Measurement Study. Test coil OD = 60.0, 70.5, 82.25 inches.

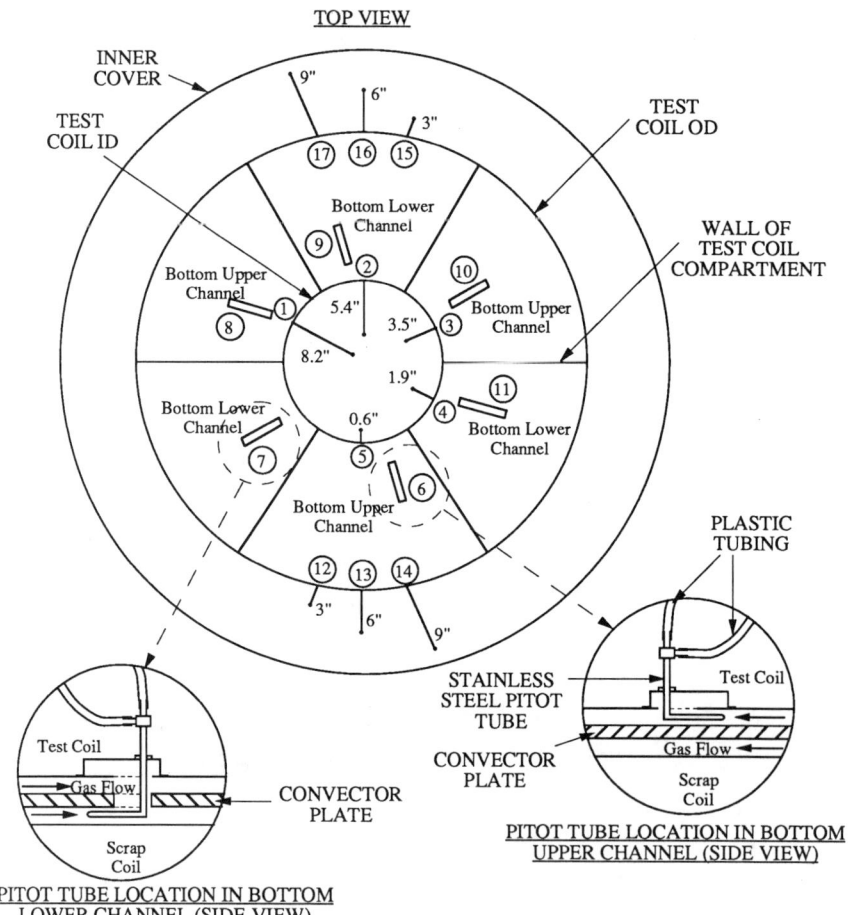

NOTE: PITOT TUBES 14 & 17 WERE REMOVED AND 13 & 16 WERE SHORTENED TO 4" FOR THE 82.25" OD TEST COIL.

Figure 4 - Top and Side Views Showing Pitot Tube Locations for Test Coil Shown in Figure 3.

The pressure drop across any Pitot tube could be read on the manometer located in the basement below the annealing base. From these pressure drop readings, the gas velocity and thus, the flow rates were calculated. Using these test coils, a series of tests were conducted on a dedicated base in the HFH annealing facility, of which some tests were conducted using different convector plate designs with and without a top plate as shown in Figures 5 to 8. Because of safety concerns and temperature limitations in the flow measuring instruments, all tests were conducted at ambient temperatures and the furnace was never used.

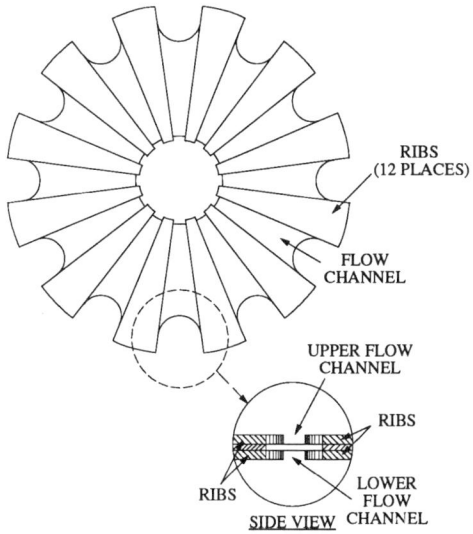

Figure 5 - Standard Convector Plate.

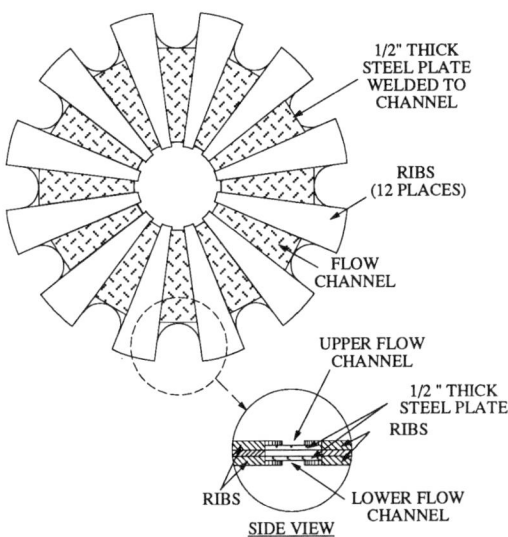

NOTE: MODIFIED CONVECTOR PLATE IS A STANDARD CONVECTOR PLATE WITH THE CHANNEL REDUCED TO 0.5-INCH DEEP BY WELDING 0.5-INCH-THICK STEEL PLATE IN THE CHANNEL.

Figure 6 - Convector Plate of Modified Design A.

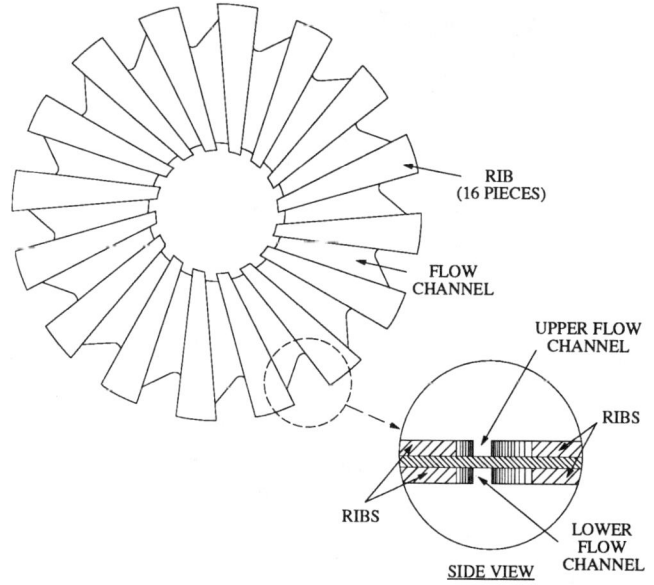

Figure 7 - Convector Plate of Modified Design B.

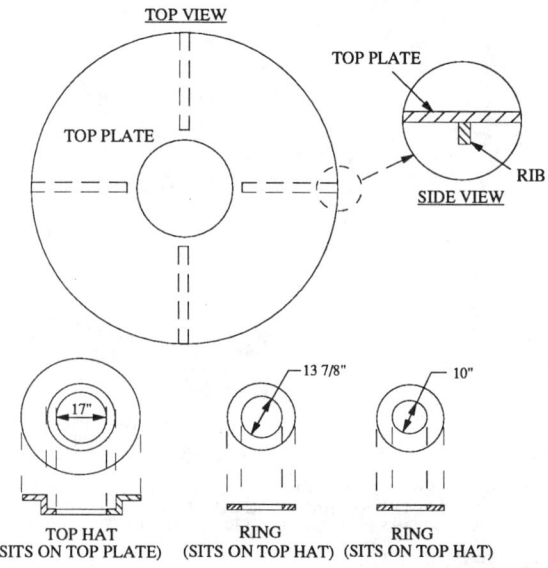

Figure 8 - Top Plate with 'Top Hat' to Constrict Top Opening.

Analysis of the results obtained revealed the following:

1. Total flow increased as the gas density decreases.
2. Consistent flow (percentage of the total flow) through the convector plates were obtained irrespective of gas compositions and densities.
3. Flow distributions through the convector plates decreased as the test coil was moved towards the top of the stack.
4. Flow distributions through the convector plates of the 82.25 inches OD test coil were higher by approximately 3 percent when compared to the 70.50 inches OD test coil.
5. The standard convector plate with a top plate showed marginal increases in the flow.

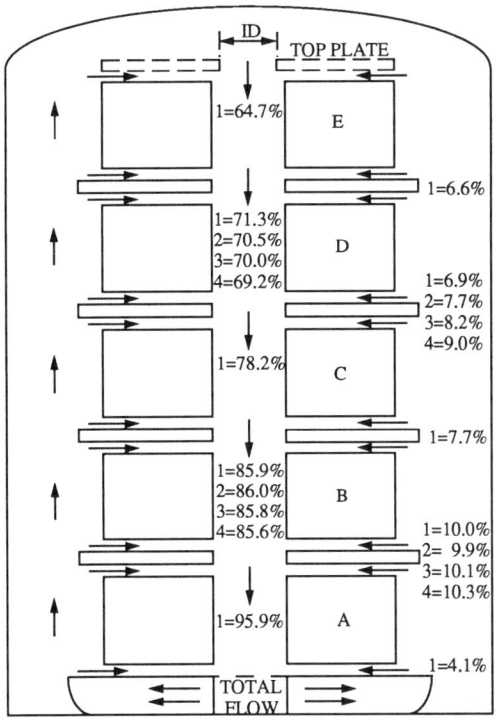

LEGEND
1. NO TOP PLATE
2. TOP PLATE, ID=17.0"
3. TOP PLATE, ID=13.9"
4. TOP PLATE, ID=10.0"

NOTE: FLOW DISTRIBUTIONS SHOWN ABOVE ARE PERCENTAGE VALUE OF THE TOTAL FLOW.

Figure 9 - Flow Distributions for 70.50 inches OD x 48.00 inches Wide Test Coil, Modified Convector Plate of Design A, With/Without Top Plate.

6. Modified convector plate of design A showed reduced flow when compared to the standard convector plate. With a top plate, marginal increases in the flow were observed, as shown in Figure 9.

7. The convector plate of design B, produced the highest flow when compared to the standard convector plate. However, this increase occurs mainly in the convector plate of the bottom coils, resulting in these coils being annealed at a higher temperature compared to the top coil and thus effecting product quality by reducing uniformity of mechanical properties within the coil stack. With a top plate, marginal increases in the flow were observed.

The results of these tests were then incorporated into the theoretical computer model so that an improved HFH batch annealing model could be obtained.

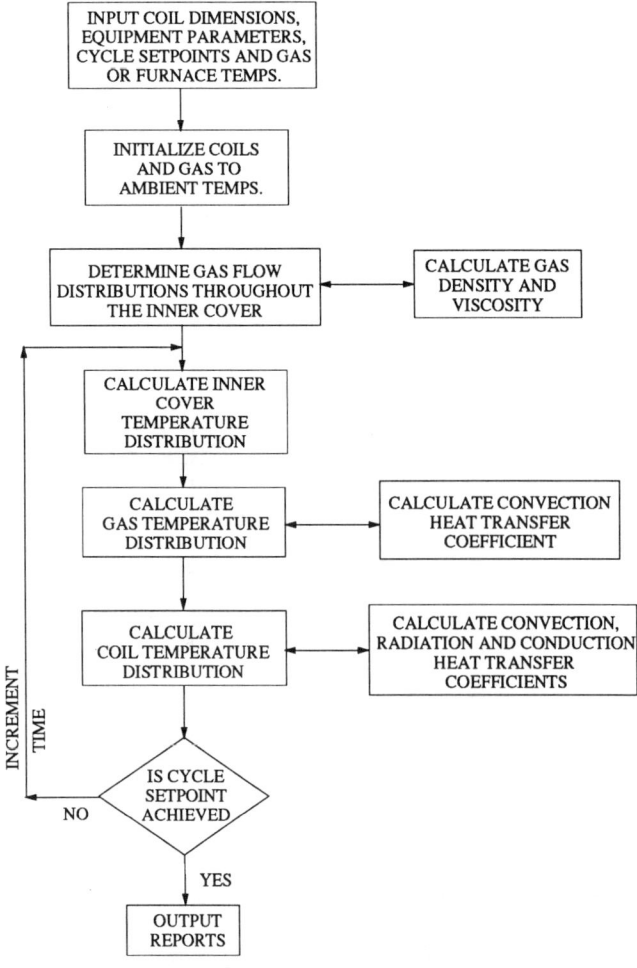

Figure 10 - Flow Chart of Theoretical Annealing Model.

Modeling

Personnel from Carnegie Mellon Research Institute's Computer Engineering Center, modified the original theoretical model developed by USS, to incorporate these latest experimental data. The model consists of approximately 20,000 lines of FORTRAN code and runs on a VAX minicomputer. Its run time is approximately 3 to 5 minutes depending on the stack geometry and number of coils in the stack.

Based on the stack geometry and coil size, the model calculates the flow distributions, and determines the convection, conduction and radiation heat transfer coefficients. Using these coefficients, the energy balance equations are solved via the finite difference technique to determine the gas and coil temperatures. This iteration is repeated until the desired cycle setpoint (cold-spot temperature) is achieved and the time required to reach this cycle setpoint is the heating or annealing time for the stack. A flow chart describing the sequence of steps that occur during execution of the model is shown in Figure 10 and its final outputs are shown in Figures 11 and 12.

This complete model was then verified using data from 40 trials with thermocouples imbedded in the coils along with data from 100 production charges. The heating-time predictions of the model were compared with the empirical model predictions, as shown in Figure 13 for CQ (Commercial Quality) and DQ (Drawing Quality) cycles. Also shown in this figure are the measured heating times obtained from the thermocouple trials with a deviation of ±1.0 hour to account for process variations. These results show that the theoretical model is capable of accurately predicting the annealing process for most of the trials to within ±1.0 hour of the measured data, thus indicating that an increase in productivity could be achieved if the empirical model was replaced by the theoretical model.

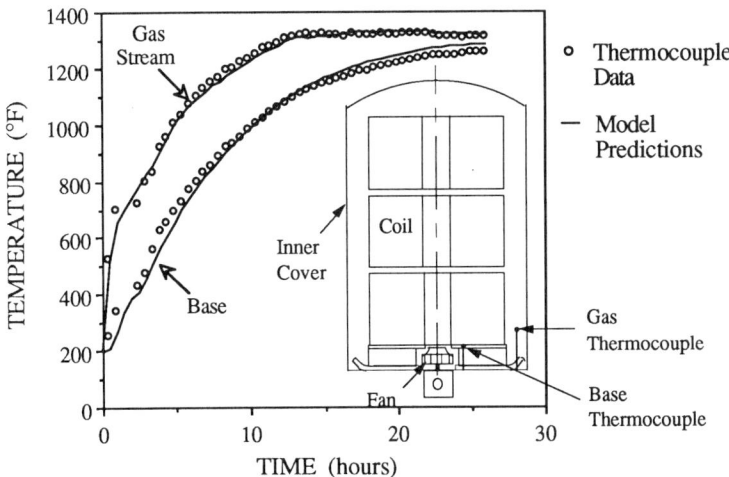

Figure 11 - Comparison of Theoretical Model Predictions Versus Thermocouple Data for Gas and Base Temperatures.

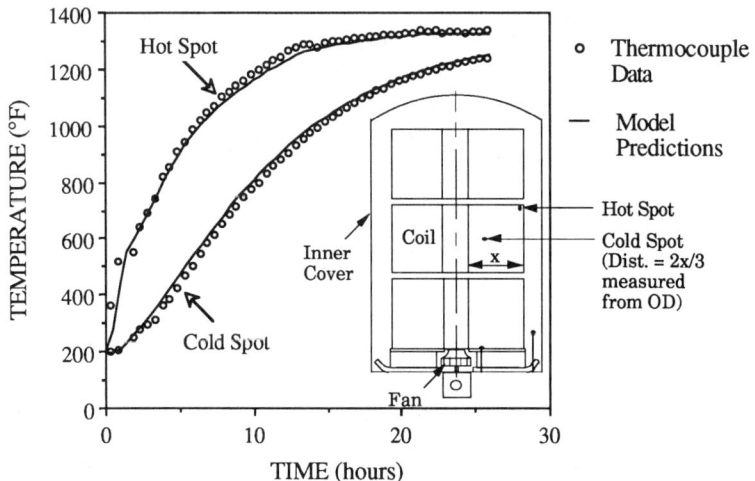

Figure 12 - Comparison of Theoretical Model Predictions Versus Thermocouple Data for Hot and Cold Spot Temperatures.

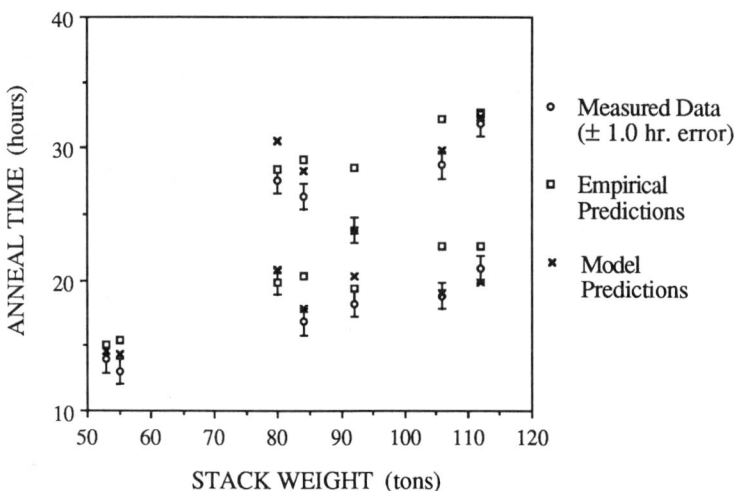

Figure 13 - Comparison of Theoretical Model Versus Empirical Heating Time Predictions for CQ and DQ Trials.

Verification of Annealing Model

Because the extent to which use of this theoretical model in determining annealing times and temperatures would affect annealing quality was not known, the performance of the model was monitored in extensive production trials starting with Commercial Quality (CQ) product in June 1990. Irvin Quality Assurance selected the test product and conducted extensive mechanical-property testing of the head, mid-width and tail from each coil for these materials. A summary of the mechanical properties obtained from these trials is shown in Table III. Table IV shows the mechanical-property summary for coils annealed using the empirical model previously used. Referring to Tables III and IV for the tail tests, the results

Table III - Property Summary for Coils Annealed Using Theoretical Model, Commercial Quality (CQ) Cycle

COIL HEAD

No. of Tests	Carbon % (Grade)		ASTM Grain Size	Yield stress (Ksi)	Tensile Stress (Ksi)	Total Elong. %	Unif. Elong. %	n Value
35	0.04-0.06	Avg.	8.8	30.8	46.4	42.3	25.2	0.23
	(1006)	σ	0.5	2.5	1.8	2.0	1.7	0.01
83	0.04 - 0.08	Avg.	8.7	30.5	46.2	42.5	25.2	0.23
	(1008)	σ	0.5	2.6	1.7	2.5	2.1	0.02
44	0.08 - 0.11	Avg.	9.0	33.0	49.6	40.6	24.6	0.22
	(1010)	σ	0.6	2.9	2.5	3.2	2.4	0.02

COIL MID-WIDTH

No. of Tests	Carbon % (Grade)		ASTM Grain Size	Yield stress (Ksi)	Tensile Stress (Ksi)	Total Elong. %	Unif. Elong. %	n Value
33	0.04 - 0.06	Avg.	8.9	32.5	48.6	40.9	24.9	0.22
	(1006)	σ	0.6	2.2	2.0	2.3	1.2	0.01
78	0.04 - 0.08	Avg.	9.0	31.8	47.5	41.7	25.0	0.22
	(1008)	σ	0.5	2.3	2.2	3.3	1.7	0.01
36	0.08 - 0.11	Avg.	9.3	34.8	51.4	40.0	24.5	0.22
	(1010)	σ	0.6	2.9	2.8	2.0	1.6	0.01

COIL TAIL

No. of Tests	Carbon % (Grade)		ASTM Grain Size	Yield stress (Ksi)	Tensile Stress (Ksi)	Total Elong. %	Unif. Elong. %	n Value
34	0.04 - 0.06	Avg.	9.1	31.7	47.8	41.8	25.4	0.23
	(1006)	σ	0.6	1.9	1.8	2.2	1.1	0.01
83	0.04 - 0.08	Avg.	8.8	31.1	46.9	42.3	25.0	0.22
	(1008)	σ	0.5	1.8	1.8	2.5	1.6	0.01
46	0.08 - 0.11	Avg.	9.4	34.3	51.3	40.1	24.3	0.22
	(1010)	σ	0.5	2.9	2.6	2.6	1.6	0.01

σ - Std. Deviation

Table IV - Property Summary for Coils Annealed Using Empirical Model, Commercial Quality (CQ) Cycle

COIL TAIL

No. of Tests	Carbon % (Grade)		ASTM Grain Size	Yield stress (Ksi)	Tensile Stress (Ksi)	Total Elong. %	Unif. Elong. %	n Value
207	0.04 - 0.06	Avg.	8.8	30.9	47.2	42.7	NA	NA
	(1006)	σ	0.7	2.6	2.1	2.0		
40	0.04 - 0.08	Avg.	8.7	30.4	47.3	42.6	NA	NA
	(1008)	σ	0.6	3.2	2.5	2.7		
75	0.08 - 0.11	Avg.	9.5	34.2	51.4	40.0	NA	NA
	(1010)	σ	0.6	3.3	2.7	2.6		

Table V - Property Summary for Coils Annealed Using Theoretical Model, Drawing Quality (DQ) Cycle

COIL HEAD

No. of Tests	Carbon % (Grade)		ASTM Grain Size	Yield stress (Ksi)	Tensile Stress (Ksi)	Total Elong. %	Unif. Elong. %	n Value
30	0.03 - 0.05	Avg.	6.8	26.2	45.1	41.7	25.8	0.23
	(7050)	σ	0.6	2.2	1.6	2.1	1.8	0.01

COIL MID-WIDTH

No. of Tests	Carbon % (Grade)		ASTM Grain Size	Yield stress (Ksi)	Tensile Stress (Ksi)	Total Elong. %	Unif. Elong. %	n Value
27	0.03 - 0.05	Avg.	7.0	26.2	45.2	40.9	25.6	0.23
	(7050)	σ	0.5	1.2	1.3	1.7	1.7	0.01

COIL TAIL

No. of Tests	Carbon % (Grade)		ASTM Grain Size	Yield stress (Ksi)	Tensile Stress (Ksi)	Total Elong. %	Unif. Elong. %	n Value
29	0.03 - 0.05	Avg.	6.9	26.9	45.4	41.3	25.4	0.23
	(7050)	σ	0.4	1.9	1.3	2.4	1.6	0.01

Table VI - Property Summary for Coils Annealed Using Empirical Model, Drawing Quality (DQ) Cycle

COIL TAIL

No. of Tests	Carbon % (Grade)		ASTM Grain Size	Yield stress (Ksi)	Tensile Stress (Ksi)	Total Elong. %	Unif. Elong. %	n Value
*	0.03-0.05	Avg.	7.0	27.0	44.6	44.2	24.7	NA
	(7050)	σ	0.6	2.0	1.8	1.8	1.3	

NA - Not Available, σ - Std. Deviation, * - Based on two years of production testing

show that for each grade, consistent average levels of mechanical properties were achieved and uniformity as measured by standard deviation was significantly improved when the theoretical model was used for annealing. Also, referring to Table III, the results indicate that consistent mechanical properties are being achieved throughout the coils when the theoretical model was used for annealing. The results also indicated that a productivity increase of 8.5 percent, due to decreased heating time, is being achieved for CQ steel sheet. Because of this good performance, a decision was made in November 1990 to anneal all CQ charges using the theoretical model.

On completion of CQ product evaluation, the performance of the theoretical model was then extended to Drawing Quality (DQ) products. Table V shows mechanical property summary for trial coils annealed using the theoretical model and Table VI for coils annealed using the empirical model. The results of Table V tail tests, show that the mechanical properties were almost identical to historical properties as shown in Table VI, but generally better uniformity was obtained for coils annealed using the theoretical model, with a productivity increase of 3 percent.

Thus, for a 50/50 mix of CQ and DQ product, an overall productivity increase of about 6 percent with good mechanical properties are being achieved at the HFH box-annealing shop of the Irvin plant.

Conclusions

Capabilities of the current theoretical model are:

1. Accounts for variations in coil gage and its effects on the radial thermal conductivity of the coils.

2. Accurately models the annealing gas flow to within 2 percent of the measured flow through the convector-plate channels.

3. Accurately predicts the temperature distribution of all coils in a wide variety of stack configurations throughout the heating and soaking phases of the annealing process.

4. Accounts for annealing process parameter variations (e.g. use of actual gas stream thermocouple temperature input, burner failures, unexpected soak time, gas stream overshoot, etc).

5. Can be used as a tool for improving annealing cycles or in developing new equipment designs (e.g. new types of convector plates, different inner cover design, use of top cover plates, etc).

6. 6 percent improvement in productivity while maintaining product quality.

Thus the theoretical annealing model predicts the heating times for different stack configuration and cycles more accurately than the empirical annealing model to provide a high quality product at improved productivity levels.

Disclaimer

The material in this paper is intended for general information only. Any use of this material in relation to any specific application should be based on independent examination and verification of its unrestricted availability for such use, and a determination of suitability for the application by professionally qualified personnel. No license under any USX Corporation patents or other proprietary interest is implied by the publication of this paper. Those making use of or relying upon the material assume all risks and liability arising from such use or reliance.

Acknowledgments

The authors would like to thank many of our associates who have contributed to the success of this project. We also wish to publicly acknowledge the efforts of Alloy Engineering of Cleveland who designed and constructed the test coils and instrumentation for the flow test.

References

1. K. G. Brickner, H. B. Kincaid, and A. E. Rudolph, " Start-up Experience with High-Flow, High-Hydrogen Box Annealing at Irvin Plant of Mon Valley Works," 30th Mechanical Working and Steel Processing Proceedings, 26 (1988), 147-154.

2. A. P. DeVito et al., "Improvements in Mechanical Properties and Surface Cleanliness using High Flow Hydrogen Annealing," 32nd Mechanical Working and Steel Processing Proceedings, 28 (1990), 249-258.

3. T. R. S. Rao, G. J. Barth and J. R. Miller, "Computer Model Prediction of Heating, Soaking and Cooling Times in Batch Coil Annealing," Iron and Steel Engineer, (9) (1983), 22-33.

4. G. F. Harvey, " Mathematical Simulation of Tight Coil Annealing," The Journal of the Australasian Institute of Metals, 22 (3) (1977), 28-37.

5. Wolfgang Potke, Rudolf Jeschar, and Georg Kehse, " Heat Transport During Annealing of Coils in a Bell-Type Furnace with Hydrogen as Protecting Gas," Stahl u. Eisen, 108 (12) (1988), 581-585.

6. A. A. Lisogor and V. I. Mitkalinnyi, " Thermal and Physical Properties of Stacks of Cold Rolled Steel," STAL in English, (12) (1970), 996-998.

V. Batch Annealing: Product Metallurgy

EXPERIENCE WITH HIGH CONVECTION HYDROGEN BATCH ANNEALING AT DOFASCO

William F. Gasse, and Steven J. Thomas

DOFASCO Incorporated
1390 Burlington Street East
P.O. Box 2460, Hamilton, Ontario L8N 3J5, CANADA.

Abstract

Start-up experience of the initial hydrogen batch annealing installation at Dofasco's Number 3 Sheet Mill is detailed. The rationale for choosing hydrogen batch annealing at Dofasco as opposed to alternatives such as continuous annealing is discussed. On-going investigations to develop optimum thermal and protective atmosphere cycles are outlined. Improvements in product uniformity and surface cleanliness results are also presented.

INTRODUCTION

Prior to the 1970's, almost all cold rolled wide strip steel production was annealed using conventional tight coil batch annealing equipment and nitrogen-based annealing atmospheres. Since that time, two new annealing technologies have emerged, continuous strip annealing and high convection hydrogen batch annealing, both driven by the need to improve final product quality. Japanese steel strip producers introduced continuous annealing of cold rolled steel for large volume production in 1971, and have since licensed and sold the technology in various forms worldwide. European steel strip producers were the first to employ hydrogen batch annealing in 1983, followed by major North American steel producers since 1987.

Compared with conventional HNX batch annealing, greater heating and cooling rates are achieved with hydrogen batch annealing, coincident with a more uniform temperature distribution throughout the coils being annealed. Approximately one-half of the productivity improvement is attributed to the use of the pure hydrogen annealing atmosphere, the other one-half to the increased convection rates attained through use of high flowrate designed equipment. Hydrogen gas has a thermal conductivity approximately seven times that of conventional nitrogen-based annealing atmospheres. Because hydrogen gas has a very low density, manufacturers of hydrogen batch annealing equipment are able to utilize larger base fans than conventional batch annealing equipment, and claim annealing atmosphere flowrates in excess of 30,000 scfm. Pure hydrogen gas also has a greater reducing potential than conventional nitrogen-based annealing atmospheres. Product quality benefits arrived at through the use of hydrogen batch annealing relative to conventional batch annealing include improved uniformity of mechanical properties, improved surface cleanliness, and the elimination of nitriding during batch annealing.

This paper describes the results obtained by Dofasco from trial investigations, commissioning of the annealing facility, and the initial six months of hydrogen batch annealing production.

INITIAL INVESTIGATIONS

Dofasco's Research Department conducted four production scale trials between September 1987 and December 1989 to evaluate high convection hydrogen batch annealing quality and performance. This work involved the hydrogen batch annealing of 24 coils and the continuous annealing of 6 coils using outside facilities.

Results of the trials indicated that with hydrogen batch annealing:

- heating and cooling of the charge is both more rapid and more uniform than with conventional batch annealing.
- heat transfer within a coil is increased, particularly in the radial direction.
- overheating is reduced to less than 10°C.
- the process is capable of achieving a temperature uniformity of 10°C throughout the charge at the end of heating.
- productivity was increased by at least 97 percent relative to conventional batch annealing at Dofasco.
- through-coil mechanical properties uniformity and surface cleanliness levels were improved relative to conventional batch annealed quality levels.
- hydrogen batch annealing is capable of achieving quality levels approaching or exceeding those observed for continuously annealed material, especially for deep-drawing qualities production.

RATIONALE FOR HYDROGEN BATCH ANNEALING

A continuous annealing line had been planned as part of a 600,000 NT per year cold rolled modernization program due to come on stream in 1992. In 1989 a comprehensive review of cold roll market trends, product types, available technology, and total production costs concluded that hydrogen batch annealing is the preferred technology for Dofasco's increase or modernization of annealing capacity. The review indicated that while continuous annealing should result in yields approximately 1.5 percent greater than with hydrogen batch annealing, and should have lower quality-associated costs, continuous annealing would overall be a more expensive process due to its much higher capital expense. Similar conclusions have been reported by other steel producers. [1,2]

Hydrogen batch annealing was favoured for the following reasons:

1. Hydrogen batch annealing is less restrictive in terms of chemistry and processing requirements than continuous annealing.

- Steelmaking - continuous annealed CQ and DQ product requires lower carbon levels, typically <.02% carbon vs .04% - .08% carbon typical for batch annealed product.
  - continuous annealed DDQ product requires vaccuum degassed ultra-low carbon grades, typically < 50 ppm carbon vs .02% - .04% carbon typical for batch annealed product.
- Hot Rolling - continuous annealed DDQ product requires higher coiling temperatures and "hot-ends" cooling practices, or stabilization using titanium or boron to minimize hot rolled microstructure and mechanical properties variation along the coil length and across the strip width.
- Pickling - continuous annealed DDQ product requires reduced linespeeds and/or scale-breaking to remove the thicker and more acid-resistant scale formed.
- Tempering - continuous annealed CQ, DQ, and HSLA product has higher yield point elongation than batch annealed product, requiring greater temper rolling reductions and consequent decrease in formability.

2. Hydrogen batch annealing offered the flexibility of modernization or incremental capacity expansion (in units of 2 bases plus 1 furnace, or approximately 20,000 tons per year) which continuous annealing does not. This same flexibility applies to capacity reduction when required.

3. Hydrogen batch annealing expansion or modernization represented a lower risk alternative to expansion via continuous annealing due to lower capital cost. Figure 1 summarizes the total cost of processing vs increase in capacity for the three possible scenarios.

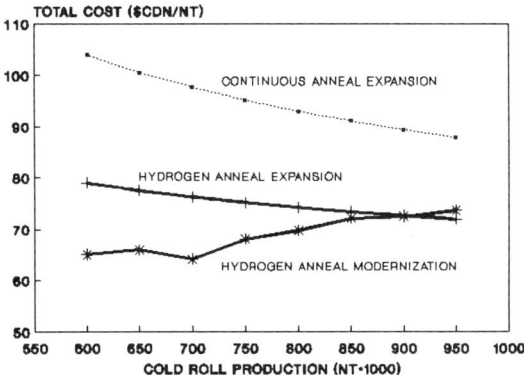

Figure 1 - Total Anneal and Temper Cost (includes added upstream process costs) vs Cold Roll Production Increase.

Hydrogen batch annealing was targeted for exposed quality DDQ and HSLA grades, where present and future customer quality restrictions are most demanding. Dofasco has traditionally produced these grades utilizing a range of chemistries, prior processing, and coil geometries to suit individual customer requirements. Hydrogen batch annealing is ideally suited to meet the quality demands for these grades, while supporting the varied product mix.

FACILITY DESCRIPTION

The two furnace, four base facility was originally proposed in 1988 for batch annealing research studies. Based on a detailed technical comparison of the equipment, and discussions with steel producers already using hydrogen batch annealing, Dofasco chose EBNER for the initial installation. The location selected was in Dofasco's Number 3 Sheet Mill, for ease of integration with existing production facilities. This shop also uses direct-fired furnaces and water-cooled annealing equipment, similar in concept to the EBNER technology. An aim throughput of 5.0 NT/hour was anticipated for the facility with a DDQ and HSLA product mix.

The design details of the EBNER equipment have been well documented elsewhere[3]; the basic components of the Dofasco facility are essentially unchanged.

The Dofasco furnaces are equipped with 12 high velocity tangential burners sized for COG fuel. The burners are arranged in three rows and are evenly spaced around the circumference of the furnace, to reduce localized heating of the inner cover. A heat shield is not used. For environmental reasons, Dofasco specified an additional small burner separate from the main combustion system to burn waste annealing atmosphere in lieu of venting. Due to the sulphur content of the COG fuel gas, Dofasco specified that the inner covers be constructed of AISI 302B stainless steel in place of the standard 309S.

The Dofasco bases can accomodate coils to 86" OD, and to a stacking height of 157" including convector plates. Maximum charge weight is approximately 116 tons. Each base is hardwired with 24 test thermocouple outputs to facilitate buried thermocouple studies. A Compaq 386 process control computer is interfaced to a General Electric Series 5 programmable logic controller equipped with Genius bus for data transfer between the master control PLC and individual base control PLC's.

COMMISSIONING TRIALS

As part of the purchase agreement, Dofasco requested that the EBNER equipment meet strict performance and utilities consumptions guarantees. DDQ and HSLA charges of 3 and 4 coil configurations were specified. Overheating was restricted to less than 10°C at any time during the test anneals, and a temperature uniformity of 10°C throughout the charge at the end of heating was also specified. Utilities consumptions were measured with all four bases in production.

A total of thirteen instrumented test anneals were run during the commissioning period. EBNER personnel made on-site equipment modifications to the convector plates and base charge plates to increase annealing atmosphere flow efficiency. As Dofasco cold rolled coils have a 20" coil bore, convector plates were originally sized with a minimum inner diameter of 18.5", to prevent telescoping of the coil eye during light gauge anneals. To increase annealing atmosphere flow down the bore while still retaining the design intentions, EBNER reduced the length of every second radial segment of the convectors at the inner diameter. To better equalize heat transfer within the charge, the annealing atmosphere passages in the base charge plate were restricted, increasing flow through the convector plates. With the modifications, the equipment met the temperature and throughput levels specified in the contract.

Table I summarizes the temperature and productivity results obtained during the commissioning trials.

The temperature data from the test anneals was also used to calibrate an off-line batch annealing simulation model for high convection hydrogen processing. Tables of predicted end-of-heating times required to achieve critical core temperature were then generated. At present these require that the operator determine the most appropriate total heating time based on the various coil width, gauge, and weight combinations in each charge. In the future, the heating times will be generated automatically by the process control computer based upon the coil information input by the operator.

PRODUCTION RESULTS

Productivity and Utilities Performance

Hydrogen batch anneal productivities are dramatically increased relative to conventional batch anneal production levels. Table 2 summarizes current hydrogen batch anneal productivity levels for CQ, DDQ, and HSLA production. Both heating and cooling productivities are increased, reflecting the benefits of the hydrogen annealing atmosphere and higher convection rates. The benefit of forced-air/water cooling of the inner cover, as opposed to forced-air cooling only, also contributes to the increase in cooling productivity noted.

Utilities consumptions had been forecast in making the recommendation to proceed with hydrogen batch annealing technology. A total utilities cost savings of more than $2.75 CDN per annealed ton was demonstrated during the commissioning trials. Figure 2 presents a comparison of utilities costs for the hydrogen batch anneal facility versus conventional batch annealing equipment.

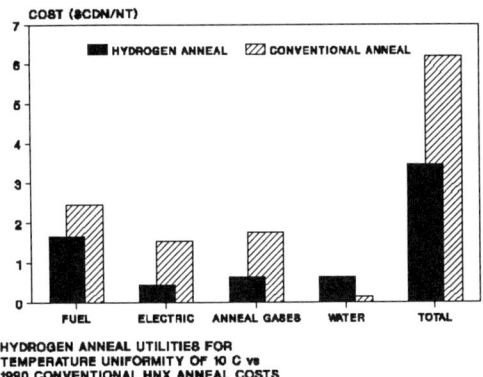

Figure 2 - Utilities Costs Comparison Between Hydrogen and Conventional Batch Annealing.

Mechanical Properties Uniformity

Uniformity of final product mechanical properties is dependent upon processing prior to, during, and subsequent to batch annealing. Table 3 provides an estimate of the magnitude of yield stress variation attributable to each process for hydrogen and conventional batch annealed HSLA, low carbon DDQ, and stabilized EDDQ products. This information is also shown graphically in Figure 3.

Figure 3 shows that:

1. The capability of hydrogen batch annealing to achieve increased temperature uniformity has greatest implications for HSLA product improvement.

2. The ability of hydrogen batch annealing to eliminate nitrogen-pickup as a source of yield stress variation has greatest implication for EDDQ product improvement.

3. Temper rolling is the predominant source of final product yield stress variation for batch annealed DDQ product.

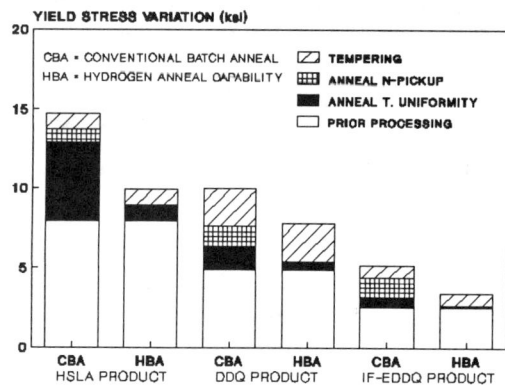

Figure 3 - Yield Stress Variations Arising from Processing: Estimated Magnitudes for Hydrogen and Conventional Batch Annealed HSLA, low-carbon DDQ, and IF-EDDQ products.

In comparison anneals using coils which received identical processing before and after annealing, hydrogen batch annealing reduced through-coil longitudinal yield stress variation by 45 percent relative to conventional batch annealing for both DDQ and HSLA products. Through-charge yield stress statistical ranges (6s) of 3.9 ksi for AKDQ, and 4.8 ksi for HSLA have been achieved for hydrogen batch annealed final product.

Figure 4 shows longitudinal yield stress distributions for hydrogen and conventional batch annealed HSLA production during the initial six month operating period. Results are reported for temper rolled lead and trail positions. The reduction in overheated coil exterior material with hydrogen batch annealing is evident in the sharp decline in frequency of yield stress less than 50 ksi relative to conventional batch annealed production. Hydrogen batch annealing to a temperature uniformity of 30°C reduced the HSLA yield stress statistical range by 4.9 ksi. Similar results have been obtained for DDQ production, where hydrogen batch annealing reduced the yield stress statistical range by 3.6 ksi.

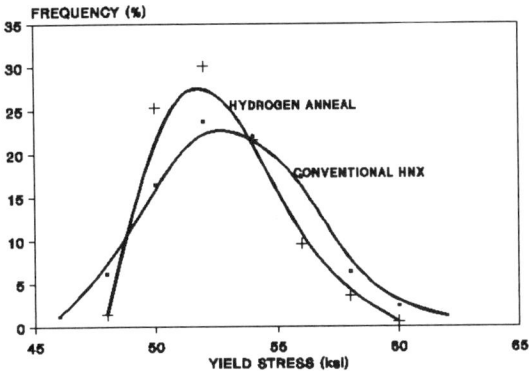

Figure 4 - Batch Annealed HSLA Yield Stress Distributions: Hydrogen and Conventional Batch Annealed Production.

Nitrogen-Pickup

Besides contributing to mechanical properties variability, nitrogen-pickup also contributes to reduced magnetic properties in motor lamination quality steels. Coreloss is increased by fine grain size, and thus is adversely affected by the fine-grained strip surface structure resulting from nitrogen-pickup, as shown in Figure 5. No nitriding is possible with the pure hydrogen annealing atmosphere, resulting in improved product quality relative to HNX batch annealing.

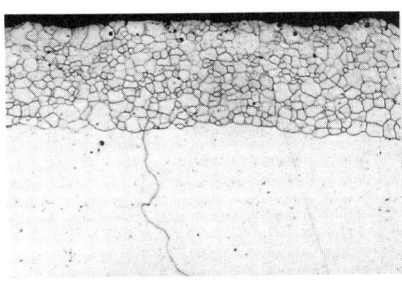 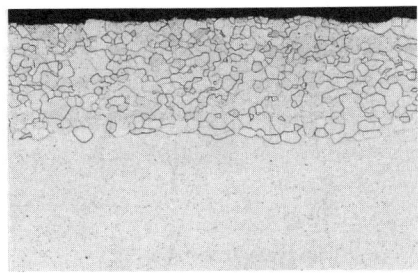

Longitudinal cross-section etched to reveal grains.
• Marshall's etch,
• Picral etch,
• Marshall's etch.

X100

Same section etched to reveal AlN precipitates.
• Picral etch
• AlN etchant,
• Ammonium hydroxide rinse.

X100

Figure 5 - Microstructures of Motor-Lamination Quality steel showing fine-grained strip surface microstructure resulting from nitrogen-pickup during HNX batch annealing.

Surface Cleanliness

Surface cleanliness of cold rolled product is characterized primarily by the levels of surface carbon, iron fines, and oxides present on the strip. One industry 'standard' for cleanliness is the FORD specification (surface carbon = 7.0 mg/m^2 maximum, after power wash).[4] Batch annealed surface cleanliness is dependent upon steel chemistry, surface chemistry, annealing temperature uniformity and time, strip geometry, and annealing atmosphere-metal reactions in the gaps between individual coil wraps.

Annealing in the pure hydrogen atmosphere facilitates the reduction of residual rolling oils and metal oxides present on the strip surface following the cold rolling operation. The capability of hydrogen batch annealing to maintain lower temperature differentials within the coils during heating allows the oils to volatilize evenly and be removed in the waste hydrogen exchange flow instead of cracking at the hot coil faces and forming the surface defect commonly referred to as "snake-edge". Surface cleanliness results from trial DDQ annealing cycles which had a two hour hold at 425°C during heating were unchanged from those achieved in production anneals which use a six hour ramped heatup.

Through-coil FORD surface carbon levels average between 2.0 mg/m^2 and 2.2 mg/m^2 for hydrogen batch annealed DDQ product, as shown in by the middle line in Figure 6. These levels represent a balance between acceptable surface cleanliness and hydrogen consumption. Best surface cleanliness results have been achieved on trial coils processed with fresh rolling solution in the final mill-stand during cold rolling and the maximum hydrogen exchange flow of 24 m^3/hour during annealing. FORD surface carbon results improve to approximately 1.4 mg/m^2 throughout the body of the coils, but coil exterior and coil bore levels remain unchanged from regular production levels. The higher levels at these positions are believed due to the continued evolution of methane during the initial portion of the cooling cycle coupled with no hydrogen exchange flow during cooling, leading to surface carbon contamination of the outside surfaces of the coils.

Relatively high surface carbon levels (to 4.5 mg/m^2) have been observed in the inner wraps of hydrogen batch annealed coils. This phenomena is also observed for conventional batch annealed material, but is not observed in coils which have been electrolytically cleaned prior to batch annealing. Investigation of the surface cleanliness after cold rolling failed to establish any relationships with batch annealed surface carbon contamination along the length of the coil. Further investigation positively correlated batch annealed surface carbon level to interwrap stress levels in the coil after cold rolling. The use of high coiling tension during initial coil buildup at the tandem cold mill results in increased interfacial pressures and decreased interwrap clearances in this position of the coil. These factors have been previously noted to increase surface carbon levels following batch annealing.[5,6,7] The results from trials to establish the effect of coiling tension on through-coil surface cleanliness levels are also shown in Figure 6.

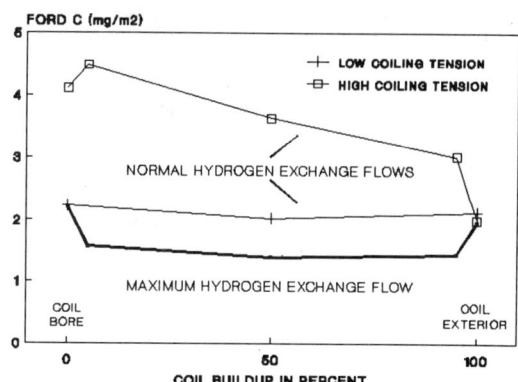

Figure 6 - Effect of Coiling Tension and Hydrogen Exchange Flow during annealing on Through-Coil Surface Carbon Levels.

Hydrogen batch annealing reduced average surface carbon levels by 23 percent for DDQ product, and by 47 percent for HSLA product relative to conventional batch annealing. Average iron fines levels were similarly reduced by 29 percent for DDQ product and by 56 percent for HSLA product. Surface contamination due to MnO formation, although not quantitatively determined, has been observed to be much less severe throughout the hydrogen annealed coils, as shown in Figure 7.

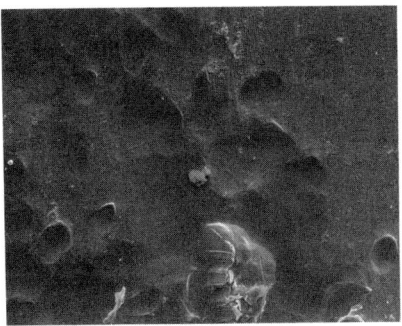

HYDROGEN ANNEALED
Few Manganese Oxides
X500

CONVENTIONAL HNX ANNEALED
Numerous manganese oxides
X500

Figure 7 - **Manganese Oxides on Batch Annealed Strip Surface: Hydrogen Annealed and Conventional HNX Batch Annealed DDQ material.**

CONCLUSION

The implementation of hydrogen batch annealing has resulted in substantially higher productivity levels, along with improved temperature uniformity in the batch annealing process relative to conventional batch annealing at Dofasco. A utilities costs savings is also realized relative to conventional batch annealing.

Results from the initial six months of operation indicate that hydrogen batch annealing has reduced DDQ and HSLA product yield stress variation. Surface cleanliness of drawing quality product is improved relative to levels obtained with conventional batch annealing. Surface cleanliness has been shown to depend on tandem mill practices, and volume of hydrogen annealing atmosphere exchanged during annealing.

REFERENCES

1. Junius, H. T., et al, Stahl und Eisen, vol.108, No.20, 1988.

2. Devito, A. P., et al, ISS-AIME Mechanical Working Conference, Cincinnati, October 1990.

3. Lochner, H., Iron and Steel Engineer, No. 67, March 1990.

4. Ford Laboratory Test Method BZ 2-3, 1979.
   Ford Engineering Material Specification ESB-M2P 117-A.

5. Inokuti, Y., Trans ISIJ, Vol.15, 1975.

6. Jenkins, R., et al, Commission of the European Communities Report eur 10132 en, 1986.

7. Chatelain, B. and Leroy, V., Commission of the European Communities Report eur 8700 en, 1987.

TABLE I: HYDROGEN BATCH ANNEALING FACILITY COMMISSIONING SUMMARY

ITEM	DDQ TEST ANNEALS		HSLA TEST ANNEALS	
	3 HIGH	4 HIGH	3 HIGH	4 HIGH
AVERAGE CHARGE WEIGHT (TONS)	78	79	78	74
CONTROL TEMPERATURE SETPOINT (°C)	704	703	627	623
MAXIMUM STEEL TEMPERATURE (°C)	710	710	631	631
TEMPERATURE UNIFORMITY (°C)	10	10	10	10
PRODUCTIVITY (TONS/HR) : HEATING	2.1	2.4	2.5	3.1
: COOLING	3.6	3.8	4.1	4.5
: OVERALL	1.33	1.46	1.55	1.84
UTILITIES CONSUMPTIONS/NT : FUEL (m3)	40.8	40.3	33.0	31.7
: ELECTRIC (KWh)	12.2	10.8	9.7	7.9
: HYDROGEN (m3)	2.0	2.1	2.2	2.5
: NITROGEN (m3)	3.0	3.1	4.1	4.2
: WATER (m3)	6.0	5.4	6.2	4.9

TABLE II: CURRENT HYDROGEN BATCH ANNEAL PRODUCTIVITY LEVELS
and
PERCENT INCREASE RELATIVE TO 1990 CONVENTIONAL HNX BATCH ANNEAL LEVELS

COMMODITY ANNEALED	PRODUCTIVITY (TONS/HOUR)		
	HEATING	COOLING	OVERALL
COMMERCIAL QUALITY PRODUCT	3.5	3.0	1.64
90°C TEMPERATURE UNIFORMITY	(+ 74%)	(+ 167%)	(+ 128%)
DEEP DRAWING QUALITY PRODUCT	2.4	2.7	1.28
30°C TEMPERATURE UNIFORMITY	(+ 40%)	(+ 121%)	(+ 78%)
HIGH STRENGTH LOW ALLOY PRODUCT	3.0	2.8	1.46
30°C TEMPERATURE UNIFORMITY	(+ 116%)	(+ 346%)	(+ 240%)

STEEL GRADE	HSLA PRODUCT			LOW-CARBON DDQ PRODUCT			I.F.- EDDQ PRODUCT		
PROCESS	SENSITIVITY	TYPICAL RANGE	ΔYS DUE TO PROCESS	SENSITIVITY	TYPICAL RANGE	ΔYS DUE TO PROCESS	SENSITIVITY	TYPICAL RANGE	ΔYS DUE TO PROCESS
STEELMAKING: CHEMISTRY	C = 32 ksi/% Mn = 5.2 Si = 12 P = 73 N = 450 Nb = 188 V = 142	12.1 - 17.5 ksi	5.4 ksi	C = 32 ksi/% Mn = 5.2 Si = 12 P = 73 N = 450	4.1 - 7.0 ksi	2.9 ksi	C = 700ksi/% Mn = 5.2 N = 450 Ti = 10	2.4 - 4.8 ksi.	2.4 ksi
HOT ROLLED COILING TEMPERATURE	.3 ksi/10°C	50°C	1.5 ksi	0 (low CT)	50°C	0.0 ksi	0.2 ksi/10°C	50°C	1.0 ksi
COLD REDUCTION	.5 ksi/1%	2%	1.0 ksi	.5 ksi/1%	2%	1.0 ksi	0	0	0.0 ksi
CONVENTIONAL HNX: BATCH ANNEAL Δ TEMPERATURE	1 ksi/10°C	50°C	5.0 ksi	.5 ksi/10°C	30°C	1.5 ksi	.13 ksi/10°C	50°C	0.7 ksi
HYDROGEN:		30°C	3.0 ksi		30°C	1.5 ksi		30°C	0.4 ksi
CONVENTIONAL HNX: BATCH ANNEAL NITROGEN PICKUP	1.2ksi/50ppm	30 ppm	0.8 ksi	1.2ksi/50ppm	50 ppm	1.2 ksi	1.2ksi/50ppm	50 ppm	1.2 ksi
HYDROGEN:		0	0.0 ksi		0	0.0 ksi		0	0.0 ksi
TEMPER ROLLING ELONGATION	2.5 ksi/1%	0.4%	1.0 ksi	6.0 ksi/1%	0.4%	2.4 ksi	4.0 ksi/1%	0.2%	0.8 ksi
CONVENTIONAL HNX: TOTAL ESTIMATED YIELD STRESS VARIATION	ESTIMATED IMPROVEMENT: ΔYS RANGE = 2.8 ksi		14.7 ksi	ESTIMATED IMPROVEMENT: ΔYS RANGE = 1.2 ksi		9.0 ksi	ESTIMATED IMPROVEMENT: ΔYS RANGE = 1.5 ksi		6.1 ksi
HYDROGEN:			11.9 ksi			7.8 ksi			4.6 ksi
HYDROGEN BATCH ANNEAL INITIAL 6 MONTHS PRODUCTION (30°C UNIFORMITY)	ACTUAL HSLA PRODUCT IMPROVEMENT: ΔYS RANGE = 4.9 ksi			ACTUAL DDQ PRODUCT IMPROVEMENT: ΔYS RANGE = 3.6 ksi			ACTUAL EDDQ PRODUCT IMPROVEMENT: ΔYS RANGE = na.		

**TABLE III: Yield Stress Variations Arising from Processing:
Estimated Magnitudes for HSLA, low-carbon DDQ, and IF-EDDQ material.**

SENSITIVITY AND TYPICAL RANGE VALUES FROM LITERATURE and/or DOFASCO PRODUCTION

GAS-METAL REACTIONS DURING 100% H_2 BATCH-ANNEALING

J.M. Mataigne, M. Lamberigts, V. Leroy

CRM, Abbaye du Val-Benoît, B - 4000 LIEGE (Belgium)

Abstract

The performances of bell-type annealing furnaces can be definitely improved by increasing heat transfer between heating cover and coil stack by using the so-called high convection (hydrogen) - high flow (large fan) technology. The advantages of this new annealing practice are nowadays well established in terms of energy saving, annealing time shortening as well as consistency of mechanical properties.
The aim of the present work is to analyze the gas-metal reactions taking place during batch-annealing in detail and to compare the efficiencies of treatments carried out under HNX and H_2. These gas-metal reactions are indeed responsible for the elimination of rolling oil residues, which warrants the steel sheet final surface cleanliness.
In order to avoid arguments about the reliability of simulation tests, the present work was based on full-scale instrumented 20-ton coils annealed in industrial furnaces.

Introduction

In order to insure easy stamping, assembling and painting operations in finishing lines, steelmakers must produce annealed and skin-passed deep drawing-quality steel sheets of excellent surface cleanliness and mechanical property consistency, all along individual coils, as well as from coil to coil. This is especially true for the automotive industry, where the smooth running of drawing presses and welding robots is essential. In this connection, metal springback after stamping must for instance be kept as constant as possible. The up-grading of continuous annealing and hot-dip galvanizing lines has certainly helped meet these requirements, particularly by the introduction of a pre-cleaning stage prior to continuous annealing and through better temperature homogeneity during annealing[1,2,3].

From this viewpoint, the recrystallization batch-annealing under N_2-5% H_2 protective atmosphere of steel sheet might look less suited to meet the same users' requirements, mainly because of temperature heterogeneities inside thus treated coils.

The performance of bell-type annealing furnaces, which otherwise exhibit attractive utilization flexibility and demand-adaptable investment costs, can only be improved by increasing the heat transfer between heating cover and coil stack[4,5,6,7].

More precisely, better temperature homogeneity must prevail inside the coils, particularly when the cold-rolled microstructure reaches the AlN precipitation and recrystallization temperature ranges. This can be more easily achieved under a pure hydrogen protective atmosphere, because of its excellent conductivity and convection properties.

As a reminder, table 1 gives comparative values of physical constants for hydrogen and nitrogen, while figure 1 shows, in a simplified geometry, how outstanding hydrogen convection heat transfer can be, provided the gas mass flow is kept constant.

Because of hydrogen low specific gravity, a high mass flow can however only be maintained inside the furnace by driving the gas with an appropriate impeller (rotor), which defines the so-called "high convection-high flow technology".

Obviously enough, the heat which is thus efficiently delivered in the coil eye then propagates by conduction throughout the whole coil. Real time temperature distribution analysis has shown that the good heat conductivity of the hydrogen gas contained in interturn gaps might alter the balance between axial and radial heat transfers to the coil's cold spot : this is further illustrated by figure 2, based on literature[8].

TABLE 1 - PHYSICAL CONSTANTS

	Thermal conduct. $mW\ m^{-1}\ K^{-1}$	Viscos. at 0°C 10^{-5} poise	Spec. heat at 24°C $kJ\ K^{-1}\ kg^{-1}$
N_2	24	16.58	1.04
H_2	163	3.42	14.32

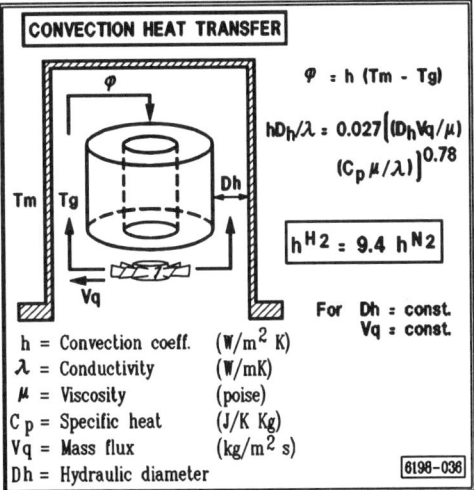

Figure 1 - Convection heat transfer in batch-annealing.

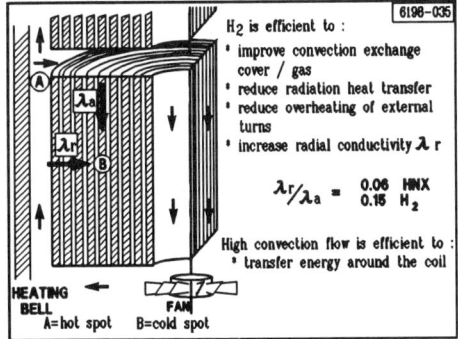

Figure 2 - Conductivity in batch-annealing

It all looks like the combined utilization of hydrogen (high convection) and gas drive technologies (high flow) should lead to a much better temperature homogeneity inside steel coils annealed in bell-type furnaces. This has received a first confirmation by measurements made on industrial furnaces, as is evidenced by figure 3, which pertains to treatments carried out under 100% H_2 and N_2-5% H_2. It should be pointed out that the external wraps of coils treated under 100% H_2 exhibit no radiation-related overheating. Furthermore, the minimized thermal gradient between "hot" and "cold" spots in the microstructure recovery and recrystallization temperature ranges should also be emphasized : this will warrant mechanical property consistency. Minimizing the thermal gradient is not so easy at the treatment outset, although it depends on how the heating cycle is conducted. This point will be addressed to later on.

Aim of the work

Although its advantages in terms of energy conservation and cycle time shortening, achieved by better heat transfer during heating and cooling, are certainly worth comments, we shall not deal with the economics of this new technology.

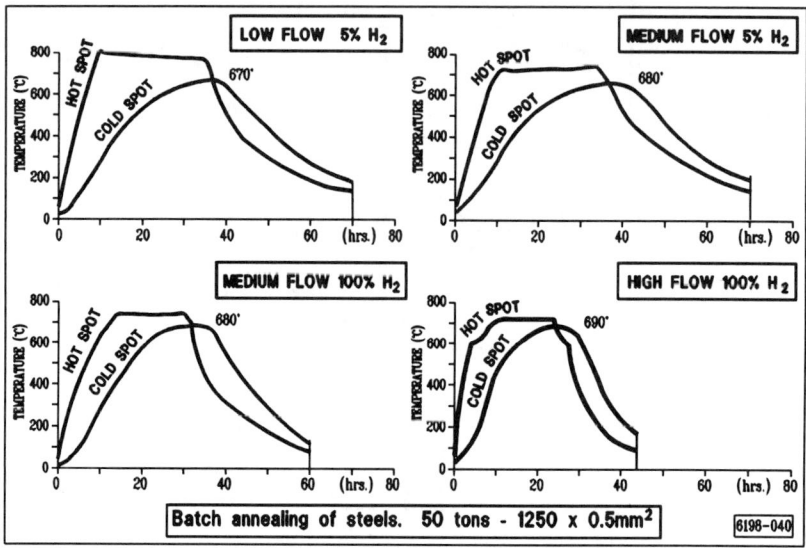

Figure 3 - Typical thermal cycles in batch annealing.

Instead, this work is intended to analyze in detail the gas-metal reactions taking place during batch-annealing and further compare treatments carried out under HNX and H_2. As a matter of fact, the gas-metal reactions under consideration are responsible for the elimination of rolling emulsion residues enclosed between successive turns of the tightly wound coil. Ultimate sheet cleanliness, which is one of the steelmaker's major objectives, depends on how efficient this organic residue elimination is.

This work is also aimed at investigating radial thermal stresses building up inside the coil due to temperature gradients. The effects of these thermal stresses on emulsion residue elimination will be discussed in detail.

In order to avoid arguments about the reliability of laboratory simulation tests, a decision was made to carry out the whole work on full-scale instrumented 20-ton coils batch-annealed in industrial bell-type furnaces.

Experimental setup

Chemical micro-reactors were created by drilling axial holes in the coil cross-section at 1/4, 1/2 and 3/4 of the steel crown thickness. Drilling was carried out dry, so that any additional organic pollution, other that pre-existent rolling emulsion residues, could be avoided. The endoscopic photography of figure 4 shows a typical as-drilled cavity : the good quality of the work is evidenced by clearly visible coil turns and interturn gaps.

The micro-reactors were then instrumented with special probes consisting of a thermocouple and a gas pick-up tube made of spectrographic quality stainless steel. Parallel to this type of drilled coil instrumentation, and with test cost reduction in view, less sophisticated sniffer-probes were also developed, so that they could be simply laid on top of the lower coil in the stack and pressed on its edge surface by the combined weight of heat convector and other coils above.

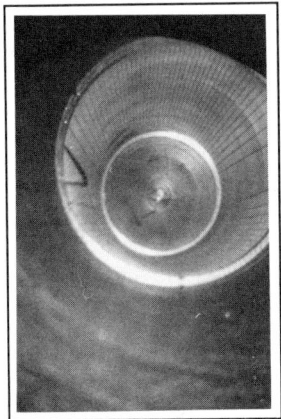

Figure 4 - Endoscopic photograph of an in-coil drilled micro-reactor.

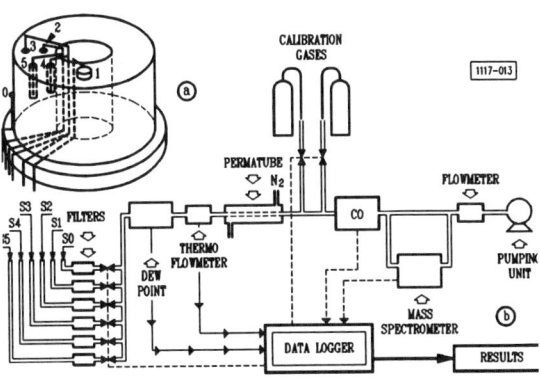

Figure 5 - Experimental set-up. a. probe location : S0 = cover - S1 = sniffer - S2-S5 = drilled probes; b. gas analysis equipments and data processing system

Corresponding gas analysis is naturally less "accurate", because it involves global interturn gap evolutions, averaged over the temperature gradient from strip axis to edge. We have therefore classified our results according to whether they had been obtained from drilled, sniffer-like or combined probes. Similar probes were also located under the protection cover, and inside the gas inlet and outlet pipes.

During each experimental annealing cycle, reaction gas was picked up sequentially from each probe and sent to a mirror dew-pointmeter and an infra-red absorption spectrometer for CO analysis. After drying, it was then analyzed in a mass spectrometer chosen for its capability for hydrocarbon analysis. The whole sequence of operations, comprising probe selection, spectrometer calibration, gas analysis, and data acquisition and processing was fully computer-controlled, which allowed for round-the-clock monitoring during the whole cycle. Figure 5 gives a schematic representation of the experimental setup. Complete description of and comments on the system have already been presented elsewhere[9,10].

Materials

The present work was carried out on sheet steel, between 0.7 and 1.0 mm in thickness. A clear distinction will be made hereafter between products rolled on a 4-stand tandem mill, characterized by a roughness index Ra of about 1.5 micron, and those rolled on a reversible mill, to a much lower roughness (Ra = 0.2 micron). In all cases, rolling emulsions were prepared from Quaker Q2023 and QN408 oils.

Experimental results

HNX batch-annealing. Concast steel sheets (1035 x 0.8 mm^2) were batch-annealed in 50-ton, 3-coil stacks under a N_2-5%H_2 protective atmosphere close to - 60°C in inlet gas dew point. Initial surface cleanliness was defined by :
- total pollution/side = 184 mg/m^2,
- iron fines pollution/side = 52 mg/m^2, and
- soluble oil pollution/side = 95 mg/m^2.

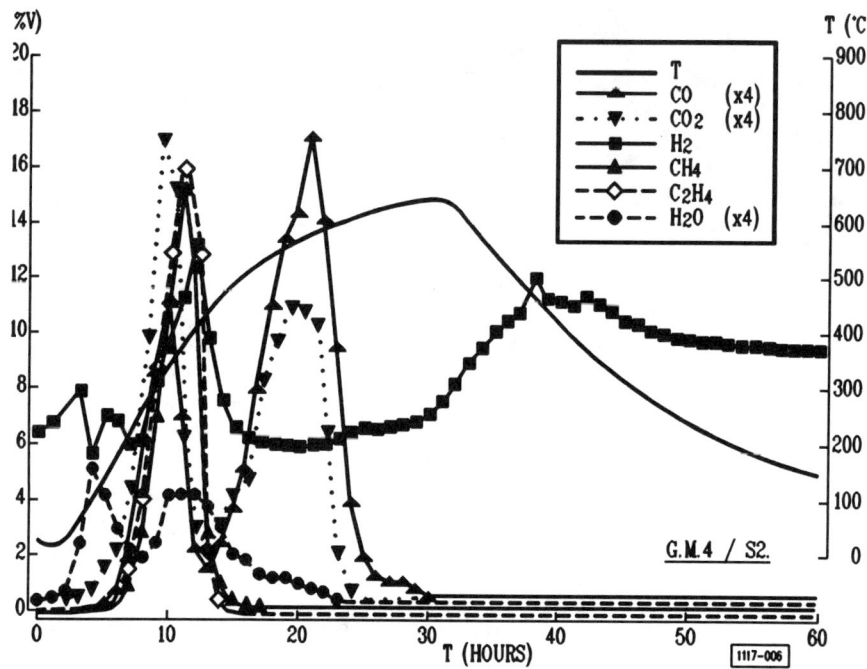

Figure 6 - Gas analysis during batch-annealing (HNX - LF drilled coil).

Based on drilled coil instrumentation data, figure 6 gives the variation of temperature and interturn gap reaction gas chemistry during the whole annealing treatment.

The probe under consideration corresponds to the chemical micro-reactor located at strip axis and 1/2 of coil cross-section thickness. A close look at the results suggests that :
- water vapour is released as soon as temperature reaches 100°C : this obviously corresponds to the initial amount of water physically entrapped in the coil,
- the occurrence of $[CO]_{LT}$ and $[CO_2]_{LT}$ species reveal the decarboxylation of esters and fatty acids in emulsion residues; we call them "low temperature" (LT) evolutions,
- the significant increase in H_2, CH_4, and C_2H_4 concentrations reflects distillation and thermocracking of the oil residue volatile fraction,
- a transient return of the protective atmosphere to its nominal 5%H_2 composition defines the so-called "neutral point"; it separates low temperature reactions, corresponding to the elimination of water and oil vapours, from high temperature reactions,
- a further evolution of $[CO]_{HT}$ and $[CO_2]_{HT}$ is observed, the interturn gap concentrations of which may reach 3 to 5 vol % (HT = high temperature),
- a very significant atmosphere hydrogen enrichment is observed during the final closed-system cooling (inlet valve closed) : this corresponds to the release of excess high temperature solute hydrogen in the ferrite.

As far as test repeatability is concerned, it should be mentioned that the peak temperature standard deviations computed for individual gas species, based on a dozen instrumented annealing cycles, is very low indeed :

$T[CO]_{LT}$, $T[CO_2]_{LT}$ = 318°C; σ = 13°C
$T[H_2]$, $T[CH_4]$ = 397°C; σ = 12°C
$T_{Neutral\ point}$ = 463°C; σ = 11°C
$T[CO]_{HT}$, $T[CO_2]_{HT}$ = 585°C; σ = 12°C.

This suggests that, for similar emulsions, metal temperature is the major parameter which actually controls the various gas-metal reactions leading to final surface de-pollution.

<u>100% H_2 batch-annealing</u>. The problem of pure hydrogen annealing protective atmospheres was first approached by means of sniffer-probe gas pick-up in an older, but revamped, furnace working under "medium flow-high convection" conditions.

Figure 7 gives the variations of temperature and gas composition, as measured from a sniffer-probe located on top of the lower coil, in the middle of the steel crown. The gas evolution sequence of the HNX cycle described above can also be observed. However, an additional phenomenon occurs, corresponding to an important release of CH_4 as metal temperature reaches 600°C, the intensity of which is clearly modulated by the inlet gas flow in the furnace (fig. 7). When the gas inlet is closed at the beginning of the cooling sequence, the CH_4 content in the atmosphere may increase to 15 vol%.

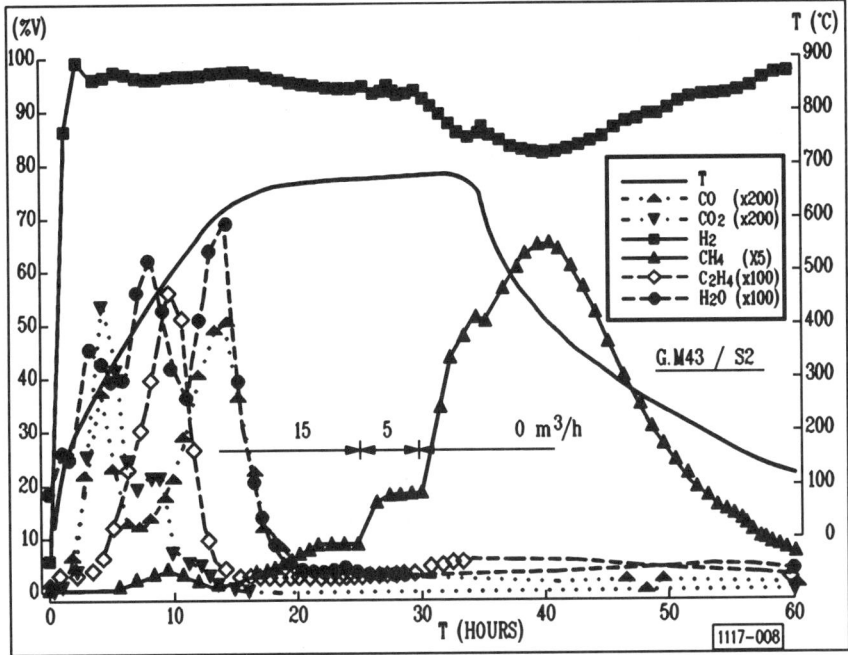

Figure 7 - Gas analysis during batch-annealing (100 H_2 - MF sniffer-probe).

100% H_2 batch-annealing was also investigated by combined drilled-coil and sniffer-probe instrumentations in a furnace working under "high convection-high flow" conditions. This made it possible to compare the relative efficiencies of both types of probe.

Figures 8.a and 8.b respectively give temperature and gas composition results obtained from drilled probe S4, located at sheet axis and 1/2 of coil cross-section thickness, and sniffer-probe S2, located at mid top edge surface.

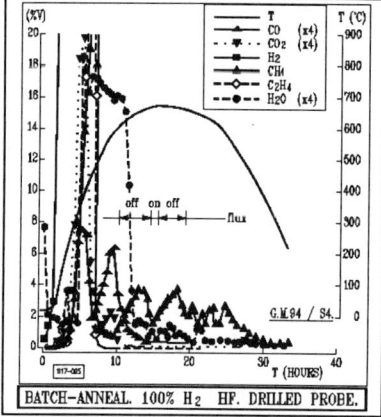
Figure 8.a - Gas analysis during batch-annealing (100 H_2 HF drilled coil).

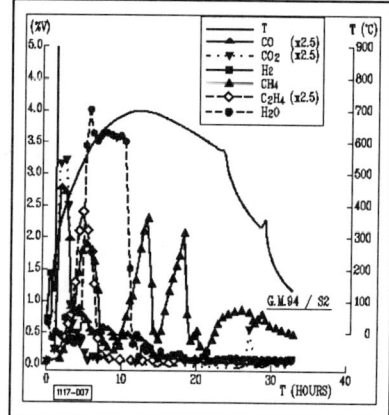
Figure 8.b - Gas analysis during batch-annealing (100 H_2 HF sniffer).

As is clearly seen, hydrogen was only fed to the furnace after some delay, for safety reasons. Besides the well known oil residue decarboxylation ($[CO]_{LT}$, $[CO_2]_{LT}$) and distillation ($[CH_4]$, $[C_2H_4]$) phenomena, important evolutions of $[CO]_{HT}$ and $[CO_2]_{HT}$ are again observed at higher temperature, along with a very significant dew point increase (fig. 8.a). The latter production of water vapour is much more intense than it is in HNX annealing cycles (fig. 6), but it should be pointed out that the corresponding H_2O/H_2 ratio is almost identical (5 10^{-2}).

At 600°C, a further CH_4 evolution is seen, the intensity of which is clearly modulated by inlet gas flow. This confirms earlier observations (fig. 7).

<u>Gas-metal reaction model</u>

The various reactions occurring during the cycle can be accounted for in four distinct stages :
1) for T = 280-400°C.
   The decarboxylation of saponifiable species in the initial oil residues leads to creating substantial amounts of $[CO]_{LT}$ and $[CO_2]_{LT}$. In this temperature range, the thus modified atmosphere is oxidizing for steel :

   $$x\ CO_2 + Fe = x\ CO + FeO_x \quad (1)$$

   On the other hand, the gas-water reaction yields

   $$CO_2 + H_2 = CO + H_2O \quad (2)$$

Thus also leading to

$$x\ H_2O + Fe = x\ H_2 + FeO_x \qquad (3)$$

2) for T = 300-450°C.
Slightly shifted to higher temperatures relative to the former stage, oil distillation is clearly revealed by the mass spectrometric detection of CH_4, C_2H_4, C_nH_m.
The intensity of these emissions obviously warrants final surface cleanliness.
After these first two stages, a thin solid hydrocarbon residue may however still remain at the steel surface, resulting from oil thermodegradation. This hydrocarbon residue will hereafter be called C_{tar}. As such, it could not be vaporized and would then resist elimination, except by chemical reaction with another species creating some volatile compound.

3) for T = 420-620°C.
In this temperature range, further evolutions of CO, CO_2 and H_2O are observed. It is worth noticing that the atmosphere then turns reducing for iron (but not for alloying elements).
Events could be schematically represented as follows :

$$FeO_x + x\ H_2 \rightarrow Fe + x\ H_2O \qquad (3')$$

which hints to some possible reaction involving the tar residue

$$H_2O + C_{tar} \rightarrow H_2 + CO \qquad (4)$$

Thus created species must then reorganize according to the Boudouard reaction

$$2\ CO = CO_2 + C_{soot} \qquad (5)$$

and again, the gas-water reaction

$$CO_2 + H_2 = H_2O + CO \qquad (2)$$

It is seen that a new allotropic form of solid carbon has been created (C_{soot}), which is clearly distinct from the tar residue mentioned earlier.
The mechanism described here then allows at least partial tar elimination while modifying the nature of the residual surface pollution. This could be important for final depollution during temper-rolling.

At this stage, model verification could be sought by comparing the actual system condition to the theoretical thermodynamic equilibria of equations (2) and (5).

Based on the analyses of gas pick-ups from probe S4 already defined in figure 8.a, figure 9 gives the time variations of theoretical and experimental thermodynamic constants K_{th} and K_{exp}.

It is seen that the Boudouard and gas-water reactions reach thermodynamic equilibrium for annealing times between 8 and 12 hours, corresponding to temperatures between 500 and 620°C. For higher temperatures, these two reactions are out of equilibrium again.

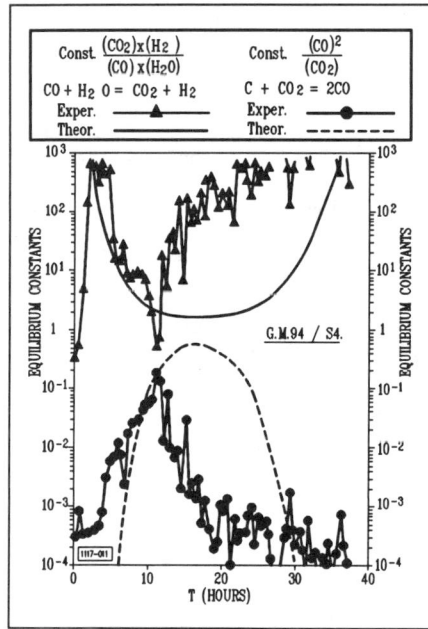
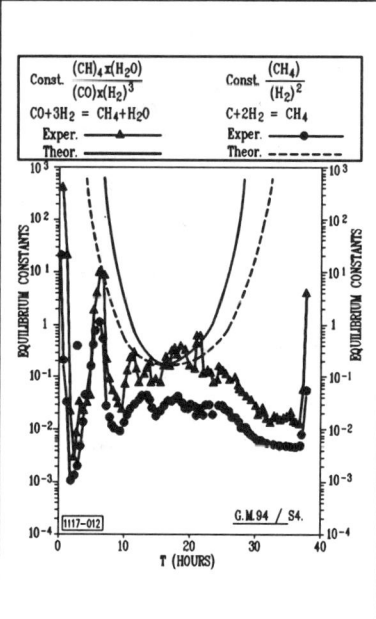

Figure 9.a - Thermodynamic analysis   Figure 9.b - Thermodynamic analysis

4) for T > 620°C.
In the case of the HNX batch-annealing of figure 6, no further reaction is observed before final cooling, except for excess high temperature solute hydrogen release from the matrix.

On the contrary, 100% H_2 batch-annealing is characterized by an additional emission of CH_4, which can be detected from end of soaking to beginning of cooling (fig. 8.a). It has already been mentioned that the amount of CH_4 increases in time when the system is closed, and rapidly decreases under the effects of either gas fluxing or purging.
In order to account for these observations, the well known metallic iron catalysed Fischer-Tropsch[1] reaction of carbon monoxide hydrogenization must be evoked :

$$CO + 3 H_2 \xrightarrow{Fe} CH_4 + H_2O \qquad (6)$$

coupled with the irreversible reaction

$$H_2O + C_{soot}, C_{tar} \rightarrow CO + H_2 \qquad (4')$$

As a matter of fact, fig. 9.b confirms that the carbon monoxide methanization actually reaches equilibrium between 12 and 21 hours of annealing time, whilst clearly modulated by inlet gas flow.

Combined equations (4') and (6) show how further residual solid carbon can be transferred to the gas phase in the form of CH_4.

In this connection, it is important to check for potential direct carbon methanization :

$$C_{soot} + 2 H_2 = CH_4 \tag{7}$$

As is clearly seen in figure 9.b, which pertains to chemistry readings from probe S4 of annealing cycle GM94 of figure 8.a, the amount of CH_4 created at this stage never reaches the equilibrium corresponding to reaction (7). This is presumably due to the on-and-off gas feed policy of this particular cycle.

On the other hand, similar measurements made during the annealing cycle of figure 7 showed that very high CH_4 contents (13 vol %) could well be attained, thus achieving thermodynamic equilibrium for reaction (7), which suggests the possibility of surface soot re-deposition. Figure 7 also confirms that closed system conditions prevailed at that stage.

It is worth reminding that the Fischer-Tropsch reaction may be seriously poisoned by sulphur present at the iron catalyst surface, thus modifying the adsorption of CO[12]. Although no experimental evidence of sulphur at the extreme surface could be reported, comments above might be used to question the composition of some sulphur-doped rolling oils currently used in tandem mills.

As far as the Fischer-Tropsch reaction is concerned, it should be mentioned that, depending on the catalyst considered, it could well lead to the creation of other C_nH_m molecules. In the temperature range of interest, we actually detected ethane traces in a C_2H_4/CH_4 ratio of $5 \cdot 10^{-3}$, which again suggests carbon monoxide hydrogenization rather than direct carbon methanization. Finally, it should also be pointed out that some uncertainty remains as regards the efficiency of de-pollution reaction (4') as it comes to the forms of residual carbon, i.e., tar and soot.

Thermal stresses induced during annealing

It is a well known fact that temperature gradients building up inside heat treated coils result in so-called radial, circumferential and axial thermal stresses. As was shown by Pawelski et al[13], radial thermal stresses σ_{rr} develop between successive coil winds and account for sticking phenomena occurring on cooling, mostly between external turns of strips cold-rolled to a low surface roughness (reversible mills).

Figure 10 shows the variations of temperature differentials ΔT_{AB} between external and mid-thickness turns, and ΔT_{CB} between internal and mid-thickness ones, all along and three annealing cycles of figure 3.

It is clearly seen that, by improving heat transfer and limiting overheating of external coil turns, "high convection-high flow" technology strongly reduces temperature differential ΔT_{AB} during heating. For the same reasons, temperature differential ΔT_{CB} is much higher for 100% H_2 annealing than for its "low convection-low flow" counterpart.

Thermal stresses generated in a coil modelled by a solid cylinder of internal and external radii r_C and r_A, can be computed from the prevailing temperature distribution.

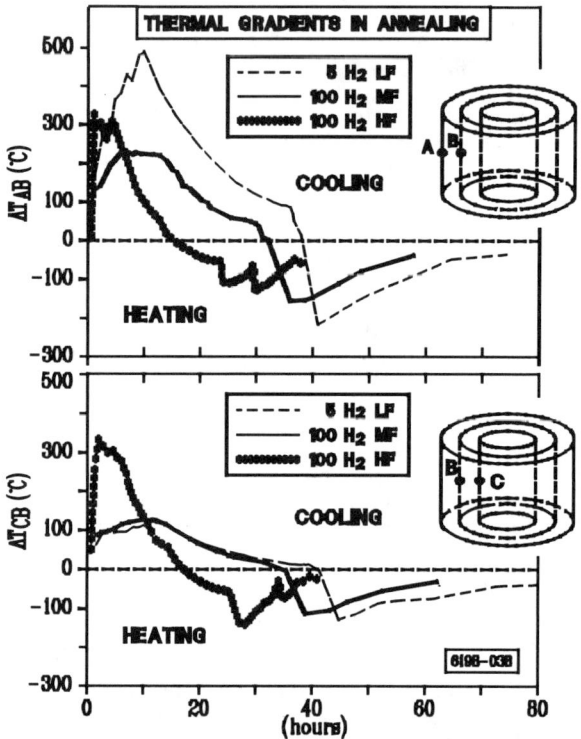

Figure 10 - Thermal analysis in batch-annealing.

As far as radial stresses are concerned, the Duhamel theory[14] leads to :

$$\sigma_{rr}(r) = \frac{E\alpha}{r^2(1-\nu)} \left[\frac{r^2-r_C^2}{r_A^2-r_C^2} \int_{r_C}^{r_A} rT(r)\,dr - \int_{r_C}^{r} rT(r)\,dr \right] \quad (8)$$

In this equation, E, ν and α respectively represent the steel's Young modulus, Poisson coefficient and coefficient of thermal expansion.

With the notations defined by figure 10, temperature radial distribution could be fairly approximated by :

$$T(r) = T_B + (T_C - T_B)\exp\frac{-(r-r_C)}{Z_C} + (T_A - T_B)\exp\frac{(r-r_A)}{Z_A} \quad (9)$$

where parameters Z_A and Z_C account for heat transfer laws.

Figure 11 shows the temperature profile computed after about 3 hours of heating, along with the corresponding radial stress distribution derived from equation (8).

As is evidenced in the same figure, a thermo-elasto-plastic finite element analysis actually confirms and generalizes the simple analytical approach described above.

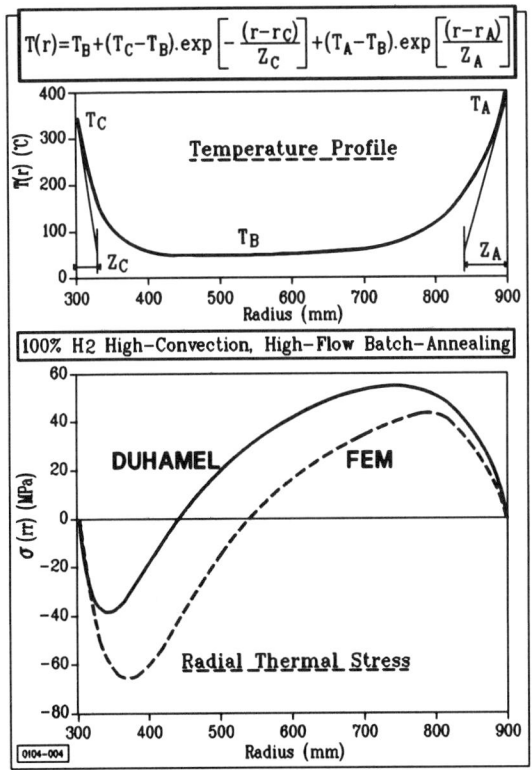

Figure 11 - Thermal and stress distributions during batch-annealing (100 H_2 HF).

It can be concluded that compressive radial stresses develop near the coil eye and combine with those inevitably introduced by the exit coiler of the tandem mill, which are themselves maximum at the internal turns (hard core).

Final surface cleanliness

The application of the "high convection-high flow" concept results in delivering more heat inside the eye, thus increasing temperature differential ΔT_{CB} between internal and mid-thickness turn leading to local compressive radial thermal stresses which prevent the coil from opening up and the gas-metal reactions described above from taking place in full.

As is shown by figure 11, important compressive radial stresses prevail near the eye at temperature between 300 and 400°C, when de-polluting preliminary decarboxylation and distillation reactions should normally proceed.

Therefore, it seems only natural to suspect that these reactions will either be slowed down or postponed, which would impair their efficiency, particularly for steelgrades undergoing short time annealing. This point was verified by measuring the surface carbon pollution at mid-thickness and close to the internal end of an annealed, but not skin-passed, ST12 steelgrade coil.

The major pollution components were identified, i.e., surface carbon residues (combustion test carried out prior to power-wash) and detachable iron fines (tape peeling and ICP analysis). The results reported in table 2 emphasise the increase in carbon pollution near the eye, which is associated with a worse back-sintering of cold rolling-induced iron fines.

TABLE 2 - SURFACE POLLUTION FEATURES
(no temper-rolling, no power-wash)

ST12 steel	Middle turns	Inner turns
total carbon (mg/m^2) combustion tests	3 - 6	20 - 25
iron fines (mg/m^2) tape tests + ICP	8 - 10	18 - 58

It seemed interesting to try and better characterize the nature of residual surface carbon pollution by X-ray photo-electron spectrometry (XPS). The technique makes it possible to detect differences between the various carbon chemical bonds. The results of table 3 were obtained after proper calibration based on several chemically defined organic compounds deposited on gold or iron substrates.

Figure 12 gives C_{1s} XPS spectra respectively recorded on cold-rolled and as-annealed (no skin-pass) specimens : in the latter case, CCl_4 degreasing was performed, in order to avoid all accidental pollutions between specimen sampling at the mill and laboratory analysis.

Figure 12.a, pertaining to the full-hard specimen, emphasizes the existence of long hydrocarbon chains, as is evidenced by the important $R-[CH_3]_n$ peak, the presence of C - OH and C = O specific radicals, together with ester and acid functions [COOR - COOH]. After annealing, the spectrum recorded at the middle of the coil is substantially different : the majority fraction now corresponds to C - C carbon bonds.
According to the gas-metal reactions commented above, this fraction should essentially result from the Boudouard reaction (C_{soot}), but the presence of some C_{tar} cannot be completely ruled out, especially in the face of RCH_3, COH and COOR, which are also detected.

For specimens taken from the internal coil turns, heavier global pollution is not only observed, but RCH_3, COH, CO and COOR components are also clearly detected, thus suggesting the presence of C_{tar}. Compressive radial thermal stress buildup near the eye then appears to have put off the oil distillation and inhibited the Boudouard and Fischer-Tropsch reactions which would normally have contributed to C_{tar} elimination.

Conclusions

The work presented here helps better understand the various gas-metal reactions occurring during recrystallisation batch-annealing under N_2 - 5% H_2 and 100% H_2 protective atmospheres.
Whenever the cold-rolled steel strip does not undergo any precleaning treatment, most of the oil residues are removed by decarboxylation and distillation. At higher temperatures, solid carbon residue elimination is achieved through the Boudouard and gas-water reactions.

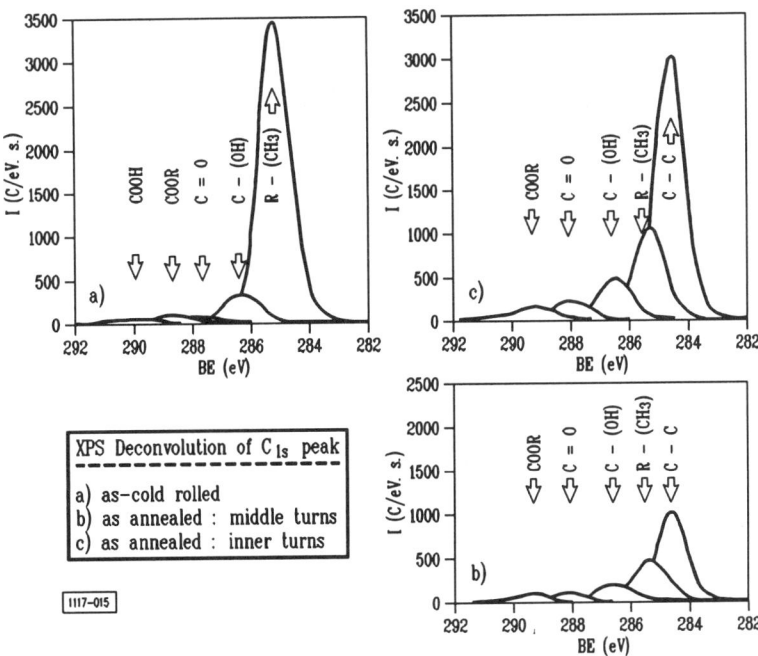

Figure 12 - Surface carbon as measured by XPS analysis.

TABLE 3 - BINDING ENERGY AND HALF-HEIGHT WIDTH OF ELEMENTAL XPS C_{1s} PEAKS

	Chemical bonding	C_{1s} Binding energy (eV)	C_{1s} Half-height width (eV)
	C - C	284.40 - 284.60	1.10 - 1.15
(M)	R - CH_3	285.10 - 285.30	1.25 - 1.30
(A)	C - O - H	286.30 - 286.60	1.33 - 1.37
(K)	$R\diagdown$ $C = O$ $R'\diagup$	287.65 - 287.95	1.35 - 1.39
(E)	$C\diagup^O_{\diagdown O-R}$	288.60 - 288.90	1.37 - 1.39
(FA)	$C\diagup^O_{\diagdown O-H}$	289.30 - 289.60	1.37 - 1.40

In 100% H_2 batch-annealing, an additional carbon residue elimination mechanism shows up, leading to methane and ethane evolutions, which can be accounted for, based on the metallic iron-catalysed Fischer-Tropsch reaction.

This invites to recommend on-and-off furnace gas feed, so that volatile CH_4 and C_2H_4 species can be extracted.

Obviously enough, combining pure hydrogen with forced gas drive ("high convection-high flow" technology) allows to accelerate the heating step, but the latter must still be kept under control, so that too high surface temperature radial gradients and associated efficiency losses of the depolluting reactions can be avoided. This grows particularly important for CQ steelgrades usually submitted to short annealing cycles, for which depollution might not be achieved in full.

A trade-off work point must then be sought for bell-like furnaces operated under "high convection-high flow" conditions, so that excellent cleanliness and high productivity are achieved.

As far as this new technology is concerned, several additional aspects should be considered, namely :
- problems related to hydrogen adsorption and desorption for ULC-IF steel grades, and
- advantage of pure hydrogen annealing practice aimed at avoiding nitrogen pick-up in HNX annealing in the case of sensitive steels[15].

Acknowledgements

This research was carried out with the joint financial support of IRSIA (Institut pour l'Encouragement de la Recherche dans l'Industrie et l'Agriculture) and ECSC (European Steel and Coal Community). The authors would like to express their appreciation to their colleagues A. Magnée and J. Wauters, for FEM computational work, and B. Chatelain, currently at Cockerill-Sambre. The technical support of COCKERILL SAMBRE (Belgium) and ARBED-Dudelange (Luxemburg) works is gratefully acknowledged.

REFERENCES

1. Ph. Paulus, V. Tusset, J. Hancart, V. Leroy, M. Economopoulos, "Howaq a new way for strip heat treating, and processing", Met. Rev. CRM, 55, 33 (1979).

2. V. Leroy, J.P. Servais, "Influence of steel surface chemistry in finishing treatment used in automotive industry", (Paper 387 presented at NACE Corrosion 85, Boston, March 1985).

3. B. Chatelain, V. Leroy, "Propreté de la bande recuite et skin-passée" (Rapport EUR 8700, ECSC Recherche Technique 1984).

4. H. Lochner, "Practical experience in annealing wide cold-rolled strip in Hicon/H_2 bell annealing, World Steel and Metallworking, 8, 132 (1956).

5. M. Daguier, "Les fours cloches à hautes performances", Traitements Thermiques, 239, 90, 19.

6. R. Eylens, W. Rausch, B. Voigt, "Einfluss von Wasserstoff als Schutzgas beim Hochkonvektionisglühen auf die Bandsauberkeit von beschichtetem Automobil-Feinblech, Stahl und Eisen, n°3, 73 (1990).

7. R. Enghofer, T.E. Mueller, J.L. Kuzdal, J.M. Mates, "Experience with 100 H_2 annealing at LTV's Indiana Harbor n°3 steel mill", Iron and Steel Engineer, 25, March (1990).

8. N. Ducrot, C. Brun, J. Caninez, G. Schollaert, "Essais de recuit sous hydrogène pur des tôles d'acier doux", Journées Sidérurgiques ATS, déc. 1987, Revue de Métallurgie CIT, 11 899 (1988).

9. B. Chatelain, V. Leroy, "Réactions gaz-métal au cours du recuit-base de la bande laminée à froid", Revue de Métallurgie CIT, 331 (avril 1986).

10. B. Chatelain, V. Leroy, F. Beco, C. Orban, R. Raisi, R. Henrion, M. Kuhn, R. Lammar, J. Redo, "Evaluation du recuit sous hydrogène des aciers doux : réactions gaz-métal", Revue de Métallurgie CIT, 173 (février 1989).

11. H.P. Bonzel, H.J. Krebs, "Surface science approach to heterogeneous catalysis : CO hydrogenation on transition metals", Surface Science, 117, 639 (1982).

12. J. Benziger, R.J. Madix, "The effects of carbon, oxygen, sulphur and potassium adlayers on CO and H_2 adsorption on Fe(100)", Surface Science, 94, 110 (1950).

13. O. Pawelski, W. Rasp, G. Martin, "Entstehung von bandklebern bei Hauben geglühtem Kaltband", Stahl und Eisen, 4, 178 (1989).

14. P.P. Penham, R.D. Hoyke, Thermal stress, Pitman London.

15. A.R. Perrin, M. Wolosiuk, A. Mc Lean, "Nitrogen absorption during batch-annealing of Al killed steels", I-SM, 45 (June 1987).
14. P.P. Penham, R.D. Hoyle, Thermal stress, Pitman London.

SELECTIVE OXIDATION OF COLD-ROLLED STEEL DURING

RECRYSTALLIZATION ANNEALING

J.M. Mataigne, M. Lamberigts, V. Leroy
C.R.M., Abbaye du Val Benoît, 4000 Liege, Belgium.

Abstract

During the recrystallisation annealing performed in N_2-H_2 or H_2 atmospheres, a complete modification of the cold-rolled steel surface chemistry can occur due to the selective oxidation of alloying elements more oxidizable than iron. This is of paramount importance for the steel strip behaviour in all finishing treatments. A theoretical model describing the selective oxidation of binary alloys has been developed some years ago by C. WAGNER. A distinction was made between internal and external selective oxidation; this paper presents an attempt to extend this model to the description of the selective oxidation of steels, taking into account the presence of several oxidizable elements and the possibility of preferential oxidation along the grain boundaries.

Introduction

Although it is reducing for iron, the protective atmosphere of either batch or continuous annealing treatments is always oxidizing for more reactive alloying or residual elements present in the steel chemistry. In the course of subsequent finishing treatments rapidly described below, the annealed steel surface may exhibit extremely variable chemical reactivity, depending on whether this selective oxidation develops in the external (as a film on the free surface) or internal (in the metallic matrix below the free surface) mode.

- In the current surface conversion treatment applied in painting lines, the phosphoric acid dissolution of iron, which governs the growth of phosphate crystals at the steel surface can only proceed satisfactorily if the annealed surface, made of metallic iron, is chemically homogeneous. The presence of insoluble surface oxide particles can indeed lead to point defects in the phosphate layer or even prevent its growth, when they pack to form a sufficiently solid cover[1].
- Modern cold-rolled sheet steel continuous galvanizing involves a recrystallization treatment prior to hot-dipping. The reduction of high temperature soaking-induced surface oxides by the aluminium contained in the liquid bath might locally accelerate Fe-Zn reactions, and thus contribute to the appearance of an irregular intermetallic interface layer[2].
- Internal oxide precipitation in the ferrite matrix can also help create preferential recombination sites for the solute hydrogen absorbed during electro-deposition. If the coating metal (Zn, Sn) is impermeable to hydrogen, gas pressure buildup at the interface can lead to blistering and later bring about pitting corrosion[3].

In order to ensure proper steel behaviour, it is therefore of the utmost importance that recrystallization annealing-induced selective oxidation phenomena be thoroughly investigated.

C. Wagner[4] has put forward the theoretical basic principles of a mathematical model describing internal selective oxidation in ideal single crystal binary alloys, where it is supposed that selective oxidation is governed by the diffusion of reactive species while precipitation front localization relative to the free surface depends on the balance between the inward oxygen-and outward oxidizable solute element-flows. External oxidation can then be treated like a particular case of internal oxidation, when selective surface oxide precipitation grows active enough as to obstruct all oxygen diffusion paths. According to theory, the internal oxidation of reactive element X in binary alloy M-X occurs during an isothermal heat treatment under constant oxygen partial pressure when its molar fraction N^0_X is less than some critical value. The latter is defined by the establishment of an outward X flow which leads to oxygen inward flow-blocking oxide precipitation near the surface : in the case of Fe-X alloys, it can be estimated by the relation below.

$$N^0_{X,crit} = \left[\frac{\pi \, g^* \, V \, N^S_0 \, D_0}{2 \, n \, V_{XO_n} \, D_X} \right]^{1/2} \tag{1}$$

Where D_0 is the coefficient of oxygen diffusion in iron :

$$D_0 = D'_0 \cdot \exp[-Q_0/RT],$$

D_X, the coefficient of X diffusion in iron :

$$D_X = D'_X \cdot \exp[-Q_X/RT],$$

N^S_O, the molar fraction of free surface-adsorbed oxygen,

V, the alloy molar volume

V_{XO_n}, the molar volume of oxide XO_n, and

g^*, the critical volume fraction of precipitated oxides which leads to obstructing all oxygen inward diffusion paths.

Under the same conditions, external selective oxidation of element X takes place if its content in the alloy is greater than critical value N^O_X defined by relation (1).

Obviously enough, actual steels are far from the ideal binary alloys of the Wagner model, due to the complexity of their chemistry (many reactive elements can take part in the selective oxidation process) and the presence of lattice defects (grain boundaries, dislocations). Reactive species diffusion paths are then more varied and oxidation products are not necessarily simple compounds like SiO_2, but may instead be more complex in composition ($MnSiO_3$). Furthermore, some of them may even dissolve iron oxide (fayalite).

Chemical balance calculations must then be thoroughly modified to take into proper account the effects of operating conditions affecting oxide stability (soaking temperature, protective atmosphere oxidizing power).

The initial Wagner model was verified particularly well for binary alloys submitted to long (several hours) and high temperature (T > 900°C) treatments which led to selective internal oxidation over a substantially thick sub-surface layer (th > 20 μm). The emergence of modern surface analysis techniques (SAM, SIMS, XPS) however made it possible to characterize the definitely shallower selective oxidation brought about by shorter heat treatments, such as the continuous annealing of low alloy DQ and DDQ steel-grades.

The present paper is aimed at verifying the relevance of the Wagner model principles in the case of actual steels submitted to either batch or continuous recrystallization annealing. In this framework, particular attention will be paid to pure hydrogen batch-annealing which is now increasingly being used in the industry. The present investigation is based on a step-by-step approach which successively investigates Fe-Si alloys and more complex steels.

Materials

Table 1 defines the chemistries of the various steels considered. It should be noted that materials S1 to S4 are industrial steels taken in the full-hard condition, after cold-rolling to thicknesses ranging from 0.6 to 1.5 mm. Material A was prepared in the laboratory, by induction melting electrolytic iron under vacuum (VIM), adding the silicon under a 500-Torr pure Ar pressure, and casting the thus obtained alloy in 100 mm-thick ingots. The latter were then forged down to 25 mm, and hot-rolled to 3.5 mm (T_{final} = 900°C). After descaling by shot blasting, the material was finally cold-rolled to 1 mm.

TABLE I. - ALLOY CHEMISTRY (10^{-3} wt %)

Steel-grade	C	P	S	Si	Mn	N	Al_{tot}	Al_{sol}	Cr	B
A	3	2	1	197	4	1.3	2	1	0	-
S1	16	22	2	3020	90	1.8	4	2	5	-
S2	77	77	7	305	730	3.5	67	64	11	-
S3	24	6	6	7	140	2.2	55	53	10	≤0.1
S4	10	15	9	5	190	2.4	37	35	10	2.3

Experimental procedures

Specimens, 3 x 5 cm^2 in surface area, were treated in a tight resistance-heating laboratory furnace, which had been instrumented so as to ensure the proper hydrogen content-and dew point-control of the protective atmosphere, the flowrate of which was kept constant (20 l/h).

Surface chemistry was assessed by SIMS (Secondary Ion Mass Spectrometry) and SAM (Scanning Auger Microprobe Analysis), while oxide particle distribution was investigated by examining polished sections under the SEM (Scanning Electron Microscopy).

For some of the specimens, surface reactivity was further evaluated by laboratory bi-cation phosphating treatments (Gr908).

Results

Fe-Si binary alloys

Alloy S1 (Fe-3% Si). The investigation was first aimed at characterizing the internal oxidation condition of coarse-grained binary alloy S1, which could be considered very close to a single crystal.

The Wagner criterion [relation (1)] is theoretically valid in this particular case. Its application leads to determining the transition limit between internal and external oxidations in an atmosphere oxidizing power-critical content diagram (fig. 1). Based on information given in the specialized scientific literature, the constants in the equation were chosen as follows:
- $g^* = 0.3$ [in accordance with Rapp's estimate][5]
- $V = 7.1$ cm^3/mol [6]
- $D_o^{'\alpha} = 3.71 \cdot 10^{-2}$ cm^2/sec [oxygen diffusion frequency factor][7]
- $Q_o^{\alpha} = 23,050$ cal/atom.gr [activation energy for oxygen diffusion in ferrite][7]
- $D_{Si}^{'\alpha} = 8.0$ cm^2/sec [silicon diffusion frequency factor][8]
- $Q_{Si} = 59,500$ cal/atm.gr [activation energy for silicon diffusion in ferrite][8]
- $R = 1.987$ cal/(atom.gr °K)
- $n = 2$ for SiO_2, and
- $V_{SiO2} = 26.12$ cm^3/mol [6].

The protecting atmosphere's oxidizing power is given in the form of partial pressure ratio P_{H2O}/P_{H2}, the value of which indeed determines the free surface-adsorbed oxygen molar fraction in relation (1) [N_o^S]:

$$H_2 \text{ [1 atm.]} + O \text{ [1 wt \% sol.]} = H_2O \text{ [1 atm.]}$$

with corresponding equilibrium constant K in ferrite given by :

$$K = [P_{H2O}/(P_{H2} \times (wt \% O))] = 0.162 \exp [22876/RT]^{(9)}$$

N_o^S can then be computed from the relation :

$$N_o^S = (1/100) \cdot (56/16) \cdot (P_{H2O}/P_{H2}) \cdot (1/K).$$

Assuming that iron and silicon form an ideal solid solution, and neglecting the possible occurrence of fayalite, the selective oxidation domain of fig. 1 is respectively limited, on its low-and high-oxidizing power sides, by the dissociation equilibria of silica and iron oxide.

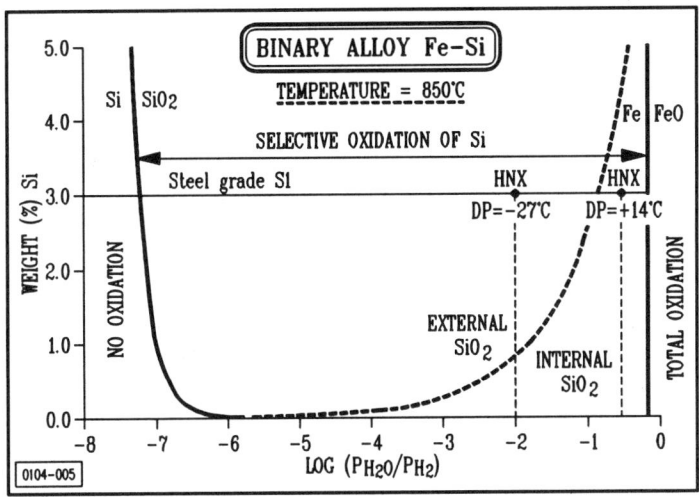

Figure 1 - Oxidation domains for binary Fe-Si alloys at 850°C.

Figure 2 gives the Auger in-depth concentration profiles recorded for Fe, O and Si on specimen S1 as after one-hour annealing at 850°C under a protective atmosphere made of 95% N_2 and 5% H_2, and characterized by a dew point of - 27°C (P_{H2O}/P_{H2}= 10^{-2}). In full accordance with theoretical predictions, external selective oxidation of silicon is shown to have taken place, which created a surface silica film of about 7 nm in thickness, thus completely insulating the substrate from the environment. This silica film was also shown to inhibit the annealed material's phosphatability.

If carried out under a 5% H_2 atmosphere characterized by a dew point of + 14°C (P_{H2O}/P_{H2} = 3.10^{-1}), the same heat treatment should have led to an internal selective oxidation of silicon. The SEM 3-D view of figure 3 characterizes the corresponding mixed surface oxidation condition. It reveals an arrangement of alternating 20 μm-deep silicon selective internal oxidation zones covered by total oxidation (fayalite was indeed detected at the surface of these areas) and patches of purely external oxidation SiO_2 surface film of about 7 nm in thickness, which prevented deeper silicon selective internal oxidation.

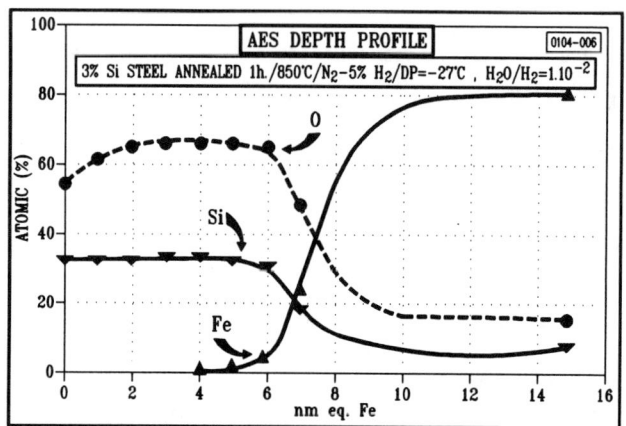

Figure 2 - Auger in-depth profiles of pure silicon external oxide on 3% Si steel.

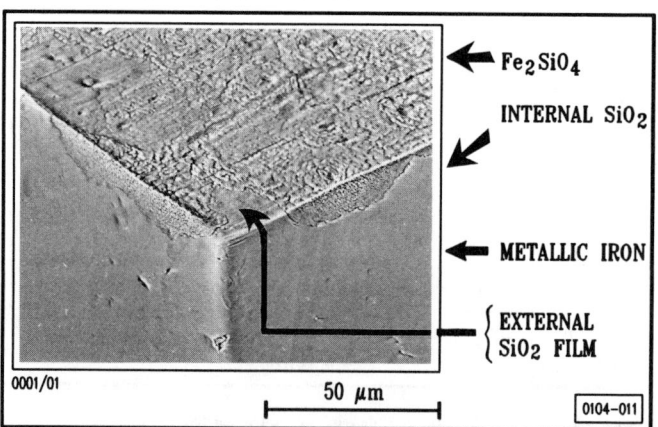

Figure 3 - Mixed oxidation pattern of silicon on 3% Si steel
(as-annealed : 850°C, 1 hour, 5% H_2, DP = + 14°C)

<u>Alloy A</u> (Fe-0.2% Si). Figure 4 defines the various oxidation domains for Fe-Si binary alloys, the silicon contents of which are closer to those of low C steelgrades. The oxidation mode, which is brought about by an annealing treatment of one hour at 850°C, under a pure H_2 protective atmosphere characterized by a dew point of - 40°C ($P_{H2O}/P_{H2} = 10^{-4}$), is described by the SEM micrographs and Auger spectrometry distribution maps of figure 5.

An external oxidation silica film is still present, in association with selective silicon oxidation along grain boundaries, down to 30 nm from the surface. As such, relation (1) cannot account for the grain boundary oxidation.

Furthermore, if a dew point of - 27°C had prevailed in the 5% H_2 atmosphere ($P_{H2O}/P_{H2} = 10^{-2}$), the same heat treatment would have led to internal silicon selective oxidation over a thickness of 2 μm or so, as is shown by the electron micrograph of figure 6.

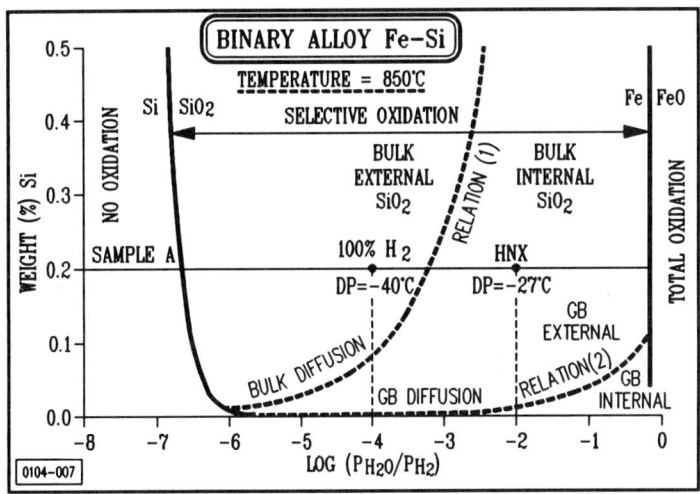

Figure 4 - Oxidation domains for low silicon content Fe-Si alloys at 850°C (relations (1) and (2)).

Contrary to the previous case, the extreme surface of this latter specimen (fig. 7) is made of metallic iron. As a matter of fact, surface analyses, carried out after the natural iron oxide, formed by simple exposure to ambient air, has been sputtered away, reveal that the free surface inside the grain is made of metallic iron, while silica is only detected along grain boundaries, down to about 15nm.

Figure 8 shows the SIMS in-depth $^{28}Si^+$ ion concentration profiles respectively recorded on specimens of material A, as heat treated for 30 seconds and one hour at 850°C, under either a pure hydrogen protective atmosphere, the dew point of which was - 40°C, or a 5% H_2 protective atmosphere, with a dew point of - 27°C. In the latter case, prolonging the treatment from 30 seconds to one hour extends the internal oxidation layer from 400 nm to 2 μm below the surface, and deepens external oxidation along grain boundaries.

On the contrary, the kinetics of purely external oxidation (100% H_2 - DP = - 40°C) is substantially slower, due to the screening effect of the external oxide film which induces ionic- instead of atomic- diffusion of moving species.

Selective oxidation along grain boundaries could be accounted for by rewriting relation (1), in such a form that room is made to accommodate the corresponding accelerated diffusion flows of oxygen and alloying elements along the grain boundaries.

Assuming that all grain boundary diffusion activation energies are half their values for bulk diffusion in a single crystal lattice, a new transition limit between internal and external oxidation modes can thus be computed, to take grain boundary diffusion phenomena into account (fig. 4) :

$$N^o_{Xcrit.GB} = \left[\frac{\pi \; g^* \; V \; N^S_o \; D'_o \; \exp \; (- Q_o/2RT)}{2 \; n \; V_{XO_n} \; D'_x \; \exp \; (- Q_x/2RT)} \right]^{1/2} \quad (2)$$

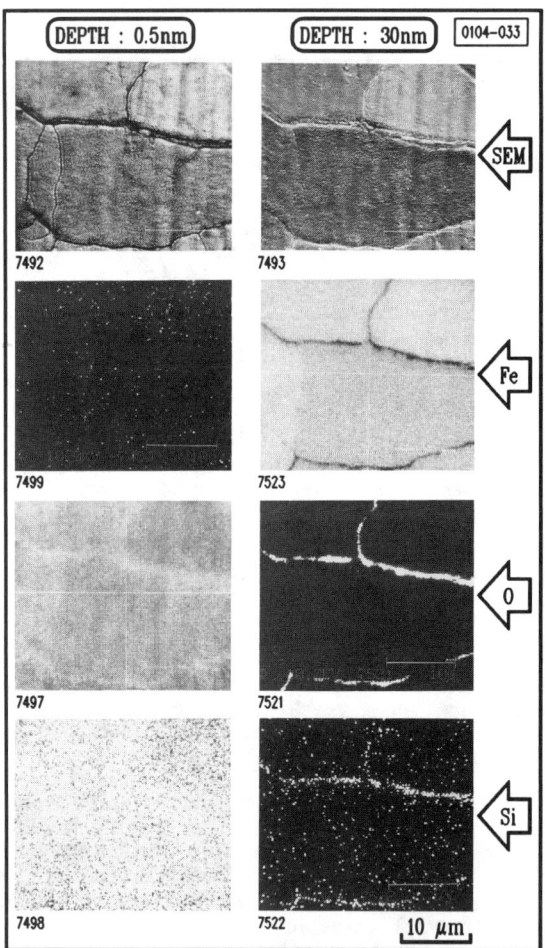

Figure 5 - SEM surface and SAM distribution maps of sample A
(as-annealed : 850°C, 1 hour, 100% H_2, DP = - 40°C)

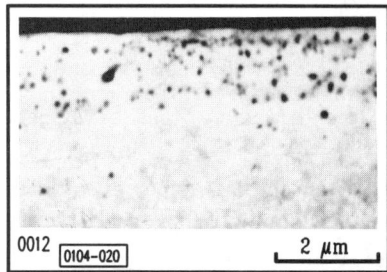

Figure 6 - BSE cross-sectional micrograph of
silicon internal oxidation of sample A
(as-annealed : 850°C, 1h, 5% H_2
DP = - 27°C).

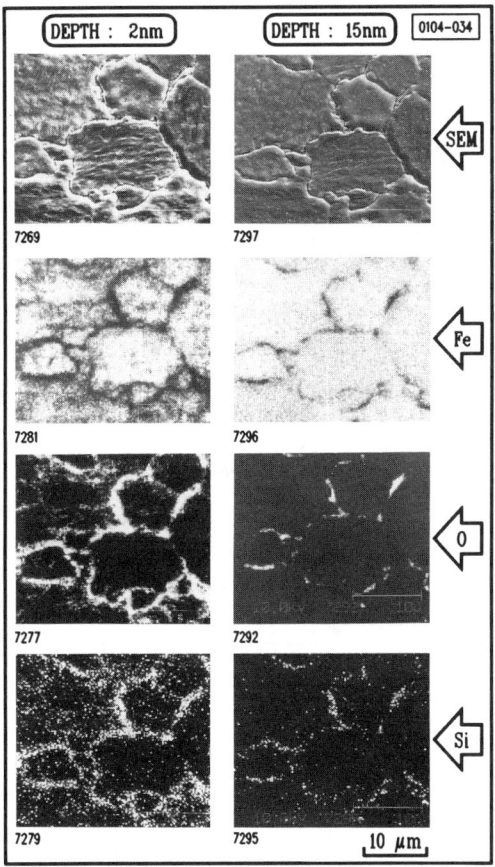

Figure 7 - SEM surface micrographs and SAM distribution of sample A (as-annealed : 850°C, 1 hour, 5% H_2, DP = - 27°C).

For the model to be quantitatively validated, this assumption should be experimentally verified, which could prove long and tedious.

As a matter of fact, what we are really seeking for here is a qualitative description of the selective oxidation behaviour of Fe-based alloys, which allows us to accept the assumption as a fair first approximation.

Both types of silicon selective oxidation described earlier are summarized in figure 9, which also schematically presents the corresponding iron and silicon in-depth SIMS profiles. It can be concluded from the evidence above that the internal oxidation of silicon in Fe-0.2% Si binary alloys annealed at 850°C coexists with some external oxidation along the grain boundaries, which is revealed by the rapid decrease in $^{28}Si^+$ secondary ion current at the extreme surface. The latter intensity increases again at depths where internal oxidation has actually developed.

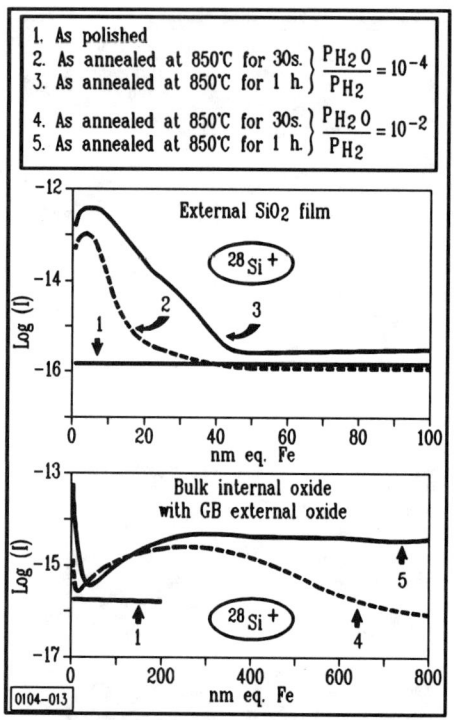

Figure 8 - Kinetics of silicon selective oxidsation on sample A as revealed by ion $^{28}Si^+$ SIMS in-depth profiles.

Steels

Predictions based on relations (1) and (2) are generally far from accurate in the case of actual steels. In fact, thus determined critical contents for the various alloying elements considered can only be as good as a steel model based on the juxtaposition of completely independent Fe-X binary alloys.

Table II summarizes such predictions computed from the physical constants reported in table III and compares them to experimental SIMS observations in the case of steel S2, as annealed for one hour at 850°C, under a 100% H_2 atmosphere, the dew point of which was - 42°C. Figure 10 illustrates the oxidation pattern of this sample by SEM and SAM distribution maps recorded at 3 nm depth below the free surface.

In agreement with thermodynamics predictions, Cr and P indeed do not take part in the oxidation. On the other hand, although MnO is thermodynamically unstable in the conditions considered, Mn actually participates, by forming complex oxides with Si and Al ($MnSiO_3 + Al_2O_3$), the stability of which is greater than that of MnO.

As a consequence, the mere presence of Si (and Al) in the steel chemistry appears to be sufficient condition for the theoretically unauthorized oxidation of Mn to take place.

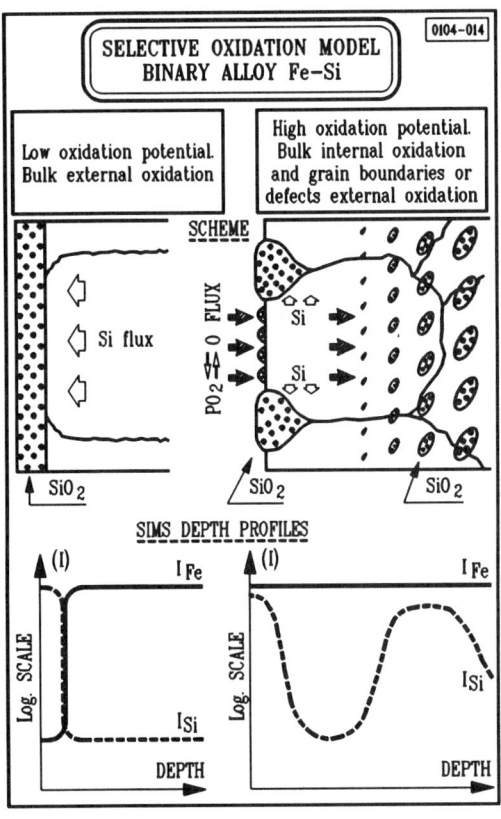

Figure 9 - Schematic view of silicon selective oxidation in Fe-Si binary alloys.

TABLE II. - PREDICTED AND OBSERVED OXIDATION MODES
STEEL S2, ANNEALED AT 850°C/1 HOUR (100% H_2/DP = - 42°C)

		Oxidation	
Element	Diffusion paths	Predicted from relations (1) and (2)	Observed and predicted from (3)
Si	Lattice & GB	external	external
Al	Lattice	internal	external
	GB	external	
Mn	Lattice & GB	non-oxidizable	external
Cr		non-oxidizable	no oxidation
P		non-oxidizable	no oxidation

TABLE III - STOECHIOMETRIC FACTORS AND MOLAR VOLUME OF OXIDES AND DIFFUSION COEFFICIENTS OF VARIOUS ALLOYING ELEMENTS IN FERRITE

Alloying elements	n	V_{XO_n} cm^3/mole	Frequency factors $D_X^{'a}$ cm^2/sec	Activation energies Q_X^a cal/at.gr
Mn	1	13.02[6]	1.49[10]	55,800[10]
Cr	1.5	14.59[6]	2.40[11]	57,300[11]
Al	1.5	12.86[6]	5.90[8]	57,700[8]
P	2.5	29.69[6]	7.1[12]	40,000[12]
B	1.5	28.3[6]	10^{-5}[13]	15,000[13]

The balance between internal and external oxidation modes then looks like being governed by the tendency of Si to external oxidation.

According to criteria (1) and (2), a somewhat more oxidizing treatment of the same material (850°C - 1 h - HNX - DP = - 2°C) should have led to the general bulk internal oxidation of Si, Mn, Cr and Al, while remaining thermodynamically unable to stabilize a pure phosphorus oxide.

SIMS analysis indicates that, for all the five elements just mentioned, internal oxidation inside the grains has actually occured alongwith grain boundary external oxidation. SEM micrograph and SAM distribution maps recorded at 3 nm below the free surface of this sample are presented at figure 11. The manganese external oxide is essentially detected along the grain boundaries. Phosphorus oxide is present both along the grain boundaries and in some particles located at the free surface. Silicon oxide could not be detected in the scanning mode but ponctual analysis revealed the presence of silicon oxide along the grain boundaries and inside the free surface particles. These particles might well have grown at the points of emergence at the surface of lattice linear defects. Table IV summarizes the corresponding predicted and observed oxidation modes.

It is clearly suggested that, in the case of complex alloys, defining the transition limit between external and internal oxidation modes requires that the oxygen inward flow be compared to the collective outward flow of all the oxidizable elements in the composition. This increases the computation complexity, because the thermodynamic stability of thus created oxides varies under the influence of test conditions : steel composition, nature of the protective atmosphere, soaking temperature, ...

A new criterion, meeting these requirements, could however be put forward :

$$\left[\sum_X N_X^0 \left[n \, D_X \, V_{XO_n} \right]^{1/2} \right] \geq \left[\frac{0.3\pi \, V}{2} \right]^{1/2} \left[N_0^S D_0 \right]^{1/2} \quad (3)$$

wherein the left-hand side represents the linear superimposition of individual contributions of the various oxidizable elements.

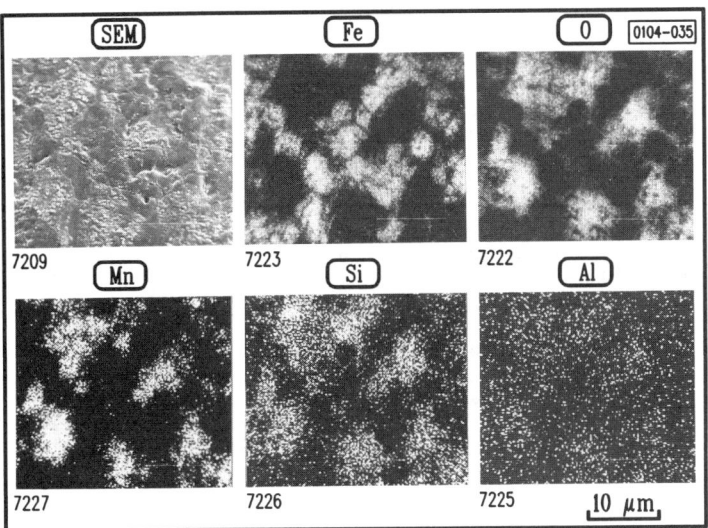

Figure 10 - SEM surface micrograph and SAM distribution maps of steel
(as-annealed : 850°C, 1 hour, 100% H_2, DP = - 42°C). Depth : 3 nm.

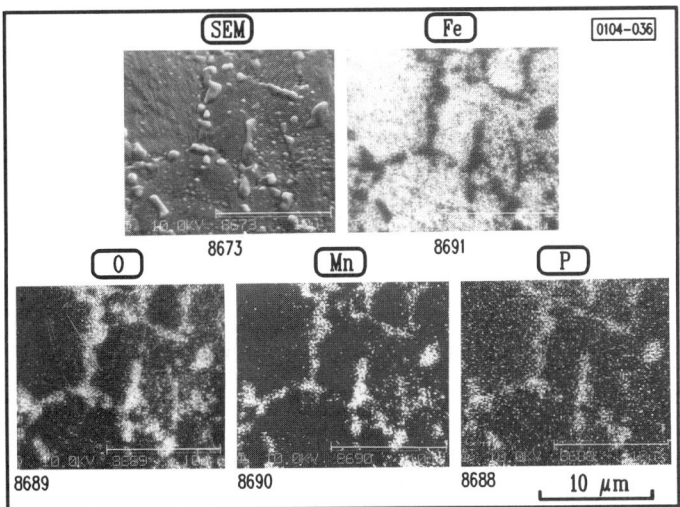

Figure 11 - SEM surface micrograph and SAM distribution maps of steel S2
(as-annealed : 850°C, 1 hour, 5% H_2, DP = - 2°C). Depth = 3 nm.

On the other hand, the right-hand side only depends on oxygen inward flow through the free surface, and is therefore independent of the steel's composition.

According to the model presented here, external oxidation should occur when inequality (3) is verified under the experimental conditions considered. In all other cases, internal oxidation should prevail.

TABLE IV. - PREDICTED AND OBSERVED OXIDATION MODES
STEEL S2, ANNEALED AT 850°C/1 HOUR (HNX/DP = - 2°C)

Element	Diffusion paths	Oxidation	
		Predicted from relations (1) and (2)	Observed and predicted from (3)
Si	Lattice	Internal	Internal
	Grain boundary	Internal	External
Al	Lattice	Internal	Internal
	Grain boundary	Internal	External
Mn	Lattice	Internal	Internal
	Grain boundary	Internal	External
Cr	Lattice	Internal	Internal
	Grain boundary	Internal	External
P	Lattice	Non-oxidizable	Internal
	Grain boundary	Non-oxidizable	External

Practically speaking, quantitative model validation is faced with two major difficulties. First of all, the number of physical constants appearing in relation (3) is very high, and most of them are only known approximately. Furthermore, the nature of reaction products is not well defined, because of possible complex oxide formation.

It must also be noted, from a theoretical point of view, that relation (3) is based on the linear combination of independent diffusion flows, neglecting the possibility of interactions, which are known to occur in practice (co-diffusion effects). Striclty speaking, it is the diffusion flow driving forces (activity gradients) which can be combined linearly, rather than the flows themselves. Deriving a new criterion from this better basis would however require much additional theoretical work. Each new steelgrade should then be considered separately and interaction effects should be accounted for on a consistent, if simplified basis, with supposed constant interference effects.

In addition, some metallurgical features affecting the diffusion kinetics of reactive species should also be taken into account : dislocation density related to the cold-rolling reduction, micro-precipitation condition, grain boundary segregations and precipitation, ferrite-austenite transformation,..

Although a fully quantitative treatment of the question looks very complicated indeed, relation (3) however provides the pieces of information that are necessary to understand some apparently paradoxical observations, such as the drastic modification in oxidation behaviour exhibited by some steels under the effect of minor additions of critical oxidizable elements.

In this respect, boron additions, initially developed in order to accelerate nitrogen precipitation during coiling of steels mostly aimed at continuous annealing, give a fascinating example of such behaviours. The simulation of continuous annealing described in table V was applied to materials S3 (ELC, AL-killed grade) and S4 (ELC, B-containing grade)..

TABLE V. - SIMULATION CONTINUOUS ANNEALING CONDITIONS

Heat. rate (°C/sec)	Soak. temp. (°C)	Soak. time (sec)	Cool. rate (°C/sec)	Overag. temp. (°C)	Overag. time (sec)	Atm.	DP (°C)
4.5	800	30	10	400	180	HNX	- 27

The distinct resulting oxidation patterns are clearly evidenced in figure 12. While classical ELC grade S3 has just suffered slight surface oxidation at the emergence of grain boundaries, its B-containing counterpart (S4) exhibits a high-density distribution of surface oxide particles, both along grain boundaries and at the free surface, inside the grains.

The regular alignment of oxide particles appearing at the free surface, inside the grains of the B-containing steel (S4) is certainly worth noticing. This feature is presumably related to the crystallographic anisotropy of the material, the strong disorientation between adjacent grains of which is clearly evidenced by the deep grain boundary thermal etching.

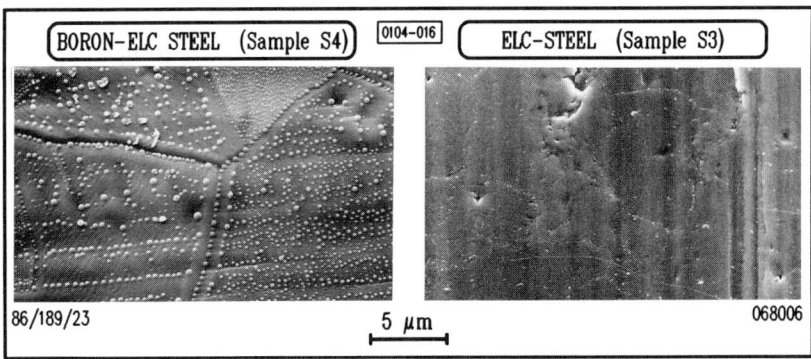

Figure 12 - SEM surface micrographs of steels S4 and S3 (as-annealed : 800°C, 30 sec, 5% H_2, DP = - 27°C, and as-overaged : 400°C, 180 sec, same atmosphere.

The numerical integration of secondary ion current profiles (presented at figure 13) limited to the sub-surface layers (200 nm) yields external surface enrichment estimates (given in Coulombs) for the various alloying elements in both steels.

Results presented in table VI emphasize the acceleration which a 23 ppm-addition of B brings about in the external oxidation kinetics of Si, Al, Mn and P. The observation can be accounted for by the rapid diffusion of boron to the free surface, where it combines with most of the incoming oxygen. For slower moving Si, Al and Mn, the inward oxygen flow is then artificially reduced, which brings them back to the conditions of external oxidation.

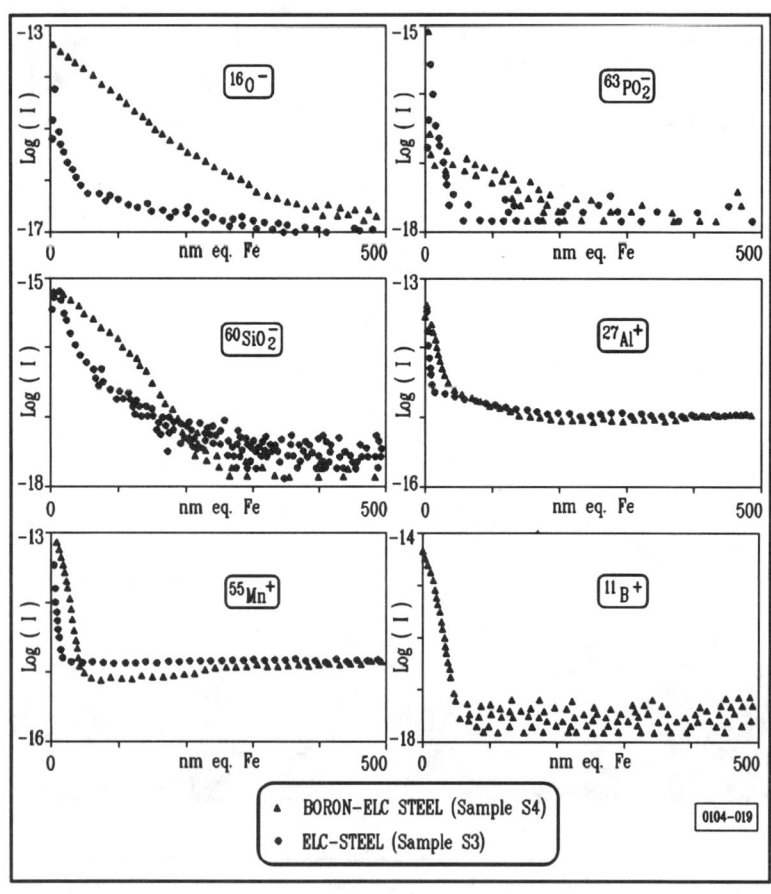

Figure 13 - SIMS in-depth profiles of steels S4 and S3 (as-annealed : 800°C, 30 sec, 5% H_2, DP = -27°C, and as-overaged : 400°C, 180 sec, same atmosphere).

TABLE VI. - INTEGRATED ION CURRENT PROFILES (10^{-15} Coulombs)

	$^{11}B^+$	$^{27}Al^+$	$^{55}Mn^+$	$^{16}O^-$	$^{60}SiO_2^-$	$^{63}PO_2^-$
S3 ELC steel	-	4433	3729	211	98	6
S4 B-ELC steel	566	5761	12960	8985	165	14

Conclusions

The present work was aimed at investigating the alloying element selective oxidation phenomena taking place during the recrystallization annealing of cold-rolled steels.

Based on the theoretical model proposed by Wagner, a criterion was put forward for the transition between internal and external oxidation modes, and it was shown to account fairly well for experimental data corresponding to binary Fe-Si alloys. The extension of the model to commercial steels proved rather complicated, due to the occurrence of synergetic external oxidation effects.

Based on the comparison of the oxygen inward flow with the combined outward diffusion flow of all oxidizable alloying elements, a qualitative simplified criterion was therefore proposed, which fits in well with experimental observations made on steels. This criterion can moreover be used to explain the accelerating effect of fast moving and highly oxidizable elements such as boron on the kinetics of external selective oxidation. The fact that it does not take the possible interactions between distinct diffusing species into proper account calls for more experimental and theoretical work.

The research reported here is now being extended to ULC-IF steels, which are increasingly used in continuous hot-dip galvanizing. In this case, the balance between the internal and external oxidations of highly oxidizable elements (Ti, Al, ...) is of the utmost importance for surface-liquid bath reactivity and associated formation of Fe-Zn intermetallics.

Acknowledgements

The authors are grateful for the financial assistance from IRSIA (Belgian Institute for the Encouragement of Scientific Research in Industry and Agriculture) and ECSC (European Coal and Steel Community).

REFERENCES

1. J.P. Servais, V. Leroy, "Application of Surface Analysis in Steel Industry", Proceedings of SIMS VI Conference, 13-18 September 1987 (Chichester : John Wiley & Sons, 1988), 551-560.

2. A. Aubry, B. Vialatte, "Reactivity of Synthetic Titanium Interstitial-free Steel in the Liquid Zinc Bath", (Paper presented at INTERGALVA 91, Barcelona, June 1991).

3. J.P. Servais et al, "New Views on the Corrosion of Tinplate", (CRM internal report S25/90, 1990).

4. C. Wagner, "Reaktionstypen bei der Oxydation von Legierungen", Zeitschrift für Elektrochemie, 63 (7) (1959), 772-782.

5. R.A. Rapp, "Kinetics, Microstructures and Mechanism of Internal Oxidation - Its Effects and Prevention in High Temperature Alloy Oxidation", Corrosion-NACE, 21 (1965), 382-401.

6. R.C. Weast Handbook of Chemistry and Physics, (CRC Press Inc., Boca Raton, Florida, 1981-1982).

7. H. Bester, K.W. Lange, "Abschätzung mittlerer Werte für die Diffusion von Kohlenstoff, Sauerstoff, Wasserstoff, Stickstoff und Schwefel in festem und flüssigen Eisen, <u>Archiv für das Eisenhüttenwesen, 43 (3) (1972), 207)213.</u>

8. A. Vignes et al, "The use of the electron microprobe for the determination of impurity diffusion coefficients", (Paper presented at the 2nd Natl Conference Electron Miroprobe Analysis, Boston, 1967),20.

9. J.H. Swisher, E.T. Turkdogan, "Solubility, Permeability an Diffusivity of Oxygen in Solid Iron", <u>Trans. of the Metallurgical Society of AIME</u>, 239 (1967), 426-431.

10. K. Nohara , K-I. Hirano, "Diffusion of Mn^{54} in iron and iron-manganese alloys", Proceedings ICSTIS, suppl. Trans. ISIJ, 11 (1971), 1267-1273.

11. P.J. Alberry, C.W. Haworth, "Interdiffusion of Cr, Mo and W in iron", <u>Metal Science</u>, 8 (1974), 407-411.

12. Y. Adda, J. Philibert, <u>La diffusion dans les solides</u>, (Presses Universitaires de France, 1966).

13. M.A. Krishtal, " <u>Diffusion Processes in Iron Alloys</u>", ed. J.J. Becker (Jerusalem, Israel Program for Scientific Translations, 1970), 124.

CHARACTERISTICS OF 100% HYDROGEN ANNEALING FURNACE

Shigeru Tajima, and Masaki Shirouzu

Sumitomo Metal Industries, Ltd.
Wakayama Steel Works
1850 Minato, Wakayama, Japan

Abstract

100% hydrogen annealer for cold rolled sheets was installed at Wakayama Steel Works in 1989 for the first time in Japan for trial purposes. The results of annealing test for two years, including special steels, are quite satisfactory in productivity, uniformity, quality and running cost. In Japan, Continuous Annealing Line (CAL) has been successfully operated and still on the increase, but 100% hydrogen annealer has excellent performance which is competitive or superior to CAL. In Wakayama Steel Works, new 100% hydrogen annealers, 11 furnaces - 22 workbases, for actual production are under construction.

Intruduction

Recently Continuous Annealing Line (CAL) has been remarkably developed for the main facility of annealing process for cold rolled sheets in Japan. However the investment and running cost for CAL are too expensive if the production amount is not so much and it is not adaptable to special steels, such as high carbon steels. Therefore, it is still very important to improve batch annealing furnace.

The new bell type batch annealing furnace, which utilized high convection in the inner cover with 100% hydrogen atmosphere, was developed in early 1970's in Europe for brass industry. This technology was adapted to seel industry at the end of 1970's.

In Wakayama Steel Works, the proportion of special steels and thin gage steels in production is fairly high, and the new annealing process had been investigated for a long time. And finally 100% hydrogen annealer, one-furnace and one-workbase, was installed in 1989 for the first time in Japan for estimation.

This paper reports that the results are quite satisifactory in production rate, running cost of hydrogen annealer and improvement of quality for mild and special steels.

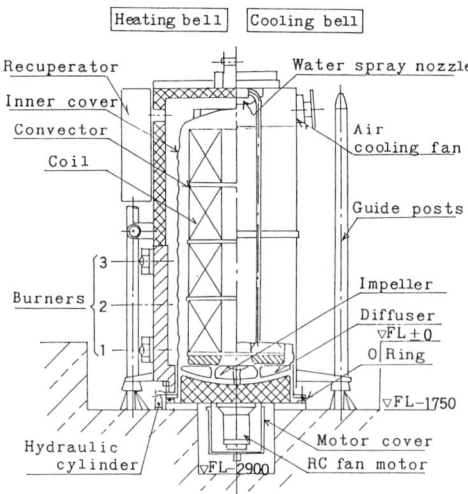

Figure 1 - General assembly of hydrogen annealer

Table I   Main Specifications of hydrogen annealer

Item	Specification
Charging measurement	2,000mm diameter
	4,300mm height
Charge weight	Maximum 100 ton
Coil temperature	850°C ∿ 250°C
Atmosphere	H_2, N_2, AX + N_2
Recirculation fan	57/32/17kW
	1,200/900/600 r.p.m.
Burners	4.8 x 10^8 J/h x 12
Clamping mechanism for inner cover	Hydraulic clamping cylinder
Cooling system	Water/Air

Table II Physical properties of hydrogen and nitrogen

Item	H_2(A)	N_2(B)	A/B
Thermal conductivity [W/(m·k)]	168×10^{-3}	24.0×10^{-3}	7
Viscosity Pa/s	8.41×10^{-6}	16.6×10^{-6}	1/2
Diffusion coefficient [cm²/s]	0.629 (H_2-CO_2)	0.160 (H_2-CO_2)	4
Density [kg·s²/m⁴]	9.04×10^{-3}	126×10^{-3}	1/14
Explosion limit [%]	4 ~ 75	-	-

Expected effects of 100% hydrogen annealer

The general assembly of a hydrogen annealer is illustrated in Fig. 1, and main specifications are summarized in Table I. The basic design of hydrogen annealer is similar to conventional annealer, but the seal of workbase is improved by using O-ring and welded stainless steel structure. These improvements enable to anneal materials in 100% hydrogen atmosphere, which realizes high convection atmosphere in the inner cover.
Physical properties of hydrogen are summarized in Table II. It is a notable property that thermal conductivity of hydrogen is 7 times higher than nitrogen. Simulation was made to know the effect of different atmosphere gas conductivity, in the assumption that there is no contact with strip surface, as shown in Fig. 2. More conductivity gas between the wrap of coils increases the radial thermal conductivity (λ radial) significantly. λ radial in hydrogen atmosphere becomes approximately 3.5 times as in pure nitrogen

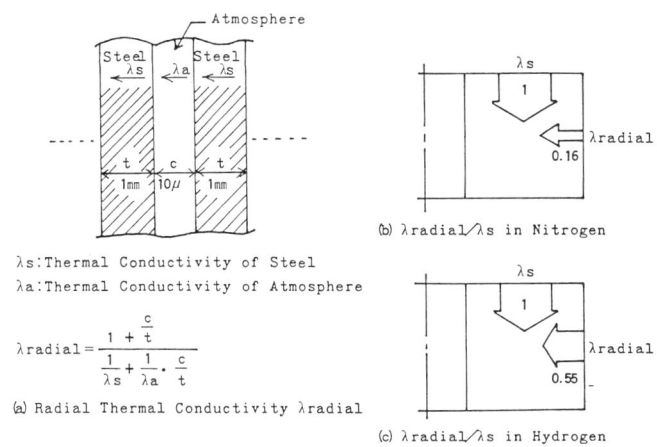

Figure 2 - Calculation of radial thermal conductivity, λ radial, for steel coil in hydrogen and nitrogen atmosphere

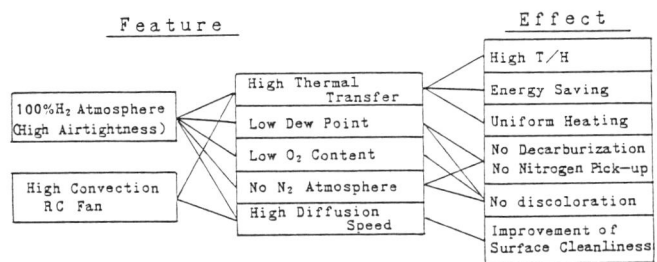

Figure 3 - Expected effects of hydrogen annealer

atmosphere as shown in Fig. 2.

The high circulation rate of the annealing atmosphere, with low density of hydrogen, increases heat transfer rate between annealing atmosphere and outside surface of coils. It is possible to get the higher heating and cooling rate, using the high firing rate burners and the water spray cooling system, with 100% hydrogen atmosphere and higher circulation rate. In addition, hydrogen annealer is expected to get uniformity of temperature and quality improvement of strip surface, with high diffusion coefficient, low dew point and low O_2 content of annealing atmosphere, as shown in Fig. 3.

Productivity

Effect of hydrogen concentration and recirculation fan revolution on productivity

Effect of hydrogen concentration and recirculation fan revolution on productive was estimated by annealing four coils of which total weight was 85 ton. Measuring points of temperature are shown in Fig. 4. Fig. 5 and Fig. 6 show the relative throughput (T/H) ratio to reach 690°C in the coldest point and 710°C in the hottest point of the charge. The standard throughput is in condition that recirculation fan revolution is 600 r.p.m. in 100% nitrogen atmosphere.

The throughput of heating increases proportionally as increasing hydrogen concentration of the HNX mixture atmosphere, as shown in Fig. 5. In pure hydrogen, heating throughput is 1.7 times higher than that in pure nitrogen. In higher recirculation fan revolution its effect becomes smaller. The throughput of cooling scarecely increases below 50% hydrogen concentration, but, in pure hydrogen cooling throughput is 1.5 times higher than that in pure nitrogen.

Fig. 6 shows effect of convection on heating cooling throughput. In the heating, effect of recirculation fan revolution is not so great as hydrogen concentration. Twice revolution of recirculation fan increases only 10% heating throughput. However in the cooling, effect of recirculation fan revolution is great and high convection (1200 r.p.m.) give 40% higher cooling throughput against low convection (600 r.p.m.).

Fig. 7 shows comparison of relative throughput ratio between water cooling and air cooling. Due to the high convection of the 100% hydrogen, water cooling is most effective and cooling throughput was twice higher than in the air cooling.

The results of this test suggest that the effect of hydrogen concentration, increasing radial direction heat transfer, is prominent for the heating throughput. Furthermore, effect of high convection, eliminating much heat from outside surface of coils, is also prominent for the cooling throughput.

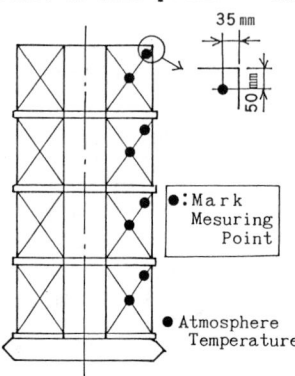

Figure 4 - Measuring points of temperature

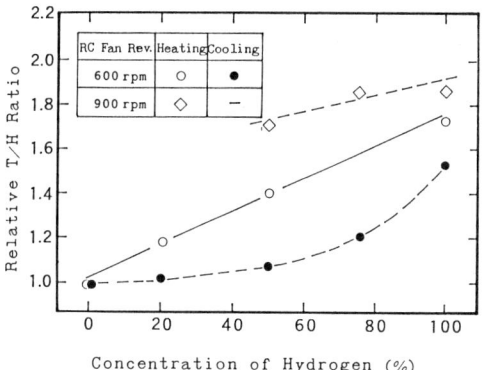

Figure 5 - Effect of concentration of hydrogen on relative throughput (T/H) ratio

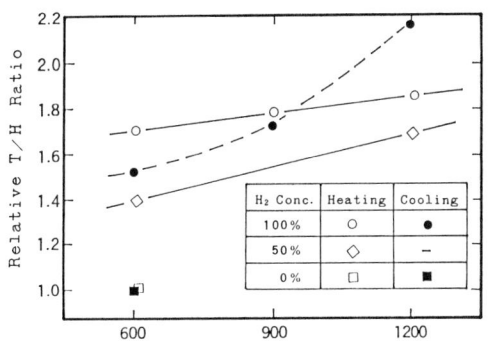

Figure 6 - Effect of recirculation fan revolution on relative throughput (T/H) ratio

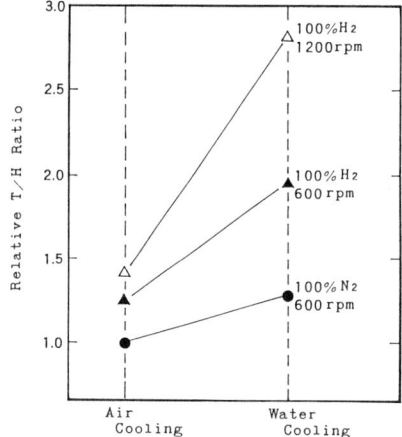

Figure 7 - Effect of atmosphere, recirculation fan revolution and water cooling on relative throughput (T/H) ratio

Table III Comparison of throughput (T/H) between hydrogen annealer and conventional annealer

	Hydrogen annealer (A)	Conventional annealer (B)	A/B
Heating T/H	3.2	2.0	1.6
Cooling T/H	4.4 (Water cooling)	1.5 (Air cooling)	2.9
Total T/H	1.8	0.9	2.0

Comparison of throughput between hydrogen annealer and conventional annealer

Throughput of actual DDQ cycle of hydrogen annealer compared with conventional annealer is shown in Table III. The throughput for 85 ton-charge weight is 1.6 times in heating and 2.9 times in cooling, and total throughput is twice higher than in conventional annealer with HNX mixture of 11%-hydrogen and 89%-nitrogen.

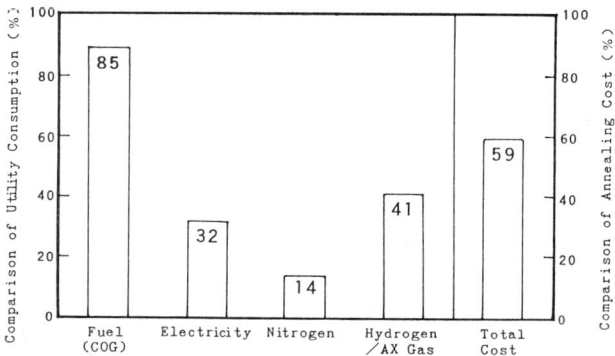

Figure 8 - Comparison of utility consumption and total cost between hydrogen annealer and conventional annealer

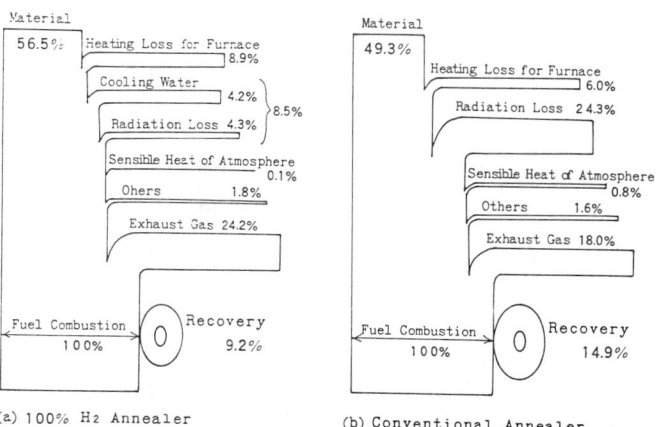

Figure 9 - Comparison of heat balance between hydrogen annealer and conventional annealer

Utility consumption

Major utility consumption and total cost of hydrogen annealer compared with conventional annealer are shown in Fig. 8.
In order to estimate fuel gas consumption, heat balance is shown in Fig. 9.
In hydrogen annealer, short annealing time and improvement of furnace insulation reduces significantly radiation loss and thermal loss by cooling water. As a result, fuel gas is used efficiently and fuel gas consumption is saved 15% against conventional annealer.
Electricity consumption is also reduced remarkably in comparison with conventioal annealer, due to low density of hydrogen. Total cost are reduced 41% against conventional annealer, including annealing atmoshpere.

Improvement of quality

Improvement of gage accuracy of strip

Mostly high carbon steels are annealed for spheroidizing (softening) before cold rolling. Gage fluctuation occurs along the circumference of the coil in cold rolling as a result of hardness non-uniformity. This is due to circumference temperature difference by conventional annealer ($\Delta t=27.0°C$) as shown in Fig. 10). The temperature distribution for outer weap of coil, 2,000mm outer diameter, during soaking stage of hydrogen annealer is controlled within $\Delta t=5.7°C$ as shown in Fig. 10. Therefore gage accuracy of strip after cold rolling is improved significantly, as shown in Table IV.

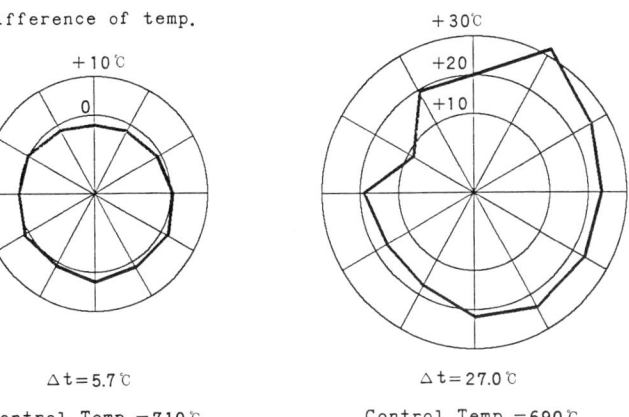

① Hrdrogen annealer   ② Conventional annealer

Figure 10 - Outer wrap temperature distribution along the circumference of the coil

Table IV Comparison of gage accuracy between hydrogen annealer and conventional annealer

Hydrogen annealer 2,000mm O.D.	Conventional annealer 2,000mm O.D.	1,500mm O.D.
±6 μm	±17 μm	±6 μm

Table V Required time to prescribed temperature difference in the charge

Temperature difference in the charge	80°C	30°C	20°C	10°C
Hydrogen annealer	19.0hr	28.0hr	32.0hr	46.5hr
Conventional annealer	27.0hr	43.0hr	52.0hr	-

Mechanical properties

Heating time for a prescribed temperature difference between the hottest and coldest point in the charge in hydrogen annealer is decreased about 40% against conventional annealer, as shown in Table V. Test annealing condition was that annealing temperature was 710°C in the hottest point of the charge and charge weight was 85 ton. Surprisingly, within only 10°C difference in the charge is possible in hydrogen annealer. Excellent temperature uniformity within the coils reduces the non-uniformity of mechanical property.
Fig. 11 shows the hardness distribution of the high carbon steel in the longitudinal direction of the strip which was annealed to spheriodize the cementite. The hardness distribution is reduced by using hydrogen annealer compared with the conventional annealer. Fig. 12 shows comparison of average values and standard deviations of mechanical properties in rolling direction for 0.8mm thickness specimens annealed by hydrogen annealer and conventional annealer. Standard deviation of all the mechanical properties for hydrogen annealing is much smaller than that of conventional annealing.

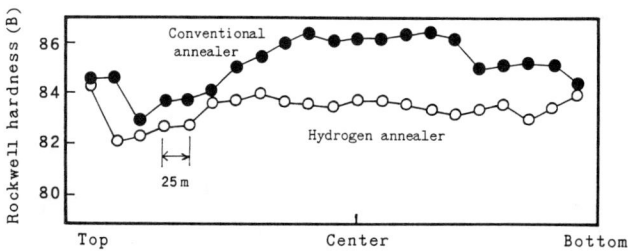

Figure 11 - Hardness distribution in the longitudinal direction of the strip

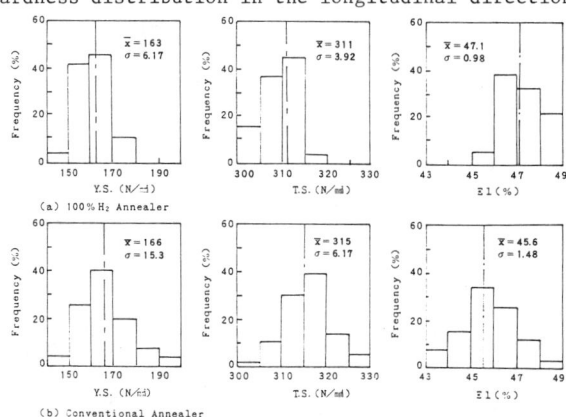

Figure 12 - Comparison of mechanical properties between hydrogen annealer and conventional annealer

Decreasing surface oxidation and decarburization

The surface of the special steels, especially on the edge such as high carbon steels, were oxidized by water vapor in annealing atmosphere of conventional annealer, due to high content of Mn and Si which have strong affinity to oxygen. In hydrogen annealer with gas-tight structure and low H_2O/H_2 ratio, width of edge discoloration and depth of grain boundary oxidation are much smaller, as shown in Fig. 13.
Furthermore surface decarburization of high carbon steel and surface nitriding of Si steel were perfectly suppressed in hydrogen annealer.

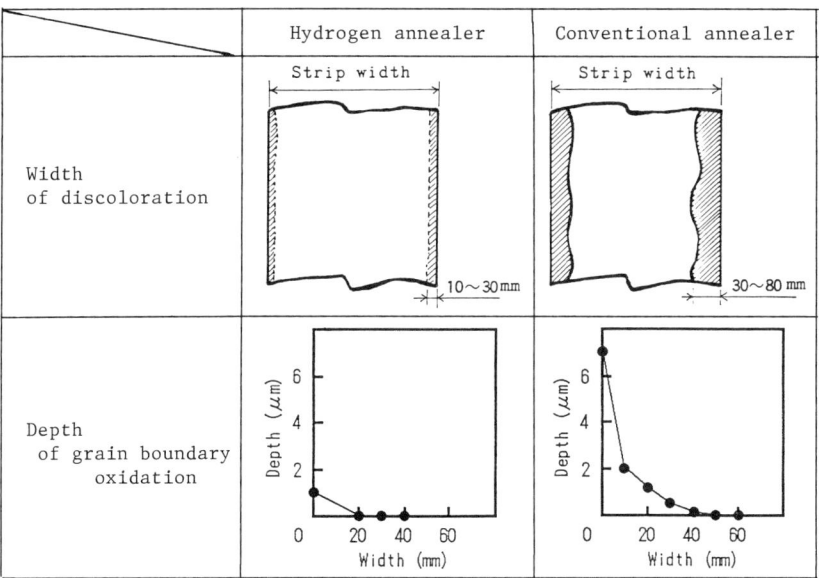

Figure 13 - Comparison of surface oxidation between hydrogen annealer and conventional annealer

Conclusion

A 100% hydrogen annealer has been estimated in Wakayama Steel Works for two years. The results are quite satisfactory in productivity, uniformity, running cost and quality especially for special steels.
Sumitomo Metal Industries decided to introduce 10 more furnaces and 21 more workbases for actual production. 6 furnaces and 12 workbases including test facility started its operation in June 1991. Another 5 furnaces and 10 workbases are under construction now and will be completed at the end of 1991.

SUBJECT INDEX

A
Aging Index, 265
    effect of annealing cycle, 270-272, 283
        MnS precipitation, 267
Alloy coating
    analyzer, 34-35
    control of Fe content, 34-35
    effect of substrate, 364-366
    powdering resistance, 39-40
AlN precipitation, 14
    effect of Al content, 279, 389-392
        hot-mill coiling temperature, 255-257, 279, 389-390
        Mn content, 255-257

B
Bake-hardening steels, 18, 49-52
    effect of C content, 249
    effect of carburising in IF, 165-171

C
Carbide dissolution, 180-186
Carbide distribution
    effect of anneal cycle, 7-8, 274-278
        Mn content, 252, 255
Coating weight control
    hot-dip galvanizing, 32-35
Continuous Annealing
    equipment, 28-32, 45-46, 54-55, 66-68, 131, 134, 406-408
    line locations, 4
    product properties, 47-48, 63, 75-76, 155
    thermal cycles, 8-9, 47, 55, 66, 134, 263, 269, 371
Cold-work embrittlement in IF
    effect of carburising, 164-165
    test procedure, 162-163
Computer control
    hot-dip galvanizing line, 35-38

D
Direct-fire furnace design, 31, 118-127
Dynamic strain-aging, 291-294

F
Ferritic hot-rolling, 295-298, 301-302
    tensile properties, 299-300
Furnace design
    Direct-fire continuous-annealing, 31, 118-127
    Hydrogen batch-annealing, 437-440
    Radiant-tube continuous-annealing, 82-90

G
Galvanizing
    effect of substrate type, 359-362
        surface condition, 354-357
    gas wiping, 32-33
    line layout, 29
    pot design, 32
Galvannealing
    analyzer, 34-35
    control of Fe content, 34-35
    effect of substrate, 364-366
    powdering resistance, 39-40

H
Heat buckle, 91-103
High-strength steels, 16-20, 38-39, 63
Hot-dip coated steels
    effect of anneal temperature, 154
        zinc-quench temperature, 154
        mechanical properties, 38-39, 155
Hot-dip galvanizing line, 29
Hydrogen batch-annealing
    chemical reactions, 430-432, 498-500
    comparison with continuous-annealing, 22, 483-484
        conventional batch-annealing, 485-487, 534-536

convector-plate design, 468-470
dew point, 433, 439
flow measurements, 467-472
furnace design, 437-440, 530
iron fines, 431
material handling, 459-461
process control, 444-454
product properties, 475-476, 485-487, 536-537
productivity, 491, 532-533
quality control, 455-458
surface cleanliness, 431, 488-489, 505-507, 537
surface reactions, 432-435, 500-503, 520-526
thermal cycle, 496

L
Laboratory simulator
continuous annealing, 375-378, 386-387

M
Microstructure of Al-killed steels
effect of C content, 274-276
hot-mill coiling temperature, 252, 276
Mn content, 252, 255, 275
MnS precipitation
effect of slab-reheat temperature, 267
sulfur content, 267
relation to aging index, 267
Multi-reflection pyrometry, 127-129, 140-141

O
Overaging
tin-plate tempers, 14

P
Precipitates in IF steels, 199-210, 222-234, 414
effect of hot-charging, 310-312, 317

Process control
continuous-annealing line, 91-103, 136-142
hot-dip galvanizing line, 35-38
Hydrogen batch-annealing, 444-454

Q
Quality control
continuous-annealing line, 71, 127-129, 140-141
hot-dip coating, 36
Hydrogen batch-annealing, 455-458

R
r-value of Al-killed steel
effect of C content, 249
hot-mill coiling temperature, 252, 257
Mn content, 252, 257-258
r-value of ferritic hot-rolled steel, 299
r-value of IF
effect of anneal temperature, 11, 316-318
carburising, 172-174
hot-charging, 316-318
Radiant-tube furnace design, 82-90
Recrystallization of IF, 192-196
activation energy, 211-216
Retained austenite in TRIP steels
effect of transformation temperature/time, 325-326
strain, 334
Roll quenching, 56-59, 66-68, 72-73, 103-105

S
Strain aging
effect of annealing cycle, 7, 270-272, 283
carbide distribution, 7
Stainless steel
formability, 348
oxidation behavior, 343-345
Strip flatness, 61-62, 75, 91-103

Strip tension control, 56-59, 143-147
Strip walking, 72
Super-formable IF, 10-12
Surface reactions in Hydrogen batch-annealing, 432-435
Surface carbon, 431

T

Temperature of continuous-annealed strip
    control system, 106-114, 135-139
    measurement, 71, 127-129, 140-141
Tensile properties of Al-killed steel
    effect of Mn content, 280
    slab-reheat temperature, 280
Texture in Al-killed steel
    effect of hot-mill coiling temperature, 253
    Mn content, 253
Texture in IF steels, 236-243, 295-298
    effect of annealing temperature, 315
    hot-charging, 315

Tin-plate tempers, 13-16, 385, 406
    effect of AlN precipitation, 14, 389-394
        annealing condition, 405, 415, 419-420
        carbon content, 400-401, 422
        hot-mill coiling temperature, 14-15, 389-394, 416
        Nb addition, 401-402, 413-414, 419-422
        overaging, 14
        Zr addition, 413-414, 419-422
Toroidal burner, 31, 118-127
TRIP steels, 16-20
    annealing cycle, 19
    chemistry, 19
    deep drawability, 19
    mechanical properties, 19, 329-332, 335
    retained austenite, 325-326, 334

W

Water Quenching, 59-62, 66-73
Weight reduction in autos, 19

AUTHOR INDEX

A
Akisue, O., 261
Asaho, R., 397

B
Beco, F., 287
Benincasa, S., 219
Bock, M., 443
Brickner, K.G., 463
Brugnera, C., 43

C
Chang, S.K., 305, 411
Chin, K.G., 305
Cook, E.A., 143

D
Deutsch, C., 43
DeVito, A.P., 463

F
Fujinaga, C., 397
Fukushima,T., 79

G
Gasse, W.F., 481
Gaulin, B.D., 219

H
Haezebrouck, D.M., 369
Harlet, Ph., 287
Hashimoto, O., 177, 339
Hashimoto, S., 159
Hayashida, T., 261
Herman, J.C., 287
Hirohata, K., 79
Honda, A., 117
Honjoh, M., 133
Hoogendoorn, Th.M., 383

I
Ida, Y., 79
Inoue, T., 159
Iwaya, J., 53
Izushi, T., 65

J
Jitsukawa, S., 65

K
Kaihara, T., 79
Kang, H.J., 305
Kato, T., 177, 397
Kishida, K., 351
Kitamura, M., 159
Koch, M., 219
Koyama, K., 261
Krauss, G., 189, 321
Kuguminato, H., 397
Kuramoto, K., 79
Kurihara, M., 117
Kwak, J.H., 411

L
Lamberigts, M., 493, 511
Lankila, A., 151
Leroy, V., 287, 493, 511
Lochner, H., 427
Louis, P., 43

M
Maeda, H., 53
Makino, H., 53
Mataigne, J.M., 493, 511
Matlock, D.K., 189, 321
Matsui, N., 65
Matsumoto, M., 159
Mega., S., 79
Messien, P., 287
Mieloo, R., 143
Miura, K., 339
Mizui, N., 247
Morita, M., 177

N
Nakagawa, T., 79
Nakamura, A., 27
Nishimura, K., 351
Nolte, K., 443

O
Odashima, H., 351
Ohno, H., 397
Okamoto, A., 247
Okura, M., 53

P
Prikryl, M., 219

R
Ramasamy, S., 463
Ranta-Eskola, A., 151
Renard, L., 287

S
Sakuma, Y., 321
Satoh, S., 177, 339
Sekine, T., 397
Shimoyama, Y., 397
Shirouzu, M., 529
Shoji, M., 117
Simmons, R.L., 463
Sinclair, J.W., 369
Subramanian, S.V., 219

T
Tajima, S., 529
Taguchi, N., 27
Takagi, K., 27
Takechi, H., 3
Tanaka, Y., 53
Taya, K., 133
Thomas, S.J., 481
Tosaka, A., 397

U
Uchino, S., 117
Ueda, I., 133
Ushioda, K., 261

V
van den Hoogen, A.J., 383

W
White, D.A., 369
Wilshynsky-Dresler, D.O., 189
Wittler, P., 443

Y
Yamazaki, M., 65
Yano, H., 27
Yoshioka, K., 339